Lecture Notes in Computer Science 15906

Founding Editors

Gerhard Goos
Juris Hartmanis

Editorial Board Members

Elisa Bertino, *Purdue University, West Lafayette, IN, USA*
Wen Gao, *Peking University, Beijing, China*
Bernhard Steffen, *TU Dortmund University, Dortmund, Germany*
Moti Yung, *Columbia University, New York, NY, USA*

The series Lecture Notes in Computer Science (LNCS), including its subseries Lecture Notes in Artificial Intelligence (LNAI) and Lecture Notes in Bioinformatics (LNBI), has established itself as a medium for the publication of new developments in computer science and information technology research, teaching, and education.

LNCS enjoys close cooperation with the computer science R & D community, the series counts many renowned academics among its volume editors and paper authors, and collaborates with prestigious societies. Its mission is to serve this international community by providing an invaluable service, mainly focused on the publication of conference and workshop proceedings and postproceedings. LNCS commenced publication in 1973.

Michael H. Lees · Wentong Cai ·
Siew Ann Cheong · Yi Su · David Abramson ·
Jack J. Dongarra · Peter M. A. Sloot
Editors

Computational Science – ICCS 2025

25th International Conference Singapore, Singapore, July 7–9, 2025 Proceedings, Part IV

Editors
Michael H. Lees
University of Amsterdam
Amsterdam, The Netherlands

Siew Ann Cheong
Nanyang Technological University
Singapore, Singapore

David Abramson
The University of Queensland
Brisbane, QLD, Australia

Peter M. A. Sloot
University of Amsterdam
Amsterdam, The Netherlands

Wentong Cai
Nanyang Technological University
Singapore, Singapore

Yi Su
Institute for High Performance Computing
A*STAR
Singapore, Singapore

Jack J. Dongarra
The University of Tennessee
Knoxville, TN, USA

ISSN 0302-9743 ISSN 1611-3349 (electronic)
Lecture Notes in Computer Science
ISBN 978-3-031-97634-6 ISBN 978-3-031-97635-3 (eBook)
https://doi.org/10.1007/978-3-031-97635-3

© The Editor(s) (if applicable) and The Author(s), under exclusive license
to Springer Nature Switzerland AG 2025

This work is subject to copyright. All rights are solely and exclusively licensed by the Publisher, whether the whole or part of the material is concerned, specifically the rights of translation, reprinting, reuse of illustrations, recitation, broadcasting, reproduction on microfilms or in any other physical way, and transmission or information storage and retrieval, electronic adaptation, computer software, or by similar or dissimilar methodology now known or hereafter developed.
The use of general descriptive names, registered names, trademarks, service marks, etc. in this publication does not imply, even in the absence of a specific statement, that such names are exempt from the relevant protective laws and regulations and therefore free for general use.
The publisher, the authors and the editors are safe to assume that the advice and information in this book are believed to be true and accurate at the date of publication. Neither the publisher nor the authors or the editors give a warranty, expressed or implied, with respect to the material contained herein or for any errors or omissions that may have been made. The publisher remains neutral with regard to jurisdictional claims in published maps and institutional affiliations.

This Springer imprint is published by the registered company Springer Nature Switzerland AG
The registered company address is: Gewerbestrasse 11, 6330 Cham, Switzerland

If disposing of this product, please recycle the paper.

Preface

Welcome to the 25th International Conference on Computational Science (ICCS - https://www.iccs-meeting.org/iccs2025/), held on July 7–9, 2025 at Nanyang Technological University (NTU), Singapore.

This 25th edition in Singapore marked our return to a fully in-person event. Although the challenges of our present times are manifold, we have always tried our best to keep the ICCS community as dynamic, creative, and productive as possible. We are proud to present the proceedings you are reading as a result.

ICCS 2025 was jointly organized by Nanyang Technological University, the A*STAR Institute of High Performance Computing, the University of Amsterdam, and the University of Tennessee.

Considered one of the most developed countries in the world, the island country of Singapore is a major aviation, financial, and maritime shipping hub in Asia. Singapore is multilingual, multiethnic, and multicultural, and as such a very popular, safe tourism destination.

NTU Singapore is a public university ranked among the world's best, with 35,000 students, and home to the world-renowned autonomous National Institute of Education and S. Rajaratnam School of International Studies. In addition to many research institutes and centers at the university, college, and school levels, NTU also hosts two National Research Foundation (NRF) and Ministry of Education (MOE) Research Centers of Excellence, namely the Singapore Center for Environmental Life Sciences Engineering (SCELSE) and the Institute for Digital Molecular Analytics & Science (IDMxS), and 11 Corporate Labs in partnership with various industries. ICCS 2025 took place on the One-north campus.

The Institute of High Performance Computing (IHPC) is a national research institute under the Agency for Science, Technology and Research (A*STAR), dedicated to advancing science and technology through computational modeling, simulation, AI, and high-performance computing. With a multidisciplinary team of scientists and engineers, IHPC drives innovation across sectors such as advanced manufacturing, microelectronics, sustainability, maritime, and biomedical sciences. It leads Singapore's national efforts in hybrid quantum-classical computing and digital twin platforms, and partners extensively with industry and government agencies to translate deep tech into real-world impact.

The International Conference on Computational Science is an annual conference that brings together researchers and scientists from mathematics and computer science as basic computing disciplines, as well as researchers from various application areas who are pioneering computational methods in sciences such as physics, chemistry, life sciences, engineering, arts, and humanitarian fields, to discuss problems and solutions in the area, identify new issues, and shape future directions for research.

The ICCS proceedings series has become a primary intellectual resource for computational science researchers, defining and advancing the state of the art in this field.

We are proud to note that this 25th edition, with 23 workshops (the Workshops on Computational Science), one co-located event (the Asian Network of Complexity Scientists Workshop), and over 300 participants, kept to the tradition and high standards of previous editions.

The theme for 2025, "**Making Complex Systems tractable through Computational Science**", highlighted the role of Computational Science in tackling the complex problems of today and tomorrow. This conference was a unique event, focusing on recent developments in scalable scientific algorithms; advanced software tools; computational grids; advanced numerical methods; and novel application areas. These innovative novel models, algorithms, and tools drive new science through efficient application in physical systems, computational and systems biology, environmental systems, finance, and others.

ICCS is well known for its lineup of keynote speakers. The keynotes for 2025 were:

- **Johan Bollen**, Indiana University Bloomington, USA
- **Jack Dongarra**, University of Tennessee, USA
- **Mile Gu**, Nanyang Technological University, Singapore
- **Erika Fille Legara**, Center for AI Research|Asian Institute of Management, Philippines
- **Yong-Wei Zhang**, Institute of High Performance Computing, A*STAR, Singapore

This year, the main track of ICCS registered 162 submissions, of which 64 were accepted as full papers, and 52 as short papers. There were on average 2.4 single-blind reviews per submission.

We would like to thank all committee members from the main track and workshops for their contribution to ensuring a high standard for the accepted papers. We would also like to thank *Springer, Elsevier,* and *Intellegibilis* for their support. Finally, we appreciate all the local organizing committee members for their hard work in preparing this conference.

We hope you enjoyed the conference and the beautiful country of Singapore.

July 2025

Michael H. Lees
Wentong Cai
Siew Ann Cheong
Yi Su
David Abramson
Jack J. Dongarra
Peter M. A. Sloot

Organization

Program Committee Chairs

Peter M. A. Sloot	University of Amsterdam, The Netherlands
Jack J. Dongarra	University of Tennessee, USA
Michael H. Lees	University of Amsterdam, The Netherlands
David Abramson	University of Queensland, Australia
Wentong Cai	Nanyang Technological University, Singapore
Cheong Siew Ann	Nanyang Technological University, Singapore
Su Yi	Institute for High Performance Computing, A*Star, Singapore

Local Program Committee at NTU Singapore

Ee Hou Yong	Nanyang Technological University, Singapore
Kang Hao	Nanyang Technological University, Singapore

Publicity Chairs

Leonardo Franco	University of Málaga, Spain
Muhamad Azfar Ramli	Institute for High Performance Computing, A*Star, Singapore

Impact Chair

Valeria Krzhizhanovskaya	University of Amsterdam, The Netherlands

Outreach Chair

Alfons Hoekstra	University of Amsterdam, The Netherlands

Program Committee Chair – Workshops on Computational Science

Maciej Paszynski　　　　　　　AGH University of Krakow, Poland

Program Committee – Workshops on Computational Science

Amanda S. Barnard　　　　　　Australian National University, Australia
Yongjie Jessica Zhang　　　　　Carnegie Mellon University, USA

Reviewers

Julen Alvarez-Aramberri	University of the Basque Country, Spain
Philipp Andelfinger	Nanyang Technological University, Singapore
Adrian Bekasiewicz	Gdańsk University of Technology, Poland
Nik Brouw	University of Amsterdam, Netherlands
Roland V. Bumbuc	University of Amsterdam, Netherlands
Wentong Cai	Nanyang Technological University, Singapore
Pedro J. S. Cardoso	Universidade do Algarve, Portugal
Eddy Caron	ENS-Lyon/Inria/LIP, France
Lock-Yue Chew	Nanyang Technological University, Singapore
Ana Cortes	Universitat Autònoma de Barcelona, Spain
Daan Crommelin	CWI Amsterdam, Netherlands
Carlo Cunha	Northern Arizona University, USA
Bartosz Czaplewski	Gdańsk University of Technology, Poland
Venkata Rupesh Kumar Dabbir	Google LLC, USA
Eric Dignum	University of Amsterdam, Netherlands
Vitor Duarte	Universidade NOVA de Lisboa, Portugal
Mariusz Dzwonkowski	Medical University of Gdańsk, Poland
Nahid Emad	Paris-Saclay University, France
Roberto R. Expósito	Universidade da Coruña, CITIC, Spain
Ruy Freitas Reis	Universidade Federal de Juiz de Fora, Brazil
Wlodzimierz Funika	AGH University of Krakow, Poland
Victoria Garibay	University of Amsterdam, Netherlands
Paweł Gepner	Warsaw University of Technology, Poland
Alex Gerbessiotis	New Jersey Institute of Technology, USA
Maziar Ghorbani	Brunel University London, UK
Konstantinos Giannoutakis	University of Macedonia, Greece
Jorge González-Domínguez	Universidade da Coruña, Spain
Yuriy Gorbachev	Soft-Impact LLC, Russia
Michael Gowanlock	Northern Arizona University, USA

George Gravvanis	Democritus University of Thrace, Greece
Derek Groen	Brunel University London, UK
Loïc Guégan	UiT the Arctic University of Norway, France
Rafiazka Hilman	University of Amsterdam, Netherlands
Cillian Hourican	University of Amsterdam, Netherlands
Neil Huynh	Institute of High Performance Computing, A*STAR, Singapore
Alireza Jahani	Brunel University London, UK
Song Jie	Institute of High Performance Computing, A*STAR, Singapore
Zhong Jin	Computer Network Information Center, Chinese Academy of Sciences, China
David Johnson	Uppsala University, Sweden
Takahiro Katagiri	Nagoya University, Japan
Sotiris Kotsiantis	University of Patras, Greece
Sergey Kovalchuk	Huawei, Russia
Valeria Krzhizhanovskaya	University of Amsterdam, Netherlands
Michael Kuhn	Otto von Guericke University Magdeburg, Germany
Jaeyoung Kwak	Nanyang Technological University, Singapore
Michael Lees	University of Amsterdam, Netherlands
Malcolm Low	Singapore Institute of Technology, Singapore
Lukasz Madej	AGH University of Science and Technology, Poland
Tomas Margalef	Universitat Autònoma de Barcelona, Spain
Paula Martins	University of Algarve, Portugal
Pedro Medeiros	Universidade Nova de Lisboa, Portugal
Isaak Mengesha	University of Amsterdam, Netherlands
Marianna Milano	Università Magna Græcia di Catanzaro, Italy
Dhruv Mittal	University of Amsterdam, Netherlands
Francisco J. Moreno-Barea	Universidad de Málaga, Spain
Marcin Paprzycki	IBS PAN and WSM, Poland
Giulia Pederzani	Universiteit van Amsterdam, Netherlands
Alberto Perez de Alba Ortiz	University of Amsterdam, Netherlands
Dana Petcu	West University of Timisoara, Romania
Jolan Philippe	IMT Atlantique, France
Dirk Pleiter	University of Groningen, Netherlands
Alexander Pyayt	EPAM Systems, Russia
Rick Quax	University of Amsterdam, Netherlands
Muhamad Azfar Ramli	Institute of High Performance Computing, A*STAR, Singapore
Amir Raoofy	Technical University of Munich, Germany

Sophie Robert	University of Orléans, France
Daniel Rodriguez	University of Alcalá, Spain
Bertil Schmidt	University of Mainz, Germany
Martin Schreiber	Université Grenoble Alpes/Inria/Laboratoire Jean Kuntzmann, France
Md. Shalihin Othman	D-SIMLAB Technologies Pte. Ltd., Singapore
Joaquim Silva	Nova School of Science and Technology - NOVA LINCS, Portugal
Mateusz Sitko	AGH University of Science and Technology, Poland
Sucha Smanchat	King Mongkut's University of Technology North Bangkok, Thailand
Alexander Smirnovsky	SPbPU, Russia
Yong Sheng Soh	National University of Singapore, Singapore
Ryszard Tadeusiewicz	AGH University of Krakow, Poland
Daisuke Takahashi	University of Tsukuba, Japan
Gary Tan	National University of Singapore, Singapore
Wen Jun Tan	Nanyang Technological University, Singapore
Vítor V. Vasconcelos	University of Amsterdam, Netherlands
Lars Wienbrandt	Kiel University, Germany
Yani Xue	Brunel University London, UK
Xin-She Yang	Middlesex University London, UK
Felix Zhu	IHPC, Singapore

Contents – Part IV

ICCS 2025 Main Track Short Papers

A Method for Handling Negative Similarities in Explainable Graph
Spectral Clustering of Text Documents 3
   *Mieczysław A. Kłopotek, Sławomir T. Wierzchoń, Bartłomiej Starosta,
Dariusz Czerski, and Piotr Borkowski*

Preliminary Comparison of Different EDs Performance, Using Simulation 11
   *Ramona Galeano, Alvaro Wong, Dolores Rexachs, Remo Suppi,
Eva Bruballa, Francisco Epelde, and Emilio Luque*

Efficient Peptide MRM Transition Prediction via Convolutional Hashing 19
   Ramon Adàlia, Gemma Sanjuan, Tomàs Margalef, and Ismael Zamora

Comparison of Crash Simulations on Two Types of Flying Cars 27
   *Hiroto Sato, Ayato Takii, Masashi Yamakawa, Yusei Kobayashi,
Shinichi Asao, Seiichi Takeuchi, and Takahiro Ikeda*

Centrality Resilience in Complex Networks 36
   *Fariba Afrin Irany, Turja Kundu, Soumya Sarakar, Animesh Mukherjee,
and Sanjukta Bhowmick*

A Customizable Agent-Based Simulation Framework for Emergency
Departments 46
   *Francisco Mesas, Manel Taboada, Dolores Rexachs, Francisco Epelde,
Alvaro Wong, and Emilio Luque*

Leveraging Positional Bias of LLM In-Context Learning
with Class-Few-Shot and Maj-Min Alternating Ordering 54
   *Aleksander Szczęsny, Maciej Markiewicz, Łukasz Radliński,
and Przemysław Kazienko*

Remote Sensing AI for Crop Planting in Wildfire Fuel Mapping 63
   *Paula Sánchez, Irene González, Carlos Carrillo, Ana Cortés,
and Remo Suppi*

Regularization Algorithm for Eliminating Singularities in the PIES
Formula for 3D Multidomain Orthotropic Problems 72
   Krzysztof Szerszeń and Eugeniusz Zieniuk

Advancing Bird Species Classification: A Fusion of Audio and Image Data 81
Jie Xie, Xueyan Dong, Zhe Wu, Zheng Lang, Yuji Wang, Chunrong He, Jinpei Song, Zhuobin Zhang, Guiqing Yu, and Jia Tang

Prototype-Pairs Decomposition for Extracting Simple and Meaningful
Rules 89
Marcin Blachnik, Mirosław Kordos, and Daniel Dąbrowski

Fast Prediction of Job Execution Times in the ALICE Grid Through
GPU-Based Inference with Quantization and Sparsity Techniques 97
Tomasz Lelek, Szymon Mazurek, Maciej Wielgosz, and Bartosz Balis

Reversible Data Hiding in Encrypted Images with Pixel Prediction
and ERLE Compression 106
Remigiusz Martyniak and Mariusz Dzwonkowski

Information Flow Between Neighboring Housing Markets: A Case
from the Seoul Metropolitan Area 114
Leehyun Jung, Minhyuk Jeong, Yena Song, and Kwangwon Ahn

A Bi-Stage Framework for Automatic Development of Pixel-Based Planar
Antenna Structures 121
Khadijeh Askaripour, Adrian Bekasiewicz, and Slawomir Koziel

Investigation of CUDA Graphs Performance for Selected Parallel
Applications 130
Oksana Diakun and Paweł Czarnul

Instance Selection by Fast Local Set Border Selector 138
Norbert Jankowski and Mateusz Skarupski

Modelling the Transient Evolution of Queues in Plugged-in Electric
Vehicles (PEV) Fast Charging Stations 146
Godlove Suila Kuaban, Tomasz Nycz, Monika Nycz, Tadeusz Czachórski, and Piotr Czekalski

Is Heterogeneous Model Soup Tasty? A Multidimensional Evaluation
of Diverse Model Soups in Language Model Alignment 154
Dawid Motyka, Paweł Walkowiak, Julia Moska, Bartosz Żuk, Karolina Seweryn, and Arkadiusz Janz

Performance Evaluation of IMS/NGN Network with SDN-Based
Transport Stratum 163
Sylwester Kaczmarek, Magdalena Młynarczuk, and Maciej Sac

Variable-Resolution Machine Learning for Rapid Multi-Criterial Antenna
Design 171
   Anna Pietrenko-Dabrowska and Slawomir Koziel

A Fast and Scalable Genomic Data Compressor for Multicore Clusters 180
   Victoria Sanz, Adrián Pousa, Marcelo Naiouf, and Armando De Giusti

Anchored Semantics: Augmenting Ontologies via Competency Questions,
Self-Attention, and Predictive Graph Learning 189
   Shengqi Li and Amarnath Gupta

Modeling Firm Birth and Death Dynamics Using Survival Fractions
and Age Distributions 198
   Yipei Guo, Huynh Hoai Nguyen, and Feng Ling

Enhancing Sentiment Analysis Through Multimodal Fusion:
A BERT-DINOv2 Approach 207
   Taoxu Zhao, Meisi Li, Kehao Chen, Liye Wang, Xucheng Zhou,
   Kunal Chaturvedi, Mukesh Prasad, Ali Anaissi, and Ali Braytee

Modeling Parallel AI Applications for Performance Analysis on Cloud
Environments 216
   Miquel Albert, Alvaro Wong, Betzabeth Leon, Dolores Rexachs,
   and Emilio Luque

Simplified Swarm Learning Framework for Robust and Scalable
Diagnostic Services in Cancer Histopathology 225
   Yanjie Wu, Yuhao Ji, Saiho Lee, Juniad Akram, Ali Braytee,
   and Ali Anaissi

A Fast MPI-Based Distributed Hash-Table as Surrogate Model for HPC
Applications 233
   Max Lübke, Marco De Lucia, Stefan Petri, and Bettina Schnor

Rockburst Forecasting Using Composite Modelling for Seismic Sensors
Data 241
   Ilia Revin, Vadim A. Potemkin, and Nikolay O. Nikitin

Accelerating LBM with C++ STL Asynchronous Parallel Model 249
   Ziheng Yuan and Takashi Shimokawabe

Accelerating Cloud-Based Transcriptomics: Performance Analysis
and Optimization of the STAR Aligner Workflow 257
   Piotr Kica, Sabina Lichołai, Michał Orzechowski, and Maciej Malawski

Adaptive Modular Housing Design for Crisis Situations 266
   Anastasiya Pechko, Katarzyna Grzesiak-Kopeć, Barbara Strug, and Grażyna Ślusarczyk

Evaluating Parameter-Based Training Performance of Neural Networks and Variational Quantum Circuits 275
   Michael Kölle, Alexander Feist, Jonas Stein, Sebastian Wölckert, and Claudia Linnhoff-Popien

Towards an Open Science—An Academic Recommendation Cloud Platform 283
   Anna Kobusińska, Damian Tabaczyński, and Victor Chang

Algorithm Selection in Short-Range Molecular Dynamics Simulations 292
   Samuel James Newcome, Fabio Alexander Gratl, Manuel Lerchner, Abdulkadir Pazar, Manish Kumar Mishra, and Hans-Joachim Bungartz

SOPMOA*: Unleashing Shared-Open Parallelism for High-Performance Multi-Objective Pathfinding 300
   Long Viet Truong, Tien Minh Dam, Tuan Anh Nguyen, Linh Thuy Thi Nguyen, and Duong Trung Dinh

From Recursion to Parallelization: Plug & Play Dynamic Programming 309
   Jiang Long

NEoN: A Tool for Automated Detection, Linguistic and LLM-Driven Analysis of Neologisms in Polish 318
   Aleksandra Tomaszewska, Dariusz Czerski, Bartosz Żuk, and Maciej Ogrodniczuk

Surrogate Models for Analyzing Performance Behavior of HPC Applications Using the RAJA Performance Suite 327
   Befikir Bogale, Ian Lumsden, Dalal Sukkari, Dewi Yokelson, Stephanie Brink, Olga Pearce, and Michela Taufer

Cattle Identification Using 2D Mask Retention Network 336
   Niraj Kumar, Sakshi Ranjan, and Sanjay Kumar Singh

Scaling Dynamics of the Electricity Utility Sector: Assessing the Role of Agglomeration Externalities and Sensitivity to Population Cutoffs in Spatial Dynamics Across European Regions 345
   Vidit Kundu, Debraj Roy, Michel Ehrenhard, and Karin Pfeffer

Pollution Simulations and in-Field Measurements Performed in March
at Longyearbyen, Spitzbergen 353
   Albert Oliver-Serra, Leszek Siwik, Natalia Leszczyńska, Maciej Sikora,
   Tomasz Maciej Ciesielski, Eirik Valseth, Jacek Leszczyński,
   Anna Paszyńska, and Maciej Paszyński

Reversed Model Verification by Inferring Conceptual Models
from Simulation Code 361
   Rumyana Neykova and Derek Groen

GPU-Accelerated Out-of-Core HMM Inference with Concurrent CUDA
Streams 369
   MohammadReza HoseinyFarahabady and Albert Y. Zomaya

Predicting Future Collaborations in a Scientific Community Using Graph
Neural Networks 377
   Nachyn Dorzhu, Tatiana Sukhomlinova, Lijing Luo, and Sergey Kovalchuk

Data-Centric Parallel Programming Abstractions for High Performance
Computations 385
   Domenico Talia

Neural Network for Evaluating the Operational Range of Antennas
with Randomly Generated Designs 394
   Bartosz Czaplewski

Optimizing U-Net Architecture Using Differential Evolution for Brain
Tumor Segmentation 403
   Shoffan Saifullah and Rafał Dreżewski

Simulation Modeling of Clinical Decision Making for Personalized Policy
Identification 412
   Ashish T. S. Ireddy and Sergey V. Kovalchuk

Verified Eigenvalue Calculation for the Laplace Operator 421
   Jijing Zhao and Shuyu Sun

A Hybrid Q-Learning Automata Routing Protocol for Wireless Sensor
Networks 429
   Jakub Gąsior

Improving Project-Level Code Generation Using Combined Relevant Context 438
  Dmitriy Fedrushkov, Denis Tereshchenko, Sergey Kovalchuk, and Artem Aliev

Author Index 447

ICCS 2025 Main Track Short Papers

A Method for Handling Negative Similarities in Explainable Graph Spectral Clustering of Text Documents[⋆]

Mieczysław A. Kłopotek(✉)[ID], Sławomir T. Wierzchoń[ID], Bartłomiej Starosta[ID], Dariusz Czerski[ID], and Piotr Borkowski[ID]

Institute of Computer Science, Polish Academy of Sciences, ul. Jana Kazimierza 5, 01-248 Warsaw, Poland
{klopotek,stw,barstar,dcz,pbr}@ipipan.waw.pl
http://www.ipipan.waw.pl

Abstract. This paper investigates the problem of Graph Spectral Clustering (GSC) with negative similarities, resulting from document embeddings different from the traditional Term Vector Space (like doc2vec, GloVe, etc.). Solutions for combinatorial Laplacians and normalized Laplacians are discussed. An experimental investigation shows the advantages and disadvantages of solutions proposed in the literature and in this research. The research demonstrates that GloVe embeddings frequently cause failures of normalized Laplacian based GSC due to negative similarities. Application of methods curing similarity negativity leads to accuracy improvement for both combinatorial and normalized Laplacian based GSC. It also leads to applicability for GloVe embeddings of explanation methods developed for Term Vector Space embeddings.

Keywords: Artificial Intelligence · Machine Learning · Graph spectral clustering · document embeddings · negative similarities

1 Introduction

Graph Spectral Clustering (GSC) is known as an effective method of clustering when data are available in the form of a similarity matrix. As the method relies on Laplacians of the similarity matrix, non-negative similarities are required. However, there exist multiple applications where non-negativity is not guaranteed, which leads to numerous formal and numerical problems, as pointed e.g. in [4]. Although solutions have been proposed for various domains, they have not been discussed for text document clustering. In this paper, we attempt to address them.

Originally, the similarity of text documents was computed as a cosine of the angle between documents embedded in the Term Vector Space (TVS for

---
[⋆] Supported by Polish Ministry of Science

short). These similarities were non-negative by definition. However, the emergence of new and more efficient embedding methods for textual documents such as Word2vec [12], Doc2Vec [7], GloVe [9], BERT [2] based on transformers and others [8] gave rise to the problem of the emergence of negative similarities. This fact causes formal, theoretical, and computational problems for GSC, as computational efficiency and accuracy deteriorate. In addition, normalized Laplacians may not be computable, and the procedure developed to explain clustering results, as described in [13] will fail.

In this paper, we address the clustering of tweets. Their sheer volume, noise, and dynamics impose challenges that hinder the effectiveness of observing clusters with high intra-cluster and low inter-cluster similarity, see e.g. [10].

The paper is organized as follows: Sect. 2 gives an overview of previous work on related topics. Section 3 contains our proposed solution to the negative similarity problem, and Sect. 4 illustrates the effectiveness of the proposed method. A summary of the article is given in Sect. 5. Due to space limitations, only relevant comments are presented here. The reader will find more details in the extended version [6].

2 Previous Work

Graph Spectral Clustering is a methodology for low-complexity approximation of graph clustering based on graph cut criteria. The best-known criteria are RCut (ratio-cut) and NCut (normalized cut) defined as follows:

$$RCut(\varGamma) = \sum_{j=1}^{k} \frac{cut(C_j, \bar{C}_j)}{|C_j|} = \sum_{j=1}^{k} \frac{1}{|C_j|} \sum_{i \in C_j} \sum_{\ell \notin C_j} s_{i\ell} \qquad (1)$$

$$NCut(\varGamma) = \sum_{j=1}^{k} \frac{cut(C_j, \bar{C}_j)}{\mathcal{V}_j} = \sum_{j=1}^{k} \frac{1}{\mathcal{V}_j} \sum_{i \in C_j} \sum_{\ell \notin C_j} s_{i\ell} \qquad (2)$$

Here $s_{i\ell}$ stands for the similarity between objects i and ℓ, (usually it is a number between 0 and 1), $\varGamma$ is the partition of objects, $\bar{C}_j$ denotes the complement of the cluster C_j, $|C_j|$ stands for the cardinality of C_j, and $\mathcal{V}_j = \sum_{i \in C_j} \sum_{\ell} s_{i\ell}$ is the volume of j-th cluster. The elements s_{ij} form a similarity matrix S, and by convention $s_{ii} = 0$ as acyclic graphs are used in GSC.

A *combinatorial Laplacian* is defined as

$$L = D - S, \qquad (3)$$

where D is the diagonal matrix with $d_{ii} = \sum_{\ell=1}^{n} s_{i\ell}$ for each $i \in [n]$. A *normalized Laplacian* $\mathcal{L}$ of the graph represented by S is defined as

$$\mathcal{L} = D^{-1/2} L D^{-1/2} = I - D^{-1/2} S D^{-1/2}. \qquad (4)$$

There exist numerous application areas where it is convenient to use negative similarity measures. They include, but are not limited to, studies based on correlations [3], investigations of electric networks [16], and others. As mentioned in the Introduction, such similarity measures constitute various problems both for the graph cut criteria and the GSC clustering methods if we want to extend them into such a realm.

To overcome the problems with negative similarities, several proposals were elaborated. One can eliminate negative similarities setting them to zero[1]

$$s_{ik}^{(pZ)} = \begin{cases} s_{ik} & \text{if } s_{ik} > 0 \\ 0 & \text{otherwise} \end{cases} \tag{5}$$

Other simple possibilities include taking absolute values, or adding a positive constant to all edge weights. Approaches depend on the application, i.e. why some weights are negative and what the negativity means.

3 Our Approach to Technical Problems

A deeper discussion of these topics can be found in the extended version [6].

3.1 The Problem of Combinatorial Laplacians

A simple transformation relies upon adding a positive constant c to all off-diagonal similarities. New matrix $\tilde{S}$ takes the form $\tilde{S} = S + c(J - I)$, where I is the identity matrix, $J = \mathbf{1}\mathbf{1}^T$ is the matrix with all elements equal to one, and $\mathbf{1}$ is the vector with all entries equal to one. Thus the entries of $\tilde{S}$ are

$$s_{ik}^{(pA)} = s_{ik} + c \tag{6}$$

and the degree matrix $\tilde{D}$ induced by $\tilde{S}$ is

$$\tilde{D} = \text{diag}(\tilde{S}\mathbf{1}) = \text{diag}(S\mathbf{1} + c(n-1)\mathbf{1}) = D + c(n-1)I \tag{7}$$

where $\text{diag}(v)$ is a diagonal matrix with v as its diagonal. Laplacian of $\tilde{S}$ is

$$\tilde{L} = \tilde{D} - \tilde{S} = L - cJ + cnI \tag{8}$$

Let (λ, v) be an eigenpair of the Laplacian L. Then

$$\tilde{L}v = Lv - cJv + cmIv = (\lambda + cn)v \tag{9}$$

since $Jv = \mathbf{0}$. This shows that $(\lambda + cn, v)$ is an eigenpair of $\tilde{L}$. So, RCut minimizing clustering remains unchanged under such an operation.

Alternatively, we can add a positive constant α to the diagonal elements of the degree matrix, that is, $\hat{D} = D + \alpha I$. Then for any eigenpair of L

$$(\hat{D} - S)v = Lv + \alpha v = (\lambda + \alpha)v \tag{10}$$

---

[1] The proposal of signed cuts in [5] ignores in fact negative weights, see [4].

Although the matrix $(\hat{D} - S)$ does not fulfill the requirements of being a combinatorial Laplacian, it still belongs to a large family of generalized graph Laplacians, [1]. Rocha and Trevisan call such matrices perturbed Laplacians and develop their theory in [11].

The similarities $s_{ik}^{(pA)}$ will get out of the range $[0, 1]$ for large enough c. To get them again into this range, we can divide them by $c+1$, leading to elements of the matrix $\bar{S} = \frac{S+c(J-I)}{1+c}$ of the form:

$$s_{ik}^{(pN)} = \frac{s_{ik} + c}{1 + c} \qquad (11)$$

which will lead to the same eigenvectors of the resulting combinatorial Laplacian $\bar{L} = \bar{D} - \bar{S}$ with $\bar{D} = \text{diag}(\bar{S}\mathbf{1})$ as for the original L.

Conclusion: The calculation of Laplacians $\tilde{L}, \hat{L}, \bar{L}$ is not necessary, because the eigenvectors of the original L will not differ. Hence, also the clustering based on the lowest eigenvectors will yield the same results.

Interestingly, the formula (11) can be assigned a geometric interpretation if we compute the similarities as cosines between the document embedding vectors in an N-dimensional space, such as the doc2vec or GloVe space, upon extending this space with an additional dimension, with a constant coordinate, see [6].

3.2 The Problem of Normalized Laplacians

Two types of problems with computation of normalized Laplacian $\mathcal{L} = D(S)^{-1/2} L D(S)^{-1/2}$ may occur. First type of problems is faced, when L contains positive off-diagonal elements while all diagonal elements are positive, so that $\mathcal{L}$ is computable, but some off-line elements remain positive. Curing this situation is analogous to combinatorial Laplacian and shall not be detailed here. Second type of problems occurs, as described among others in [4], when some elements of D are negative. This is a more profound problem than just square rooting negative numbers, [4]. The NCut criterion refers to the cluster volume that may turn out to be negative. A cluster with negative volume – that is with strongly dissimilar documents – has a chance to minimize the NCut criterion. Instead of clusters with strongly similar documents one gets ones with strongly dissimilar ones. This issue strongly resembles the problems with kernel k-means which may not reach the minimum of k-means criterion. Therefore, the NCut criterion must be addressed at the very beginning.

We will consider several proposals.

- Adding a constant to the diagonal of the matrix D,
- Adding a constant to each element of the similarity matrix S,
- Manipulating similarity computation by taking not the cosine of the angle between documents, but half of this angle.
- Replacing similarity with the exponent of the negated distance between documents on a unit sphere.

Consider adding a positive number to all similarities. The clustering taking similarities into account, that is RCut, will not change. Adding a sufficiently large constant will make Normalized Laplacian computable. But will the NCut change under such an operation? Define

$$s_{ik}^{(pD)}(x) = \frac{s_{ik}}{\sqrt{d_{ii}+x}\sqrt{d_{kk}+x}} \quad (12)$$

$s_{ik}^{(pD)}(0)$ is the negated off-diagonal element of normalized Laplacian.

It can be shown that if $s_{ik}^{(pD)}(0) > s_{i\ell}^{(pD)}(0)$, then $s_{ik}^{(pD)}(c) > s_{i\ell}^{(pD)}(c)$, for $c > 0$ and $s_{ik} > s_{i\ell}$. Consider three documents, i, k, ℓ and let

$$\frac{s_{ik}}{\sqrt{d_{kk}+c}} > \frac{s_{i\ell}}{\sqrt{d_{\ell\ell}+c}}$$

If we are in the realm of non-negative similarities (other cases can be handled similarly)

$$\frac{s_{ik}^2}{s_{i\ell}^2} > \frac{d_{kk}+c}{d_{\ell\ell}+c}$$

With an increase of c, the expression on the right-hand side grows/decreases towards one. Therefore, if originally $s_{ik} > s_{i\ell}$, then the expression is true.

This means that adding a constant to the normalized Laplacian diagonal keeps to a great extent the ordering of similarities, so that the results of clustering may be similar, unless the normalization changes proportions between similarities in the original Laplacian. In case of some negative d_{ii}, adding an appropriate constant may turn the Laplacian into a computable one, resulting in clustering similar to the one originally intended.

However, the problem is that this solution tends to be in fact a version of the newly introduced NRCut [14], and not NCut. So another approach is needed.

Another solution would be adding a constant c to each similarity ($s_{ik}^{(pA)}(c) = s_{ik} + c$. Again, no warranty that the ordering of all normalized similarities will be the same and hence that clustering result is the same.

One solution could be to transform the similarity matrix S into a positive one $S^{(pQ)}$ as follows:

$$s_{ik}^{(pQ)} = \cos\left(\frac{\pi}{2}\frac{\arccos s_{ik}}{\max_{i,k\in[n], i\neq k}\arccos s_{ik}}\right) \quad (13)$$

whereby max is computed over all off-diagonal elements of the S matrix. cos is non-negative in the range $[0, \frac{\pi}{2}]$ while it is negative for greater angles. By dividing the actual angles between documents by the maximal angle, and multiplying with $\frac{\pi}{2}$ we scale all the angles into the non-negative cosine range. Now, the traditional normalized Laplacian is applicable. The ranking of similarities of combinatorial Laplacian is preserved completely, but again no warranty for the normalized similarities. The above formula can be generalized to:

$$s_{ik}^{(pC)} = \cos\left(\frac{\arccos s_{ik}}{1+c}\right) \quad (14)$$

$c = 1$ is for sure a reasonable choice, because *arccos* returns values in the range $[0, \pi]$ and dividing this result by 2 scales them into the required range $[0, \frac{\pi}{2}]$.

If the graph has isolated nodes, we already get into trouble with normalized Laplacian because of division by zero; the same applies to the NCut criterion. Therefore, a change in understanding NCut is needed in such a way that a cluster with all nodes isolated has a non-zero volume. So, the similarity needs to be transformed. As the Euclidean distance between two normalized vectors x_i, x_j equals to $\|x_i - x_j\|^2 = 2(1 - s_{ij})$, where $s_{ij} = cos(x_i, x_j)$, our proposal is

$$s_{ik}^{(pE)} = e^{-(1-s_{ik})/2} \quad (15)$$

This should be applied to get a new similarity matrix S' as well as a redefinition of NCut to NCut$^{(pE)}$ (based on the new similarities). There is no need to worry about isolated nodes. If we generalize the transformation $s_{ik}^{(pE)}$ to

$$s_{ik}^{(pE)}(c) = e^{-(1-(s_{ik}+c))/2} \quad (16)$$

the normalized Laplacian will remain the same for all values $c \geq 0$ because adding a constant in the exponent is the same as multiplying the similarity with another constant.

3.3 Negativity Versus Explainability

GSC result explanation procedure elaborated in [13] encounters serious problems as it is based on the products of word embedding vectors and cluster center vectors which would lead to meaningless negative word importance. The correction proposed for combinatorial Laplacian based GSC keeps the spirit of [13]. As normalized Laplacian is concerned, we show in [13] that additive corrections of similarity measure does not disturb the explanation bridge.

4 Experiments

We conducted experiments on the effectiveness of GSC methods to predict hashtags for a large set of **short** tweets using different methods to deal with negative similarities, as mentioned in the formulas (6), (11), (13), (14), (16), (5) for $c = 0, 1, 2, 3$. Note that $c = 0$ means that no correction of negative similarity was performed. For the modified similarity matrices, both combinatorial and normalized Laplacians were used in GSC. The computations were performed for the traditional Term Vector Space (TVS, tf, tfidf) as well as for the GloVe based embeddings: TweetGlove (trained on Twitter data) and WikiGlove (trained on Wikipedia Data). The results can be accessed via the link https://github.com/ipipan-barstar/ICCS25.MfHNSiEGSCoTD.

As expected, the Term Vector Space embeddings have no negative similarity problems. TweetWiki embedding leads to numerous negative similarity matrix entries, but no problem with row sums occurs for our samples. The most difficult problems occur for the WikiGloVe embedding, as there are many more negative similarities and there are multiple rows with negative entries in three of the samples. When the correction of negative similarities is based on zeroing them, Normalized Laplacian based clustering could be executed. We see that GloVe based do not have a big advantage over TVS based embeddings. The results are the worst compared to other methods. When the correction of negative similarities is based on adding a constant to all off-diagonal similarities, with or without dividing for normalization, Normalized Laplacian-based clustering could be executed except for $c = 0$ in WikiGlove embedding because the diagonal of D contained negative entries. GloVe based GSC does not have any TVS based embeddings. At the same time, adding the constant $c = 1$ significantly improves the performance, while higher constants do not contribute much to the results. When normalizing over the largest angle between document vectors, the results are worse for TVS embeddings, and slightly worse for GloVe embeddings. When dividing the angle between document vectors, the results constitute an improvement when dividing by at least two, but dividing by higher values does not contribute anything. When replacing primary similarities with their exponential variants, the variants do not differ much, but replacement of negative similarities with exponential ones helps the GloVe based embeddings, and also the TVS embeddings benefit from this transformation. The GSC results for combinatorial Laplacians are significantly worse, and the effects of transformations are generally marginal, as expected.

Detailed results for all samples are available at https://github.com/ipipan-barstar/ICCS25.MfHNSiEGSCoTD.

5 Conclusions

In this paper we discussed the issues in graph spectral clustering of documents resulting from growing popularity of embeddings different from the traditional Term Vector Space. The major problem is the negative cosine similarities between documents under these embeddings. We have studied six different methods for overcoming negative similarities. Essentially, the combinatorial Laplacian-based clusterings seem to be unaffected by negative similarities, as demonstrated by theoretical arguments. In case of normalized Laplacians, the method of setting negative similarities to zero yields the worst results. The other methods perform similarly. Interestingly, it turns out that for Term vector Space embeddings there may be an improvement of performance when the similarity correction is applied. We were also able to provide a geometric interpretation of one of the studied methods [6]. Note however other approaches to use similarities, e.g. [15]. This study was limited to two GloVe type embeddings, based on Wiki training data and Tweeter training data.

References

1. Biyikoglu, T., Leydold, J., Stadler, P.F.: Laplacian eigenvectors of graphs. Perron-Frobenius and Faber-Krahn type theorems. In: Lecture Notes in Mathematics, vol. 1915. Springer, Heidelberg (2007). https://doi.org/10.1007/978-3-540-73510-6

2. Devlin, J., Chang, M.W., Lee, K., Toutanova, K.: Bert: Pre-training of deep bidirectional transformers for language understanding (2019). https://doi.org/10.48550/ARXIV.1810.04805
3. Knyazev, A.: Edge-enhancing filters with negative weights. In: 2015 IEEE Global Conference on Signal and Information Processing (GlobalSIP), pp. 260–264 (2015). https://doi.org/10.1109/GlobalSIP.2015.7418197
4. Knyazev, A.: Signed Laplacian for spectral clustering revisited. arXiv (2017). https://arxiv.org/abs/1701.01394
5. Kunegis, J., Schmidt, S., Lommatzsch, A., Lerner, J., Luca, E.W.D., Albayrak, S.: Spectral analysis of signed graphs for clustering, prediction and visualization, pp. 559–570. Society for Industrial and Applied Mathematics (2010). https://doi.org/10.1137/1.9781611972801.49
6. Kłopotek, M.A., Wierzchoń, S.T., Starosta, B., Czerski, D., Borkowski, P.: A method for handling negative similarities in explainable graph spectral clustering of text documents – Extended Version. CoRR **abs/2504.12360** (2025). https://arxiv.org/abs/2504.12360
7. Lau, J.H., Baldwin, T.: An empirical evaluation of doc2vec with practical insights into document embedding generation (2016). https://doi.org/10.48550/ARXIV.1607.05368
8. Lin, T., Wang, Y., Liu, X., Qiu, X.: A survey of transformers. AI Open **3**, 111–132 (2022). https://doi.org/10.1016/j.aiopen.2022.10.001
9. Pennington, J., Socher, R., Manning, C.: GloVe: global vectors for word representation. In: Moschitti, A., Pang, B., Daelemans, W. (eds.) Proceedings of the 2014 Conference on Empirical Methods in Natural Language Processing (EMNLP), pp. 1532–1543. Association for Computational Linguistics, Doha, Qatar (2014). https://doi.org/10.3115/v1/D14-1162
10. Ravi, J., Kulkarni, S.: Text embedding techniques for efficient clustering of Twitter data. Evol. Intel. **16**, 1667–1677 (2023). https://doi.org/10.1007/s12065-023-00825-3
11. Rocha, I., Trevisan, V.: A Fiedler-like theory for the perturbed Laplacian. Czech Math. J. **66**, 717–735 (2016). https://doi.org/10.1007/s10587-016-0288-4
12. Rong, X.: word2vec parameter learning explained (2014). https://doi.org/10.48550/ARXIV.1411.2738, arXiv:1411.2738 [cs.CL]
13. Starosta, B., Kłopotek, M.A., Wierzchoń, S.T., Czerski, D., Sydow, M., Borkowski, P.: Explainable graph spectral clustering of text documents. PLoS ONE **20**(2), e0313238 (2025). https://doi.org/10.1371/journal.pone.0313238
14. Starosta, B., Kłopotek, M.A., Wierzchoń, S.T.: Approaches to explainability of output of graph spectral clustering methods. to appear in monograph "Design and Implementation of Artificial Intelligence Systems". published by University of Siedlce (2025)
15. Stec, H., Ekanadham, C., Kallus, N.: Is cosine-similarity of embeddings really about similarity? arXiv (2024). https://doi.org/10.1145/3589335.3651526
16. Zelazo, D., Buerger, M.: On the definiteness of the weighted Laplacian and its connection to effective resistance. In: 53rd IEEE Conference on Decision and Control, pp. 2895–2900. IEEE (2014). https://doi.org/10.1109/cdc.2014.7039834

Preliminary Comparison of Different EDs Performance, Using Simulation

Ramona Galeano[1,2](✉), Alvaro Wong[1], Dolores Rexachs[1], Remo Suppi[1], Eva Bruballa[3], Francisco Epelde[4], and Emilio Luque[1]

[1] Computer Architecture and Operating Systems Department, Universidad Autónoma de Barcelona (UAB), Barcelona, Spain
RamonaElizabeth.Galeano@autonoma.cat,
{Alvaro.Wong,Dolores.Rexachs,Remo.Suppi}@uab.cat, Emilio.luque@uab.es
[2] Facultad Politécnica, Universidad Nacional de Asunción, San Lorenzo, Paraguay
[3] Escoles Universitaries Gimbernat (EUG), Computer Science School, Universitat Autónoma de Barcelona, Sant Cugat del Valles 08174, Spain
eva.bruballa@eug.es
[4] Medicine Department, Hospital Universitari Parc Tauli, Barcelona, Spain
fepelde@tauli.cat

Abstract. A reference model is a model of something that contains a fundamental objective or idea of something and can be established as a reference for multiple purposes. We performed an analysis based on a reference model following the Spanish Ministry of Health's document of standards and recommendations to achieve better care, efficiency, and uniformity in emergency services. This standard describes the guidelines and resources needed in hospital emergency services for better patient care. We also analyzed the standards of emergency services in the following countries: the United States, the United Kingdom, Germany, Canada, Paraguay, Argentina, the United Arab Emirates, and Turkey. The objective of the research is to analyze the efficiency of the Spanish model following its reference standard compared to the standards of other countries, to explore how the Spanish emergency service would work using the parameters of the emergency service standards compared to the standards of different countries, specifically the KPI we analyzed is the Door to the Doctor (DtD), through simulation. It was concluded that in some countries, DtD times improve compared to the Spanish reference standard, and in some cases, they worsen.

Keywords: simulation · agent-based model · KPI · standard

1 Introduction

The standards outline the guidelines and resources required for hospital Emergency Departments (EDs) to care for patients. As the ED is a central clinical unit of a hospital, it aims to achieve greater efficiency, more transparent communication, and uniformity of services across the health system. Spain has standards and recommendations for EDs. The objective of this document of standards and recommendations is to make criteria available to the administrators of hospital EDs for the organization and management of these units, for improving safety, efficiency, and quality in service provision, as well as for their design and equipment [1].

Some studies carried out on the analysis of standards in EDs are the following: The study conducted by Bermejo et al. [4], the objective is to determine which triage systems are used in Spain, specifically in the hospital emergency services of the national public health network, define their characteristics, and evaluate the degree of compliance with the recommendations regarding the study. The study conducted by Arco et al. [3] compares the physical structure, human resources, and quality indicators of emergency services in public hospitals in the Spanish regions of Madrid and Catalonia. The study conducted by Jones et al. [7] compares emergency nursing practice standards across five countries(United States, Canada, United Kingdom, New Zealand, and Australia). The study conducted by Moon et al. [12] studied triage accuracy using the Korean triage, they used Korean Triage and Acuity Scales (KTAS) based on the Canadian Triage and Acuity Scale (CTAS) to classify patients according to their symptoms. The nurses determine the triage level based on the symptoms. The study conducted by Pines et al. [13] provides an international perspective on ED crowding, including data and trends from 15 countries outside the United States.

In this paper, we present a simulation study using an Agent-Based Model (ABM) grounded in the standards and recommendations for hospital EDs in Spain [1]. The model incorporates key parameters outlined in the Spanish guidelines, including triage procedures, staffing, treatment, boxes, and patient volume. Using these specifications, we configured our ED simulator and conducted a comparative analysis with healthcare systems in Germany, the United Kingdom, the United States, Turkey, the United Arab Emirates, Canada, Argentina, and Paraguay. The objective was to evaluate the performance of the Spanish ED system when simulated under the parameters defined by these other countries. A key Performance Indicator (KPI) in this study is the Door-to-Doctor (DtD) time. DtD is a metric that measures the length of time from when the patient enters the Emergency Department until the medical screening exam begins by the physician. The simulator employed in this research was developed in earlier work by the High-Performance Computing for Efficient Applications and Simulation (HPC4EAS) research group at the UAB [8]. Section 2 describes The emergency department model. Section 3 presents the Results and discussion, and Sect. 4 presents the Conclusions and Future Work.

2 The Emergency Department Model

In our research, we present a simulation of an ABM taking as a reference the document of standards and recommendations for hospital emergencies of Spain [1], using the parameters specified in that standard: staffing of doctors and nurses, staffing of triage, number of boxes, number of patients. The other unmodified parameters, such as patient flow distribution, doctor and nurse patient attention time, triage time, imaging and labs services, are those validated by the Parc Tauli Hospital in Sabadell. The patient entrance used is from the Parc Tauli Hospital, as can be seen in Fig. 1 (a); on the x-axis, the hours of the day in which the patients arrive are observed, and on the y-axis, the percentage of patient arrivals per hour can be observed. As can be seen in the graph, the hours with the lowest number of patients arriving are between 1 a.m. and 8 a.m., after which patients come in a more significant proportion. The time when the peak of patients increases considerably is between 11 a.m. and 12 a.m.

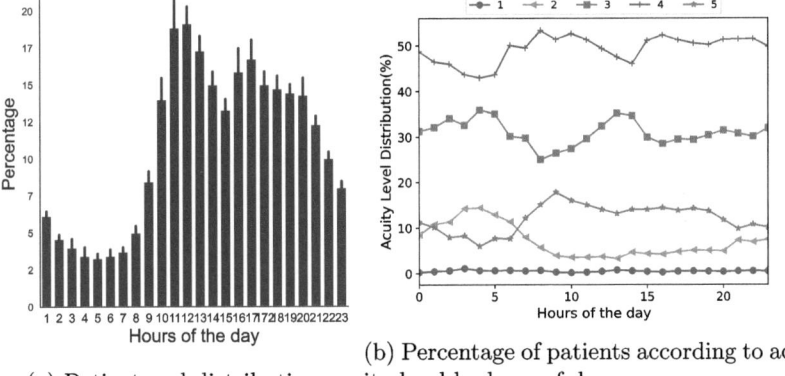

(a) Patient and distribution.

(b) Percentage of patients according to acuity level by hour of day.

Fig. 1. Percentage of arrivals and distribution of patients by time of day and percentage of patients according to acuity level by hour of day.

In the Fig. 1 (b) shows the percentage of acuity level of patients arriving per hour. There are five acuity levels used in triage, where patients with acuity levels I, II, and III are the most serious, and IV and V are the mildest. It can be observed in the graph that the lowest percentage of patients who arrive at the ED are patients with severity level I, which is represented by the blue lines. Acuity level I represents 1% of patients who arrive at the hospital, acuity level II represents 10% of patients, acuity level III represents 30%, acuity level IV represents 49%, and acuity level V approximately 15%. On the x-axis, you can see the hours of the day, and on the y-axis, the percentage.

In Table 1 compares triage times in each country according to the triage standard used by that country, grouped by patient severity level. United Kingdom (UK) uses the MTS [10], United States (US) uses Emergency Severity Index (ESI) [6], Germany uses the MTS [5], Canada uses the CTAS [1], United Arab Emirates (UAE) uses the ESI [2], Turkye uses the ESI [14], Argentina uses the MTS [9], Spain uses the CTAS and Spanish Triage System (SET) [1], and Paraguay uses the SET [11].

Table 1. Comparison of triage time used in each country according to the acuity level. At level I, for all countries, attention must be immediate. Level II-V: Standard time of medical care from arrival at the hospital to the visit (DtD), expressed in minutes. Triage type, input, and parameters.

Country	Type	Triage time (minutes)				Input	Parameters		
		Level II	Level III	Level IV	Level V	Patient	Nurse	Doc	Box
Spain	CTAS/SET	15	30	60	120	300	9	8	22
US	ESI	10	60	120	240	100	20	10	30
UK	MTS	10	60	120	240	200	20	10	30
Germany	MTS	10	60	120	240	150	10	6	15
Canada	CTAS	15	45	120	240	70	10	8	20
UAE	ESI	10	60	120	240	200	15	10	30
Turkye	ESI	10	60	120	240	250	15	8	25
Argentina	MTS	10	60	120	240	50	10	6	20
Paraguay	SET	15	30	60	120	20	5	4	10

3 Results and Discussion

Using our emergency simulator, we ran several simulations, each lasting nine months. We modified the patient, doctor, nurse, and box parameters established in the standards and recommendations document for Spain [1] and the standards of the other countries analyzed. This data can be found in Table 1. We substituted this data into our simulator as input parameters for each country. The severity acuity level distribution and other data, such as patient flow distribution, doctor and nurse patient attention time, triage time, imaging and labs services are the real data from the Parc Tauli Hospital in Sabadell. In all the figures analyzed, the x-axis shows the hours of the day, and the y-axis shows the average time of medical care from arrival at the hospital to the visit (DtD), expressed in minutes. The value in each Level column is the "maximum" DtD time for each acuity level (I to V) in each country. At level I for all countries, attention must be immediate. The value in the patient column is the total number of patients arriving at the ED per day in each country. The patient, nurse, doctor, and box columns are parameters for simulation according to each country's standards. Severity acuity level I patients, neither pass admission nor triage, they go directly to doctor. Patients with severity level I are very few and do not have significant differences, which is why we selected patients with severity levels II and III.

The first set of simulation results can be seen in Fig. 2; we group 4 countries, which are (a) Spain, (b) the United States, (c) Canada, and (d) the UK. The objective is to analyze how hospitals in Spain would behave using the parameters of standards in other countries. We analyze the acuity level II for Spain, the United States, Canada, and the UK. For this level, the standard triage time for these countries is 15 min for Spain, 10 min for the United States, 15 min for Canada, and 10 min for the UK, as shown in Table 1.

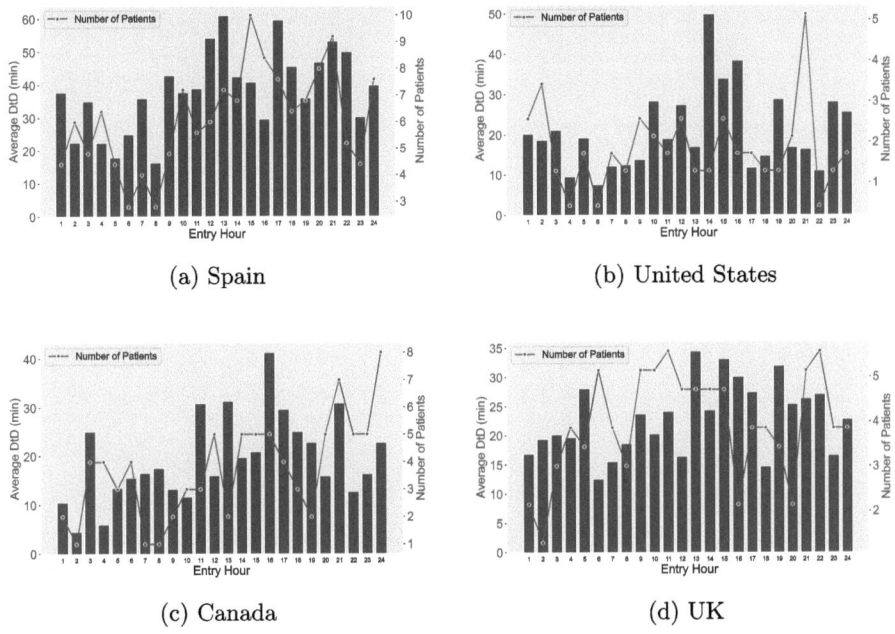

Fig. 2. Comparison of the average DtD for patients with acuity level II from Spain, United States, Canada and UK.

We can see in Fig. 2 (b) that in the United States, the DtD times improve much more than those of Spain; on average, they reach more than the time required by the triage standard for that acuity level. We can see in Fig. 2 (c) that in Canada it shows that the times for DtD improve much more than those of Spain, and on average, they manage to reach more than the time required by the triage standard for acuity level II. We can see in Fig. 2 (d) For the UK, using the data from the UK standard, the times are better than those of Spain, and the average maximum DtD time is shorter than that of Spain.

In Fig. 3, we group 6 countries, which are (a) Spain, (b) Argentina, (c) Paraguay, (d) Germany, (e) United Arab Emirates and, (d) Turkey. We analyzed the acuity level III for Spain, Argentina, Paraguay, Germany, United Arab Emirates and, Turkey. For this acuity level, the standard triage time for these countries is 30 min for Spain, 60 min for Argentina, 30 min for Paraguay, 60 min for Germany, 60 min for United Arab Emirates and, 60 min for Turkey as shown in Table 1.

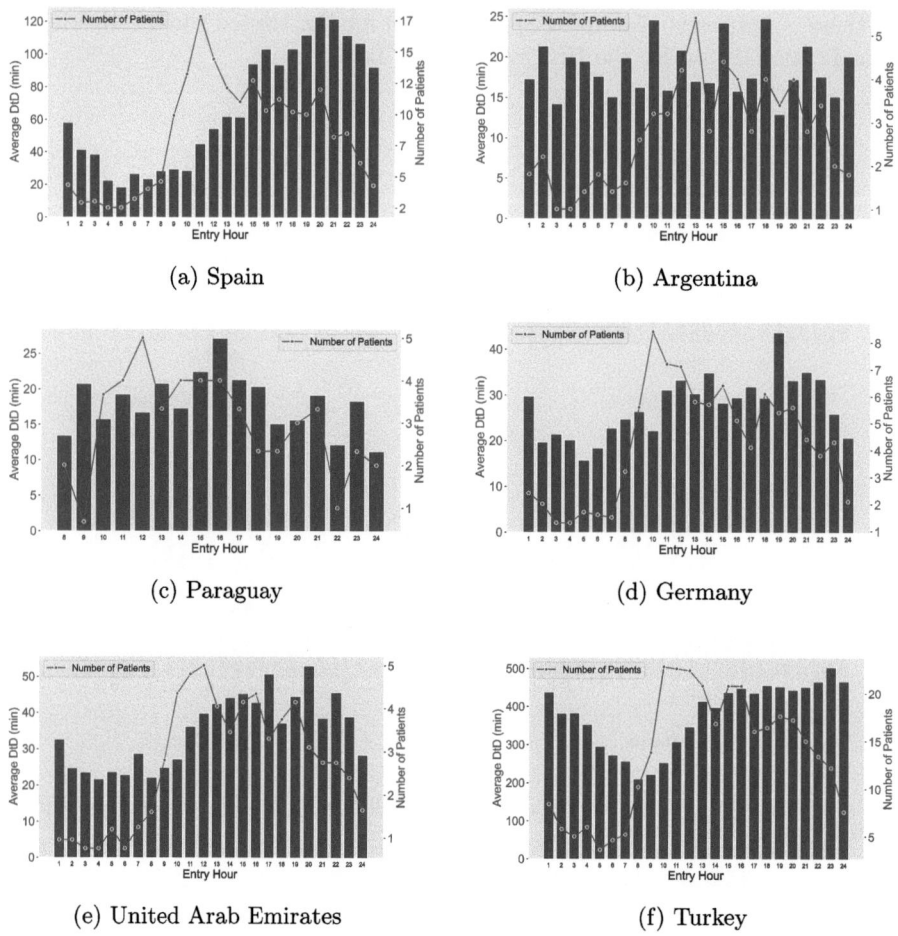

Fig. 3. Comparison of the average DtD for patients with acuity level III from Spain, Argentina, Paraguay, Germany, United Arab Emirates and, Turkey.

We can see in Fig. 3 (b) that in Argentina, the DtD times improve much more than those of Spain, and on average, they manage to reach more than the time required by the triage standard for that acuity level. We can see in Fig. 3 (c) that in Paraguay, the DtD times improve much more than those of Spain, and on average, they manage to reach more than the time required by the triage standard for acuity level III. In the case of Argentina and Paraguay, they always meet the time required by the standard of those countries for acuity level III, as can be seen in Fig. 3. It can be observed that between 4 and 10 a.m. Spain reached the 30-minute time required by the standard. It can be seen that (d) Germany achieves the DtD time required by its standard, which is 60 min, and its times are better than those of Spain.

In Fig. 3 (e), it is observed that in the United Arab Emirates, the DtD times behave better than those of Spain for severity level III. It reaches the DtD time required by its standard for all hours. In Fig. 3 (f), with the data from the Turkish standard, it shows that the DtD times are worse than those of Spain in all cases. The triage system used by the United Arab Emirates is the ESI, and the one used by Turkey is also the ESI, as shown in Table 1 as well as the parameters used for the simulation, such as the number of resources of patients, nurses, doctors, and boxes according to each country.

4 Conclusion and Future Work

In this research, we present a simulation of an ABM, taking the reference standard and recommendations for ED in Spain as a reference, using the parameters specified in said standard, such as triage, staffing, boxes, and number of patients. Based on the parameters of said standard, we substitute these data from our ED simulator. We compare it with the parameters of the standards of other countries such as Germany, UK, the US, Turkey, the United Arab Emirates, Canada, Argentina, and Paraguay through simulations. We analyze how the Spanish system works using the parameters of the standards of the other countries. Specifically, the KPI we analyze is the DtD. We concluded that with some countries, the DtD time in Spain can be reduced, and it worsens by replacing it with other countries' standards. This research could help to re-evaluate the standard that is being used in that country so that it helps to reduce patient waiting time. Hospital administrators or government health managers could benefit from our simulator to evaluate if there are standards that could be applied to the country to help reduce patient waiting time and use them in hospitals. In future work, we intend to expand this research further and include the parameters of the population pyramid and more KPIs in the study, such as the Length of Patient Stay (LoS).

Acknowledgments. This research has been supported by the Agencia Estatal de Investigacion (AEI), Spain and the Fondo Europeo de Desarrollo Regional (FEDER) UE, under contract PID2020-112496GB-I00 and PID2023-146978OB-I00, and partially funded by the Fundacion Escuelas Universitarias Gimbernat (EUG).

Disclosure of Interests. The authors have declared that no competing interests exist.

References

1. Unidad de urgencias hospitalaria estándares y recomendaciones (sf). https://www.sanidad.gob.es/areas/calidadAsistencial/excelenciaClinica/docs/UUH.pdf. Accessed 21 Apr 2025
2. AlSerkal, Y., et al.: Triage accuracy and its association with patient factors using emergency severity index: findings from united Arab emirates. Open Access Emerg. Med., pp. 427–434. (2020). https://doi.org/10.2147/OAEM.S263805, https://www.tandfonline.com/doi/full/10.2147/OAEM.S263805

3. del Arco Galán, C., et al.: Physical structure, human resources, and health care quality indicators in public hospital emergency departments in the autonomous communities of Madrid and Catalonia: a comparative study. Emergencias **29**(6), 373–83 (2017). https://revistaemergencias.org/wp-content/uploads/2023/09/Emergencias-2017_29_6_373-383-383_eng.pdf
4. Bermejo, R.S., Fadrique, C.C., Fraile, B.R., Centeno, E.F., Cueva, S.P., María, E.: Triage in spanish hospitals. Emergencias **25**(1), 66–70 (2013). https://revistaemergencias.org/wp-content/uploads/2023/09/Emergencias-2013_25_1_66-70_eng.pdf
5. Christ, M., Grossmann, F., Winter, D., Bingisser, R., Platz, E.: Modern triage in the emergency department. Deutsches Ärzteblatt Int. **107**(50), 892 (2010). https://pmc.ncbi.nlm.nih.gov/articles/PMC3021905/
6. Gilboy, N., Tanabe, P., Travers, D., Rosenau, A.M.: Emergency severity index (ESI): a triage tool for emergency department care, Version 4. Implementation Handbook, 2012 Edition. Agency for Healthcare Research and Quality, Rockville, MD (2011). https://www.ahrq.gov/sites/default/files/wysiwyg/professionals/systems/hospital/esi/esihandbk.pdf
7. Jones, T., Shaban, R.Z., Creedy, D.K.: Practice standards for emergency nursing: an international review. Australas. Emerg. Nurs. J. **18**(4), 190–203 (2015). https://www.sciencedirect.com/science/article/abs/pii/S1574626715000828
8. Liu, Z., Rexachs, D., Luque, E., Epelde, F., Cabrera, E.: Simulating the micro-level behavior of emergency department for macro-level features prediction. In: 2015 Winter Simulation Conference (WSC), pp. 171–182. IEEE (2015). https://ieeexplore.ieee.org/abstract/document/7408162
9. Londoño, V.C.C.: Triage in the intoxicated patient. Revista Argentina de Medicina **11**(4), 345–348 (2023). https://revistasam.com.ar/index.php/RAM/article/view/901
10. Mackway-Jones, K., Marsden, J., Windle, J.: Emergency Triage: Manchester Triage Group. Wiley (2013). http://dickyricky.com/books/medical/Emergency%20Triage%203e%20-%20ALSG%202014.pdf
11. Mesquita, M., Pavlicich, V., Luaces, C.: El sistema español de triaje en la evaluación de los neonatos en las urgencias pediátricas. Revista chilena de pediatría **88**(1), 107–112 (2017). https://www.scielo.cl/pdf/rcp/v88n1/en_art08.pdf
12. Moon, S.H., Shim, J.L., Park, K.S., Park, C.S.: Triage accuracy and causes of mistriage using the Korean triage and acuity scale. PLoS ONE **14**(9), e0216972 (2019)
13. Pines, J.M., et al.: International perspectives on emergency department crowding. Acad. Emerg. Med. **18**(12), 1358–1370 (2011)
14. Çinar, O., Çevik, E., Salman, N., Cömert, B.: Emergency severity index triage system and implementation experience in a university hospital. Turk. J. Emerg. Med. **10**(4), 156–164 (2010). https://turkjemergmed.com/abstract/292

Efficient Peptide MRM Transition Prediction via Convolutional Hashing

Ramon Adàlia[1,2](✉), Gemma Sanjuan[1], Tomàs Margalef[1], and Ismael Zamora[2]

[1] Universitat Autònoma de Barcelona, Cerdanyola del Vallès, Spain
Ramon.Adalia@autonoma.cat, {Gemma.Sanjuan,Tomas.Margalef}@uab.cat
[2] Lead Molecular Design, S.L., Sant Cugat del Vallès, Spain
{ramon.adalia,ismael.zamora}@leadmolecular.com

Abstract. We present a novel method for predicting multiple reaction monitoring (MRM) transitions for peptides in targeted proteomics. Our approach employs a hash-based representation inspired by convolutional neural networks, efficiently encoding peptide fragments as sparse count vectors that capture local sequence context. Using gradient-boosted decision trees, our method achieves mean Hits@5 scores of 3.4318 (hash-based) and 3.5405 (hybrid model with target frequency), significantly outperforming baselines. Transpiling trained models into Zig enables exceptional computational efficiency, with low memory usage (1180 kB) and a throughput of 388–451 peptides/second even on mobile devices, enabling lightweight, high-speed processing for scalable peptide MRM transition prediction in high-throughput proteomics workflows.

Keywords: MRM transitions · Peptide quantification · Edge computing

1 Introduction

Multiple Reaction Monitoring (MRM) is a mass spectrometry technique enabling precise peptide quantification in applications ranging from biomarker discovery to pharmaceutical development. Mass spectrometry identifies compounds by measuring the mass-to-charge ratio (m/z) of ions, with MRM specifically employing a three-step process: (1) filtering ions with the desired m/z, (2) fragmenting these filtered ions via collision with inert gas, and (3) filtering fragments by m/z for highly-specific detection. Determining appropriate m/z values for the final filtering step requires analyzing pure samples with the second filter deactivated—a time-intensive and resource-demanding process [9], underscoring the value of computational prediction methods.

Peptides, amino acid chains linked by peptide bonds (typically < 50 residues), fragment into characteristic **b ions** (N-terminal) and **y ions** (C-terminal) during mass spectrometry. Amino acids contain an amino group, a carboxyl group, hydrogen, and a variable side chain determining their properties, all connected to a central alpha carbon atom. The 20 standard amino acids use single-letter codes (A, C, D, E, F, G, H, I, K, L, M, N, P, Q, R, S, T, V, W, Y), with the fragment notation indicating the amino acid count (e.g., in CYIQNCPLG, b_3=CYI and y_2=LG), as shown in Fig. 1.

Existing approaches have significant limitations: small-molecule models [1] are unsuitable due to structural differences; simple representations such as amino acid composition [2] miss positional information despite wide use [3,4]; and neural networks like ESM [8] capture sequence features effectively but impose excessive computational demands.

Our approach addresses these challenges by replacing CNN [6] convolutions with hashing to efficiently encode peptide fragments as sparse vectors, capturing local sequence context with minimal overhead. Trained with LightGBM [5] and transpiled to Zig, our method achieves both high accuracy and computational efficiency for real-world proteomics applications.

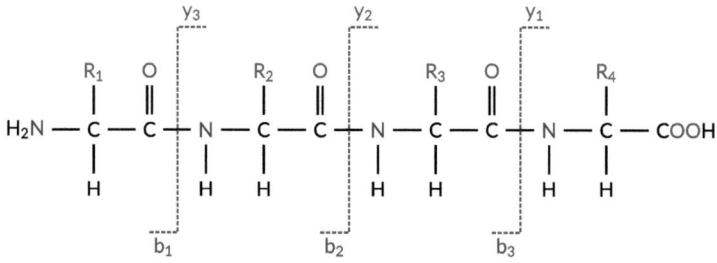

Fig. 1. Peptide fragmentation notation for a peptide with four amino acids.

2 Methodology

2.1 Data

The dataset used in this study was obtained from *Multiple Reaction Monitoring Assays for Large-Scale Quantitation* [7] and comprises 2,965 unique peptide sequences (ranging from 6–25 amino acids, median 11) with experimentally optimized MRM transitions. The majority of peptides (2,922) have five transitions, while the remainder have between 3–9, totaling 14,881 transitions represented as b/y ions with specific charge states (e.g., b9++). Charge distributions are predominantly +1 (11,592), +2 (3,177), and +3 (111). Figure 2 illustrates key dataset characteristics, highlighting the prevalence of smaller ions.

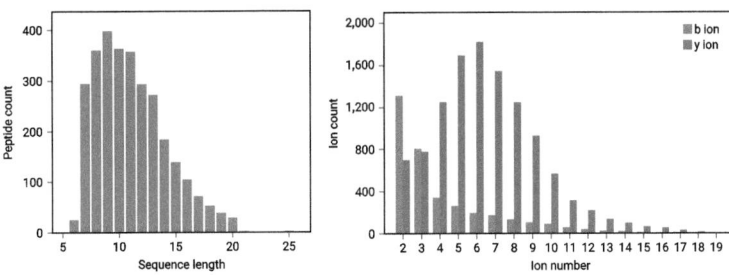

Fig. 2. Dataset composition showing sequence length distribution and b/y ion counts.

2.2 Peptide Fragment Representation

Algorithm 1. Peptide fragment sequence representation with neighborhood hashing

---
Require: Input sequence s of length $n \geq 3$, radius $R \geq 0$
1: Initialize empty array H for hashes
2: **for** $j \leftarrow 1$ to n **do**
3: $\quad v_j^0 \leftarrow [0] \parallel description(s_j)$ $\quad\quad\quad\quad\quad\quad\quad\quad\quad\quad\quad\quad\quad$ ▷ Initial encoding
4: $\quad h_j^0 \leftarrow hash(v_j^0)$; Append h_j^0 to H
5: **end for**
6: **for** $i \leftarrow 1$ to R **do**
7: $\quad h_1^i \leftarrow hash([i, h_1^{i-1}, h_2^{i-1}])$; Append to H $\quad\quad\quad\quad\quad\quad\quad$ ▷ First AA
8: $\quad$ **for** $j \leftarrow 2$ to $n - 1$ **do**
9: $\quad\quad h_j^i \leftarrow hash([i, h_{j-1}^{i-1}, h_j^{i-1}, h_{j+1}^{i-1}])$; Append to H $\quad\quad\quad$ ▷ Middle AAs
10: $\quad$ **end for**
11: $\quad h_n^i \leftarrow hash([i, h_{n-1}^{i-1}, h_n^{i-1}])$; Append to H $\quad\quad\quad\quad\quad\quad$ ▷ Last AA
12: **end for**
13: **return** H

Our peptide fragment representation (Algorithm 1) draws inspiration from 1D CNNs. Each amino acid is mapped to a two-element integer vector encoding: an ordinal position (e.g., A = 1, C = 2, Q = 3, etc.) and a binary inclusion indicator (1 if part of the fragment, 0 otherwise). For instance, in the b_2 fragment of peptide ACQA, the first amino acid 'A' encodes as [1,1] (included) whereas the final 'A' encodes as [1,0] (excluded).

Instead of standard convolutions, we leverage hashing to capture neighborhood context, aggregating local information through a hash function that processes neighboring amino acids. For a vector $[c_1, c_2, \ldots, c_n]$, we compute its hash as:
$$X_{i+1} = (a \cdot X_i + c_i) \mod m, \quad \text{for } i \in [1, n-1]$$
with parameters $m = 2^{32}$, $X_1 = 1013904223$, and $a = 1664525$.

Hashes are computed iteratively for radius R, transforming the resulting list H into a sparse count vector that comprehensively represents the fragment. Finally, charge states are incorporated as an additional integer feature.

2.3 Model Training

We employ Gradient Boosting Decision Trees via LightGBM, selected for their numerous advantages with high-dimensional sparse data, invariance to monotonic transformations, robustness to correlated features, and straightforward inference paths. Models are trained with default hyperparameters (100 trees, maximum 31 leaves) using LambdaMART with NDCG objective and a radius parameter of $R = 2$.

To enhance computational efficiency, we preprocess the sparse matrix by eliminating constant columns, reducing the feature count based on the selected R. For the simplest case where $R = 0$, only 41 features remain: one for the charge state and 40 for the amino acid presence/absence patterns. This dimensional reduction necessitates consistent feature selection between the training and inference phases.

2.4 Model Evaluation and Inference

We assess performance through 5-fold cross-validation on the dataset (593 peptides per fold), ranking candidate transitions for held-out peptides. Performance is quantified using the Hits@5 metric, which counts correct transitions appearing in the top five predictions.

For efficient inference, we transpile LightGBM's model output into optimized Zig functions that take hashmap input (column indices with corresponding counts) and produce prediction scores. This approach creates compact standalone binaries with Zig's advanced compiler optimizations, eliminating LightGBM dependencies while ensuring cross-platform compatibility and processing efficiency.

3 Results and Discussion

3.1 Baseline Methods

For a random baseline model, we compute the exact Hits@5 distribution from empirical data rather than relying on simulations. For a peptide of length n, total number of possible transitions is $6(n-1)$ (derived from $n-1$ possible b/y-ions and 3 charge states). When randomly sampling 5 transitions, the number of correct predictions follows a hypergeometric distribution with the probability mass function:

$$P(X = k) = \frac{\binom{T}{k}\binom{6(n-1)-T}{5-k}}{\binom{6(n-1)}{5}},$$

where T represents experimentally identified transitions (typically 5). Averaging probabilities across all peptides yields a probability model with an expected Hits@5 value of 0.3396 for the random model.

A more effective baseline is the *target model*, which ranks transitions by their frequency in the dataset, always predicting [y6+, y5+, b2+, y7+, y4+] in the top 5. This straightforward approach achieves an average Hits@5 score of 2.2405.

3.2 Model Performance

We evaluated random, target, hash-based, and *hash+target* models using cross-validation. The hybrid model integrates target encoding with convolutional hashing, incorporating ion frequency within training folds to mitigate overfitting. This is implemented as an additional integer feature (e.g., for ion y_8^+, the feature represents how frequently y_8^+ appears among the top-5 fragment ions in the training set). Figure 3 illustrates performance distributions via Hits@5.

The results demonstrate marked improvement from the baseline models to hash-based models. The hash-based model achieves a mean Hits@5 of 3.4318 compared with 2.2405 for the target model (representing a 58% improvement, $p < 10^{-5}$, assessed with a one-sided paired permutation test). The *hash+target* model further enhances the performance to 3.5405 ($p = 0.00381$).

While all the models achieve comparable rates of at least one correct prediction, the hash-based approaches demonstrate substantially higher frequencies of multiple correct predictions, with the hybrid model producing 19.28× more perfect predictions than the target model does. This highlights how our approach excels particularly when maximizing correctly predicted transitions is critical for downstream applications.

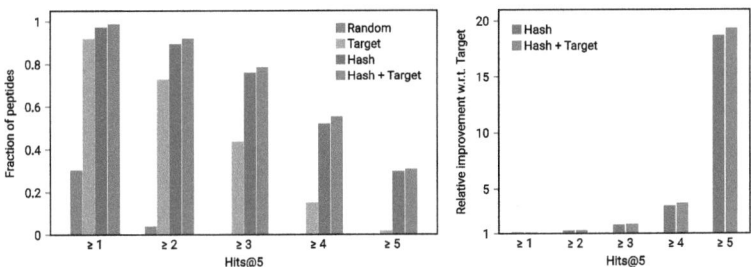

Fig. 3. Distribution of Hits@5 scores across evaluated models, showing hash-based models achieve higher frequencies of multiple correct predictions.

4 Ablation Study and Radius Effects

We conducted a detailed investigation into the impact of the radius parameter R on model performance, with $R = 0$ representing the complete removal of the convolutional effects. Figure 4 reveals substantial performance improvement when R increases from 0 to 1, with gains plateauing at higher values, suggesting diminishing returns from incorporating larger neighborhood contexts.

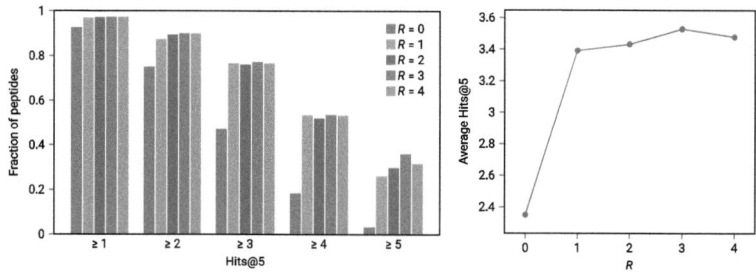

Fig. 4. Effect of radius R on Hits@5 scores, showing significant improvement from $R = 0$ to $R = 1$ and diminishing returns at higher values.

Table 1 presents computational metrics across different R values (Intel i7-8750H for training, Android Snapdragon 695 for inference). As R increases, the feature count grows exponentially (39 to 106,735), with the training time increasing from 18 s to over 14 min. Notably, memory usage at inference remains constant (1180 kB) across all configurations, whereas inference throughput decreases marginally (14% from $R = 0$ to $R = 4$), demonstrating excellent scalability.

Table 1. Training and inference performance for different radius values R.

R	Features	Training Time	Inference		
			Mem. (kB)	Time (s)	Peptides/sec
0	39	17.68 s	1180	6.564 ± 0.074	451.71
1	12,750	1:23.89 min	1180	6.789 ± 0.012	436.74
2	50,147	6:45.87 min	1180	6.989 ± 0.016	424.24
3	82,040	11:05.74 min	1180	7.205 ± 0.034	411.52
4	106,735	14:32.50 min	1180	7.632 ± 0.078	388.50

5 Conclusion

We have introduced a novel approach for predicting MRM transitions using a hash-based representation inspired by convolutional neural networks. By encoding peptide fragments as sparse count vectors with gradient boosting trees, our method achieves significant improvements over established baselines. The hybrid model integrating target frequencies with convolutional hashing achieves a mean Hits@5 of 3.5405, a 58% improvement over the baseline approach.

Our ablation study reveals the importance of the local sequence context, with peak performance gains occurring between $R = 0$ and $R = 1$. This demonstrates the value of neighborhood information while suggesting diminishing returns from larger contexts. The approach shows exceptional efficiency through Zig transpilation, maintaining consistent memory usage (1180 kB) across configurations while processing 388–451 peptides/second on mobile devices.

This combination of accuracy and efficiency makes our method practical for resource-constrained environments. The compact representation enables training on large datasets with minimal overhead. Future work could extend this methodology to other biomolecule types or incorporate domain knowledge to enhance prediction accuracy, streamlining integration into proteomics workflows.

Acknowledgments. This study was supported by Pla de Doctorats Industrials del Departament de Recerca i Universitats de la Generalitat de Catalunya (grant 2023-DI-00006).

Disclosure of Interests. The authors have no competing interests relevant to this article.

References

1. Adàlia, R., et al.: Development of a predictive multiple reaction monitoring (MRM) model for high-throughput ADME analyses using learning-to-rank (LTR) techniques. J. Am. Soc. Mass Spectrom. **35**(1), 131–139 (2023). https://doi.org/10.1021/jasms.3c00363
2. Chou, K.C.: A novel approach to predicting protein structural classes in a (20–1)-d amino acid composition space. Proteins **21**(4), 319–344 (1995). https://doi.org/10.1002/prot.340210406
3. Du, Q.S., Jiang, Z.Q., He, W.Z., Li, D.P., Chou, K.C.: Amino acid principal component analysis (aapca) and its applications in protein structural class prediction. J. Biomol. Struct. Dyn. **23**(6), 635–640 (2006). https://doi.org/10.1080/07391102.2006.10507088. pMID: 16615809
4. Jahandideh, S., Abdolmaleki, P., Jahandideh, M., Asadabadi, E.B.: Novel two-stage hybrid neural discriminant model for predicting proteins structural classes. Biophys. Chem. **128**(1), 87–93 (2007). https://doi.org/10.1016/j.bpc.2007.03.006. https://www.sciencedirect.com/science/article/pii/S0301462207000749
5. Ke, G., et al.: Lightgbm: a highly efficient gradient boosting decision tree. In: Guyon, I., Luxburg, U.V., Bengio, S., Wallach, H., Fergus, R., Vishwanathan, S., Garnett, R. (eds.) Advances in Neural Information Processing Systems. vol. 30. Curran Associates, Inc. (2017). https://proceedings.neurips.cc/paper_files/paper/2017/file/6449f44a102fde848669bdd9eb6b76fa-Paper.pdf
6. Kiranyaz, S., Avci, O., Abdeljaber, O., Ince, T., Gabbouj, M., Inman, D.J.: 1d convolutional neural networks and applications: a survey. Mech. Syst. Signal Process. **151**, 107398 (2021). https://doi.org/10.1016/j.ymssp.2020.107398. https://www.sciencedirect.com/science/article/pii/S0888327020307846
7. Michaud, S.A., et al.: Multiple reaction monitoring assays for large-scale quantitation of proteins from 20 mouse organs and tissues. Commun. Biol. **7**(1), 6 (2024). https://doi.org/10.1038/s42003-023-05687-0

8. Rives, A., et al.: Biological structure and function emerge from scaling unsupervised learning to 250 million protein sequences. Proc. Natl. Acad. Sci. **118**(15), e2016239118 (2021). https://doi.org/10.1073/pnas.2016239118. https://www.pnas.org/doi/full/10.1073/pnas.2016239118
9. Shou, W.Z., Zhang, J.: Recent development in high-throughput bioanalytical support for in vitro admet profiling. Expert Opin. Drug Metabol. Toxicol. **6**(3), 321–336 (2010). https://doi.org/10.1517/17425250903547829547829. pMID: 20163321

Comparison of Crash Simulations on Two Types of Flying Cars

Hiroto Sato[1(✉)], Ayato Takii[2], Masashi Yamakawa[1], Yusei Kobayashi[1], Shinichi Asao[3], Seiichi Takeuchi[3], and Takahiro Ikeda[1]

[1] Kyoto Institute of Technology, Matsugasaki, Sakyo, Hashikami 606-8585, Kyoto, Japan
m4623016@kit.edu.ac.jp
[2] Kobe University, 1-1 Rokkodaicho, Nada-Ku, Kobe 657-8501, Hyogo, Japan
[3] College of Industrial Technology, Nishikoya, Amagasaki 661-0047, Hyogo, Japan

Abstract. Numerical simulations are being used to ensure safety in the development of flying cars, which are expected to solve all kinds of traffic problems. In order to reproduce flight in a virtual environment, this research has been conducting numerical simulation research using the moving computational domain method and the multi-axis sliding mesh method to calculate the coupling of fluid and rigid body motion. This method enables rotational motion of the rotor and flight in an infinite region, and visualization of the flow field around the aircraft and its behavior including control. In this study, a comparison of the behavior during sudden rotor stops was performed on two different aircraft: the coaxial contra-rotating octorotor eVTOL treated in a previous study and a newly modeled domed dodecarotor eVTOL. The increased number of rotors and the wider circular arrangement of the rotors allowed for a wider range of measures to be taken in the event of rotor stoppage, and the crash risk was reduced by a factor of approximately 1/100. The differences in crash conditions and changes in aircraft attitude based on a comparison of the two models indicate that this calculation method allows for design improvements and behavior prediction based on relative evaluation.

Keywords: Computational Fluid Dynamics · Advanced Air Mobility · Crash

1 Introduction

In recent years, increasing attention has been directed toward utilizing low-altitude airspace—currently characterized by relatively low usage density—as a new mode of personal transportation. This has led to the growing interest in so-called "flying cars," also known as Advanced Air Mobility (AAM). Among them, electric vertical takeoff and landing aircraft (eVTOL), which require no runways and offer potential advantages in terms of convenience and environmental impact [1], are especially noteworthy. However, since these vehicles are intended to operate over urban areas, safety remains the most critical concern. In fact, mechanical failures account for approximately 15% of all accidents involving small aircraft and helicopters, with over 50% of such incidents resulting in fatalities [2], highlighting the serious risks posed by component failures.

To improve development efficiency, a variety of numerical simulation techniques have been proposed. Prior studies have made significant progress in this field: Jun-Young An et al. [3] analyzed aircraft performance using dynamic modeling, while Okazaki et al. [4] investigated control strategies for small drones. Further research has explored turning maneuvers and attitude changes during abrupt rotor stoppage [5–7], demonstrating the importance of coupling dynamic and fluid models. However, these studies have primarily focused on a single vehicle model, limiting the ability to evaluate differences in aircraft characteristics or to validate simulation results against real-world behavior.

To address this gap, the present study conducts simulations of abrupt rotor stoppage in a larger eVTOL model with increased rotor count, size, and weight. By comparing the flight behavior and attitude response during crash scenarios across different vehicle configurations, this research aims to quantitatively assess the impact of rotor count on crash risk and attitude controllability.

2 Numerical Approach

2.1 Governing Equation

The Reynolds number at the tip of the propeller during hovering of the aircraft considered in this study is approximately 5.98×10^8, and the maximum Mach number is approximately 0.52. Therefore, the fluid is treated as a compressible fluid, and the 3-D Euler equation for a non-viscous compressible fluid and the equation of state for an ideal gas are used to prioritize computational efficiency and to avoid consideration of viscosity. The governing equations for the fluid are shown in Eqs. (1)-(3).

$$\frac{\partial \boldsymbol{q}}{\partial t} + \frac{\partial \boldsymbol{E}}{\partial x} + \frac{\partial \boldsymbol{F}}{\partial y} + \frac{\partial \boldsymbol{G}}{\partial z} = 0, \tag{1}$$

$$\boldsymbol{q} = \begin{bmatrix} \rho \\ \rho u \\ \rho v \\ \rho w \\ e \end{bmatrix}, \boldsymbol{E} = \begin{bmatrix} \rho u \\ \rho u^2 + p \\ \rho uv \\ \rho uw \\ u(e+p) \end{bmatrix}, \boldsymbol{F} = \begin{bmatrix} \rho v \\ \rho uv \\ \rho v^2 + p \\ \rho vw \\ v(e+p) \end{bmatrix}, \boldsymbol{G} = \begin{bmatrix} \rho w \\ \rho uw \\ \rho vw \\ \rho w^2 + p \\ w(e+p) \end{bmatrix}, \tag{2}$$

$$p = (\gamma - 1)\left\{e - \frac{1}{2}\rho\left(u^2 + v^2 + w^2\right)\right\}. \tag{3}$$

where $\boldsymbol{q}$ is the conserved quantity vector, $\boldsymbol{E}$, $\boldsymbol{F}$, and $\boldsymbol{G}$ are the inviscid flux vectors in the x, y, and z directions. ρ is the density of the fluid, u, v, and w are velocity components in the x, y, and z directions, p is the pressure of the gas, e is the total energy per unit volume, and γ is the specific heat ratio. All variables are dimensionless quantities. Moreover, to reproduce the motion of a flying car through a weakly coupled fluid-rigid-body calculation, the three-dimensional Newton-Euler equations of motion for translation and rotation are used.

2.2 Computational Conditions

To simulate the flight of a flying car under complex conditions, we coupled flight dynamics and fluid dynamics as in Takahashi et al. [7], and used the MCD (Moving Computational Domain) method [8, 9] and the multi-axis sliding mesh method [10]. Table 1 shows the characteristic values used to nondimensionalize the calculations. The characteristic length is the total length of the aircraft, the characteristic density is the air density, and the characteristic velocity is the speed of sound. The boundary conditions are slip wall boundary for the airframe surface, a Riemann boundary condition for the outer region, and sliding mesh interface boundary for the split interface.

Table 1. Characteristic values

Density of the air	1.247 kg/m^3
Characteristic velocity	340.29 m/s
Characteristic length	13.0 m

3 Flight Simulation of Flying Car

3.1 Computational Model

Figure 1 shows the computational model of the octorotor eVTOL with four pairs of coaxial contra-rotating rotors used in the previous study and the dodecarotor eVTOL with 12 rotors arranged in a dome shape that will be compared in this study. Table 2 shows the characteristics of the newly introduced model. The new model is considerably larger than the model used in the previous study. One rotor has three blades, and the smallest grid is in front of the rotor blade blades, with a minimum grid width of 6 mm as in the previous study. The computational model used in this study is an unstructured tetrahedral mesh and was created using MEGG3D [11, 12]. The total number of elements is approximately 5.8 million, and these are computed in an OpenMP parallel environment [13]. The time step size for each step was set to 0.0005, and the total simulated duration was approximately 6 s. Using an Intel Core i9-14900K processor, the computation time required for simulating approximately 5 s of real time was about 18 days per case.

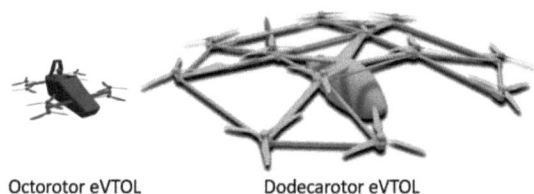

Fig. 1. Computational Model for octorotor eVTOL and dodecarotor eVTOL

Table 2. Comparison of Aircraft Specifications

	octorotor eVTOL	dodecarotor eVTOL
Overall length and width [m]	4 × 4	13 × 13
Airframe weight [kg]	400	1400
Number of rotors [-]	8	12

3.2 Flight Simulation Conditions

To outline this simulation, rotors are named as shown in Fig. 2. First, the 12 rotors are roughly classified into three groups of three rotors each, FL (Front Left) FR, RL, and RR. The rotors with the greatest influence in the pitch direction are classified as 1, those with the greatest influence in the roll direction are classified as 2, and those with the greatest influence in the altitude direction are classified as 3. The direction of rotation is distinguished by color.

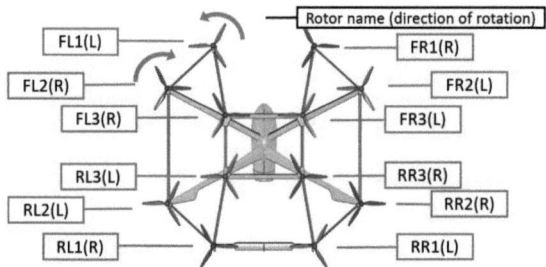

Fig. 2. Rotor name of dodecarotor eVTOL

We will compare crash conditions and changes in aircraft attitude for two different aircraft models. For this purpose, a simulation was first conducted in which some of the rotors of the new aircraft were stopped suddenly while hovering. The combinations of rotors to be stopped are shown in Table 3. After that, we maintain hovering by PID control and mixing. The results will be compared with those of previous studies [7].

Table 3. Experimental conditions

terms	Rotor name to be stopped	Number of rotors to be stopped
(1)	FL1	1
(2)	FL	3
(3)	FL, RR	6
(4)	FL, FR3, RR	7
(5)	FL, FR3, RR, RL3	8
(6)	FL, FR1	4

4 Calculation Results

4.1 Comparison of One Rotor Stopped

A comparison is made between the FLU rotor stop in the coaxial contra-rotating octorotor eVTOL and the FL1 rotor stop in the dodecarotor eVTOL. Figure 3 shows the aircraft attitudes and torques over time after rotor stoppage. Solid lines represent octorotor eVTOL 1, and dotted lines represent dodecarotor eVTOL 2. The dodecarotor exhibits attitude changes reduced by a factor of eight or less but experiences greater vibration during hovering. This is likely due to its larger fuselage and longer propellers, which cause higher torque and blade tip speeds. The coaxial octorotor reduces vibration by lowering its rotational speed, made possible by thrust recovery from the lower propeller utilizing the upper propeller's wasted energy [14]. Furthermore, the rotor stop positions differ—about 1:1 from the center of gravity for the octorotor and 2:1 for the dodecarotor—which clearly affects the resulting moment and aircraft motion. These results demonstrate that aircraft structure and rotor layout significantly influence stability and vibration. Moreover, the effect of rotor placement is clearly reflected in the generated moments and aircraft behavior, indicating that the outcomes are reproducible even across different aircraft types.

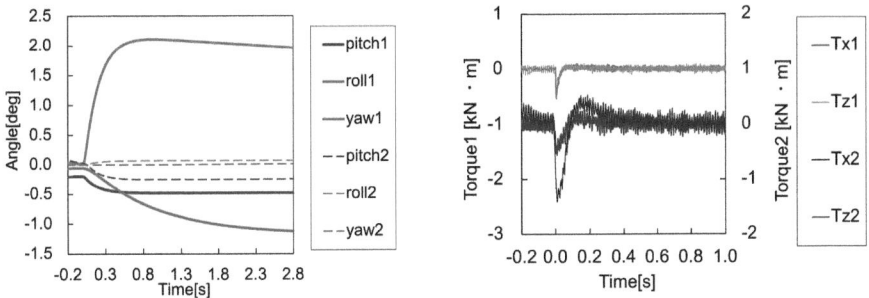

Fig. 3. Comparison of fuselage attitude and torque around fuselage

4.2 Comparison of Crashes Due to Sudden Rotor Stoppage

In the octocopter-type eVTOL, the aircraft crashed when the FLU and FLL rotors were suddenly stopped. Based on this, two failure conditions are examined for the dodecacopter-type eVTOL: (2) simultaneous stoppage of three front-left (FL) rotors, and (3) stoppage of an additional rotor on top of Condition (2). Figure 4 shows the time histories of attitude changes and overlaid snapshots during descent for all three conditions. The solid line represents the octocopter (1), and the dashed line the dodecacopter (2). Q-criterion isosurfaces (second invariant of the velocity gradient tensor) are shown for both configurations: octorotor (left) and dodecarotor (right). These are colored by velocity magnitude, while the aircraft surfaces are colored by pressure. All values are non-dimensional.

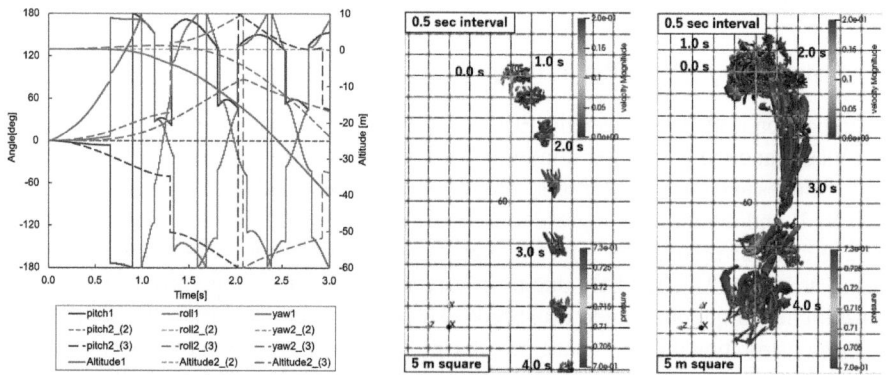

Fig. 4. Time history of aircraft attitude changes and Sequential images

In the dodecarotor eVTOL, Condition (2) maintained stable altitude, indicating successful hover, while the other conditions resulted in crashes. This suggests that even with front-left rotors stopped, attitude control remains possible due to four surrounding rotors (FR1, FR2, RL1, RL2) generating compensating moments near the center of gravity.

Comparing crash behaviors, the dodecarotor showed less attitude disturbance. Within 3 s, the octorotor rotated 3.5 times and dropped 40 m, whereas the dodecarotor rotated only 0.75 times and dropped 25 m. Assuming that all rotors fail independently with equal probability ppp, the crash probabilities are $\frac{1}{7}p^2$ and $\frac{8}{495}p^2$ for the octorotor and dodecarotor, respectively. These results demonstrate that increasing rotor count and distributing them circularly improves stability and greatly reduces crash risk in the event of rotor failures.

4.3 Summary of Rotor Sudden Stop in Dodecarotor eVTOL

From conditions (1)-(6), we summarize the possibility of crashes depending on the number and location of rotor stops. First, it was found that up to three rotor stoppages, no combination of rotors would cause a crash, while four to six rotor stoppages would lead to a crash if the stoppages were unevenly located. When more than seven rotors were stopped, the aircraft was able to maintain its attitude; however, it could not sustain its altitude, leading to a crash.

From these results, we calculate the combination of rotor stoppages that lead to crashes and the probability of this aircraft crashing due to rotor stoppage by assuming that the probability of one rotor stoppage is p and constant. The results are summarized in Table 4.

Table 4. Combination of stop rotor arrangement to crash

Number of stopped rotors	0–3	4	5	6	7	8	9	10	11	12
combination	0	8	60	78	792	495	220	66	12	1
Probability of a crash	0	$\frac{8}{495}p^4$	$\frac{60}{792}p^5$	$\frac{78}{924}p^6$	p^7	p^8	p^9	p^{10}	p^{11}	p^{12}

The likelihood of a crash due to rotor stoppage can be estimated based on rotor-generated lift. Although four rotors can theoretically maintain attitude control—as seen in conventional quadcopters—greater lift is needed for heavier eVTOLs. Analysis showed that with four rotors operating at 169% of their hover speed, lift could balance aircraft weight within the control range. However, experiments revealed that even five functioning rotors failed to maintain hover. This discrepancy is attributed to rotor deceleration variations, strut interference, and aerodynamic effects that reduce actual lift. Once the aircraft tilts beyond a certain angle, sufficient lift can no longer be produced, revealing the limitations of multirotor systems under sudden changes. These findings validate the experiment. From Table 4, the probability that the dodecarotor eVTOL crashes due to rotor failures corresponds to the sum of the probabilities in the bottom row. Given that all terms are positive and the individual probability p is sufficiently small compared to 1, the arithmetic-geometric mean inequality can be applied. As a result, the total crash probability is obtained as $0.361p^8$.

4.4 Comparison of Crash Probabilities

Similarly, for octorotor eVTOL, let p denote the probability of a single rotor failure. Since a crash occurs when both rotors in a coaxial pair fail, the total crash probability is calculated as $0.658p^5$. While coaxial contra-rotating rotors offer compact and efficient assembly, failure in a coaxial pair significantly increases crash risk. In the dodecarotor eVTOL, even with a front-left rotor failure, attitude control is possible due to the symmetrical placement of four other rotors around the center of gravity. However, circular rotor arrangements tend to be larger and heavier, presenting a tradeoff between safety and performance. To contextualize risk, we compare with commercial airplanes, where the fatal accident rate is 1 in 13.7 million [15]. Assuming 10-h flights, the accident rate is 7.3×10^{-6} per 1,000 h, with 21% due to mechanical failures [16]. Given that airplanes carry 100 times more passengers, the acceptable crash probability due to rotor failure in flying cars is 1.53×10^{-8}. To match this safety level, dodecarotor rotor failure probability must be under $0.03[/1000\,h]$, and $0.00033[/1000\,h]$ for octorotors. These values serve as rotor design targets to ensure safety equivalent to that of airplanes.

5 Conclusions

Using coupled fluid–rigid body simulations with MCD and sliding mesh methods, we analyzed the effects of sudden rotor stoppages in a domed dodecarotor eVTOL, incorporating attitude control. Various rotor failure combinations were evaluated under hovering conditions, and comparisons were made with other aircraft, including a coaxial octorotor model. The simulations showed that hovering was maintainable with up to three rotors stopped, and in some cases, even with four to six, depending on their positions. This enabled a quantitative evaluation of crash probability. Theoretical lift calculations and experimental results also highlighted the limitations of multi-rotor systems in responding to sudden lift loss, particularly under aerodynamic interference.

Compared to the coaxial octorotor model from prior research, the dodecarotor's greater number of rotors and circular layout enabled more failure-tolerant configurations, significantly reducing crash risk by about 1/100. The study also revealed differences in crash dynamics and control behavior stemming from rotor placement and fuselage geometry. This simulation approach proved effective in reproducing emergency conditions and provided a practical tool for improving design reliability and predicting behavior under failure scenarios.

Acknowledgments. This paper is based on results obtained from a project, JPNP14004, subsidized by the New Energy and Industrial Technology Development Organization (NEDO).

References

1. Ison, D.C.: Consumer willingness to fly on advanced air mobility (AAM) electric vertical takeoff and landing (eVTOL) aircraft. Collegiate Aviat. Rev. Int. **42**(1), 103–118 (2024). https://doi.org/10.22488/okstate.24.100223
2. Kunimitsu, I., Tatsuo, O.: Statistical data analyses on aircraft accidents in japan: occurrences, causes and countermeasures. Am. J. Oper. Res. **5**(3), 222–245 (2015). https://doi.org/10.4236/ajor.2015.53018
3. An, J.-Y. et al.: Practical approach to mission autonomy of multiple unmanned air mobilities. Int. J. Control, Autom. Syst. **22**(8), 2513–2536 (2024). https://doi.org/10.1007/s12555-023-0662-6
4. Okazaki, H., Isogai, K.: Motor failure issues in flying drones. IEICE Fundam. Rev. **14**(1), 44–59 (2020). https://doi.org/10.1587/essfr.14.1_44
5. Gomi, R., et al.: Flight simulation from takeoff to yawing of eVTOL airplane with coaxial propellers by fluid-rigid body interaction. Adv. Aerodyn. **5**(1) (2023)
6. Magata, T., Takii, A., Yamakawa, M., Takeuchi, S., Chung, Y.M.: Forward and turning flight simulation of flying cars and comparative evaluation of flight dynamics models considering wind disturbances. J. Comput. Sci. **85**, 102519 (2025)
7. Takahashi, N., et al.: Numerical Simulation of the Octorotor Flying Car in Sudden Rotor Stop. In: ICCS 2023, pp.33–46 (2023)
8. Yamakawa, M., et al.: Numerical simulation for a flow around body ejection using an axisymmetric unstructured moving grid method. Comput. Therm. Sci. 4(3), 217–223 (2012)
9. Asao, S., Matsuno, K., Yamakawa, M.: Simulations of a falling sphere with concentration in an infinite long pipe using a new moving mesh system. Appl. Therm. Eng. **72**(1), 29–33 (2014)

10. Takii, A., et al.: Six degrees of freedom flight simulation of tilt-rotor aircraft with nacelle conversion. J. Comput. Sci. **44**, 101164 (2020)
11. Ito, Y., Nakahashi, K.: Surface triangulation for polygonal models based on CAD data. Int. J. Numer. Meth. Fluids **39**(1), 75–96 (2002)
12. Ito, Y.: Challenges in unstructured mesh generation for practical and efficient computa- tional fluid dynamics simulations. Comput. Fluids **85**(1), 47–52 (2013)
13. Yamakawa, M., et al.: Domain decomposition method for unstructured meshes in an OpenMP computing environment. Comput. Fluids **45**(1), 168–171 (2011)
14. Theys, B., et al.: Influence of propeller configuration on propulsion system efficiency of multi-rotor unmanned aerial vehicles. In: 2016 ICUAS, pp. 1–9. IEEE (2016)
15. Barnett, A., et al.: Airline safety: still getting better? J. Air Transp. Manage. **119**, 102641 (2024). https://doi.org/10.1016/j.jairtraman.2024.102641
16. Khadivar, H., et al.: Flying safe: the impact of corporate governance on aviation safety. J. Air Transp. Manage. **124**, 102743 (2025)

Centrality Resilience in Complex Networks

Fariba Afrin Irany[1], Turja Kundu[1](✉), Soumya Sarakar[2], Animesh Mukherjee[3], and Sanjukta Bhowmick[1]

[1] Department of Computer Science and Engineering, University of North Texas, Denton, TX, USA
[2] Microsoft India, Hyderabad, West Bengal, India
[3] Department of Computer Science and Engineering, IIT Kharagpur, Kharagpur, West Bengal, India

Abstract. In this paper we study centrality resilience, that is, how well closeness and betweenness centralities are maintained under attacks. We propose efficient attack models to disrupt the rank of the top k centrality vertices. To develop our attack models, we extend the concept of rich clubs of influential vertices to the more general framework of scattered rich clubs–dense subgraphs of high-centrality vertices that are spread across the network. To improve computational efficiency, we use snowball sampling to identify these important substructures.

Our results over real-world networks demonstrate that our algorithm can identify the single or scattered rich clubs efficiently and is more effective in disrupting the centrality rankings of the network, compared to other baseline methods.

Keywords: Attack Algorithm · Snowball Sampling · Scattered Rich Club

1 Introduction

Networks (or graphs) are mathematical models of complex systems of interacting entities that occur across diverse disciplines including cyber-security, bioinformatics, and mobile networks. The entities in the complex systems are represented as vertices, and their dyadic interactions as edges. Resiliency to attacks is an important property of networks. Most research focus on connectivity, that is how attacks can disconnect the network [4]. In this paper, we develop attack models to study network resilience with respect to path-based centralities. We term this type of resilience as *centrality resilience.* as opposed to the connectivity resilience of the earlier studies.

Motivation: Centrality resilience is a powerful tool for insidiously disrupting the functioning of a network without a drastic change to its structure. When one part of a network cannot communicate with the rest of the system, it is easy to infer that the cause is due to disconnectivity. Attack on centrality may not disconnect the network, but result in longer distances and more time to transmission when traversing the network. That the increased length of the distances is due to the change in the ranking of the high centrality vertices may not be immediately apparent until the centralities of the system are re-computed. Such techniques can be applied for both malicious (stealth attacks in cybersecurity, where the location of attack cannot be immediately known) and benign (reducing the load on bottlenecks, under limited resources) attacks.

Our goal is to develop edges attack models that will disrupt the centrality distribution of the network, i.e. high centrality nodes will no longer be of high centrality.

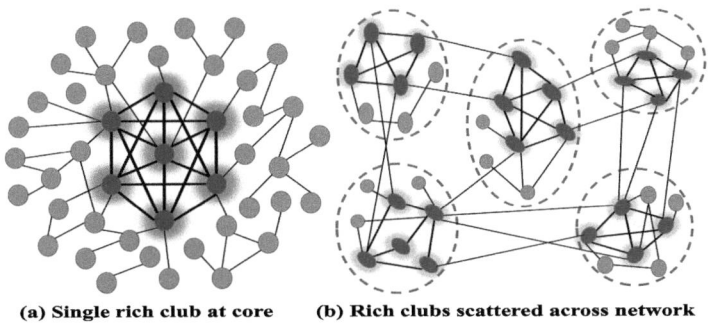

(a) Single rich club at core **(b) Rich clubs scattered across network**

Fig. 1. Rich clubs in complex networks. (a) Network with a single rich club. (b) Network with rich clubs scattered across the network.

Key Steps and Contributions: Our steps to achieve this objective are as follows.
Step 1: Identify the structural properties that affect centrality resilience. (Sect. 2).

We identify substructures that affect centrality resilience based on the observation of [13] that path-based centralities form dense clusters or "rich clubs". A *rich club*, is an assortative subgraph, where all the vertices have high value of a vertex-based property p, in this case, betweenness or closeness centrality. Breaking these rich clubs will affect the ranking of the high centrality vertices. We extend the concept of a single rich club [13] to *scattered rich clubs*, i.e. clusters that contain high centrality vertices that may be spread across the network. Figure 1 compares the structure of networks with a single (a) and multiple scattered (b) rich club(s). Here, "high centrality vertices" refers to the union of the top-k (here set to 20) high betweenness and closeness vertices.

Contribution: We demonstrate that rich clubs of high centrality vertices can be spread across multiple clusters of a network and develop a metric to measure the *degree of scatteredness* based on the distribution of high centrality vertices across the clusters.

Step 2: Develop efficient algorithms to extract scattered rich clubs. (Sect. 3).

To compute the degree of scatteredness, we need to find the high centrality nodes as seeds and construct dense clusters around them. This approach is very computationally intensive. Further, several iterative steps are required to find the appropriate sets of nodes that form the clusters. We observe due to their high centrality, the vertices can reach out to many neighbors in a few hops, and thus the rich clubs have expander graph like properties. We use snowball sampling [11], which exploits the expander property of graphs, to identify the regions containing high centrality vertices.

Contribution: We develop an efficient algorithm to find scattered rich clubs using snowball sampling. We demonstrate that our sampling method can find most of the high centrality nodes and with much lower complexity than the naive method of finding high centrality vertices and then forming clusters.

Step 3: Develop attack models based on scattered rich clubs (Sect. 4).

Our final step is to develop attack models by removing edges that belong to the rich clubs. Attacks based on removing the vertices are equivalent to removing multiple edges. We therefore posit that edge removal is a more fine grained operation where the attack is spread strategically across the networks. We use snowball sampling to find the single/scattered rich clubs, and then select edges to delete from these rich clubs.

Contribution: We develop sample-based attack strategies for disrupting the rank of high centrality vertices, based on single and scattered rich clubs. We quantify how much the ranking of the vertices have been perturbed, as per the Jaccard index. Our results show that our attack models are more effective than other baseline attack methods.

Related Work. Understanding network structure that governs robustness is an important research area. Adiga *et al.* [2] studied the robustness of the top cores under sampling and in noisy networks and Laishram *et al.* [8] developed core based approaches to increase network resilience. Here, we focus on resilience with respect to centrality.

There have been several papers on attack models that perturb the centrality of top k high centrality nodes. The attack models proposed in [7] target nodes with high importance and the average time complexity of the three models proposed in the paper is $O(E)$. A heuristic algorithm in [3] balances the centrality measures by link addition. [5] proposes an attack model targeting nodes with high eigenvector centrality.

Table 1. Test suite of networks, along with their properties.

Network	Type	Nodes	Edges	Avg Clus Co-eff	Max Core
dmela	Biological	7393	25569	0.01	11
euroraod	Infrastructure	1174	1417	0.016	2
HepPh	Citation	34546	420877	0.284	30
CondMat	Collaboration	23133	93439	0.63	25
as20000102	Autonomous System	6474	12572	0.25	12
caida	Autonomous System	26476	53383	0.21	22
HepTh	Citation	27770	352285	0.312	37
email-univ	Communication	1133	5451	0.22	11
AstroPh	Collaboration	18772	198050	0.63	56
grid-fission-yeast	Biological	2026	12637	0.221	34

2 Scattered Rich Clubs

We evaluate our approach on ten real-world networks from SNAP (Stanford Large Network Dataset) [9] and Network repository [1], as given in Table 1. We show that high centrality vertices can be distributed across the network cores and propose a new metric, the *degree of scatteredness*, to quantify this distribution.

Distribution of Rich Clubs of High Centrality Vertices. In several networks, the high betweenness and closeness centrality vertices are located in the innermost cores [12,13]. Since the innermost cores form a dense subgraph, therefore this becomes a rich club. However, in many networks the high centrality vertices can be distributed across cores. This phenomenon requires a more general definition, that of *scattered rich clubs*.

Scattered Rich Club: Given a graph $G(V, E)$, a vertex property f, and a threshold value, p, *scattered rich clubs* are a *set of disjoint* subgraphs, $\{S_1, S_2, \ldots, S_n\}$, where $S_i(V_i, E_i)$, such that $V_1 \cap V_2 \cap \ldots \cap V_n = \phi$ and $\forall v \in (V_1 \cup V_2 \cup \ldots \cup V_n), f(v) \geq p$.

The implicit expectation of a single subgraph in rich clubs is relaxed in scattered rich clubs. This generalizes the definition of rich clubs, because every network will have scattered rich clubs, even if the number of vertices in the rich clubs is one. Each scattered rich club includes at least one high-centrality node and may be augmented with a few neighboring nodes to form a dense, non-trivial cluster.

Identifying Clusters Forming the Scattered Rich Clubs. The work of Estrada *et al.* [6] shows that real-world networks fall in two categories. Either they have a dense core with sparsely connected periphery (single rich club) or are composed of subgraphs which are individually densely connected but have sparse inter-connection (scattered rich club). We identify scattered rich clubs using these steps, as shown in Fig. 2.

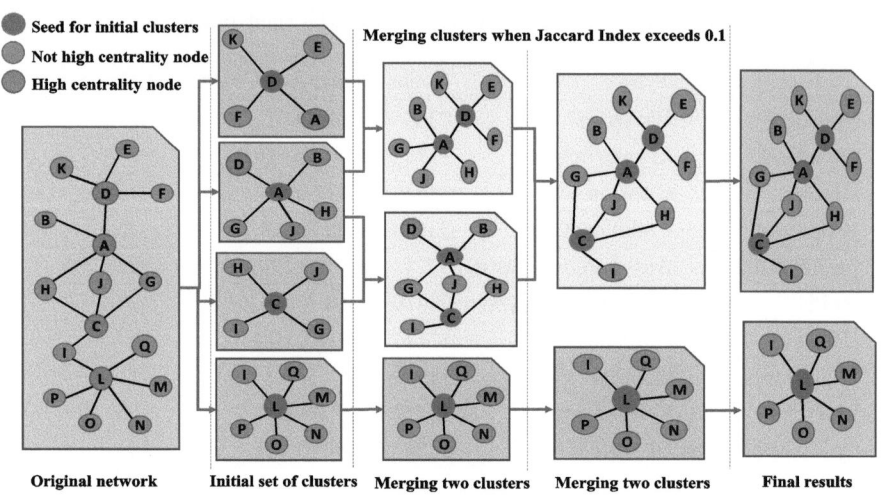

Fig. 2. Step-by-step illustration of the clustering algorithm.

First, we create the union of the top k ($k=20$) high betweenness (N_{hbc}) and closeness (N_{hcc}) centrality vertices of the network to create a unified set of high centrality vertices $N_{hc} = N_{hcc} \cup N_{hbc}$. *Second*, we form a cluster comprising of each node in N_{hc} and its neighbors. The total number of clusters formed be equal to the number of nodes ($|N_{hc}|$). *Third*, we merge overlapping clusters. For each cluster pair C_i, C_j, we compute the Jaccard Index as the ratio of their intersecting nodes to their total nodes. If it exceeds a given threshold (set to 0.1) we merge them as $C_i = C_i \cup C_j$ and repeat until no further merges are possible between any cluster pairs. This ensures that the maximum possible number of high-centrality nodes are put together in a cluster. At the end of these steps, we will have the disjoint clusters containing the high centrality vertices and their neighbors. If the network has a single rich club, then there will be one cluster, otherwise, there will be multiple clusters.

Quantifying the Degree of Scatteredness. In scattered rich clubs, the number of clusters, and the number of high centrality nodes in each cluster vary. We quantify this distribution using a new metric, *the degree of scatteredness*, as follows;

Let H be the total number of high centrality nodes and K the total number of clusters. Let H_i be the number of high centrality nodes in cluster C_i, ordered such that $H_i \geq H_j$ for $i < j$. For each cluster C_x, the ratio of the number of high-centrality vertices in the cluster to the number of clusters so far seen, $R_x = \frac{H_x}{x}$. The degree of scatteredness is then the mean of these ratios over the clusters; $\frac{1}{K} \left(\sum_{i=1}^{K} R_i \right)$

Using this formula, a single cluster will give the degree of scatterdness 1. When every cluster has one vertex, the value will be $\frac{1}{K} \left(\sum_{x=1}^{K} \frac{1}{x} \right)$, which will tend to zero as K becomes large. Table 2 shows the degree of scatteredness of our test networks. The more the high centrality nodes scatter into clusters, the value of scatteredness is lower.

Table 2. Degree of scatteredness and high-centrality nodes prediction via sampling. Multiplicity of clusters is shown as, $K(M) = K$ clusters with M high-centrality nodes.

Degree of scatteredness of Networks and high-centrality nodes prediction						
Dataset	High-centrality nodes	Number of Clusters	Distribution of High Centrality Nodes	Degree of Scatteredness	Precision	Recall
dmela	25	25	25(1)	0.152	0.07	0.88
euroroad	33	31	29(1), 2(2)	0.167	0.09	0.55
HepPh	28	21	17(1), 2(2), 1(4), 1(3)	0.293	0.04	0.96
CondMat	28	21	19(1), 1(2), 1(7)	0.362	0.14	0.79
as20000102	24	15	12(1), 2(5), 1(2)	0.402	0.70	0.79
caida	25	12	9(1), 1(11), 1(3), 1(2)	0.577	0.39	0.96
HepTh	26	10	6(1), 2(3), 1(2), 1(12)	0.609	0.15	0.77
email-univ	26	10	7(1), 2(2), 1(15)	0.683	0.15	0.85
AstroPh	31	9	6(1), 2(2), 1(21)	0.763	0.15	0.74
grid-fission-yeast	33	6	4(1), 1(26), 1(3)	0.862	0.30	0.45

3 Identifying Scattered Rich Clubs Using Snowball Algorithm

While the steps in Sect. 2 can locate single and scattered rich clubs, in practice, finding high centrality vertices for large networks is computationally intensive. Moreover, in real-world applications, the entire network may not be available for analysis.

To address these challenges, we propose identifying the rich clubs by sampling the network, using snowball sampling. *Snowball sampling* was presented in [10], where the authors conjectured that samples with higher expansion factors are more likely to be representative of the community structure of the network. We posit that the rich clubs are good expanders, since the high centrality vertices embedded in them can, in a few hops, reach a wide set of vertices. Therefore, we modify the snowball sampling to find the high centrality vertices. The unique features of our algorithm are as follows; We set the threshold, k, as 10% of the nodes in the network and used high-degree, high-clustering coefficient nodes as seed nodes. After each run of the sampling we obtain a sampled subgraph. We analyze the core periphery structure of this subgraph and designate the nodes in the innermost and second innermost core as high centrality nodes. We continue obtaining new snowball samples, until the set of high centrality nodes does not change, or a maximum threshold of runs (here set to 40) has been executed. At the end of this process, the sampled subgraphs are the *predicted rich clubs* and the nodes in the inner and second inner cores are the *predicted high centrality vertices*.

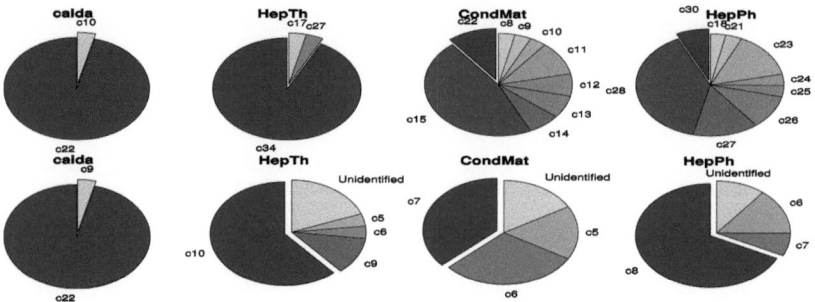

Fig. 3. Distribution of high-centrality nodes across network cores. Top: original networks. Bottom: networks sampled using the snowball in a single run with 10% of nodes.

The algorithm's time complexity is dominated by the size of the subgraph S. Each edge in S is accessed once during snowball sampling and again during core-periphery computation. The complexity per sample is $O(T \cdot E(S))$; T is the maximum iterations.

Results and Discussion. Table 2 shows the effectiveness of the sampling algorithm in finding high centrality nodes using high degree and high clustering coefficient nodes as seed. Our ground truth is the union of the set of top 20 high betweenness centrality vertices and the top 20 high closeness centrality vertices. The precision values of the predicted set are generally low. This is because the total predicted set size can be higher than the ground truth. The recall, whether all the nodes in the ground truth were obtained is high, more that 0.70, for most of the networks with single and scattered rich clubs.

4 Attack Models for Disrupting Centrality

We develop the attack model, using the subgraphs obtained through snowball sampling, to remove edges such that the ranking of the high-centrality nodes is disrupted. As shown in Fig. 3 graphs sampled using one instance of snowball sampling have most high-centrality nodes concentrated in the inner two cores. Further, the distribution of high centrality nodes of the sampled graph roughly mimic their distribution in the original graph. Thus, we deem a node to have a high core number if it is in the inner or second innermost core of the sampled subgraph, and select an edge for deletion if both its endpoints have a high core number in the sampled graph.

We test the centrality resilience by removing 2%, 4%, 6%, and 8% of the total edges. If we reach a limit of edges to choose, we relax the condition and select an edge if at least one of its endpoints has a high core number in the sampled graph. We stop the process if the required number of edges are removed or no more edges left for removal. After edge removal, we identify the top 20 high betweenness and closeness centrality vertices of the perturbed network and compare these high centrality vertices with the ones in the original network using the Jaccard index. The closer the value is to 0, the more perturbed the network. The time complexity of the attack algorithm is $E(S)$, dominated by the cost of a single snowball sampling run that produces subgraph S.

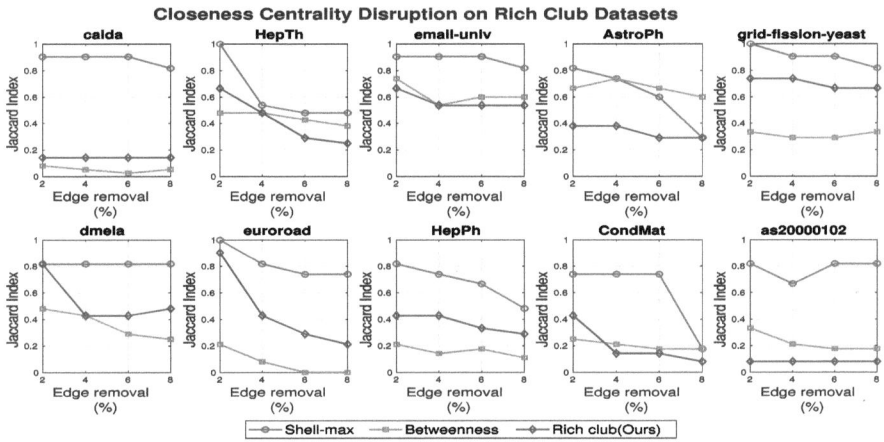

Fig. 4. Closeness centrality disruption under shell-max (red), betweenness (green), rich club (blue) attacks. Top: scatteredness > .5. Bottom: scatteredness < .5. (Color figure online)

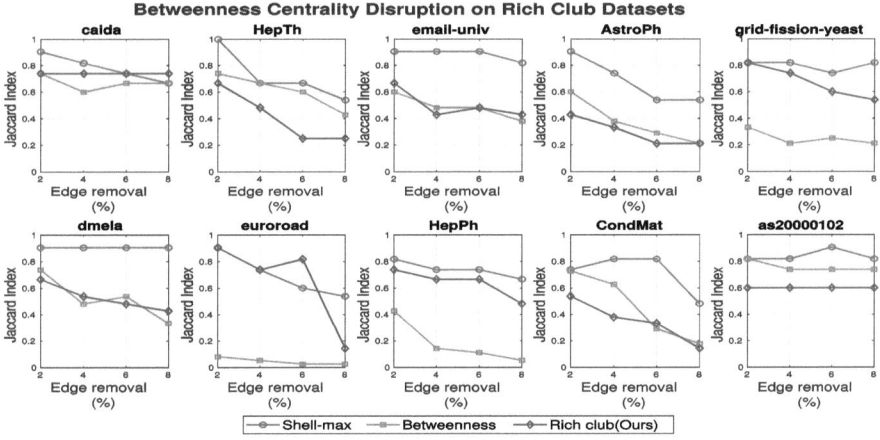

Fig. 5. Betweenness centrality disruption under shell-max, betweenness, rich club attacks. Top: scatteredness > .5. Bottom: scatteredness < .5.

Results. We compare our scattered rich club attack method (i) Shell-Max, [14] that removes edges with high k-core value ; (ii) Betweenness centrality, where edges are removed according to the order of high edge betweenness centrality and use Jaccard Index to measure the change in top 20 high centrality vertices. The lower the Jaccard index the higher the disruption. Figures 4 and 5 show that in most of the cases our method outperforms the baseline methods. Our attack model, in general, creates higher disruption than Shell-Max, and is comparable to betweenness centrality based attacks. However, the time complexity of computing edge betweenness centrality is $O(VE)$, which is an order of magnitude higher than the scatterred rich club based method.

5 Conclusion

We demonstrate that rich club of central nodes can be scattered across the network as opposed to being concentrated at the core. We discuss the implications of scattered rich clubs in terms of network centrality resilience and develop a predictive approach for discovering the scattered rich clubs using snowball sampling. Scattered rich club provide more detailed insights into how change occurs in complex networks. In future, we will study how scattered rich clubs affect other properties such as communities, and whether they can be used to predict high centrality nodes in dynamic networks.

Acknowledgments. This work was supported by NSF award CCF-1956373.

References

1. Network data repository. https://networkrepository.com/index.php
2. Adiga, A., Vullikanti, A.K.S.: How robust is the core of a network? In: ECML-PKDD, pp. 541–556. Springer (2013)
3. Alenazi, M.J., Cetinkaya, E.K., Sterbenz, J.P.: Cost-constrained and centrality-balanced network design improvement. In: 2014 6th International Workshop on Reliable Networks Design and Modeling (RNDM), pp. 194–201. IEEE (2014)
4. Chan, H., Akoglu, L., Tong, H.: Make it or break it: manipulating robustness in large networks. In: SDM, pp. 325–333. SIAM (2014)
5. Chiluka, N., Andrade, N., Gkorou, D., Pouwelse, J.: Personalizing eigentrust in the face of communities and centrality attack. In: 2012 IEEE 26th International Conference on Advanced Information Networking and Applications, pp. 503–510. IEEE (2012)
6. Estrada, E.: Topological structural classes of complex networks. Phys. Rev. E **75**(1), 016103 (2007)
7. Georgiadis, G., Kirousis, L.: Lightweight centrality measures in networks under attack. Complexus **3**(1–3), 147–157 (2006)
8. Laishram, R., Sariyüce, A.E., Eliassi-Rad, T., Pinar, A., Soundarajan, S.: Measuring and improving the core resilience of networks. In: Proceedings of the 2018 World Wide Web Conference on World Wide Web, pp. 609–618 (2018)
9. Leskovec, J.: Stanford Large Network Dataset Collection. http://snap.stanford.edu/data/
10. Maiya, A.S., Berger-Wolf, T.Y.: Online sampling of high centrality individuals in social networks. In: Pacific-Asia Conference on Knowledge Discovery and Data Mining, pp. 91–98. Springer (2010)

11. Maiya, A.S., Berger-Wolf, T.Y.: Sampling community structure. In: Proceedings of the 19th International Conference on World Wide Web, pp. 701–710 (2010)
12. Meyer, P., Siy, H., Bhowmick, S.: Identifying important classes of large software systems through k-core decomposition. Adv. Complex Syst. **17** (2014)
13. Sarkar, S., Bhowmick, S., Mukherjee, A.: On rich clubs of path-based centralities in networks. In: Proceedings of the 27th ACM International Conference on Information and Knowledge Management, pp. 567–576 (2018)
14. Sun, S., Liu, X., Wang, L., Xia, C.: New link attack strategies of complex networks based on k-core decomposition. IEEE Trans. Circ. Syst. II Express Briefs **67**(12), 3157–3161 (2020). https://doi.org/10.1109/TCSII.2020.2973668

A Customizable Agent-Based Simulation Framework for Emergency Departments

Francisco Mesas[1]([✉]), Manel Taboada[1], Dolores Rexachs[2], Francisco Epelde[3], Alvaro Wong[2], and Emilio Luque[2]

[1] Escuelas Universitarias Gimbernat (EUG), Computer Science School, Universitat Autonoma de Barcelona, Sant Cugat del Vallès, Barcelona, Spain
{francisco.mesas,manel.taboada}@eug.es
[2] Computer Architecture and Operating System Department, Universitat Autonoma de Barcelona, Barcelona, Spain
{dolores.rexachs,alvaro.wong,emilio.luque}@uab.cat
[3] Consultant Internal Medicine, University Hospital Parc Tauli, Universitat Autonoma de Barcelona, Sabadell, Barcelona, Spain
fepelde@tauli.cat
https://webs.uab.cat/hpc4eas/

Abstract. Emergency Departments (EDs) face increasing complexity due to rising patient demand, resource constraints, and the need for efficient service coordination. Traditional simulation models, while useful, cannot be easily adapted to a different hospital environments, making it difficult to transfer and scale solutions. Based on previous work, with a simulator working in a hospital, this work describes a modular Agent-Based Modeling and Simulation (ABMS) approach for increasing flexibility adaptation and reuse in ED simulations. The proposed technique, which deconstructs existing models into individual components, will allow hospitals with different workflows and operational constraints to construct customized simulations. To validate this methodology, we develop a structured, modular framework using NetLogo and Python. The suggested metasystem enables adaptive simulation-based decision assistance for emergency departments, which improves resource allocation and operational planning.

Keywords: Agent Based Modeling · Emergency Departments · Modular Simulation Framework · Resource Optimization · Adaptive Healthcare Modeling

1 Introduction

Emergency Departments (EDs) face challenges such as growing patient demand and complexity of service, which requires careful coordination of staff and resources [6,11]. Simulation helps analyze complex systems under predictable and unpredictable conditions. This article presents the key principles and one experimental design of a basic ED using an Agent-Based Modeling and Simulation (ABMS) framework. Unlike the previous works, which obtain a tool that have been designed to be applied in a specific ED, and it is difficult to adapt for being applied in others, the proposal is based on a modular system, which improves the design and allows reuse of the different elements, enabling the computational models can be applied in multiple EDs that have different operation.

Simulation becomes a useful Decision Support System that allows the analysis of different types of situation and also obtains data that are difficult to obtain in reality, answering 'what if' questions to predict real-life system outcomes [8]. There are several techniques to model and simulate a system, but when we talk about ED, in the literature, we find references mainly of two of them, Discrete Event Simulation (DES) and ABMS. For system flow analysis, DES is the most commonly used method, while ABMS offers a more dynamic and detailed perspective by modeling the behavior and interactions between multiple individual agents, such as patients, doctors, nurses, and their environment. One of the important characteristics of ABMS is the "emergent properties", in other words, "the higher-level system properties emerge from the interactions of lower-level subsystems (Agents)", making it the ideal choice according to various studies [9,14]. This adaptability makes it possible to create environments that are customized to the requirements of the system under study.

The variability in ED operations results in differences in regulatory systems and certifications, e.g., in the field of phlebotomy, we observe a regulatory divergence between the United States and Spain [4,10]. When considering the implementation of simulation techniques to improve EDs, these structural and regulatory variations must be taken into account to the specific characteristics of each emergency system. Once a simulator is operational and validated in a hospital, different studies show that its adaptability is adequate for hospitals with similar operations. However, it becomes more challenging for hospitals with different regulations or operational approaches due to the monolithic nature of current models [1,5]. A monolithic system is one in which different components of the software are strongly integrated and can complicate its adaptation to new contexts. Taking into account this situation, two initial solutions are presented: modify the existing monolithic model to adapt it to new needs, despite the difficulties this may entail, or develop a new simulator from scratch.

This article presents an alternative approach inspired by the modularity and versatility of Lego® blocks, allowing us to transform the monolithic approach. Before starting, it is necessary to define the methodology to prevent it from becoming a monolithic structure again. Using a validated simulator of an ED, the objective is to identify and extract some agents to encapsulate them into a reusable agent box.

The remainder of this article is structured as follows: Sect. 2 provides an overview of the concepts and limitations of the existing simulation models to understand the objective of the article; Sect. 3 analyzes a validated ED simulator by the HPC4EAS group to identify and decompose the monolithic system; Sect. 4 explains the design of the experimental validation, the structure needed for make the 'agent box' and the results of the experiments, and Sect. 5 describes the conclusions and future plans for the research work.

2 Theoretical Framework

Simulation approaches are essential tools for explore complicated systems in EDs and ABMS allows for a deeper and more complete understanding of how a system works, making it the ideal choice according to various studies [9,14]. The HPC4EAS group, a research group of the Universitat Autònoma de Barcelona (UAB), collaborating with the ED staff of Sabadell Hospital (Corporació Sanitària Parc Taulí), a reference center in the Catalan health System, has developed a conceptual model and a computational model for ED that utilize the ABMS technique, distinguishing between active and passive agents. Active agents are capable of making decisions, and passive agents do not take initiative on their own, but are essential to execute predetermined processes and enable interactions. The interaction between these agents and the modeled environment allows replication of the particularities of a real ED [12].

The research begins with the development of a conceptual model derived from a meticulous analysis of the elements of the ED, including the triage system that classifies urgency into five severity levels, specifically the Manchester Triage System [15] with level I being the most critical and level V the least [16].

After establishing the conceptual model and understanding the mechanisms of the ED operation, the next step was the creation of the computational model. The actual model becomes a sophisticated tool to predict the behavior of the ED, implemented with NetLogo. Regarding the software, although other alternatives such as Mesa, Repast, or AnyLogic have been studied [13], the choice of NetLogo powered by Python, combines the familiarity of the group and the ease of prototyping of NetLogo with the advanced analytical capabilities of the Python ecosystem, offering a pragmatic and tailored hybrid solution.

The group's validated simulator has been previously applied to various studies, including resource optimization [3], analysis of rare scenarios [2], and modeling disease transmission like MRSA within the ED [7].

After different studies in the same hospital with good results, it is extremely difficult to adapt to other ED with many different behaviors. It can be seen that the models and simulators developed to date by the HPC4EAS research group and other researchers operate in a monolithic manner, creating certain limitations in terms of adaptability, and some of them propose some alternative [1,5]. ABMS provides a modular structure in which individual agents interact on the basis of straightforward rules. This makes it an excellent solution for tackling this issue and easing adaptation to various scenarios. First, it is necessary to define how to get the definitions for an ED in a standard way for the conceptual model and how to transform it reusing the components of the computational model.

As can be seen in Fig. 1, after generating the conceptual model with the data collected from the experts, the objective is to be able to define a modular system, which we will call a metasystem, which already has the basic components based on the work previously carried out and the specifications of different hospitals.

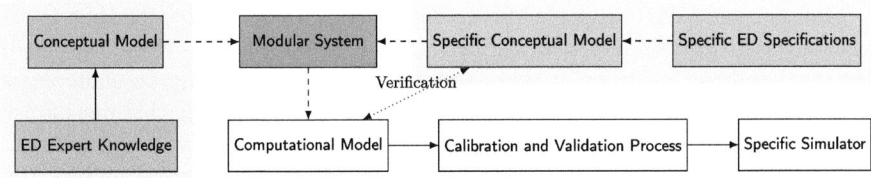

Fig. 1. Diagram of the design process of a simulator using a modular system for a specific ED.

The need for a metasystem makes sense, and in order to have it, it is first necessary to analyze what we currently have and break down those agents into a modular system. In the following section, analyze the current structure of the computational model from the HPC4EAS group to see which parts of the conceptual model can be easily linked and perform an experimental validation with short requirements to see how it can be possible.

3 Breaking Down the Current Simulator

In the ED simulator of the Hospital of Sabadell we find several fundamental agents. In this section, the different elements of the system are broken down, and each of the properties that they have in the current version is analyzed to understand the design of the metasystem. On the one hand, within the active elements of our system, we can discern between active agents and locations that act as active agents.

Currently, different agents are clearly identified in the actual ED Simulator: Admissions staff, Triage Nurses, Patients, Doctors (in Area A), Doctors (in Area B), Nurses (in Area A), Nurses (in Area B), Auxiliary Staff, the Information System, Careboxes, Test Rooms, Waiting Rooms, Ambulances, and Hospitals. The first eight agents listed have exclusively human roles. The last six agents are physical spaces (e.g., triage rooms, Area A/B) and should instead be modeled as passive environmental entities with properties like; Capacity limits, resource availability (e.g., beds, equipment) and spatial coordinates for agent navigation. This distinction ensures that agents interact with their environment rather than treating locations as autonomous actors, preserving the focus of ABMS on agent-environment interaction.

The system allows to measure different information like waiting times, workflow, and manage interactions which help to create a system that is organized. Nevertheless, some of the agents have inadequate simplification of the model, for example, admission agents has an incoming-queue attribute inside the agent, which specifies a fixed list of agents awaiting service. In real-world scenarios, admissions staff typically do not operate with a static queue; agents (like patients) assess their environment, organize themselves by observing the behavior of others, or sometimes ask questions to determine their turn.

This approach introduces a direct dependency between agents that is not consistent with the principles of the ABMS, where interactions should be mediated by decentralized rules or emergent features of the system. In some cases, a ticketing system is used to establish the order of service, but this mechanism should be explicitly modeled as a separate environment entity or rule. Details such as these can affect the usefulness of the model, as an overly simplified representation may not capture the real dynamics of the system. If we want to see a truly decentralized system in action, it requires a bit of rethink in how we build these agents, especially regarding how they sense and react to their environment. Doing so will allow us to tap into more realistic and surprising emergent behaviors, which is really the essence of ABMS.

4 Experimental Validation

This section covers the experiment conducted to validate this initial approach to a new fully agent-based methodology that will allow us to create our metasystem. The proposed case study presents a simplified version of Sabadell's current hospital concept, designed to verify that the methodology can be implemented as proposed. The work is early, so it requires this simplification for analyze each individual part.

In the simplified version, ED is organized into two main zones and a triage area that serves as an entry point to manage medical care according to urgency. Zone A is dedicated to critical care, serving patients with priority levels I to III (the most urgent). On the other hand, Zone B focuses on non-critical care (levels IV and V) with a waiting area to manage patient flow. The system involves various agents with different roles. Patients enter through the triage area, where the triage nurse assesses their urgency depending on the priority level, and care nurses provide essential care in CareBoxes or treatment rooms while updating the patient's status. The computer system tracks patient data, supports decision-making processes, and monitors resource use. The workflow is: patients enter → triage → routing (Zone A/B) → care (CareBox/treatment) → discharge.

An experimental design has been proposed to analyze the methodology to have our metasystem basis predefined. In this experimental design, NetLogo was used and combined with the Python programming language to allow the creation of a structure for each of the agents independently in independent boxes. This has been a significant challenge, as NetLogo does not have a way to organize modules into different files or structures, making it a tedious task to have a highly complex system since everything is intertwined.

To manage modularity, we organized the NetLogo project hierarchically. Directories separate UI elements, agent definitions (subdivided per agent type with files for setup, loop, actions, etc.), global variables, and environment setup. A specific file naming convention aids clarity. The complication is that NetLogo does not accept this structure by default, taking advantage of the integration it has with Python a script in this language is added for processing and sorting this structure into a format that NetLogo can understand.

By organizing the project, each component of the system is managed in isolation. This means that the definitions, behaviors, and configurations of each agent are located in individual files, easy to identify. Thanks to this, it is easier to update or modify an agent without affecting others, which facilitates the maintenance of the model over time. To validate its operation, a simplified conceptual and computational model was performed.

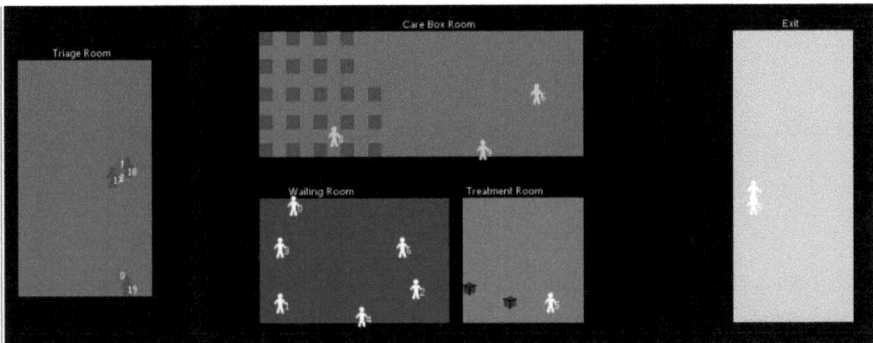

Fig. 2. Basic example for ED created with the modular structure.

Figure 2 shows the experimental work using our methodology. It consists of a Triage Room on the left, a care box for the most urgent patients at the top, waiting and treatment rooms for less urgent patients, and, on the right, the agents who have already completed all the steps. In addition, the integration of the files into a single executable file (main) is automated. Using a Python script, all the code needed for NetLogo has been prepared with the modular parts, which has allowed them to be synchronized and made to work during the simulation. Having tested this part, we now need to adapt the current monolithic model to improve its adaptability to validate our metasystem.

5 Conclusions

The use of simulation in ED is becoming increasingly important, given the increasing level of saturation they face. Through simulation, various challenging situations can be examined, allowing emergency services to adequately prepare and respond to adverse circumstances, especially in critical contexts. Simulation plays a crucial role in the strategic planning of these services. After analyzing the advantages of ABMs and seeing the performance of current simulation systems with rigid and poorly modular behaviors, a solution has been proposed that allows separating the behavior of the agents and allowing new agile implementations by effectively reusing previous work.

The proposed structure allows the individual behavior of each agent to be analyzed in the most realistic way, avoiding excessive coupling and the complexity of monolithic models. Although the initial model incorporates severe simplifications (for example, in queue management and static behaviors of some agents), the proposed structure will allow future improvements aimed at incorporating dynamics previously implemented in a monolithic manner with mechanisms for decentralized coordination. This task required a combination of NetLogo and Python, as the project can be structured into independent modules, with each agent defined in separate files. This modular approach facilitates system updates based on the specific context of each ED. However, NetLogo has limitations that do not allow the inclusion of modules in this manner, which is improved by integrating Python to make the metasystem.

The experimental validation in Sect. 4 has been useful for verifying the methodology, but is insufficient for full integration. The next steps will need to focus on defining how ED information is collected and structured for connection with the metasystem modules. In addition, it will be necessary to finish decomposing the current simulator into completely independent modules, ensuring that each one can interact in a flexible and scalable manner within the new environment. Following the defined structure will improve the adaptability of the system to different hospital contexts and facilitate future updates, but real-world testing would be needed across different EDs to fully confirm it.

In conclusion, defining the structure of a metasystem to simulate ABMS in the ED will allow for improved management of multiple hospitals. By allowing each agent to be modeled independently at the computational level, we can focus on future changes and not on restructuring monolithic systems. This approach opens new possibilities for emergency departments to prepare for future challenges. The change to modular systems and the possibility of collaborative development of these modules can significantly improve simulation capabilities. This advancement will allow us to improve not only the ED, but also other systems that need to adapt to various environments. This concept can also be applied to the effective management of urban areas or public transportation networks, where reusing components and flexible adaptation are important to optimize functions.

Acknowledgments. This research has been supported by the Agencia Estatal de Investigación (AEI), Spain and the Fondo Europeo de Desarrollo Regional (FEDER) UE, under contracts PID2020-112496GB-I00 and PID2023-146978OB-I00.

Disclosure of Interests. The authors have no competing interests to declare that are relevant to the content of this article.

References

1. Abo-Hamad, W., Arisha, A.: Simulation-based framework to improve patient experience in an emergency department. Eur. J. Oper. Res. **224**(1), 154–166 (2013). https://doi.org/10.1016/j.ejor.2012.07.028

2. Bruballa, E., Taboada, M., Cabrera, E., Rexachs, D., Luque, E.: Simulation of unusual or extreme situations of hospital emergency departments. In: ICCS Procedia Computer Science, pp. 209–212. IARIA, Nice, France (2014)
3. Cabrera, E., Taboada, M., Iglesias, M.L., Epelde, F., Luque, E.: Optimization of healthcare emergency departments by agent-based simulation. Procedia CS **4**, 1880–1889 (2011). https://doi.org/10.1016/j.procs.2011.04.204
4. Fidler, J.R.: The role of the phlebotomy technician: skills and knowledge required for successful clinical performance. Eval. Health Prof. 20(3), 286–301 (1997)
5. Godfrey, T., et al.: Supporting emergency department risk mitigation with a modular and reusable agent-based simulation infrastructure. In: Winter Simulation Conference, pp. 162–173. IEEE (2023). https://doi.org/10.1109/WSC60868.2023.10407894
6. He, J., Hou, X.Y., Toloo, S., Patrick, J.R., Fitz Gerald, G.: Demand for hospital emergency departments: a conceptual understanding. World J Emerg Med **2**(4), 253–261 (2011). https://doi.org/10.5847/wjem.j.1920-8642.2011.04.002
7. Jaramillo, C., Taboada, M., Epelde, F., Rexachs, D., Luque, E.: Agent based model and simulation of MRSA transmission in emergency departments. Procedia Comput. Sci. **51**, 443–452 (2015). https://doi.org/10.1016/j.procs.2015.05.267
8. McGuire, F.: Using simulation to reduce length of stay in emergency departments. In: Proceedings of Winter Simulation Conference, pp. 861–867 (1994). https://doi.org/10.1109/WSC.1994.717446
9. Monks, T., Currie, C.S.M., Onggo, B.S., Robinson, S., Kunc, M., Taylor, S.J.E.: Strengthening the reporting of empirical simulation studies: introducing the stress guidelines. J. Simul. **13**(1), 55–67 (2019). https://doi.org/10.1080/17477778.2018.1442155
10. Piazza, J., et al.: It's not just a needlestick: exploring phlebotomists' knowledge, training, and use of comfort measures in pediatric care to improve the patient experience. J. Appl. Lab. Med. **3**(5), 847–856 (2019). https://doi.org/10.1373/jalm.2018.027573
11. Samadbeik, M., et al.: Patient flow in emergency departments: a comprehensive umbrella review of solutions and challenges across the health system. BMC Health Serv. Res. **24**(1), 274 (2024). https://doi.org/10.1186/s12913-024-10725-6
12. Taboada, M., Cabrera, E., Epelde, F., Iglesias, M.L., Luque, E.: Agent-based emergency decision-making aid for hospital emergency departments. Emergencias **24**, 189–195 (2012)
13. Wrona, Z., et al.: Overview of software agent platforms available in 2023. Information **14**(6) (2023). https://doi.org/10.3390/info14060348
14. Yousefi, M., Yousefi, M., Fogliatto, F.S.: Simulation-based optimization methods applied in hospital emergency departments: a systematic review. Simulation **96**(10), 791–806 (2020). https://doi.org/10.1177/0037549720944483
15. Zachariasse, J.M., et al.: Validity of the manchester triage system in emergency care: a prospective observational study. PLoS ONE **12**(2), e0170811 (2017). https://doi.org/10.1371/journal.pone.0170811
16. Zhengchun, L., Rexachs, D., Epelde, F., Luque, E.: A simulation and optimization based method for calibrating agent-based emergency department models under data scarcity. Comput. Ind. Eng. **103**, 300–309 (2017). https://doi.org/10.1016/j.cie.2016.11.036

Leveraging Positional Bias of LLM In-Context Learning with Class-Few-Shot and Maj-Min Alternating Ordering

Aleksander Szczęsny(✉), Maciej Markiewicz, Łukasz Radliński, and Przemysław Kazienko

Department of Artificial Intelligence, Wrocław University of Science and Technology,
Wybrzeże Stanisława Wyspiańskiego 27, 50-370 Wrocław, Poland
aleksander.szczesny@pwr.edu.pl

Abstract. Selecting appropriate examples for in-context learning significantly impacts the performance of Large Language Models. In this paper, we show that leveraging LLMs' positional biases and incorporating knowledge of class distribution can improve classification outcomes, especially for underrepresented classes. We introduce *Class-few-shot*, a method that balances class representation among few-shot examples. To investigate this, we conduct almost 10,000 experiments on four datasets and three models, cross-checking how different biases affect models' performance and how they interact. We show that presenting classes from the most to least numerous using an alternating pattern leads to better results than standard few-shot prompting with the same number of examples. Additionally, we compare the general few-shot and *Class-few-shot* results, outlining the strengths of both approaches. All of our raw experiment results, prompts and codes are publicly available on GitHub (https://github.com/olorules/Class-few-shot).

Keywords: Few-shot · Class-few-shot · In-context learning · Large Language Models · Positional bias · Alternating ordering · Maj-Min

1 Introduction

Few-shot in-context learning [5] is one of the key approaches to increasing the performance of Large Language Models (LLMs) [10], and one that can be applied with almost no additional effort. Over the years, the identification of factors that contribute to the effectiveness of few-shot has become an area of research. This paper aims to verify how positional bias and class distribution influence the results and proposes **Class-few-shot** – a method designed to increase performance, especially the recall for minority classes, by balancing the class distribution. The key contributions of the paper include: (1) A Class-few-shot *Maj-Min* alternating order method for selecting in-context examples that improves performance. (2) Analysis of class balancing influence on the most numerous class *Maj* and the least numerous class *Min* results. (3) Nearly 10,000 publicly available experiments for further study.

2 Related Work

In-context learning [5] is an important alternative to fine-tuning [19] and can be used in a wide range of applications, from simple classification [10] to ad hoc personalization [18]. Initial research on in-context learning quickly identified some factors that influenced performance, such as recency bias and majority label bias [21], or the number and quality of examples [13]. These discoveries led to further exploration, particularly in two areas central for this work: the impact of information conveyed in the prompt, and the impact of order and example distribution on model performance.

2.1 Example Order, Distribution and Demonstration

One of the key biases demonstrated in [21] was models' tendency to prioritize the majority label and examples closer to the end of the prompt, which was present even in many-shot in-context learning [1]. Multiple strategies have been suggested to overcome the order of examples issue [12, 16, 20]. There has not been much research comparing the ordering bias with other biases [21]. Most works related to class distribution focus on developing methods, for example, selection [2, 11, 17], or dataset balancing [9] and expansion [8]. However, recent research confirms the significant impact of dataset imbalance on in-context learning [6].

3 *Class-Few-Shot* and Positional Bias

In the standard few-shot approach, the examples presented in a prompt are usually chosen randomly or manually, and the number of those examples is fixed. While sampling makes the class distribution of few-shot examples resemble the class distribution of the dataset, it may result in some classes being missing or overrepresented. Additionally, random few-shot sampling does not fully take advantage of the positional bias, as it cannot guarantee consistent class presence.

3.1 *Class-Few-Shot*: A New Approach to Balancing Few-Shot Class Distribution

We present a new method called *Class-few-shot*, Fig. 1. Its purpose is to ensure that when sampling few-shot examples, each class is represented n times. This

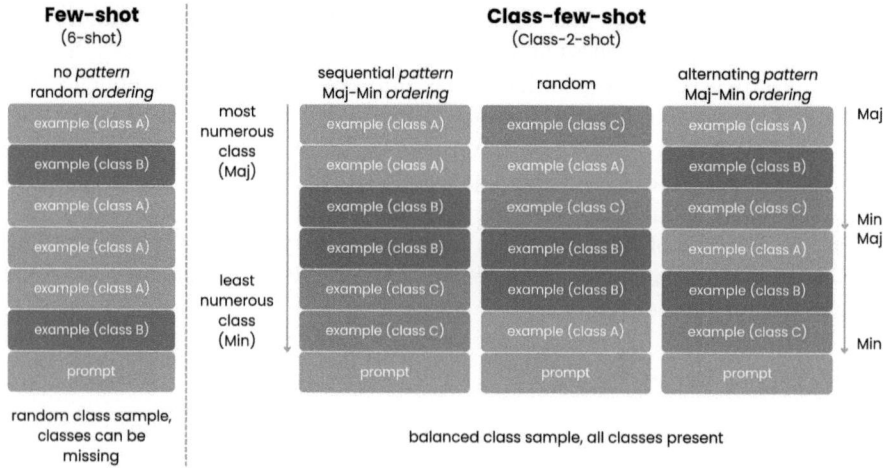

Fig. 1. *Class-few-shot* and few-shot in three classes scenario.

means that the final *shot* number depends on the class number of the specific task. The precise class presentation is controlled by parameters described below:

- **Class order** (*Ordering*) determines the order of examples representing each class in prompts, based on the number of examples belonging to the class:
  - **Maj-Min** – Sequence from the most to the least numerous class (first majority class *Maj*, last minority class *Min*),
  - **Min-Maj** – Sequence from the least to the most numerous class.
- **Class-few-shot pattern** (*Pattern*) defines a pattern of the *class order* in a prompt. It is only relevant for $n > 1$ shots per class:
  - **Sequential** (*Seq*) – all examples from a class are presented together, then all examples from the next class, etc., following given *Ordering*,
  - **Alternating** (*Alt*) – examples are given one at a time, presenting different classes sequentially.
- **Random** - a special case of Class-few-shot with a fully random example order. Class balance is maintained with n examples per class, but no specific *ordering* or *pattern* is applied. This serves as a baseline where only class balance influences results, isolating the effect from positional bias.

3.2 Positional Bias

The suggested approach aims to exploit the positional bias of LLMs. Prior research indicated bias towards information near the end of the prompt [21] and suggested that models are prone to favoring a specific position of the label [1,12,16,20]. We test various orderings and patterns to identify which one can be leveraged with *Class-few-shot* to provide better example selection for unbalanced datasets, as the bias effects vary by task and model.

We test various orderings and patterns to identify which one can be leveraged with *Class-few-shot* to provide better example selection for unbalanced datasets, as the bias is proven to work differently depending on the task and model.

4 Experimental Setup

We would like to compare the impact of *Class-few-shot* to all other factors influencing the performance, as it is difficult to isolate the impact of one specific factor. We decided to choose four different datasets, described in Sect. 4.1, as well as three representative models of different sizes: **Llama-3.1-8B-Instruct** (small), **Mixtral-8x7B-v0.1-Instruct** (medium, MoE), and **Llama-3.3-70B-Instruct** (large). To perform the experiments, we used the *lm-evaluation-harness* [7] framework with custom samplers implementing Class-few-shot. Models were hosted using the *vLLM* library.

4.1 Datasets

We chose 4 single-label classification datasets of different type, balance, class number, and difficulty (based on [10]). Dataset statistics are presented in Table 1. For the SNLI [4] dataset, we used the train set for few-shot samples and have sampled a 10% (5k) stratified random subset from the set to increase performance. The test set has not been altered. For Sarcasmania [15] we built the train and test sets by splitting the whole dataset with a train/test ratio of 9:1, with stratification.

4.2 Experiment Plan

Our experiments considered testing all permutations of parameters against each other, for all models and datasets. *Class-few-shot* parameters included *Class*

Table 1. Class distribution of datasets used in the study (test/validation set). Percentages are calculated relative to the total number of samples in each dataset.

Dataset	Class	Count	Percentage	Total Samples	Difficulty	Balance
SNLI [4]	Entailment	3368	34.3%	9824	Easy	Well-balanced
	Contradiction	3237	32.9%			
	Neutral	3219	32.8%			
TweetEval [3]	Anger	558	39.3%	1421	Medium	Imbalanced
	Sadness	382	26.9%			
	Joy	358	25.2%			
	Optimism	123	8.7%			
Sarcasmania [15]	No-Sarcasm	2129	53.5%	3978	Hard	Well-balanced
	Sarcasm	1849	46.5%			
Word Context [14]	Yes (Same Meaning)	319	50.0%	638	Medium	Perfect
	No (Other Meaning)	319	50.0%			

Order and *Class-few-shot Pattern*. Performance in standard *few-shot* was also evaluated as a reference. Each experiment has been repeated 10 times, using a fixed seed for each run: *1234*, *1235*, *1236*, etc.

To ensure reproducibility, we separated examples sampling from order and pattern. Thanks to that, all experiments were carried out with identical sets of examples, which were constant between configurations. For example, a *Min-Maj Alt* run on seed *1234* had the same few-shot examples as *Maj-Min Alt*, *Min-Maj Seq*, or *random* runs on the same seed. Class list in the prompts was presented in the same way across all experiments.

Experiments with *class-n-shot* were carried out in three scenarios in which the parameter $n \in \{1, 2, 3\}$, and the model was given n examples representing each class. The number of examples used in reference *few-shot* experiments was the same as in corresponding *class-n-shot* configurations. E.g., for the TweetEval dataset (4 classes), the model was given either $n = 4$, $n = 8$, or $n = 12$ examples. This enables a direct comparison of the results obtained by the two approaches.

4.3 Class-Few-Shot as an Experimental Methodology

To test the positional and class bias, we have to make sure that the test environment provides as much stability and reproducibility as possible. This is why we use the *Class-few-shot* method to perform all of the experiments. In standard few-shot, class distribution of examples will reflect the class distribution of the dataset. In the case of a strongly imbalanced dataset, there might be some runs with no possibility of creating a consistent *pattern*, or *class order*, as few-shot could only consist of one class. This could potentially influence the results. Therefore, we use *Class-few-shot* to assess biases, and perform standard few-shot experiments only as a reference.

5 Results

The following section presents a selection of experiments and observed trends, along with a thorough analysis.

5.1 Positional Bias in *Class-Few-Shot*

The results presented in Table 2 illustrate the impact of *ordering* and *pattern* on the performance of LLMs in classification tasks across selected datasets. While the performance differences between various ordering strategies are not striking, the consistently better results of the *Maj-Min alternating ordering* across all tested models suggest that structured variation in example presentation plays a role in optimizing model performance. Additionally, the random scenario tends to provide stable, above average results.

Furthermore, it is worth highlighting the variability in the impact of these parameters on specific models. For both Llama models, despite the gap between

Table 2. Comparison of different *Class-few-shot* scenarios. The results presented above are averaged from Class-2-shot and Class-3-shot (Class-1-shot does not enable *pattern* usage).

Dataset	Model	Maj-Min		Min-Maj		Random
		Seq	Alt	Seq	Alt	
SNLI	Llama-3.1-8B	0.750	**0.770**	0.740	0.760	0.764
	Mixtral-8x7B	**0.815**	0.802	0.778	0.791	0.799
	Llama-3.3-70B	**0.821**	**0.820**	0.794	0.805	0.811
TweetEval	Llama-3.1-8B	0.773	0.778	**0.781**	0.774	0.777
	Mixtral-8x7B	0.728	**0.740**	**0.738**	0.731	0.734
	Llama-3.3-70B	0.767	**0.775**	0.772	0.771	**0.774**
Sarcasmania	Llama-3.1-8B	0.721	**0.778**	0.709	0.706	0.732
	Mixtral-8x7B	**0.945**	**0.945**	0.912	0.910	0.928
	Llama-3.3-70B	0.822	**0.847**	**0.848**	0.840	0.839
Word Context	Llama-3.1-8B	0.613	0.638	**0.651**	0.621	0.631
	Mixtral-8x7B	0.684	0.679	0.684	**0.692**	0.685
	Llama-3.3-70B	0.725	**0.728**	0.718	0.722	0.721
Average	Llama-3.1-8B	0.714	**0.741**	0.720	0.715	0.726
	Mixtral-8x7B	**0.793**	**0.791**	0.778	0.781	0.786
	Llama-3.3-70B	0.784	**0.793**	0.783	0.784	0.786

sizes, the differences observed across various parameter configurations were significantly bigger than those seen for Mixtral-8x7B. This suggests that model architecture may play a critical role in leveraging positional bias.

Table 3 shows the mean F1 score for the least numerous class, *optimism* (9%), in the most imbalanced dataset, TweetEval. The results indicate that any *Class-n-shot* setting significantly outperforms standard few-shot for the minority class, suggesting that *Class-few-shot* helps balance output distribution and increases sensitivity to minority classes. In contrast to the outcomes detailed in Table 2, here the *Seq* pattern performs best, placing rare class instances closer to the prompt, which is consistent with prior findings [21].

5.2 *Class-N-Shot* vs. Few-Shot Comparison

Figure 2 shows the average performance differences between standard few-shot and *Class-few-shot* across all experiments, highlighting the potential gains of using *Class-few-shot*.

For Llama-8B, both *class-2-shot* and *class-3-shot* outperform in the *random* and *Maj-Min Alternating* scenario, with the latter improving the results by 2% points compared to standard few-shot. In the case of Mixtral 8 × 7B, *Class-few-shot* consistently outperforms standard few-shot across all tested configurations. Llama-70B exhibits the least gain from employing *Class-few-shot*.

Table 3. F1 score for the minority class of the TweetEval dataset (*optimism*).

Model	Few-Shot	Random	Min-Maj		Maj-Min	
			Seq	Alt	Seq	Alt
Llama3.1_8B	0.568	0.618	0.597	0.604	**0.633**	0.621
Mixtral_8x7B	0.439	0.497	0.496	0.502	**0.505**	0.481
Llama3.3_70B	0.619	0.628	0.616	0.628	**0.637**	0.632

Performance generally improves with more examples, and while the *Maj-Min Alternating* strategy yields the best results, it is worth noticing that the *random* strategy results in better performance, suggesting that class balancing alone can boost in-context learning. Note that results for Sarcasmania were excluded for Llama-70B due to extremely poor 0-shot performance, with the model often failing to produce coherent responses.

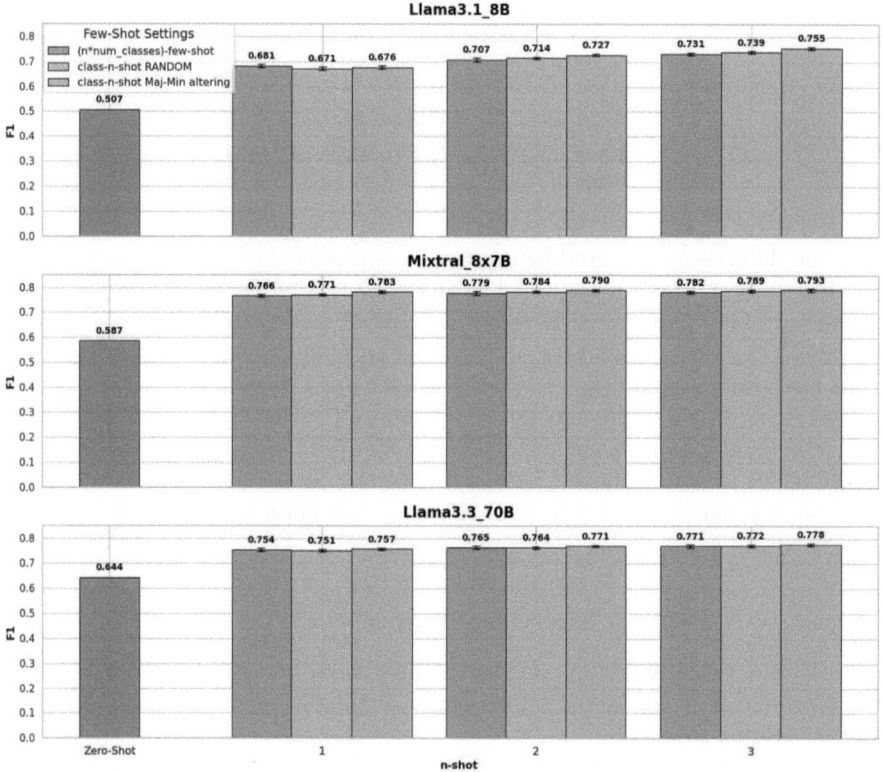

Fig. 2. F1 comparison between 0-shot, few shot and *Class-few-shot*, averaged across all the experiments with $n \in \{1, 2, 3\}$.

6 Conclusions and Future Work

Our findings reveal that *Maj-Min Alternating Class-few-shot* improves the performance of LLMs over the standard few-shot approach. Furthermore, the results confirm that *Class-few-shot* can improve the F1 score on minority classes by approximately 5 p.p.

Potential future work could explore how *Class-few-shot* affects the class distribution, as minority class results have improved significantly. Other directions include analysing positional bias with *Class-few-shot* in a many-shot [1] context, as well as expanding the number of datasets and studying the role of class order in prompts beyond few-shot examples.

References

1. Agarwal, R., et al.: Many-shot in-context learning (2024)
2. Baldassini, F.B., Shukor, M., Cord, M., Soulier, L., Piwowarski, B.: What makes multimodal in-context learning work? (2024)
3. Barbieri, F., Camacho-Collados, J., Neves, L., Espinosa-Anke, L.: Tweeteval: unified benchmark and comparative evaluation for tweet classification (2020)
4. Bowman, S.R., Angeli, G., Potts, C., Manning, C.D.: A large annotated corpus for learning natural language inference (2015)
5. Brown, T.B., et al.: Language models are few-shot learners (2020)
6. Gao, H., Zhang, F., Zeng, H., Meng, D., Jing, B., Wei, H.: Exploring imbalanced annotations for effective in-context learning (2025)
7. Gao, L., et al.: A framework for few-shot language model evaluation (2023). https://doi.org/10.5281/zenodo.10256836
8. Kaszyca, O., Kazienko, P., Kocoń, J., Cichecki, I., Kochanek, M., Szydło, D.: Is it possible for ChatGPT to mimic human annotator? Authorea Preprints (2023)
9. Kochanek, M., et al.: Improving training dataset balance with ChatGPT prompt engineering. Electronics **13**(12), 2255 (2024)
10. Kocoń, J., et al.: ChatGPT: jack of all trades, master of none. Inf. Fusion 101861 (2023)
11. Liu, J., Shen, D., Zhang, Y., Dolan, B., Carin, L., Chen, W.: What makes good in-context examples for GPT-3? (2021)
12. Lu, Y., Bartolo, M., Moore, A., Riedel, S., Stenetorp, P.: Fantastically ordered prompts and where to find them: overcoming few-shot prompt order sensitivity (2021)
13. Min, S., et al.: Rethinking the role of demonstrations: what makes in-context learning work? (2022)
14. Pilehvar, M.T., Camacho-Collados, J.: WIC: the word-in-context dataset for evaluating context-sensitive meaning representations (2019)
15. Siddiqui, R.: Sarcasmania dataset (2019)
16. Sorensen, T., et al.: An information-theoretic approach to prompt engineering without ground truth labels (2022)
17. Wang, S., Yang, C.H.H., Wu, J., Zhang, C.: Bayesian example selection improves in-context learning for speech, text, and visual modalities (2024)
18. Woźniak, S., Duszenko, J., Kocoń, J., Kazienko, P.: Improving LLM-based recommender systems with user-controllable profiles. In: The 1st Workshop on Human-Centered Recommender Systems at TheWebConf 2025 - The ACM Web Conference 2025. ACM (2025)

19. Woźniak, S., Koptyra, B., Janz, A., Kazienko, P., Kocoń, J.: Personalized large language models (2024)
20. Wu, Z., Wang, Y., Ye, J., Kong, L.: Self-adaptive in-context learning: an information compression perspective for in-context example selection and ordering (2022)
21. Zhao, T.Z., Wallace, E., Feng, S., Klein, D., Singh, S.: Calibrate before use: improving few-shot performance of language models (2021)

Remote Sensing AI for Crop Planting in Wildfire Fuel Mapping

Paula Sánchez(✉), Irene González, Carlos Carrillo, Ana Cortés, and Remo Suppi

Computer Architecture and Operating Systems Department,
Universitat Autònoma de Barcelona, Barcelona, Spain
{paula.sanchez.gayet,irene.gonzalez.fernandez,
carles.carrillo,ana.cortes,remo.suppi}@uab.cat

Abstract. Accurate wildfire prediction requires updated, high-resolution fuel maps that account for seasonal vegetation variations. The flammability of crops varies by season, affecting the behavior of wildfires. This study combines remote sensing indices and machine learning to dynamically update fuel models in cropland zones. Using Sentinel-2 data, the status of the cropland is classified as "planted" or "unplanted", achieving 80% accuracy. Applied to a 2019 wildfire in Catalonia (Spain), the updated fuel map closely matched the observed fire spread. The methodology outperforms traditional approaches and is efficient, allowing for real-time updates based on seasonal changes.

Keywords: Remote Sensing indices · Machine Learning · Seasonal fuel map · Wildfire · Cropland

1 Introduction

To accurately forecast the behavior of a wildfire, having access to updated high-resolution fuel maps is crucial. However, obtaining maps that correspond to the specific year of a fire and accurately reflect the seasonal fuel characteristics of the affected area is often challenging. Fuel maps classify vegetation characteristics into fuel models, which are required for forest fire spread models. One of the most commonly used sets is the Scott and Burgan one [13] which includes the NB3 class for agricultural fields, treating cultivated land as a firebreak. However, depending on the season, some croplands can become highly flammable, contributing to fire spread. Current fuel maps lack this seasonal variability, so a tool to assess and update cropland flammability would enhance wildfire simulation accuracy. The availability of long-term open data, particularly from remote sensing instruments like Sentinel-2, has facilitated the creation of machine learning models to address information gaps in complex systems. Sentinel-2 provides periodic, high-resolution (10 m pixel) data, which is valuable for agricultural monitoring. This data provides new opportunities to collect more precise information about croplands to improve fuel map descriptions. In fact, this capability has already been used to monitor various crops, such as maize crop growth and rice field detection [2], and it has also been used to determine the extent of croplands [12]. This study proposes using remote sensing indices to train a machine learning model to determine if cropland is planted or unplanted. With remote sensing data updated every five days, this methodology allows fuel maps to be updated according to the current season.

2 Materials

The study conducted in this work focuses on the agricultural areas of Catalonia (Spain), using data from 2021, 2022, and 2023. Remote sensing indices from this period have been used as features to train an AI model, along with cropland data, such as the type of crop and the months they are planted or unplanted, to create the ground truth. The training data can be divided into two categories: *Crop Yield Data* and *Remote Sensing Indices Data*.

2.1 Crop Yield Data

Catalonia's DACC [5] collects crop data in the Unique Agrarian Declaration (DUN) [5]. Each year, all crop fields declared in DUN are available in a shapefile, where each crop field is represented as a polygon. In this study, only a subset of fields (polygons) was selected based on specific criteria. The selected fields met the following conditions: the same crop cultivated from 2021 to 2023, a minimum area of $100\,\text{m}^2$, a single crop per year, and no interruptions like houses or structures. Polygons with invalid geometries, such as self-intersections, degenerated shapes, or other topological issues, were discarded. The selected crops included soft wheat, barley, oats, and corn.

2.2 Remote Sensing Indices Data

After obtaining crop yield data, remote sensing indices were processed using multispectral Sentinel-2 data (Level-2A product). The data was downloaded from 2021 to 2023 through the openEO API [1]. The vegetation indices-NDVI [3], MSAVI [3], GNDVI [6], and EVI [3]-were computed using bands 2, 3, 4, and 8 at a spatial resolution of 10 m. These indices were selected for their widespread use in vegetation monitoring, and among them, preference was given to those without free parameters. To ensure Sentinel-2 data quality, the Scene Classification Layer (SCL) [11] was used to identify and exclude clouds and cloud shadows that make some pixels unreliable for analysis. These pixels were assigned NaN values, excluding them from the analysis.

3 Methodology

This section describes the proposed methodology to determine whether a cropland should be considered planted or not, consisting of the following 4 steps:

Image Segmentation and Preprocessing (ISP), *Data Cleaning (DC)*, *Model Development and Performance Evaluation (MDPE)* and *Fuel Map Updating (FMU)*. Subsequently, a more detailed explanation of these steps is included.

3.1 Image Segmentation and Preprocessing (ISP)

After downloading all bands for the whole Catalonia area and for all available days in 2021, 2022, and 2023, the remote sensing indices were computed. For each cropland field, only the pixels from the remote sensing index that were entirely contained within the field boundaries were identified. A margin was intentionally left when selecting the pixels, as cropland fields are often not fully planted up to their borders. Figure 1 exemplifies this process using three pictures. The left picture shows the borderline of a given cropland with a white line. In the central image, the pixels used as a representation of the cropland are white colored and, finally, the right image illustrates the value of a certain remote sensing index for the representative pixels of the study field. For each remote sensing index, the median of the selected pixels was computed to reduce the influence of outliers and better represent the crop field, excluding anomalies caused by equipment, objects, or soil variability. Crop fields that contained only a single pixel (while accounting for edge constraints) were accepted as valid. However, fields with no interior pixels were excluded from the analysis.

Fig. 1. Steps for selecting representative pixels for a given cropland.

After evaluating the median of remote sensing indices for each cropland and day over the three years, a database was created with the following fields:

- **Crop ID**: This value corresponds to a unique identifier for each cropland.
- **NDVI, EVI, GNDVI, MSAVI**: Indices with a theoretical range between -1 and 1. In practice, bright (e.g., clouds) and dark areas (e.g., shadows) may produce anomalous values, which are treated as noise and discarded.
- **Crop**: This field describes the type of crop planted in the corresponding cropland. The possible values are *Soft wheat, Barley, Oats* and *Corn*.
- **Date**: A day corresponding to the date on which the data was recorded.
- **Planting status**: Will be 1 if the cropland is planted or 0 if it is unplanted.

To determine the planting status, fields have been selected where the typical crop cycles are known (see Table 1). For these fields, the months in which they are certainly planted and the months in which they are without crops (as no second crop is grown) are known. Transitional months (those involving planting, growth onset, or harvesting) were intentionally excluded from the table, as crop presence during these periods can be variable [8].

Table 1. Planted and Unplanted months for selected crops in Catalonia [8].

Crop	Planted Months	Unplanted Months
Soft wheat	1, 2, 3, 4, 5	8, 9, 10
Barley	1, 2, 3, 4, 5	7, 8, 9, 10
Oats	1, 2, 3, 4, 5	8, 9, 10
Corn	6, 7, 8	11, 12, 1, 2, 3

3.2 Data Cleaning (DC)

Despite removing clouds and shadows from the SCL layer, some outliers may still exist. To address this, we consider that the selected crops follow planting cycles characterized by regular seasonal patterns. This seasonal signal creates wave-like time series data, which can be approximated by a sinusoidal curve. Therefore, the Fast Fourier Transform is used to clean the data effectively [9]. In remote sensing, such as with vegetation indices, it helps separate low-frequency patterns (e.g., seasonal vegetation growth) from high-frequency noise (e.g., interference from undetected clouds). A high-frequency filtering process is applied to detect and remove outliers for each crop in the dataset, filtering out high frequencies while keeping lower ones. An inverse Fourier Transform then reconstructs the "cleaned" data, removing unwanted fluctuations. By comparing the original and smoothed data, outliers are identified and removed.

3.3 Model Development and Performance Evaluation (MDPE)

The present study uses AI to classify cropland data from 2021 to 2023 based on planting status and four vegetation indices as predictive features. All variables were standardized to have a mean of zero and a standard deviation of one, ensuring a uniform contribution to the model. The dataset was split into 80% for training and 20% for testing. Multiple classification algorithms were evaluated, including *Logistic Regression, Decision Trees, Random Forest, XGBoost,* and *CatBoost,* with *CatBoostClassifier* chosen for its superior accuracy. The model's performance was evaluated using metrics such as *Accuracy, Precision, Recall,* and *F1-Score* [10]. The obtained results are shown in Table 2.

Table 2. CatBoost performance metrics.

Target	Precision	Recall	F1-Score
Unplanted	0.79	0.82	0.81
Planted	0.82	0.79	0.80
Weighted Avg	0.81	0.80	0.80
Accuracy	0.80		

3.4 Fuel Map Updating (FMU)

When a forest fire occurs, the pre-trained *CatBoost* model updates the fuel map to reflect the current cropland status. This process begins by identifying the area of interest, downloading the necessary satellite bands, and computing remote sensing indices. The analysis focuses on up to 5 days of pre-fire data, assuming optimal weather conditions. The median values of remote sensing indices are calculated for each agricultural field, following the procedure described in the *ISP* step. These median values will be used as input for the AI, which will classify the fields as either planted or unplanted. Unplanted fields are assigned the Burgan NB3 fuel model [13], acting as fire barriers, while planted fields are assigned the GR4 model [13], which represents 2-foot-deep vegetation in arid to semi-arid climates such as Catalonia. The pixels of the polygons that represent planted crop fields will be assigned this fuel model, which best represents the behavior of a forest fire in agricultural areas, overwriting the original fuel map data. It is important to note that in the pixels being overwritten, the original fuel map might have classified the area as NB3 (croplands, not-burnable) if it was correctly identified as cropland. It is also possible that the fuel map does not recognize the area as cropland, especially if it is an older fuel map where the area might have been classified as something else, like grass or shrub. Regardless of the previous classification, the new information will overwrite the data of the old pixel. This fuel map update scheme is illustrated in Fig. 2.

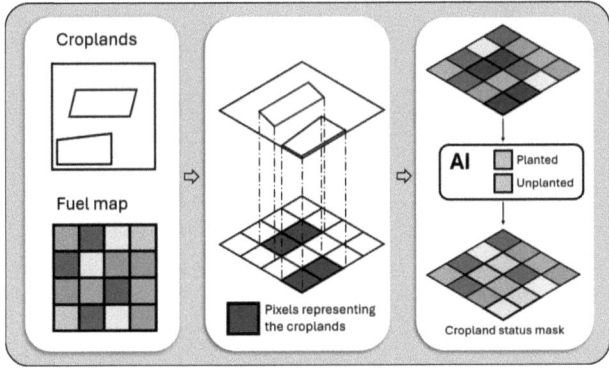

Fig. 2. Fuel map update process using the AI model to modify croplands status.

4 Experimental Study and Results

4.1 Case Study

To evaluate the effect of updating cropland fuel data using an AI model, a real wildfire in Nalec, Catalonia (Spain), from June 2019, was selected as a case study. This wildfire was chosen because 60% of the burned area was agricultural fields. Figure 3a shows the ignition point of the fire with a triangle (June 24 at 17:56) and the orange shape indicates the final burned area (June 24 at 21:05). To locate agricultural fields in the area, crops declared in the DUN of 2019 were used (see Fig. 3b). Multispectral data from Sentinel-2 for June 17, 2019 (7 days before the wildfire occurred), was used. Wildfire simulations were conducted using FARSITE [4], with a LCP file from PREVINCAT [7]. While results depend on them, the methodology remains independent, ensuring broad applicability.

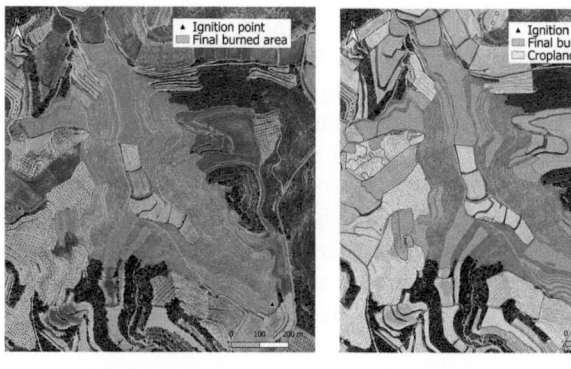

(a) *Nalec* fire.   (b) Croplands in the fire area.

Fig. 3. Ignition point and final burned area (a) and croplands in the region (b). (Color figure online)

4.2 Experimental Results

This section presents the simulation forecasts for the study case when updating cropland status using the proposed AI model. To assess the impact of classifying cropland as planted or unplanted, two extreme scenarios were analyzed: all croplands planted (scenario *P*) and all unplanted (scenario *UP*). The scenario using the AI-updated fuel map is referred to as scenario *AI*. In all three simulations, only the fuel map was modified; all other input data remained unchanged. The simulation covered a total fire duration of 3 h and 9 min. The orange area in all images indicates the real final fire perimeter.

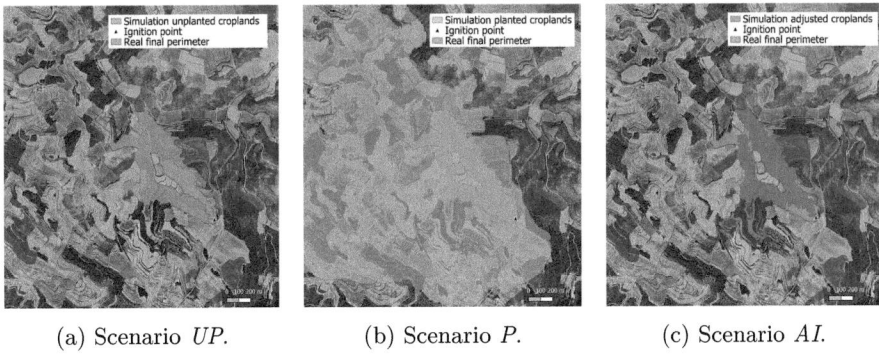

(a) Scenario UP.   (b) Scenario P.   (c) Scenario AI.

Fig. 4. Wildfire simulation using scenarios: UP (a), P (b) and AI (c). (Color figure online)

In Fig. 4a (Scenario UP), the fire spread is significantly underestimated because unplanted croplands act as barriers, preventing the fire from advancing. This results in the simulated fire being confined to a much smaller area compared to the real perimeter (blue shape in Fig. 4a). In contrast, Fig. 4b (Scenario P) removes this barrier effect, leading to a substantial overestimation of the fire spread. The simulated area is much larger than the actual fire perimeter, extending beyond the available landscape data (LCP). Finally, Fig. 4c (Scenario AI) shows the final burned area using the seasonal fuel map from the AI model. This simulation is the most accurate, as it incorporates detailed cropland information, distinguishing between planted and unplanted areas at the time of the fire.

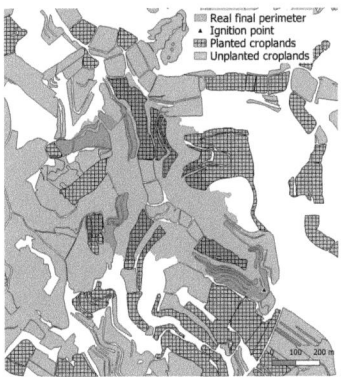

Fig. 5. Croplands in the fire area classified as planted or unplanted by the AI.

Figure 5 presents a zoomed-in view of the fire area and includes information about the croplands classified by the AI model as planted or unplanted. This is the classification used to update the fuel map for Scenario *AI*. The white areas in the figure are non-cropland areas, therefore, the AI model does not update the fuel information associated with them. It is worth noting that the "hole" in the real burned area corresponds to areas identified by the AI model as unplanted cropland, which aligns with the fire behavior that avoids burning this zone. Additionally, the AI model correctly classifies the croplands to the "head of the fire" (north) as unplanted, consistent with the firefighter report describing these areas as non-burnable. Furthermore, most of the croplands that fall within the real final perimeter are accurately marked by the AI model as planted, demonstrating its ability to capture the conditions that facilitated the spread of the fire.

5 Conclusions

This study emphasizes the importance of integrating remote sensing data into fire behavior modeling to improve wildfire spread predictions. The results demonstrated the impact of cropland status (planted vs. unplanted) on fire dynamics. For that reason, an AI system has been proposed to train a *CatBoost* model, which uses several remote sensing indices as training features along with seasonal information about crop planting status extracted from the DUN database. The AI model was tested on the *Nalec* fire, being capable of identifying unplanted crops, aligning with firefighter reports, and confirming the model's potential as a reliable tool for managing wildfires. This methodology is fast and easy to implement, making it suitable for near-real-time applications.

Acknowledgments. This work is part of the project CPP2021-008762 financed by the Spanish Ministry of Science and Innovation, and European Union-*NextGenerationEU*/ PRTR. It has also been granted by MCIN AEI/10.13039/501100011033 under contract PID2020-113614RB-C21 and PID2023-146193OB-I00, and by the Catalan government under grant 2021 SGR-00574. We thank *CTFC* for its guidance.

References

1. OpenEO API. https://openeo.org/. Accessed 24 Jan 2025
2. Ali, A.M., et al.: Integrated method for rice cultivation monitoring using sentinel-2 data and leaf area index. Egypt. J. Remote Sens. Space Sci. **24**(3, Part 1), 431–441 (2021). https://doi.org/10.1016/j.ejrs.2020.06.007
3. Antunes Daldegan, G., Gonzalez-Roglich, M., Noon, M., Zvoleff, A.I.: Assessing the performance of NDVI, 2-band EVI and MSAVI vegetation indices for land degradation monitoring across variable biomass cover at global scale. In: AGU Fall Meeting Abstracts, vol. 2020, pp. SY003–0008 (2020)
4. Finney, M.A.: FARSITE: fire area simulator-model development and evaluation (1998). https://doi.org/10.2737/rmrs-rp-4
5. Generalitat de Catalunya: Mapa de cultivos - sigpac (2023). https://agricultura.gencat.cat/ca/ambits/desenvolupament-rural/sigpac/mapa-cultius/. Accessed 26 Jan 2025

6. Gitelson, A., Kaufman, Y., Merzlyak, M.: Use of a green channel in remote sensing of global vegetation from EOS-MODIS. Remote Sens. Environ. **58**, 289–298 (1996). https://doi.org/10.1016/S0034-4257(96)00072-7
7. González-Olabarria, J., Piqué, M., Busquets, E.: Cartografia de vegetació per la simulació d'incendis forestals. Servidor PREVINCAT (2019)
8. Ministerio de Agricultura, Pesca y Alimentación: Calendario de siembra, recolección y comercialización 2014–2016 (2016). disponible en: https://servicio.mapama.gob.es/tienda/
9. Oberst, U.: The fast Fourier transform. SIAM J. Control Optim. **46**, 496–540 (2007). https://doi.org/10.1137/060658242
10. Powers, D.: Evaluation: from precision, recall and f-measure to ROC, informedness, markedness & correlation. J. Mach. Learn. Technol. **2**, 2229–3981 (2011). https://doi.org/10.9735/2229-3981
11. Sanchez, C., Mena, F., Charfuelan, M., Nuske, M., Dengel, A.: Assessment of sentinel-2 spatial and temporal coverage based on the scene classification layer. In: 2024 IEEE International Geoscience and Remote Sensing Symposium, pp. 4099–4103. IEEE (2024). https://doi.org/10.1109/igarss53475.2024.10642213
12. Savitha, C., Talari, R.: Mapping cropland extent using sentinel-2 datasets and machine learning algorithms for an agriculture watershed. Smart Agric. Technol. **4**, 100193 (2023). https://doi.org/10.1016/j.atech.2023.100193
13. Scott, J.H., Burgan, R.E.: Standard fire behavior fuel models: a comprehensive set for use with Rothermel's surface fire spread model. Technical report, U.S. Department of Agriculture, Forest Service, Rocky Mountain Research Station (2005). https://doi.org/10.2737/rmrs-gtr-153

Regularization Algorithm for Eliminating Singularities in the PIES Formula for 3D Multidomain Orthotropic Problems

Krzysztof Szerszeń(✉) and Eugeniusz Zieniuk

Institute of Computer Science, University of Bialystok, Konstantego Ciołkowskiego 1M, 15-245 Białystok, Poland
{k.szerszen,e.zieniuk}@uwb.edu.pl

Abstract. This paper presents an algorithm designed to eliminate the direct evaluation of both strongly and weakly singular boundary integrals in the Parametric Integral Equation System (PIES) for the analysis of three-dimensional multidomain orthotropic problems. The proposed method regularizes PIES by incorporating an auxiliary regularization functions with coefficients that effectively remove singularities. Consequently, the regularized PIES eliminates the need to explicitly evaluate singular integrals, enabling all integrals to be computed using standard Gaussian quadrature.

Keywords: Regularized PIES · Singular Boundary Integrals · 3D Boundary Value Problems · Subdomains · Orthotropy · Bézier Surface Patches

1 Introduction

The computational analysis of various engineering problems often relies on well-established methods such as the Finite Element Method (FEM) [1] and the Boundary Element Method (BEM) [2]. In the authors' research, an alternative method to FEM and BEM was developed to eliminate the need for discretizing both the domain and its boundary. This was achieved through an analytical modification of the traditional Boundary Integral Equation (BIE) and the formulation of a Parametric Integral Equation System (PIES) [3]. The PIES enables direct incorporation of the boundary shape into the BIE formulation through parameterized functions, thereby eliminating the need for boundary discretization in numerical computations. This approach enables a more efficient and continuous representation of the problem domain. In PIES, the boundary is globally defined using a small set of control points, and in 3D problems, it is represented by parametric surface patches.

Although the analytical modification of BIE and the development of PIES introduced significant improvements, they did not eliminate the presence of singular integrals. As in BIE, the integrals in PIES still exhibit both strong and weak singularities in their integrands, arising from the characteristics of the integrand functions. Over the years, various techniques have been developed to address these challenges within the Boundary

Integral Equation (BIE) framework, including subdivision methods [4–6], analytical evaluation of nearly singular integrals [7–9], and specialized quadrature rules [10, 11]. Despite their widespread application, identifying a universal approach for handling a broad range of singular integrals remains a significant challenge.

The paper proposes an algorithm that eliminates the need for directly computing boundary singular integrals in the PIES formulation through regularization. Extending previous work on homogeneous media [12, 13], this study generalizes the approach to three-dimensional problems involving subdomains characterized by orthotropic material properties. The proposed regularization involves the use of auxiliary functions containing regularization coefficients designed to eliminate both weakly and strongly singular terms. Notably, this approach is independent of the boundary shape and specified boundary conditions, enhancing its versatility. To model subdomain boundaries, Bézier surface patches are used, providing a smooth and flexible representation. Test results demonstrate a significant improvement in solution accuracy after applying the proposed regularization.

2 PIES for Subdomains with Piecewise Homogeneous Properties of an Orthotropic Medium

We consider a three-dimensional potential problem in a piecewise homogeneous orthotropic medium, governed by the partial differential equation

$$k_{11}^{(i)} \frac{\partial^2 u}{\partial x_1^2} + k_{22}^{(i)} \frac{\partial^2 u}{\partial x_2^2} + k_{33}^{(i)} \frac{\partial^2 u}{\partial x_3^2} = 0. \tag{1}$$

Figure 1 illustrates the domain Ω, consisting of three homogeneous subdomains Ω_i ($i = 1, 2, 3$), each characterized by orthotropic coefficients $k_{11}^{(i)}, k_{22}^{(i)}, k_{33}^{(i)}$.

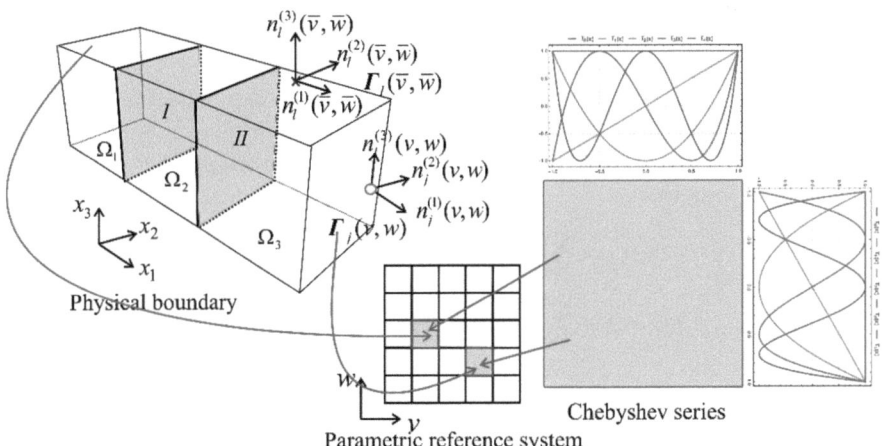

Fig. 1. Mapping of the boundary Γ of the domain Ω, consisting of homogeneous subdomains $\Omega_1, \Omega_2, \Omega_3$, onto a parameterized plane with boundary functions $u_j(v, w)$ and $p_j(v, w)$ approximated using Chebyshev series.

PIES facilitates a one-dimensional mathematical reduction of the problem by evaluating it on the boundary, similar to BEM. However, unlike BEM, it maps the physical boundary onto a parameterized reference domain. In 3D, this involves projecting the boundary onto a parameterized plane, as shown in Fig. 1. The PIES formulation for each subdomain is given by

$$0.5u_l(\bar{v},\bar{w}) = \sum_{j=1}^{N} \int_{v_{j-1}}^{v_j} \int_{w_{j-1}}^{w_j} \{\overline{U}_{lj}^*(\bar{v},\bar{w},v,w)p_j(v,w) - \overline{P}_{lj}^*(\bar{v},\bar{w},v,w)u_j(v,w)\} J_j(v,w) dv dw, \quad (2)$$

where

$v_{j-1} < \bar{v}, v < v_j, w_{j-1} < \bar{w}, w < w_j, l = 1,2,3,\ldots,N$.

In Eq. (2), $u_j(v,w)$ and $p_j(v,w)$ represent the boundary functions, while $J_j(v,w)$ denotes the Jacobian of the transformation from the parametric domain to the Cartesian coordinate system. The boundary of each subdomain is defined using N parametric Bézier surface patches $\Gamma_j(v,w)$. Figure 1 depicts the boundary represented by 18 first-degree Bézier patches, with six patches per subdomain and two patches positioned along each shared interface. For more complex geometries, higher-degree patches can be employed. Each Bézier patch is directly embedded in the integrand functions of Eq. (2), as given by

$$\overline{U}_{lj}^*(\bar{v},\bar{w},v,w) = \frac{1}{4\pi\sqrt{k_{11}^{(i)}k_{22}^{(i)}k_{33}^{(i)}}} \frac{1}{\left(\eta_1^2/k_{11}^{(i)} + \eta_2^2/k_{22}^{(i)} + \eta_3^2/k_{33}^{(i)}\right)^{0.5}}, \quad (3)$$

$$\overline{P}_{lj}^*(\bar{v},\bar{w},v,w) = \frac{1}{4\pi\sqrt{k_{11}^{(i)}k_{22}^{(i)}k_{33}^{(i)}}} \frac{\eta_1 n_j^{(1)}(v,w) + \eta_2 n_j^{(2)}(v,w) + \eta_3 n_j^{(3)}(v,w)}{\left(\eta_1^2/k_{11}^{(i)} + \eta_2^2/k_{22}^{(i)} + \eta_3^2/k_{33}^{(i)}\right)^{1.5}}. \quad (4)$$

Here, η_g for $g = 1,2,3$ are defined as

$$\eta_g = \Gamma_l^{(g)}(\bar{v},\bar{w}) - \Gamma_j^{(g)}(v,w), \quad (5)$$

where l and j denote the Bézier patches containing the source point (load point) and the field point (integration point), respectively. The terms $n_j^{(g)}(v,w)$ represent the components of the boundary normal vector, and the index g corresponds to the Cartesian coordinate directions, while the index i denotes the subdomain Ω_i. Moreover, we assume that the boundary functions $u_j(v,w)$ and $p_j(v,w)$ in (2) are expressed as Chebyshev series, taking the following form

$$u_j(v,w) = \sum_{p=0}^{P-1}\sum_{r=0}^{R-1} u_j^{(pr)} T_j^{(p)}(v) T_j^{(r)}(w). \quad (6)$$

$$p_j(v,w) = \sum_{p=0}^{P-1}\sum_{r=0}^{R-1} p_j^{(pr)} T_j^{(p)}(v) T_j^{(r)}(w), \quad (7)$$

where $u_j^{(pr)}$ and $p_j^{(pr)}$ represent the successive coefficients in these series. After obtaining $u_j(v,w)$ and $p_j(v,w)$ along the boundary using Eq. (2) by determining the values of the coefficients $u_j^{(pr)}$ and $p_j^{(pr)}$, the solution at any point within the computational domain can be determined by applying the integral identity within the subdomain Ω_i, as outlined in [3].

3 Singularity Removal in the PIES Formulation via Regularization

An analysis of Eq. (2) reveals a special case when $l = j$, meaning the load point and the integration point lie on the same patch. In such a case, as $v \to \bar{v}$ and $w \to \bar{w}$, the integrand function (3) becomes weakly singular, while (4) becomes strongly singular. This behavior arises from the fact that, under these conditions, the denominator in (3) and (4) approaches zero, hence the entire formulas tend toward infinity. To address this, Eq. (2) is transformed into an equivalent regularized form that eliminates these singularities. The procedure begins by replacing the boundary functions $u_j(v, w)$ and $p_j(v, w)$ in (2) with the auxiliary regularization functions (8) and (9), as described below

$$\overset{\vee}{u}_j(v, w) = A_l(\bar{v}, \bar{w})$$

$$\left[\frac{\Gamma_l^{(1)}(\bar{v}, \bar{w}) - \Gamma_j^{(1)}(v, w)}{k_{11}^{(i)}} + \frac{\Gamma_l^{(2)}(\bar{v}, \bar{w}) - \Gamma_j^{(2)}(v, w)}{k_{22}^{(i)}} + \frac{\Gamma_l^{(3)}(\bar{v}, \bar{w}) - \Gamma_j^{(3)}(v, w)}{k_{33}^{(i)}} \right],$$

$$+ B_l(\bar{v}, \bar{w}), \tag{8}$$

$$\overset{\vee}{p}_j(v, w) = A_l(\bar{v}, \bar{w}) \left[n_j^{(1)}(v, w) + n_j^{(2)}(v, w) + n_j^{(3)}(v, w) \right]. \tag{9}$$

These functions satisfy (1) and depend on the geometry, boundary normals, material properties and the unknown regularization coefficients $A_l(\bar{v}, \bar{w})$, $B_l(\bar{v}, \bar{w})$. Next, subtracting Eq. (2) with $u_j^\vee(v, w)$ and $p_j^\vee(v, w)$ from the original version with $u_j(v, w)$ and $p_j(v, w)$, we obtain

$$0.5\{u_l(\bar{v}, \bar{w}) - B_l(\bar{v}, \bar{w})\} = \sum_{j=1}^{N} \int_{v_{j-1}}^{v_j} \int_{w_{j-1}}^{w_j} \{\bar{U}_{lj}^*(\bar{v}, \bar{w}, v, w)$$

$$\left\{ p_j(v, w) - A_l(\bar{v}, \bar{w}) \left[n_j^{(1)}(v, w) + n_j^{(2)}(v, w) + n_j^{(3)}(v, w) \right] \right\}$$

$$- \bar{P}_{lj}^*(\bar{v}, \bar{w}, v, w) \{u_j(v, w)$$

$$- A_l(\bar{v}, \bar{w}) \left[\frac{\Gamma_l^{(1)}(\bar{v}, \bar{w}) - \Gamma_j^{(1)}(v, w)}{k_{11}^{(i)}} + \frac{\Gamma_l^{(2)}(\bar{v}, \bar{w}) - \Gamma_j^{(2)}(v, w)}{k_{22}^{(i)}} + \frac{\Gamma_l^{(3)}(\bar{v}, \bar{w}) - \Gamma_j^{(3)}(v, w)}{k_{33}^{(i)}} \right]$$

$$+ B_l(\bar{v}, \bar{w}) \} J_j(v, w) dv dw. \tag{10}$$

At the singular point ($l = j$, $v = \bar{v}$, $w = \bar{w}$) the singularities in (3) and (4) are removed by choosing

$$A_l(\bar{v}, \bar{w}) = \frac{p_l(\bar{v}, \bar{w})}{n_l^{(1)}(\bar{v}, \bar{w}) + n_l^{(2)}(\bar{v}, \bar{w}) + n_l^{(3)}(\bar{v}, \bar{w})}, \tag{11}$$

$$B_l(\bar{v}, \bar{w}) = u_l(\bar{v}, \bar{w}). \tag{12}$$

By substituting Eqs. (11) and (12) into (10), the final regularized PIES formulation is obtained, as presented below

$$\sum_{j=1}^{N} \left\{ \int_{v_{j-1}}^{v_j} \int_{w_{j-1}}^{w_j} \bar{U}_{lj}^*(\bar{v},\bar{w},v,w) \left[p_j(v,w) - \frac{n_j^{(1)}(v,w)+n_j^{(2)}(v,w)+n_j^{(3)}(v,w)}{n_l^{(1)}(\bar{v},\bar{w})+n_l^{(2)}(\bar{v},\bar{w})+n_l^{(3)}(\bar{v},\bar{w})} p_l(\bar{v},\bar{w}) \right] - \right.$$

$$\int_{v_{j-1}}^{v_j} \int_{w_{j-1}}^{w_j} \bar{P}_{lj}^*(\bar{v},\bar{w},v,w) \left[-\frac{\frac{\Gamma_l^{(1)}(\bar{v},\bar{w})-\Gamma_j^{(1)}(v,w)}{k_{11}^{(i)}} + \frac{\Gamma_l^{(2)}(\bar{v},\bar{w})-\Gamma_j^{(2)}(v,w)}{k_{22}^{(i)}} + \frac{\Gamma_l^{(3)}(\bar{v},\bar{w})-\Gamma_j^{(3)}(v,w)}{k_{33}^{(i)}}}{n_l^{(1)}(\bar{v},\bar{w})+n_l^{(2)}(\bar{v},\bar{w})+n_l^{(3)}(\bar{v},\bar{w})} p_l(\bar{v},\bar{w}) + \right.$$

$$\left. u_j(v,w) - u_l(\bar{v},\bar{w}) \right] \right\} J_j(v,w) dv dw = 0. \tag{13}$$

Thanks to the regularization, all integrals in Eq. (13) become nonsingular.

4 Subdomain Assembly

For computational purposes, the boundary functions are approximated using Chebyshev series (6,7). The use of Chebyshev series allows for systematic improvement in accuracy by increasing the number of terms. The collocation method is then applied to solve Eq. (13) to obtain the values of the coefficient $u_j^{(pr)}$ and $p_j^{(pr)}$. In handling multiple interconnected subdomains, matrix elements must be defined on both external and interface boundaries. For each of the three subdomains shown in Fig. 1, separate systems of algebraic equations can be formulated based on the PIES in the following form

$$\text{for } \Omega_1 \; [H_{\Gamma_1} \; H_{\Gamma_{1I}}] \begin{bmatrix} p_{\Gamma_1} \\ p_{\Gamma_{1I}} \end{bmatrix} = [G_{\Gamma_1} \; G_{\Gamma_{1I}}] \begin{bmatrix} u_{\Gamma_1} \\ u_{\Gamma_{1I}} \end{bmatrix}, \tag{14}$$

$$\text{for } \Omega_2 \; [H_{\Gamma_{2I}} \; H_{\Gamma_2} \; H_{\Gamma_{2II}}] \begin{bmatrix} p_{\Gamma_{2I}} \\ p_{\Gamma_2} \\ p_{\Gamma_{2II}} \end{bmatrix} = [G_{\Gamma_{2I}} \; G_{\Gamma_2} \; G_{\Gamma_{2II}}] \begin{bmatrix} u_{\Gamma_{2I}} \\ u_{\Gamma_2} \\ u_{\Gamma_{2II}} \end{bmatrix}, \tag{15}$$

$$\text{for } \Omega_3 \; [H_{\Gamma_{3II}} \; H_{\Gamma_3}] \begin{bmatrix} p_{\Gamma_{3II}} \\ p_{\Gamma_3} \end{bmatrix} = [G_{\Gamma_{3II}} \; G_{\Gamma_3}] \begin{bmatrix} u_{\Gamma_{3II}} \\ u_{\Gamma_3} \end{bmatrix}. \tag{16}$$

Here, $\Gamma_1, \Gamma_2, \Gamma_3$ denote external boundaries, while Γ_{1I}, Γ_{2I} and $\Gamma_{2II}, \Gamma_{3II}$ refer to interface boundaries I and II, respectively. To assemble the global system, continuity and equilibrium conditions must be enforced on the shared boundaries, as outlined below

$$u_{\Gamma_{1I}} = u_{\Gamma_{2I}} = u_{\Gamma_I}, p_{\Gamma_{1I}} = -p_{\Gamma_{2I}} = p_{\Gamma_I}, u_{\Gamma_{2II}} = u_{\Gamma_{3II}} = u_{\Gamma_{II}}, p_{\Gamma_{2II}} = -p_{\Gamma_{3II}} = p_{\Gamma_{II}}. \tag{17}$$

Combining (14–16) with the compatibility conditions (17) yields the global system.

$$\begin{bmatrix} H_{\Gamma_1} & H_{\Gamma_{1I}} & -G_{\Gamma_{1I}} & 0 & 0 & 0 & 0 \\ 0 & H_{\Gamma_{2I}} & -G_{\Gamma_{2I}} & H_{\Gamma_2} & H_{\Gamma_{2II}} & -G_{\Gamma_{2II}} & 0 \\ 0 & 0 & 0 & 0 & H_{\Gamma_{3II}} & -G_{\Gamma_{3II}} & H_{\Gamma_3} \end{bmatrix} \begin{bmatrix} p_{\Gamma_1} \\ u_{\Gamma_I} \\ p_{\Gamma_I} \\ p_{\Gamma_2} \\ u_{\Gamma_{II}} \\ p_{\Gamma_{II}} \\ p_{\Gamma_3} \end{bmatrix} = \begin{bmatrix} G_{\Gamma_1} & 0 & 0 \\ 0 & G_{\Gamma_2} & 0 \\ 0 & 0 & G_{\Gamma_3} \end{bmatrix} \begin{bmatrix} u_{\Gamma_1} \\ u_{\Gamma_2} \\ u_{\Gamma_3} \end{bmatrix}. \tag{18}$$

5 Results

5.1 Example 1

We return to the domain shown in Fig. 1, consisting of three subdomains $\Omega_1, \Omega_2, \Omega_3$. We assume that all three subdomains share identical parameters for the orthotropic medium $k_{11} = k_{22} = 0.5$ and $k_{33} = 2$. Defining the boundaries of these subdomains requires specifying 18 interconnected Bézier patches of the first degree, defined by 16 corner points. We assume that the expected field distribution on the boundary and inside the domains is described by the following function depending on the Cartesian coordinates $x = \{x_1, x_2, x_3\}$ and satisfying the the Eq. (1).

$$u_A(x_1, x_2, x_3) = 2x_1^2 + 2x_2^2 - x_3^2. \tag{19}$$

Dirichlet boundary conditions are specified on each surface patch defining the boundary, based on this function. The problem is solved using both the singular (2) and regularized (13) PIES formula on the boundary. The singularity appearing in (2) results from its mathematical formula, as indicated at the beginning of Sect. 3 of the paper. Accuracy is assessed at internal points, with relative errors shown in Table 1.

Table 1. Relative error [%] at selected points for Example 1.

(x_1, x_2, x_3)	Analytical (19)	PIES (2) [%]	PIES (13) [%]
(5.5,0.25,1.0)	60.125	0.903981	0.061675
(4.5,0.5,1.0)	40.500	0.303981	0.024603
(3.5,0.25,1.0)	24.125	0.180077	0.011574
(2.5,0.25,1.0)	12.125	0.129525	0.009263
(1.5,0.25,1.0)	4.125	0.102096	0.008490
(0.25, 0.25,1.0)	-0.250	0.106148	0.000548

The obtained results indicate a significant improvement in the accuracy of the solutions obtained using the proposed regularization approach compared to the singular variant of PIES.

5.2 Example 2

In the next example, we extend the analysis of the field distribution defined by (19) to a new domain, shown in Fig. 2, described by both flat and curved Bézier patches. This domain is divided into two subdomains, Ω_1 and Ω_2. The subdomain boundaries are represented by 7 cubic Bézier patches for the curvilinear segments and 9 linear Bézier patches. Defining the complete boundary with 16 patches required specifying 112 control points.

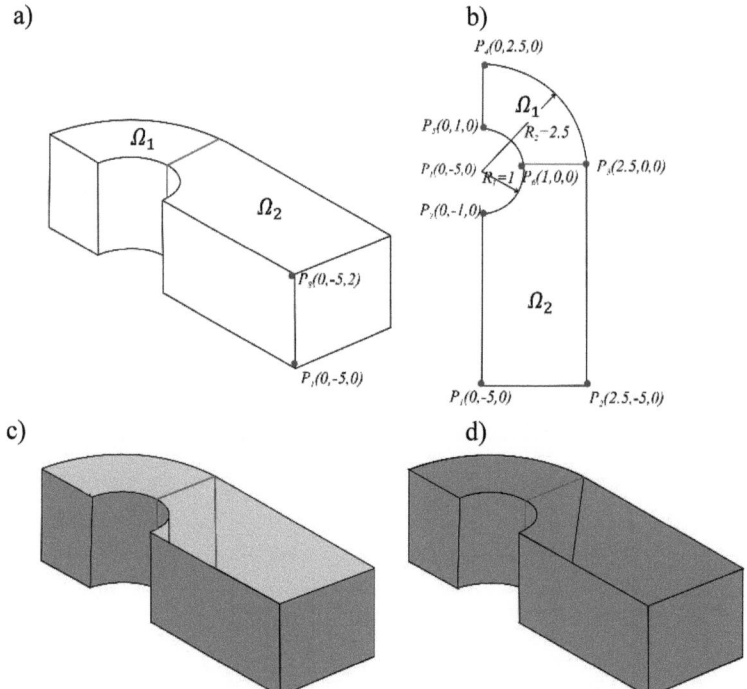

Fig. 2. Modeling of subdomains using Bézier patches: the domain contour in a 3D view (a) and the projection of the lower base, with the subdomains Ω_1, Ω_2 (b); the defined base of the subdomains using 2 cubic Bézier patches and 1 linear patch, along with the side walls of the domains modeled with 3 cubic Bézier patches and 8 linear patches (c); and the complete boundary definition of the subdomains (d).

It is assumed that both subdomains Ω_1, Ω_2, have identical properties with the following parameters for the orthotropic medium $k_{11} = k_{22} = 0.5$ and $k_{33} = 2$. The Dirichlet boundary conditions are again obrained from (19). Table 2 presents a comparison of the relative errors for solutions obtained at selected points within the subdomains Ω_1 and Ω_2 with help of (2) and (13).

Table 2. Relative error of the solutions at selected points within the subdomains for Example 2.

(x_1, x_2, x_3)	Analytical (19)	PIES (2) [%]	PIES (13) [%]
(0.862,1.690,0.926)	6.34442	0.03376	0.00673
(0.536,-2.136,0.409)	9.53990	0.05628	0.00215
(0.949,-3.754,1.528)	27.6615	0.07986	0.00145
(1.237,-2.989,1.656)	18.1899	0.01701	0.00162
(0.573,-1.741,1.278)	5.08737	0.01459	0.00268
(1.491,-2.870,1.648)	18.21070	0.01211	0.00677
(0.779,-2.090,0.460)	9.73909	0.02113	0.00232

Once again, this example demonstrates a clear improvement in the solutions obtained using the regularized formula (13).

6 Conclusions

This paper presents an algorithm that eliminates the need for direct evaluation of both strongly and weakly singular boundary integrals within the PIES, enabling efficient analysis of three-dimensional, multidomain orthotropic problems. Notably, all integrals are computed numerically, removing the requirement for analytical evaluation. Once regularized, these integrals can be accurately evaluated using standard Gaussian quadrature. The proposed method preserves all the established advantages of PIES in homogeneous regions. These benefits include a simplified subdomain representation - achieved with minimal input data and the decoupling of boundary geometry, defined via surface patches, from the approximation of boundary functions using Chebyshev series. The preliminary verification of the proposed approach, as presented in this work, will be extended in future research through a comparative analysis with other numerical methods, focusing on computational efficiency, accuracy and execution time.

References

1. Zienkiewicz, O.C., Taylor, R.L., Zhu, J.Z.: The Finite Element Method: its Basis and Fundamentals. Elsevier (2005)
2. Brebbia, C.A., Telles, J.C.F., Wrobel, L.C.: Boundary Element Techniques: Theory and Applications in Engineering. Springer, Cham (2012)
3. Zieniuk, E., Szerszen, K.: The PIES for solving 3D potential problems with domains bounded by rectangular Bézier patches. Eng. Comput. **31**(4), 791–809 (2014)
4. Scuderi, L.: On the computation of nearly singular integrals in 3D BEM collocation. Int. J. Numer. Meth. Eng. **74**, 1733–1770 (2008)
5. Gao, X., et al.: An adaptive element subdivision technique for evaluation of various 2D singular boundary integrals. Eng. Anal. Boundary Elem. **32**, 692–696 (2008)
6. Zhang, J., et al.: A binary-tree subdivision method for evaluation of singular integrals in 3D BEM. Eng. Anal. Boundary Elem. **103**, 80–93 (2019)

7. Niu, Z., et al.: Analytic formulations for calculating nearly singular integrals in two-dimensional BEM. Eng. Anal. Boundary Elem. **31**, 949–964 (2007)
8. Zhou, H., et al.: Analytical integral algorithm applied to boundary layer effect and thin body effect in BEM for anisotropic potential problems. Comput. Struct. **86**, 1656–1671 (2008)
9. Ren, Q., Chan, C.L.: Analytical evaluation of the BEM singular integrals for 3D Laplace and Stokes flow equations using coordinate transformation. Eng. Anal. Boundary Elem. **53**, 1–8 (2015)
10. Lutz, E.: Exact Gaussian quadrature methods for near-singular integrals in the boundary element method. Eng. Anal. Boundary Elem. **9**, 233–245 (1992)
11. Tausch, J.: Adaptive quadrature rules for Galerkin BEM. Comput. Math. Appl. **113**, 270–281 (2022)
12. Zieniuk, E., Szerszeń, K.: Near corner boundary regularization of the parametric integral equation system (PIES). Eng. Anal. Boundary Elem. **158**, 51–67 (2024)
13. Szerszeń, K., Zieniuk, E.: Elimination of computing singular surface integrals in the PIES method through regularization for three-dimensional potential problems. International Conference on Computational Science ICCS 2024, Lecture Notes of Computer Science, vol. 14833, Part II, pp. 325–338 (2024)

Advancing Bird Species Classification: A Fusion of Audio and Image Data

Jie Xie[1(✉)], Xueyan Dong[2], Zhe Wu[1], Zheng Lang[1], Yuji Wang[1], Chunrong He[3], Jinpei Song[3], Zhuobin Zhang[3], Guiqing Yu[3], and Jia Tang[3]

[1] School of Computer and Electronic Information/School of Artificial Intelligence, Nanjing Normal University, Nanjing, China
xiej8734@gmail.com
[2] Beijing Union University, Beijing 100101, China
[3] Hunan Hupingshan National Nature Reserve Administration Bureau, Changde 415300, China

Abstract. Automated classification of bird species is crucial for large-scale environmental monitoring, providing valuable insights into temporal and spatial changes in ecosystems. Previous studies have primarily focused on using either acoustic or visual data for bird species recognition. However, few studies have explored the simultaneous use of both acoustic and visual data to improve classification performance. In this study, we propose a dual branch network based on pre-trained models to enhance bird species classification by integrating acoustic and visual information. Specifically, ResNet50 is used for visual data, while CNN14 is employed for acoustic data. The extracted feature embeddings are then fused, and attention mechanisms are applied to further improve classification performance. Experimental results demonstrate that our proposed model achieves significantly higher accuracy compared to using audio or image data alone. The best-performing model achieved an accuracy of 96.44%, precision of 96.62%, recall of 94.30%, and F1-score of 95.01%. This study highlights the potential of combining acoustic and visual data for bird species classification and suggests that attention mechanisms can further enhance model performance.

Keywords: Bird species classification · Fine-grained classification · Fusion of information · Transfer learning

1 Introduction

Although birds are widely recognized as excellent indicators of biodiversity due to their crucial ecosystem services, a global decline in bird populations has been observed [3]. To enhance protection policies and boost bird populations, the first step is to understand the current status of birds, which can be achieved through regular monitoring. Furthermore, large-scale monitoring of bird diversity is crucial for gaining a better understanding of bird populations. Traditional bird monitoring methods, which require ecologists to conduct censuses, are both time-consuming and expensive. Therefore, the monitoring scale is limited in both temporal and spatial scales. Recently, the declining prices of sensor technology, coupled with the simultaneous development of artificial intelligence (AI) technology, have made it possible to monitor bird diversity in larger spatial and temporal scales. Increasingly, efforts are relying on sensors to collect data, which is then analyzed using AI techniques. Based on the analysis output, knowledge of bird diversity can be obtained.

Previous studies have proposed various approaches for bird species classification, which are either based on acoustic or visual information. For the recent work on bioacoustic signals, lots of deep learning architectures have been proposed including deep neural network [18], convolutional neural network [4,8,15,23–25], recurrent neural networks [17], and transformer [19,22,27]. In addition, several techniques have been investigated to further improve classification performance, such as self-supervised learning [6,16,26] and data augmentation [10,12,21]. Similar to acoustic classification of bird species, deep learning methods are widely explored in classifying bird species using images [1,2,11,13].

Classifying bird species by their calls or images alone is challenging due to factors like similarity between species, background noise, varying acquisition conditions, and variations in background, lighting, and capture angles in images [14]. Thus, combining acoustic and visual information may enhance classification performance. In this study, we propose a dual-branch network based on pre-trained models for bird sound classification. Specifically, ResNet50 is used as the image encoder, while CNN14 is used as the audio encoder. Then, extracted feature embeddings are fused to improve the classification performance. In addition, the attention mechanism is used to improve the classification performance. The main contribution is to deal with the optimal attention mechanism selection and combination of all aforementioned steps to improve bird species classification performance.

2 Related Work

Researchers have explored the integration of acoustic and visual information for the classification of bird species. In a study by [14], the Scale-Invariant Feature Transform was employed to detect local features in bird images, which were then used to train a support vector machine classifier. When instances were not classified with sufficient certainty, they were rejected and reclassified using Mel frequency cepstral coefficients extracted from bird songs when available. Another approach introduced by [20] involved automatically identifying bird species from video recordings. This method applied image and audio processing and classification techniques, constructing models using pre-trained neural networks like ResNet50V2 and EfficientNetB0. A novel method proposed by [5] integrated visual and auditory data for species identification, improving accuracy and robustness. A deep CNN was used to extract features from bird images, while an LSTM network analyzed bird calls. By integrating these modalities early in the classification process rather than relying solely on either data type or performing late fusion, they achieved a significant performance improvement. However, the aforementioned studies have some limitations. Firstly, relying on handcrafted features may hinder performance. Secondly, using pre-trained models originally intended for ImageNet may not be ideal for processing audio data. Lastly, all methods assume an equal number of training and testing data for each species, which is often not the case in real-world scenarios.

3 The Proposed Method

A dual branch network based on pre-trained models is proposed for bird species classification (Fig. 1). First, the issue of mismatched quantities of sound and image training data within the same category is processed to having the same quantities. Then, audio data is converted to a log-Mel spectrogram by a short-time Fourier transform. Next, two pre-trained models based on AudioSet and ImageNet are applied to acoustic and visual data separately for extracting embeddings. Finally, five different attention mechanisms are applied to audio and image embedding separately to improve the final classification performance.

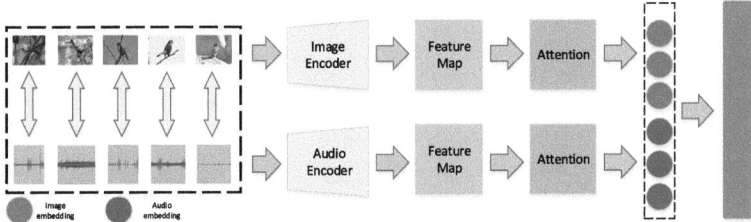

Fig. 1. Flowchart of our proposed dual branch network based on pre-trained models for bird sound classification

3.1 Datasets Description

In this study, we select 20 common bird species in Huping Moutain, Hunan Province, China as the experimental target. The audio data is collected from Xeno-canto website[1], while the images are obtained from eBird website[2]. For the image, a maximum cap of 300 images has been established for each bird species to maintain uniformity within our image dataset. Consequently, the upper limit for the number of images in any avian category is restricted to 300. For the audio, we first download the data from Xeno-canto, which is then segmented into 10 s with a fixed sampling rate of 16 kHz. Finally, the number of audio and images for each bird species is described in Table 1. Since we process the audio

---

[1] https://xeno-canto.org/.
[2] https://ebird.org/home.

and image data simultaneously, the number of audio and image samples should be identical. For each bird species, the number of audio and image samples is first compared, and the modality having more samples will be decreased.

Table 1. Number of image and audio samples for all bird species

Category	Audio Count	Image Count	Category	Audio Count	Image Count
Black-streakedScimitarBabbler	166	248	Large-billedCrow	314	300
BrownDipper	47	300	Light-ventedBulbul	447	300
ChestnutBulbul	98	300	ManipurFulvetta	37	300
CollaredFinchbill	79	300	PygmyCupwing	350	300
CrestedKingfisher	13	300	Radde'sWarbler	318	300
Elliot'sLaughingthrush	129	300	Red-breastedFlycatcher	551	300
Fork-tailedSunbird	112	300	Rufous-facedWarbler	239	300
GreatTit	565	300	Streak-breastedScimitarBabbler	241	300
Green-backedTit	258	300	WarblingWhite-eye	326	300
Grey-headedWoodpecker	611	300	Yellow-browedBulbul	112	300

3.2 Pre-trained Models

For audio classification tasks, large-scale pre-trained audio neural networks (PANNs) have been employed to extract feature embeddings [9]. PANNs represent a diverse range of convolutional neural networks designed to classify 527 distinct sound classes. Specifically, a 14-layer CNN was transferred and fine-tuned for several audio pattern recognition tasks. This CNN, pre-trained on the AudioSet dataset, has demonstrated robust generalization across numerous audio pattern recognition tasks [9]. In this study, we utilized the CNN14 architecture from [9], which comprises five blocks of 3×3 convolutional filters, followed by batch normalization and ReLU activation [9].

To extract feature embeddings from bird images, we utilized the ResNet50 model, which was pre-trained on the ImageNet dataset [7]. This pre-trained model capitalizes on its deep learning architecture to effectively capture the nuanced visual features of bird images, thereby providing a robust foundation for subsequent analysis and classification tasks.

3.3 Attention Mechanism

To further enhance feature representation in deep neural networks, Convolutional Block Attention Module (CBAM), Squeeze-and-Excitation (SE), Criss-Cross Attention (CCA), Efficient Channel Attention (ECA), and Shuffle Attention (SA) mechanisms are inserted into pre-trained models for improving bird sound classification performance. In this study, the attention mechanism is inserted into the generated feature maps of both image and audio encoders separately, and then fused together for bird sound classification.

4 Experiments

4.1 Comparison of Single Modality and Double Modalities

Figure 2 shows the performance of bird sound classification using single- and double-mode methods. From the table, we can observe that the performance of double modalities is better than both acoustic and visual modality, where the highest accuracy, precision, recall, and F1-score are 94.91%, 94.57%, 90.94%, 92.06%. For image classification, ResNet50 achieves better performance than EfficientB0. In contrast, the use of pre-trained AudioSet models (CNN14) is better than pre-trained ImageNet models (EfficientNetB0), which is often the case in previous studies [20]. One reason is that the domain mismatch between AudioSet and bird sounds is smaller than between ImageNet and bird sounds.

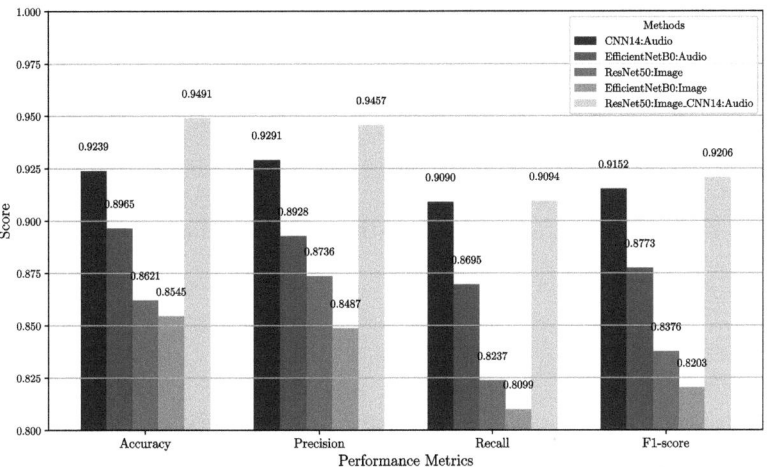

Fig. 2. Classification performance of different single modality and one double modality.

4.2 Comparison of Various Attention Mechanisms

Since ResNet50 and CNN14 achieve the best performance for both image and audio data, they are selected and combined for subsequent analysis. To further improve the classification performance, several attention mechanisms are investigated. Table 2 shows the performance comparison of attention mechanisms. All attention mechanisms can improve classification performance, indicating the necessity of using the attention mechanism. Among these attention mechanisms, ECA achieves the best performance, where precision, precision, recall, and F1 score are 96.44%, 96.62%, 94.30%, and 95.01%, respectively. Given that we apply attention to the deep-level feature map with a small spatial size, simply employing channel-wise attention yields better performance.

Table 2. Comparison of various attention mechanisms for our proposed dual branch network

Method	Accuracy	Precision	Recall	F1-score
ResNet50+CNN14	0.9491	0.9457	0.9094	0.9206
ResNet50+CNN14+CBAM	0.9517	0.9569	0.9272	0.9384
ResNet50+CNN14+CCA	0.9499	0.9555	0.9310	0.9393
ResNet50+CNN14+SE3	0.9550	0.9540	0.9274	0.9355
ResNet50+CNN14+ECA	**0.9644**	**0.9662**	**0.9430**	**0.9501**
ResNet50+CNN14+SU	0.9580	0.9595	0.9309	0.9411

Furthermore, we plot the confusion matrix of the best-performing model (Fig. 3). The highest confusion is between *Grey-headed woodpecker* and *Light-vented bulbul*. Previous studies have explored the use of audio and image data for bird sound classification, which often used a pre-trained ImageNet-based model [20]. However, the difference between nature image and spectrogram makes the pre-trained AudioSet-based model a better fit for processing audio. Compared to [20], the classification accuracy is improved by 0.38% without attention mechanisms. Using ECA, the accuracy can be improved by 2.39% which verifies the effectiveness of information fusion and attention mechanisms.

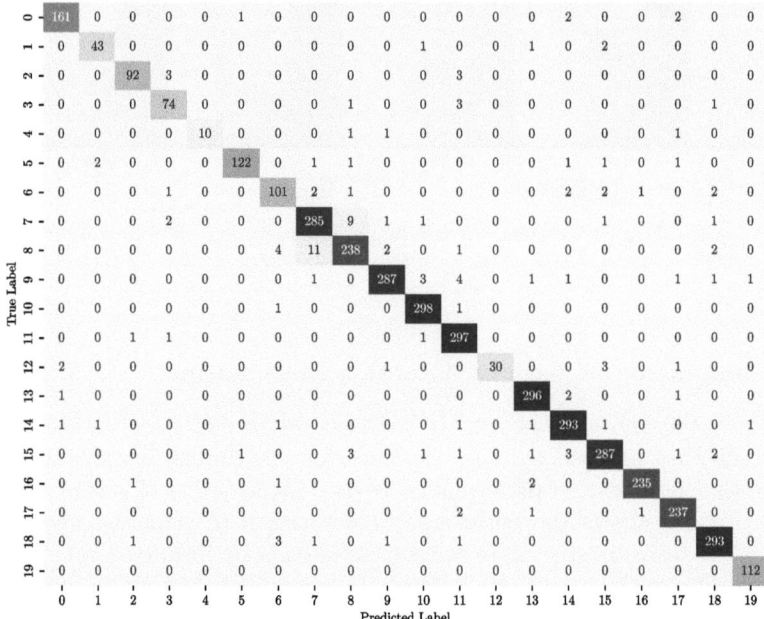

Fig. 3. Sum of confusion matrix of the best-performing model. Here, the index from 0 to 19 corresponds to the bird species from *Black streaked Scimitar Babbler* to *Yellow-browed bulbul*.

5 Conclusion and Future Work

In this study, we proposed a dual branch network based on pre-trained models to improve bird species classification by integrating acoustic and visual data. The proposed method uses ResNet50 for image data and CNN14 for audio data, with feature fusion and attention mechanisms to improve classification performance. Experimental results in the HPS dataset demonstrated that the combined use of acoustic and visual data significantly outperformed using either modality alone. The best performing model achieved an accuracy of 96. 44%, a precision of 96. 62%, a recall of 94. 30%, and a F1 score of 95. 01%. 95. 01%. Future work includes exploring advanced models, sophisticated data augmentation, generalizing to other species, and incorporating other modalities such as text or environmental data for bird species classification.

Acknowledgment. This work is supported by National Natural Science Foundation of China (Grant No: 32371556, 61902154 and 72004092). This work is also supported by the 2019 Science and Technology Plan of Beijing Municipal Education Commission (Grant No. KM201911417005).

References

1. Chen, T., Li, Y., Qiao, Q.: Fine-grained bird image classification based on counterfactual method of vision transformer model. J. Supercomput. **80**(5), 6221–6239 (2024)
2. Chen, X., Zhang, H., Song, J., Guan, J., Li, J., He, Z.: Micro-motion classification of flying bird and rotor drones via data augmentation and modified multi-scale CNN. Remote Sens. **14**(5), 1107 (2022)
3. Fraixedas, S., Lindén, A., Piha, M., Cabeza, M., Gregory, R., Lehikoinen, A.: A state-of-the-art review on birds as indicators of biodiversity: advances, challenges, and future directions. Ecol. Ind. **118**, 106728 (2020)
4. García-Ordás, M.T., Rubio-Martín, S., Benítez-Andrades, J.A., Alaiz-Moretón, H., García-Rodríguez, I.: Multispecies bird sound recognition using a fully convolutional neural network. Appl. Intell. **53**(20), 23287–23300 (2023)
5. Gavali, P., Banu, J.S.: Visual-acoustic fusion techniques for accurate identification of Indian bird species. Int. J. Comput. Digit. Syst. **17**(1), 1–22 (2024)
6. Hagiwara, M.: Aves: animal vocalization encoder based on self-supervision. In: ICASSP 2023-2023 IEEE International Conference on Acoustics, Speech and Signal Processing (ICASSP), pp. 1–5. IEEE (2023)
7. He, K., Zhang, X., Ren, S., Sun, J.: Deep residual learning for image recognition. In: Proceedings of the IEEE Conference on Computer Vision and Pattern Recognition, pp. 770–778 (2016)
8. Hu, S., Chu, Y., Wen, Z., Zhou, G., Sun, Y., Chen, A.: Deep learning bird song recognition based on MFF-ScSEnet. Ecol. Ind. **154**, 110844 (2023)
9. Kong, Q., Cao, Y., Iqbal, T., Wang, Y., Wang, W., Plumbley, M.D.: PANNs: large-scale pretrained audio neural networks for audio pattern recognition. IEEE/ACM Trans. Audio Speech Lang. Process. **28**, 2880–2894 (2020)

10. Kumar, A.S., Schlosser, T., Kahl, S., Kowerko, D.: Improving learning-based birdsong classification by utilizing combined audio augmentation strategies. Eco. Inform. **82**, 102699 (2024)
11. Kumar, M., Yadav, A.K., Kumar, M., Yadav, D.: Bird species classification from images using deep learning. In: International Conference on Computer Vision and Image Processing, pp. 388–401. Springer, Cham (2022)
12. Lauha, P., et al.: Domain-specific neural networks improve automated bird sound recognition already with small amount of local data. Methods Ecol. Evol. **13**(12), 2799–2810 (2022)
13. Liu, H., Zhang, C., Deng, Y., Xie, B., Liu, T., Li, Y.F.: Transifc: invariant cues-aware feature concentration learning for efficient fine-grained bird image classification. IEEE Trans. Multimed. (2023)
14. Marini, A., Turatti, A.J., Britto, A., Koerich, A.L.: Visual and acoustic identification of bird species. In: 2015 IEEE International Conference on Acoustics, Speech and Signal Processing (ICASSP), pp. 2309–2313. IEEE (2015)
15. Morales, G., et al.: Method for passive acoustic monitoring of bird communities using UMAP and a deep neural network. Eco. Inform. **72**, 101909 (2022)
16. Moummad, I., Farrugia, N., Serizel, R.: Self-supervised learning for few-shot bird sound classification. In: 2024 IEEE International Conference on Acoustics, Speech, and Signal Processing Workshops (ICASSPW), pp. 600–604. IEEE (2024)
17. Noumida, A., Rajan, R.: Multi-label bird species classification from audio recordings using attention framework. Appl. Acoust. **197**, 108901 (2022)
18. Rajan, R., Johnson, J., Abdul Kareem, N.: Bird call classification using DNN-based acoustic modelling. Circuits Syst. Signal Process. **41**(5), 2669–2680 (2022)
19. Rauch, L., Schwinger, R., Wirth, M., Sick, B., Tomforde, S., Scholz, C.: Active bird2vec: towards end-to-end bird sound monitoring with transformers. arXiv preprint arXiv:2308.07121 (2023)
20. Sharma, N., Vijayeendra, A., Gopakumar, V., Patni, P., Bhat, A.: Automatic identification of bird species using audio/video processing. In: 2022 International Conference for Advancement in Technology (ICONAT), pp. 1–6. IEEE (2022)
21. Sun, Y., Maeda, T.M., Solís-Lemus, C., Pimentel-Alarcón, D., Buřivalová, Z.: Classification of animal sounds in a hyperdiverse rainforest using convolutional neural networks with data augmentation. Ecol. Ind. **145**, 109621 (2022)
22. Tang, Q., Xu, L., Zheng, B., He, C.: Transound: hyper-head attention transformer for birds sound recognition. Eco. Inform. **75**, 102001 (2023)
23. Xiao, H., Liu, D., Chen, K., Zhu, M.: Amresnet: an automatic recognition model of bird sounds in real environment. Appl. Acoust. **201**, 109121 (2022)
24. Xie, J., Zhu, M.: Sliding-window based scale-frequency map for bird sound classification using 2D-and 3D-CNN. Expert Syst. Appl. **207**, 118054 (2022)
25. Xie, S., Xie, J., Zhang, J., Zhang, Y., Wang, L., Hu, H.: MDF-Net: a multi-view dual-attention fusion network for efficient bird sound classification. Appl. Acoust. **225**, 110138 (2024)
26. Zhang, C., Li, Q., Zhan, H., Li, Y., Gao, X.: One-step progressive representation transfer learning for bird sound classification. Appl. Acoust. **212**, 109614 (2023)
27. Zhang, S., Gao, Y., Cai, J., Yang, H., Zhao, Q., Pan, F.: A novel bird sound recognition method based on multifeature fusion and a transformer encoder. Sensors **23**(19), 8099 (2023)

Prototype-Pairs Decomposition for Extracting Simple and Meaningful Rules

Marcin Blachnik[1]($\boxtimes$), Mirosław Kordos[2], and Daniel Dąbrowski[1]

[1] Department of Industrial Informatics, Silesian University of Technology, ul. Krasińskiego 8, 40-019 Katowice, Poland
{marcin.blachnik,daniel.dabrowski}@polsl.pl
[2] Department of Computer Science and Automatics, University of Bielsko-Biała, ul. Willowa 2, 43-309 Bielsko-Biała, Poland
mkordos@ubb.edu.pl

Abstract. We present a preliminary study of a model-agnostic method called prototype pair decomposition that generates simple and accurate decision rules from datasets. The research focuses on its application to decision trees. It starts by selecting representative prototypes obtained by a prototype construction method, then pairs of prototypes from opposite classes are determined. These pairs define subspaces containing a fragment of the decision boundary in which a shallow decision tree is applied to extract simple decision rules consisting of a few premises. The results indicate that the proposed solution allows the extraction of locally competent simple rules that are comparable in terms of classification accuracy to a large and complex set of global rules obtained from standard decision trees.

Keywords: Explainable AI · Decision trees · Prototype-Based Learning

1 Introduction

The explanation of data and the decisions taken by machine learning models have been investigated since the beginning of AI research [12]. The problem is constantly gaining importance as the data get bigger and more complex. It is important that the user not only knows the prediction result, but also understands why a given result was reached by the model. This allows users to trust the models, which in many cases is a condition to apply them in practice [10].

Machine learning methods are often categorized as black boxes or white (glass) boxes. Black box models, like kernel methods, neural networks, and ensembles (e.g., random forest), typically offer high accuracy but limited interpretability. In contrast, white box models-such as decision trees, sequential covering, case-based reasoning (CBR) [4], and prototype-based rules [2]-provide interpretable decisions, often at the cost of accuracy.

To interpret black-box models, two main approaches are used [11]: model-independent methods and those tailored to specific models. Model-independent methods extract rule-based explanations by labeling data with predictions from black-box models. For instance, [7] approximates neural network activations with linear fragments to extract rules. Knowledge extraction from deep networks using CBR is shown in [8]. While kernel-based models interpretation as prototype-based rules can be found in [1].

Another strategy is local interpretation, as in LIME [10], which fits a linear model near the query point—later extended with autoencoders [13]. Graphical explanations like SHAP [6] and saliency maps [14] offer local insights, particularly for convolutional networks.

In this paper, we propose for the first time a preliminary study on a model-agnostic method for rule extraction that leverages problem decomposition through a technique called prototype pair decomposition (PPD) to improve the incorporability of decision boundaries in classification problems. The proposed algorithm integrates case-based reasoning with classical crisp rules by constructing locally competent and interpretable rules. This is achieved by identifying a pair of adversarial (enemy) prototypes, which are prototypes of opposite classes. This pair partitions the decision boundary of the classification problem into regions by assigning each training sample to it nearest pair of prototypes. Since the prototypes belong to different classes, the resulting pair defines a fragment of the decision boundary that is locally approximated by a simple set of rules. From a system-wide perspective, the proposed solution decomposes the classification problem into regions formed by subsets of the original dataset. Each region is identified by a pair of prototypes, and in each region, an independent model can be trained.

In our study we use a decision tree as a reference model for measuring the benefits of the proposed solution. We show that the rules obtained from decision trees with the PPD algorithm are much simpler that those generated by standard decision trees trained on the entire training data.

2 The Prototype-Pairs Decomposition Algorithm

The PPD algorithm (see sketch 1) requires a training set $\mathbf{T} = \{\mathbf{X}, \mathbf{y}\}$ and the number of prototypes k as input. It is assumed that $y_i \in \{P, N\}$, that is a binary classification problem.

PPD starts by selecting representative prototypes obtained by any instance selection or prototype construction method [3]. We use a clustering-based prototype selection, clustering each class separately. Then the set of possible pairs $\mathbf{U}$ is determined by identifying the nearest positive and negative prototype for each training sample (adversarial prototypes). Thus, each pair consists of one instance from a positive and one from a negative class. The regions are encoded using the Cantor pairing function which allows encoding a pair by a single integer value. Each pair defines the decision boundary of the nearest neighbor classifier. The pairs define subspaces also called regions, which are the key concept of the

Algorithm 1. A PPD algorithm

```
function PPD(X,y,k)
    (P_p, P_n) ← FINDPROTOTYPES(X, y, k)
    r ← zeros(1, n) //Create vector to store region identifiers
    u = ∅
    for all i ∈ {1...n} do
        a ← argmin (D(p_j, x_i)) //Get nearest positive samples
            ∀p_j in P_P
        b ← argmin (D(p_j, x_i)) //Get nearest negative samples
            ∀p_j in P_N
        z ← CANTORPAIRINGFUNCTION(a, b)
        u = u ∪ z //Set of unique regions
        r_i ← z //Assign i'th vector to region z
    end for
    s ← GETREGIONSTATISTICS(X, y, r)
    z ← GETINCORRECTREGION(s)
    while z ≠ ∅ do
        g[g = z] ← ∅ //Set all values in g equals z to None
        u ← u \ z
        X' ← UNASSIGNEDSAMPLES(X, r)
        r' ← ASSIGNREGION(X', u, P_P, P_N)
        r ← UPDATE(r, r')
        s ← GETREGIONSTATISTICS(X, y, r)
        z ← GETINCORRECTREGION(s) //Get region which doesn't fulfil given statistics or ∅
    end while
    return r,u //Return array assigning each training vector to a particular region
    //and a set of regions.
end function
```

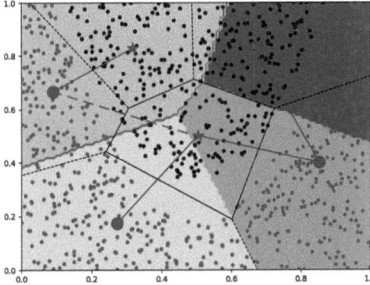

Fig. 1. A visualization of the PPD algorithm showing training samples (purple and dark blue dots), prototypes (red markers), pairs (linked prototypes), and the obtained regions encoded by the background color. (Color figure online)

algorithm. Next the regions with the fewest samples and highest unbalanced ratio are pruned. The procedure is repeated until all regions satisfy the given properties.

Next, the process of training the decision trees for each region begins. Each of the decision trees is trained on the subset (region) of the original training set. The size of the regions scales inversely with the number of regions K. The total computational complexity of the proposed method is the sum of the complexities of the PPD algorithm ($O(n)$) and of the decision trees trained on subsets of the samples. On average it is $O\left(n + n \log\left(\frac{n}{K}\right)\right)$.

Algorithm 2. An AssignRegion algorithm used for prediction and reassigning regions which do not fulfill statistical requirements.

```
function ASSIGNREGION(X, u, P_P, P_N)
    n ← ||X||
    for all i ∈ {1...n} do
        for all u ∈ u do
            (p_P, p_N) ← CANTORUNPAIRINGFUNCTION(u) //Get prototypes defining region u
            d ← D(p_P, x_i)² + D(p_N, x_i)² //Get distance to pair of prototypes
            if d < d_min then //Find smallest distance
                d_min ← d
                z ← u
            end if
        end for
        g_i ← z //Assign i to region z
    end for
    return g
end function
```

Visualization of the PPD algorithm is presented in Fig. 1. After defining the regions that are stored in $\mathbf{u}$ (here we have $K = \|\mathbf{u}\|$ regions), a small decision tree is trained using samples within the region. This can be parallelized and executed for each region independently. As the regions contain a relatively small subset of samples, a fragment of the decision boundary which is contained inside a region can be approximated with a simple set of rules. This ensures that the rules are clear and easy to interpret, usually consisting of just a few premises. Since a decision tree is one of the most popular methods for rule extraction, supporting very good scalability, it becomes the first choice for our experiments, but other rule extraction methods can also be used.

During prediction, PPD starts by identifying a region for each test sample using the *AssignRegion* function presented in sketch 2. The arguments of this function are: a set of test samples $\mathbf{X}$, a set of region identifiers $\mathbf{u}$, and a position of prototypes $\mathbf{P}_P, \mathbf{P}_N$. Then, the main procedure starts by iterating over all test samples. Next, it iterates over a set of possible region identifiers $\mathbf{u}$. Each region identifier is then decoded to the corresponding prototypes $\mathbf{P}_P, \mathbf{P}_N$ using the inverse Cantor function. Then, the nearest pair is determined by the minimum sum of squared distances to the positive and negative prototypes of the pair. Finally, when all samples in $\mathbf{X}$ have a pair of prototypes assigned, the final classifier starts. In a loop, all samples that share the same region are selected and the decision tree related to that particular region is applied to that samples.

3 PPD with the Decision Trees. Toy Example

To better explain the PPD algorithm, a toy example is presented, where the PPD algorithm is applied to a binary classification problem.

Figure 2d shows the obtained decision boundary where the input space is divided into 3 regions marked with straight lines. Within each region, a simple decision tree is constructed. In this example, we set the maximum depth of the tree to 2. As a result, three trees are created, as shown in Fig. 2a, 2b, 2c. To show the benefits of the proposed algorithm over the classical decision tree, the

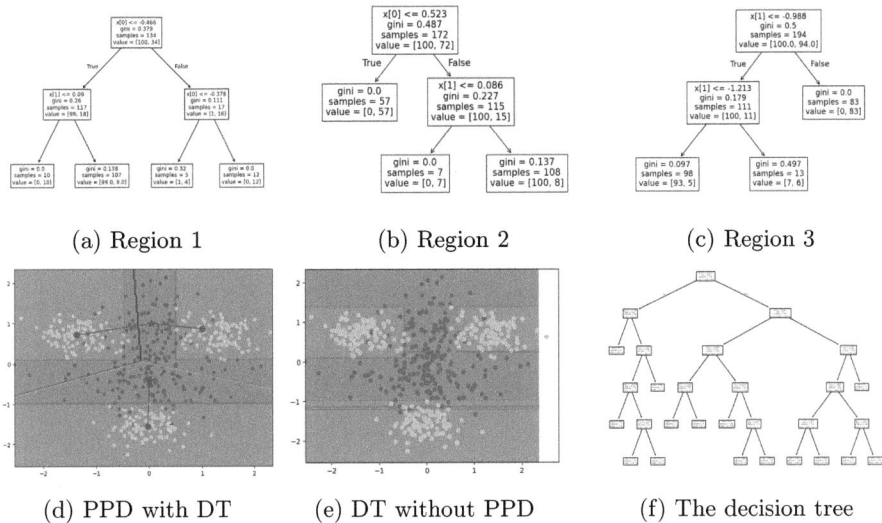

Fig. 2. A comparison of the decision borders and the trees obtained with PPD (a),(b),(c),(d) algorithm and without PPD algorithm (e),(f)

decision tree that achieves a comparable prediction performance is shown in Fig. 2e and its decision border is visualized in Fig. 2f. Comparison of these two examples clearly shows that the rules generated by PPD regions with decision trees are much simpler and easier to understand.

4 Experiments

To verify the performance of the PPD algorithm with internal decision trees, we evaluated its prediction accuracy as a function of the depth of internal PPD trees vs. the accuracy of classical decision trees and the influence of PPD parameters on its prediction accuracy using a 10-fold cross-validation.

The software used for the experiments was created in Python, Scikit-learn [9], imbalanced-learn libraries [5] and our own PPD library. It is available at https://github.com/mblachnik/Prototype-Pair-Ensemble/tree/24_PPD_Tree. The datasets obtained from https://www.openml.org present complex two-class classification problems with a non-linear decision boundaries.

4.1 PPD vs Standard Decision Tree. Tree Depth Comparison

In this section we evaluate the dependencies between size and accuracy of standard decision trees vs. of PPD with small trees inside the PPD regions.

Table 1 shows that for the same accuracy of both methods, standard decision trees on average required 8.56–9.88 premises of a rule, while PPD with trees required only 3–6 rule premises. Moreover, the number of rules in standard trees

Table 1. Prediction performance of the decision tree and the PPD-trees along with the average depth of the tree and the number of leaves. For the PPD-trees also the number of regions is given.

dataset	samples	attr.	unb. ratio	Decision tree				PPD(Tree)				
				Acc	Std	depth	leaves	Acc	Std	depth	leaves	# reg.
codrnaNorm	488565	8	0.5	95.48	0.08	9.88	882	95.05	0.11	4.00	15.18	33
electricity-normalized	45312	8	0.738	83.77	0.30	9.50	550	83.67	0.59	5.00	26.20	25
covtype	581012	54	0.952	81.25	0.13	9.73	691	81.23	0.71	5.00	26.83	12
banana	5300	2	0.813	88.77	1.20	8.92	237	88.75	1.27	3.00	7.33	6
ring	7400	20	0.981	85.07	0.79	8.56	185	84.31	1.77	6.00	23.71	14
twonorm	7400	20	0.998	84.61	1.45	8.86	344	83.31	1.98	5.00	25.70	10

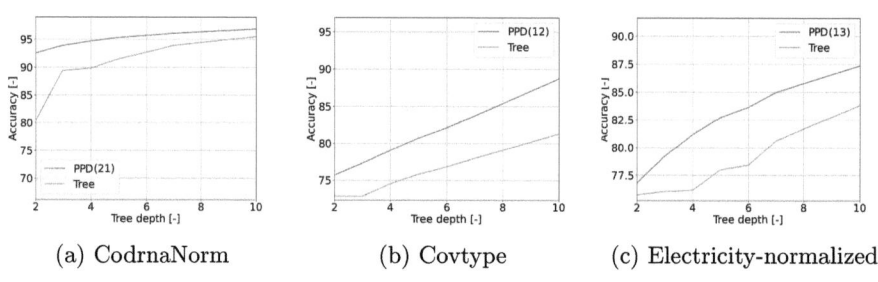

(a) CodrnaNorm (b) Covtype (c) Electricity-normalized

Fig. 3. Relation between the prediction performance and depth of the tree. The number of regions for PPD is given in brackets

varies between 185 and 882, while for PPD with trees only 7 to 26 rules are required, with additional prototype-based rule for region identification.

To obtain accuracy for various tree depths PPD was configured as follows: minimum samples per region: 400, unbalanced rate: 0.2, and prototypes obtained with k-means clustering with 10 clusters per class (in total 20 prototypes were selected). The obtained results are shown in Fig. 3. It can be seen that PPD allows for a significantly simpler representation of the rule base. As trees are constructed inside the PPD regions, they can be much simpler. The maximum number of leaves in a decision tree is $2^{depth+1}$. As can be observed, PPD with local trees of depths 2 to 3 and 10 regions allows for an equivalent accuracy to standard trees of depth 6 to 7.

4.2 Influence of PPD Parameters on Its Performance

First, the influence of the number of clusters on the number of regions was analyzed. The obtained results can be divided into two sets of similar behavior: large datasets and small datasets. For large datasets (CodrnaNorm, Electricity-normalized, Covtype) the number of clusters scales linearly with the number of regions, almost independently of the value of $min\_support$. An example for the CordaNorm dataset is shown in Fig. 4a. For smaller datasets low $min\_support$

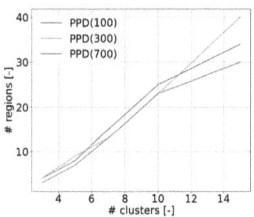 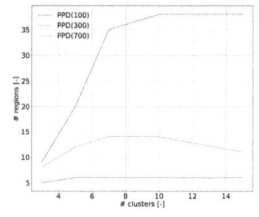 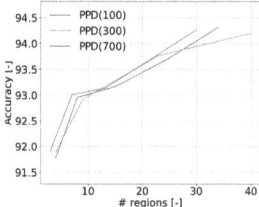

(a) Clusters-Regions large   (b) Clusters-Regions small   (c) Regions-Accuracy

Fig. 4. The influence of model parameters on its performance. (a) Relation between the number of clusters and the number of obtained regions for CordaNorm dataset - an example of a large dataset. (b) the influence of the number of clusters on the number of regions for the Ring dataset - an example of a small dataset. (c) relation between the prediction performance and the number of regions for the CordaNorm (large) dataset. In the legend, in brackets is given the value of $min\_support$.

leads to a higher number of regions, while large $min\_support$ merges the smallest regions into the larger ones, and the final number of obtained regions gets fixed as shown in Fig. 4b for Ring dataset.

Second, the influence of the number of regions on model prediction performance was analyzed. Similarly, here, different behavior was observed with respect to the size of the dataset.

For larger datasets, we observed an increase in prediction performance with the growth of the number of regions, as this allows better prototypes to be found and small regions to be removed, as shown in Fig. 4c. Consequently, a higher accuracy can be achieved. For smaller datasets, the relation between the number of regions and the prediction performance is less regular and leads to classification performance fluctuations because the number of regions is almost constant.

5 Conclusions

The proposed prototype-pairs decomposition (PPD) algorithm combines case-based reasoning with local rule extraction via in-region decision trees. It produces simpler rule sets than standard decision trees and balances global and local interpretability. This hybrid approach offers an advantage over purely global methods, especially on large datasets with complex decision boundaries.

The application of PPD to multiple class problems, to high dimensional spaces and to models other than decision trees will be the subject of our future research.

Acknowledgments. The research was supported by the Excellence Initiative - Research University program implemented at the Silesian University of Technology, year 2024, project number 11/040/SDW/10-21-01 and the research project BK-227/RM4/2025 funded by the Silesian University of Technology.

Disclosure of Interests. The authors have no competing interests to declare that are relevant to the content of this article.

References

1. Blachnik, M., Duch, W.: LVQ algorithm with instance weighting for generation of prototype-based rules. Neural Netw. **24**(8), 824–830 (2011)
2. Blachnik, M., Kordos, M., Duch, W.: Extraction of prototype-based threshold rules using neural training procedure. In: Villa, A., Duch, W., Érdi, P., Masulli, F., Palm, G. (eds.) ICANN 2012. LNCS, vol. 7553, pp. 255–262. Springer, Heidelberg (2012). https://doi.org/10.1007/978-3-642-33266-1_32
3. Garcia, S., Derrac, J., Cano, J., Herrera, F.: Prototype selection for nearest neighbor classification: taxonomy and empirical study. IEEE Trans. Pattern Anal. Mach. Intell. **34**(3), 417–435 (2012)
4. Kolodner, J.: Case-Based Reasoning. Morgan Kaufmann (2014)
5. Lemaître, G., Nogueira, F., Aridas, C.K.: Imbalanced-learn: a python toolbox to tackle the curse of imbalanced datasets in machine learning. J. Mach. Learn. Res. **18**, 1–5 (2017). http://jmlr.org/papers/v18/16-365.html
6. Lundberg, S.M., Lee, S.I.: A unified approach to interpreting model predictions. In: Advances in Neural Information Processing Systems, vol. 30 (2017)
7. Chakraborty, M., Biswas, S.K., Purkayastha, B.: Rule extraction from neural network trained using deep belief network and back propagation. Knowl. Inf. Syst. **62**(9), 3753–3781 (2020). https://doi.org/10.1007/s10115-020-01473-0
8. Ma, C., Zhao, B., Chen, C., Rudin, C.: This looks like those: illuminating prototypical concepts using multiple visualizations. In: Advances in Neural Information Processing Systems, vol. 36 (2024)
9. Pedregosa, F., et al.: Scikit-learn: machine learning in Python. J. Mach. Learn. Res. **12**, 2825–2830 (2011)
10. Ribeiro, M.T., Singh, S.: Why should I trust you?: explaining the predictions of any classifier. In: 22nd ACM SIGKDD (2016)
11. Mekkaoui, S.E., Benabbou, L., Berrado, A.: Rule-extraction methods from feedforward neural networks: a systematic literature review (2023). https://arxiv.org/html/2312.12878v1
12. Saranya, A., Subhashini, R.: A systematic review of explainable artificial intelligence models and applications: recent developments and future trends. Decis. Anal. J. **7**, 100230 (2023)
13. Shankaranarayana, S.M., Runje, D.: ALIME: autoencoder based approach for local interpretability. In: Yin, H., Camacho, D., Tino, P., Tallón-Ballesteros, A.J., Menezes, R., Allmendinger, R. (eds.) IDEAL 2019. LNCS, vol. 11871, pp. 454–463. Springer, Cham (2019). https://doi.org/10.1007/978-3-030-33607-3_49
14. Shrikumar, A., Greenside, P., Kundaje, A.: Learning important features through propagating activation differences. In: International Conference on Machine Learning, pp. 3145–3153. PMLR (2017)

Fast Prediction of Job Execution Times in the ALICE Grid Through GPU-Based Inference with Quantization and Sparsity Techniques

Tomasz Lelek[1], Szymon Mazurek[2,3]( ), Maciej Wielgosz[2,3],
and Bartosz Balis[1]

[1] Faculty of Computer Science, AGH University of Krakow, al. Mickiewicza 30, 30-059 Krakow, Poland
{tlelek,balis}@agh.edu.pl
[2] Faculty of Computer Science, Electronics and Telecommunications, AGH University of Krakow, al. Mickiewicza 30, 30-059 Krakow, Poland
{smazurek,wielgosz}@agh.edu.pl
[3] ACC Cyfronet AGH, Nawojki 11, 30-072 Krakow, Poland

Abstract. We propose a latency-optimized neural network model to dynamically predict job execution times for the ALICE experiment at CERN, replacing static Time-To-Live (TTL) allocations. Utilizing Nvidia A100 GPUs, we optimize inference latency via FP16 and INT8 quantization, 2:4 sparsity, quantization-aware training, and graph compilation. Results show that FP16 and sparsity reduce latency for larger batches, while INT8 is optimal for single-sample predictions. For single-sample online inference, static INT8 quantization achieves a median 0.38 ms prediction time, a 1.8x improvement over the 0.71 ms baseline. The model achieves a 1.9-hour RMSE, improving on the 14.23-hour RMSE of current TTL assignments. With sub-40ms inference latency on GPU hardware, this work demonstrates how NN optimization can help achieve performance demands of large-scale distributed computing systems.

Keywords: Deep learning · ALICE · latency optimization · high-energy physics · real-time inference · job scheduling

1 Introduction

AI is increasingly important in large-scale computing environments, where vast amounts of system data are continuously generated. Effectively analyzing this data is crucial for efficient administration, maintenance, and resource management [7]. Large-scale computing is essential in high-energy physics experiments, such as ALICE at CERN. The ALICE Grid, with 60 global computing clusters, runs 500k daily jobs running simulations and analyzing data. With up to 200k concurrent jobs, efficient scheduling is crucial. Currently the job scheduling algorithm uses Time-To-Live (TTL) values (expected execution time), but

---

T. Lelek and S. Mazurek—Equal contribution.

© The Author(s), under exclusive license to Springer Nature Switzerland AG 2025
M. H. Lees et al. (Eds.): ICCS 2025, LNCS 15906, pp. 97–105, 2025.
https://doi.org/10.1007/978-3-031-97635-3_12

these are often overestimated, leading to suboptimal resource allocation. While accurate execution time estimation could improve job scheduling, a prediction service needs to be very fast due to the high job throughput. Neural network inference and training acceleration through pruning, quantization, and other optimizations is well-studied [9,13]. However, research often emphasizes memory footprint and performance degradation resistance [3]. Inference speed is frequently evaluated with fixed network architectures, data modalities, and batch sizes, limiting insight into technique effectiveness [3,11]. Evaluations also tend to focus on complex CNNs or transformers with many parameters [9].

In this work, we develop a neural network model to predict execution times of ALICE Grid jobs, based on their input parameters and machine characteristics. We then analyze inference acceleration techniques, including quantization, semi-structured sparsity, and graph-based model compilation to reduce the inference latency, critical for optimizing job scheduling in the ALICE large-scale computing infrastructure. Our key contributions include demonstrating that INT8 quantization achieves a 1.8× speedup for single-sample inference, while semi-structured sparsity improves large-batch processing by up to 4×. We also highlight the trade-offs in quantization strategies, noting that weight quantization is highly effective, whereas activation quantization requires careful application. Finally, we present a scalable framework which can be considered as a blueprint for deploying deep learning optimizations, enhancing efficiency at large scale. These contributions provide a guidance for future large-scale scientific projects, outlining how deep learning optimizations can be integrated into mission-critical workflows where latency, scalability, and accuracy are essential.

The paper is organized as follows. Section 2 outlines research context and methods. Section 3 presents experimental results. Section 4 contains discussion and concluding remarks.

2 Methods

2.1 Architecture of the ALICE System

The architecture of the ALICE system, extended with the *Prediction Service* which is the subject of this research, is shown in Fig. 1. This service should efficiently predict the job runtime (TTL) based on its submission parameters specified by the user, and the target machine. In addition to the obvious requirement of minimizing the TTL prediction error, the model deployed within the prediction service must perform the predictions as fast as possible. In our scenario, the high job arrival rate requires that the total time from job submission to receiving the TTL prediction from the service be under 40 ms. While the latency will be influenced by each component of the new service, here we focus solely on the deep learning model inference time.

2.2 Hardware Accelerators in AI

In modern DL models, most computations fall into the matrix multiplication and addition category (MMA). The nature of these operations allows for mas-

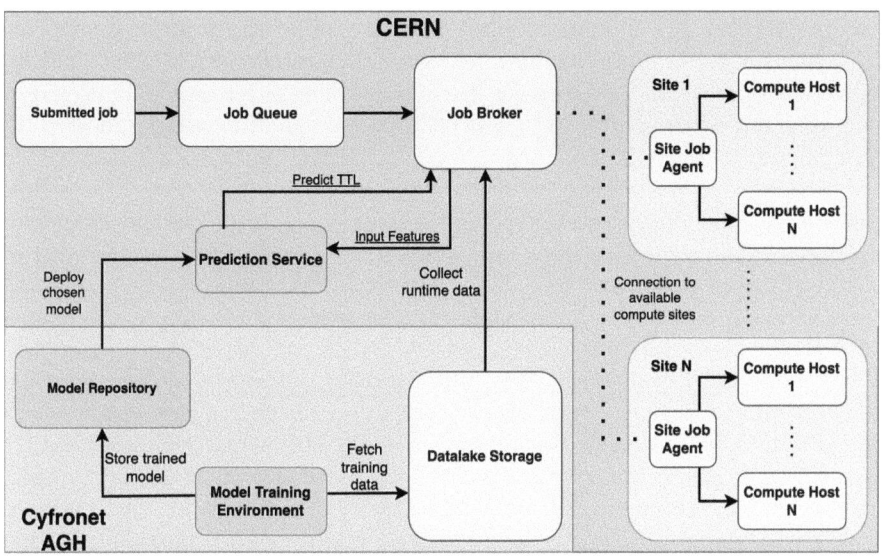

Fig. 1. The architecture of the ALICE computing framework, integrating CERN's infrastructure and the proposed AI-based prediction service. Elements of the architecture discussed in this paper are marked in green. (Color figure online)

sively parallel execution, as most atomic operations on separate elements of the matrices can be done independently. This fact led to the development of specialized hardware accelerators focusing on parallel-vector operations. Among those, graphics processing units (GPUs) are the most widespread. They are rapidly evolving, with each generation introducing new, faster architectures with support for additional features, such as operations on floating-point values with reduced precision or higher I/O bandwidth.

The Nvidia Ampere GPU series are one of the most recent developments in this domain. They introduce numerous improvements compared to the previous ones, such as the new generation of tensor cores; extended support for numeric precision formats, and cache memory expansion [2]. Our experiments utilize this type of GPU along with the following techniques: *semistructured sparsity* [11], *quantization* [4], and *model compilation* with the Pytorch native compiler, Inductor [1].

2.3 Neural Network Architecture and Experiment Design

To solve the TTL prediction problem posed, we created a simple feedforward neural network. The architecture consisted of blocks of linear and batch normalization layers followed by ReLU activation [8,12]. The last block was an exception as it contained only the linear layer.

In the sparse runs, pruning was applied to all linear layers except the last one, as the single output neuron did not meet the divisibility criterion for 2:4

sparsity patterns. The number of input neurons was set to 19840 for sparse runs and 19809 for dense ones. The difference arises from the requirement that 2:4 sparse tensors must have dimensions divisible by 32 or 64 for the FP32 and BF16 precisions, respectively. To meet these criteria, zero-padding was applied to the input feature vector in sparse runs.

To assess model latency, forward pass execution times were measured after applying the optimization techniques discussed earlier. The measurements included median and P99 values across 1000 trials, with each trial preceded by a GPU warm-up period not included in the recorded results. Latency measurements were conducted for batch sizes ranging from 1 to 16384, each being a consecutive power of 2.

For final model evaluation, the optimizations yielding the highest inference time reduction for given batch size ranges were selected. These models were trained and validated using the 10-fold cross-validation method, assessing the mean squared error of the average root (RMSE), the maximum prediction error and the median error across all folds. The network training process continued for up to 100 epochs, with early stopping implemented when no validation loss improvement occurred for 5 consecutive epochs. The AdamW optimizer [10] was used with a learning rate and weight decay set at 10^{-3}. The validation subset was extracted as 10% from the training data for each fold, ensuring robust model evaluation.

2.4 Experiments Setup and Used Hardware

The experiments were carried out within the HPC cluster, using a single Nvidia A100 GPU, up to 16 cores of AMD EPYC 7742 CPU and up to 200 GB of RAM. The experimental code was created with the Pytorch 2.5.1 framework [1], using the CUDA 12.1 toolkit for GPU integration. For sparse matrix operations, we used CUTLASS [6] 3.4.0 as a backend for sparse matrix operations.

3 Results

3.1 Analysis of Available Features and Job Execution Times in the Past ALICE Workloads

In the existing ALICE scheduling system, the job execution times are assigned static TTL values, determined manually by the operators. These values are often highly overestimated to ensure that jobs are completed successfully, even on the slowest CPUs in the grid. Although this conservative approach prevents job failures due to insufficient time, it is obviously suboptimal from the perspective of effective resource utilization.

To understand the extent of the inefficiencies, we calculated the RMSE between operator-defined TTLs and actual job execution time. Analysis has shown that the mean RMSE was equal to 14.23 h, the median to 15.12 h, and the maximal RMSE observed was 23.89 h. Most jobs are assigned TTLs that exceed 20 h, while their actual execution time in most cases is considerably shorter.

Thus, it is clearly visible that there exists vast room for improvement to achieve more accurate execution time predictions.

Based on the analysis of historical job data, we have identified 20 categorical and numerical features that describe the job when it is submitted. Categorical features were encoded into one-hot vectors, with the missing categorical values being treated as a separate category. For the numerical features, missing data was filled with the median value calculated across all jobs in the data set. Next, each numerical feature was independently z-score standardized. Lastly, we combined categorical and numerical features into a single one-dimensional vector that was used as input for the network.

3.2 Evaluating Inference Latency for Chosen Optimization Techniques

We first evaluated the forward pass time of the constructed network for a given batch size. The median and P99 latency are shown in Fig. 2. The results show some visible trends. It can be seen that using BF16 precision leads to a lower inference latency in all scenarios. Even when weights or activations were quantized with lower precision and BF16 was used only to represent activations, it still provided performance advantages. Interestingly, in runs using low-precision quantization, the latency was usually drastically higher than in the unoptimized baseline model. An exception is a single sample performance of a model with statically quantized INT8 weights. These two phenomena can be explained by the overhead introduced with the quantization techniques. In the case of static weight quantization, the activations of each layer are computed in the input precision (either FP32 or BF16 in our case). To perform such operation, weights have to be casted into the corresponding precision to compute activations, thus introducing additional overhead. However, when the forward pass is bound by memory transfer between GPU vRAM and streaming multiprocessor cache, copying of data in lower precision is faster than in higher ones. As the number of samples is small, the overhead caused by additional computation does not yet impact the processing time in a significant way. When the batch size increases, the overhead of quantization starts to manifest, showing lower inference speeds. Similar phenomena occur in dynamic quantization. As activations are in INT8 precision and weights are INT4, casting is also required. Furthermore, dynamic quantization of activations requires computing quantization parameters on the fly, adding another source of delay. Sparse runs perform worse than baseline in small batch sizes; however, they outperform all other methods when the batch size starts to grow. It is caused by additional operations required to properly process compressed sparse tensors [5]. Lastly, we observe that graph compilation leads to slight speed-ups in each case. This is in contrast to what could be expected, although we speculate that the cause lies in the lack of utilization of CUDA graphs, which we purposely disabled.

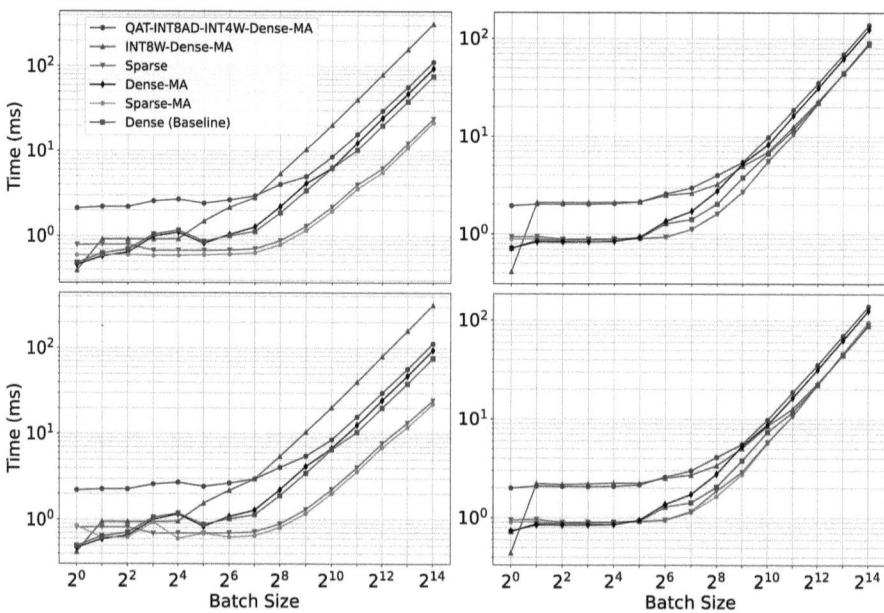

Fig. 2. Forward pass latency with different optimization techniques applied depending on an input batch size. Charts in the top show median forward pass time, while bottom ones present P99 values. The left and right columns refer to measurements in BF16 and FP32 activation precision, respectively. QAT denotes quantization aware training; AD refers to dynamic activation quantization in corresponding precision, and W refers to static weight quantization with corresponding precision. In both BF16 and FP32 activation precisions, we refer to dense run with no optimizations as a baseline.

3.3 Evaluating the Performance of the Obtained Optimized Models

We chose to evaluate the performance of only specific models based on the previous latency measurements, each showing an advantage for specific batch size ranges. The optimization techniques chosen were as follows: BF16 activation precision, BF16 activation precision combined with semistructured weight pruning, and static weight quantization to INT8 format, with activations remaining in BF16. The FP32 dense network served as a baseline. In each case (including baseline), the compilation of the model was added, as it was safe to assume that it will not affect the qualitative performance of the model. In previous experiments, it was also shown that it consistently provided a speed-up of inference. The results, along with brief descriptions of advantages in latency reduction, are shown in Table 1.

Table 1. Average metrics summarizing 10-fold cross validation evaluation of the networks with different latency reduction techniques applied, ± SEM. Each run included model compilation and dense weights, unless stated otherwise. Best results are highlighted in **bold**. We also note the advantages of each technique for inference speedup.

Metric	Method			
	FP32	BF16	BF16-Sparse	BF16-INT8W
RMSE	**1.93** ± 0.01	1.92 ± 0.01	1.91 ± 0.01	1.91 ± 0.1
Avg. max error	19.11 ± 0.67	19.14 ± 0.71	**18.92** ± 0.06	19.1 ± 0.67
Avg. median error	0.51 ± 0.01	0.48 ± 0.01	**0.47** ± 0.01	0.48 ± 0.01
Advantage in latency reduction	Baseline	Simple to use, moderate speed-up	Significant speed-up for larger batches	Speed-up when processing single sample

4 Discussion and Conclusions

In this study, we have presented a latency-optimized neural network to predict workload execution time in the ALICE computing infrastructure. We have examined different combinations of optimization techniques, including semi-structured sparsity, mixed-precision training, and inference, static and dynamic quantization into low-bit representations of weights and activations, quantization-aware training, and model compilation algorithms. We performed our analyses on a wide range of input batch sizes, spanning far beyond our use case, to provide insight for researchers looking for different use cases in the future. We find that the optimal latency reduction technique varies depending on the input batch size. Semi-structured sparsity is suitable for larger batch sizes, while for smaller ones, it can introduce additional overhead. Quantization has to be applied with caution, as it can introduce additional overhead. This is especially true for dynamic quantization techniques, where the number of additional operations is even higher. Despite the obvious benefits of reducing the memory footprint, improper use of quantization can lead to a large decrease in processing speed.

During the final evaluation, our model greatly reduced the wall-time prediction error compared to the baseline obtained manually assigned by the infrastructure operators. The average prediction RMSE on test subsets was at the level of 1.9 h, while for manual assignment, it was 14.23 h across the available historical data. Importantly, the evaluated latency reduction methods did not result in the degradation of the final model performance. Based on the experimental results, we conclude that the proposed model is suitable as the backbone of the TTL estimation system within the ALICE architecture.

This work can be expanded in several directions. From the perspective of the predictive system design, extended evaluations will be made in terms of latency measurement and reduction, seeking to find further optimization targets. Secondly, a comprehensive evaluation of newer hardware can be performed. With the GPU market rapidly evolving, it is possible that using newer ones alone would yield better speedups. Other methods of latency reduction should also be explored, i.e. structured pruning or different quantization techniques. The model architecture or feature processing could be expanded to reduce the prediction errors even more. Evaluations on more extended datasets are also needed to further prove the robustness of the model.

Acknowledgments. This work is co-financed by the Polish Ministry of Science and Higher Education under Agreement No. 2022/WK/01 and through the PMW program. We gratefully acknowledge Polish high-performance computing infrastructure PLGrid and the Academic Computer Centre Cyfronet AGH for providing computer facilities and support within computational grant no. PLG/2024/017775 and PLG/2024/017612. The research presented in this paper received partial financing from the funds assigned by Polish Ministry of Science and Higher Education to AGH University of Krakow. Research project supported/partly supported by program "Excellence initiative - research university" for the AGH University of Krakow.

Disclosure of Interests. Authors declare no conflicts of interest.

References

1. Ansel, J., et al.: Pytorch 2: faster machine learning through dynamic python bytecode transformation and graph compilation. In: Proceedings of the 29th ACM International Conference on Architectural Support for Programming Languages and Operating Systems, vol. 2 (2024)
2. Choquette, J., Gandhi, W., Giroux, O., Stam, N., Krashinsky, R.: NVIDIA A100 tensor core GPU: performance and innovation. IEEE Micro **41**(2), 29–35 (2021)
3. Danhofer, D.A.: Inducing semi-structured sparsity by masking for efficient model inference in convolutional networks. arXiv (2024)
4. Gholami, A., Kim, S., Dong, Z., Yao, Z., Mahoney, M.W., Keutzer, K.: A survey of quantization methods for efficient neural network inference. CoRR (2021)
5. Goumas, G.I., Kourtis, K., Anastopoulos, N., Karakasis, V.P., Koziris, N.: Performance evaluation of the sparse matrix-vector multiplication on modern architectures. J. Supercomput. **50**, 36–77 (2009)
6. Huang, X., Zhang, X., Yang, P., Xiao, N.: Benchmarking GPU tensor cores on general matrix multiplication kernels through cutlass. Appl. Sci. (2023)
7. Ilager, S., Muralidhar, R., Buyya, R.: Artificial intelligence (AI)-centric management of resources in modern distributed computing systems. In: 2020 IEEE Cloud Summit, pp. 1–10 (2020)
8. Ioffe, S., Szegedy, C.: Batch normalization: accelerating deep network training by reducing internal covariate shift. In: Proceedings of the 32nd International Conference on International Conference on Machine Learning, ICML 2015, vol. 37, pp. 448–456. JMLR.org (2015)

9. Liang, T., Glossner, J., Wang, L., Shi, S., Zhang, X.: Pruning and quantization for deep neural network acceleration: a survey. Neurocomputing **461**, 370–403 (2021)
10. Loshchilov, I., Hutter, F.: Decoupled weight decay regularization. In: International Conference on Learning Representations (2017)
11. Mishra, A.K., et al.: Accelerating sparse deep neural networks. arXiv (2021)
12. Nair, V., Hinton, G.E.: Rectified linear units improve restricted Boltzmann machines. In: International Conference on Machine Learning (2010)
13. Wrobel, K., Karwatowski, M., Wielgosz, M., Pietroń, M., Wiatr, K.: Compression of convolutional neural network for natural language processing. Comput. Sci. **21**(1) (2020). https://doi.org/10.7494/csci.2020.21.1.3375

Reversible Data Hiding in Encrypted Images with Pixel Prediction and ERLE Compression

Remigiusz Martyniak[1(✉)] and Mariusz Dzwonkowski[1,2]

[1] Department of Teleinformation Networks, Faculty of Electronics, Telecommunications and Informatics, Gdansk University of Technology, Gabriela Narutowicza 11/12, 80-233 Gdańsk, Poland
remigiusz.martyniak@pg.edu.pl
[2] Department of Radiology Informatics and Statistics, Faculty of Health Sciences, Medical University of Gdansk, Tuwima 15, 80-210 Gdańsk, Poland

Abstract. Reversible Data Hiding in Encrypted Images (RDHEI) is a technique that enables the embedding of additional information into encrypted carrier images, facilitating data extraction and exact restoration of the original image upon decryption. In this paper, enhancements to an RDHEI algorithm utilizing a block-wise pixel value prediction scheme have been analyzed and proposed. To better exploit spatial pixel correlations, seven additional prediction models have been introduced, along with the identification of reference pixels to improve the variable block-wise reconstruction of the carrier data. Furthermore, the integration of Huffman coding within the RDHEI scheme has been evaluated and compared with Extended Run-Length Encoding. Benchmark results against other RDHEI methods from the literature are presented at the end of the paper.

Keywords: Encrypted images · ERLE · lossless scheme · RDHEI

1 Introduction

The evolution of data security has become a critical aspect in contemporary digital communication. Digital images can function not only as a medium for visual information but also as vessels for securely embedded data, accessible exclusively through designated decryption keys. However, most data embedding methods introduce permanent losses in the utilized carrier upon extracting the embedded data, making it impossible to restore its original form. In fields such as medicine, forensics, or military systems, such losses are deemed unacceptable. Hence, a specialized data embedding method known as Reversible Data Hiding (RDH) was developed, ensuring the recovery of embedded data and the data carrier without any information loss. To achieve a high embedding capacity, various embedding techniques have been employed. These include approaches based on histogram shifting [1, 2], difference expansion [3], and pixel value ordering [4].

Over the years, a growing number of RDH algorithms in encrypted domain (RDHEI) have been developed, providing the capability for lossless recovery of carrier data and embedded data while ensuring security. Advancements in the RDHEI field can be particularly crucial in medical imaging, e.g. when embedding patient information and metadata into medical images such as X-rays, MRIs, or CT scans, ensuring that the authenticity and integrity of the images can be verified without compromising their quality and usability for diagnosis [5]. Several notable methods have contributed to the development of Reversible Data Hiding in Encrypted Images. Yi and Zhou [6] introduced a Parametric Binary Tree Labeling (PBTL) approach that divides and labels pixels for data embedding, using parameters stored within the encrypted image. The method involves a detailed process of labeling, block permutation, and pixel restoration to recover both the image and embedded data.

The Tang method [7] focuses on block-wise data hiding, utilizing a logistic map for encryption and compression to create embedding space. This method requires auxiliary information for successful data extraction and image recovery. In the Yin method [8], a median edge detector is employed to generate a label map, which plays a crucial role in determining the data embedding capacity. The embedding process involves label maps, Huffman coding, and multiple MSB substitutions, offering a balance between data capacity and security.

Mohammadi, Nakhkash, and Akhaee [9] introduced a high-capacity RDHEI technique that uses a local difference predictor. This approach increases embedding capacity by locally predicting pixel differences, allowing for the concealment of larger amounts of data while ensuring the complete recovery of the original image. Huiqi Zhang, Lin Li, and Qingyan Li [10] introduced a reversible data embedding algorithm based on block-wise multi-prediction, where the original image is divided into blocks of 8×8 px. The value of each pixel in every block is computed based on one or more adjacent pixels with known values.

This paper builds upon an existing RDHEI approach based on specific block-wise pixel value prediction models [10]. The considered approach efficiently embeds data within encrypted images, ensuring both high-capacity data embedding and the integrity of the original image. The contributions presented in this paper include:

- Expanding the block-wise pixel prediction scheme by incorporating seven additional prediction models to better exploit spatial pixel correlations;
- Identifying reference pixels to aid in the reconstruction of the carrier data and formulating the structure of the auxiliary information binary sequence to ensure independent extraction of the embedded data and reconstruction of the cover image;
- Conducting a detailed comparative analysis of the RDHEI scheme implemented with Huffman coding and fine-tuned ERLE compression for various block sizes of the segmented cover image.

2 Considered RDHEI Scheme

The RDHEI algorithm comprises three stages:

- Preprocessing (content owner's side).
- Data hiding (data hider's side).
- Data extraction and image recovery (receiver's side).

Preprocessing consists of three sequentially performed operations on the original image: block-wise pixel prediction, compression, and encryption with key K_e. Next, the embedding entity (i.e., data hider) embeds secret data using the data hiding key K_h in the previously encrypted image. On the receiver's side, the embedded data is recovered, and the original image is reconstructed using the corresponding data hiding key and the image encryption key.

2.1 Preprocessing

The content owner conducts operations on the carrier image, allowing the data hider to embed data on their end and facilitating the receiver in recovering the embedded data and reconstructing the original image. Additionally, the outcome of the preprocessing is the acquisition of auxiliary information (AI), which is a binary sequence containing the necessary information for embedding and data recovery. AI is passed to the data hider within an encrypted placeholder of the carrier image.

Block-wise Pixel Prediction. The carrier image (i.e., algorithm's input) is divided into smaller blocks of size $b \times b$. Pixel prediction is performed on these prepared blocks. Each pixel value is predicted based on one or more neighboring pixels, whose values are already known. The value of pixel is predicted based on neighboring pixels: X_{ul} (upper-left), X_u (upper), X_{ur} (upper-right), X_l (left).

In this work, 23 prediction models were used for pixel prediction. The initial 16 models were integrated based on the methodology described in [9], while the additional models introduced in this paper are presented in Table 1. This approach leaves room for further improvements, e.g. by increasing the number of prediction models even further to a total of 32, while still maintaining their binary representation at 5 bits.

Table 1. Prediction models added to the block-wise pixel prediction step.

Model	Predicted value of a pixel	Model	Predicted value of a pixel
17	Round($(X_{ul} + X_{ur}) / 2$)	21	Max(X_u, X_l, X_{ul}, X_{ur})
18	Round($(X_{ul} + X_{ur} + X_u) / 3$)	22	Min(X_u, X_l, X_{ul}, X_{ur})
19	Round($(X_l + X_{ul} + X_u) / 3$)	23	Max(X_{ul}, X_{ur}), if $X_u \leq$ min(X_{ul}, X_{ur})
20	Round($(X_l + X_u + X_{ur}) / 3$)		Min(X_{ul}, X_{ur}), if $X_u \leq$ max(X_{ul}, X_{ur})

Reference pixels are essential for the accurate reconstruction of the image on the recipient's side. Due to their significant importance, they constitute a part of the AI sequence. For each block, prediction is carried out using each of the 23 models. To identify the model with the smallest error, the SAD function is employed [10]. The model characterized by a SAD function value close to zero for a specific block is utilized for predicting pixel values within that block. After conducting predictions for each block, the difference between the original image and the image obtained through prediction is calculated, thus creating an error map. Additionally, a map of models used to predict pixel values within each block is generated. Due to the presence of negative values in the error map, it is essential to apply suitable encoding for their representation. For this purpose, sign-magnitude encoding is employed.

Compression and Encryption. To maximize embedding capacity, the error map is compressed using Huffman coding and Extended Run-Length Encoding (ERLE). ERLE, an advanced form of Run-Length Encoding (RLE) proposed by Chen and Chang [11], optimizes lossless compression for consecutive symbol sequences, common in image data. RLE compresses runs by recording the symbol and its run length, requiring additional markers for run transitions in RDHEI algorithms. ERLE enhances this with fixed-length codewords (prefix 0 plus L_{fix} bits) for short runs (<4 symbols) and variable-length codewords for longer runs (≥4 symbols), featuring a prefix (L_{pre} bits: $L_{\text{pre}} - 1$ ones, then 0), a length symbol encoding value of $L - 2^{L_{\text{pre}}}$ (L being the length of the run), and a tail bit (indicating symbol of the run). After error map compression, the AI sequence—which stores the data for image reconstruction—is formulated. It comprises reference pixel bits (first row and column of the carrier), prediction models bits, sign-magnitude encoded module bits, and the compressed error map. This AI data is placed at the beginning of a placeholder sequence of length $m \cdot n \cdot 8$, where m and n are the width and height of the image, respectively. The remaining space in the placeholder is designated for data embedding. The placeholder is encrypted using a pseudo-random bit sequence (key K_e) of equal length via bitwise XOR operation, yielding an encrypted sequence for the data hider.

2.2 Data Hiding

The empty space in the encrypted placeholder is designated for the data hider, who can embed additional, secret data into it. To help identify the starting point for the data embedding process, the length of the data hiding key K_h, used for encrypting the additional information, is associated with the length of the compressed error map. Thus, knowing the key K_h, the data hider can locate the end of AI in the encrypted placeholder and start embedding additional information without interfering with previously embedded auxiliary data. To obtain the encrypted secret sequence, an XOR operation is performed between the bits of additional information and the key K_h. The encrypted sequence, obtained in this manner, is then embedded in the encrypted placeholder immediately after the encrypted AI sequence.

2.3 Data Extraction and Image Recovery

After receiving the encrypted message with embedded data, the recipient is able to extract the necessary information to reconstruct the original carrier image and the embedded data according to the image encryption key K_e and the data hiding key K_h.

The first step is to extract the embedded AI sequence from the encrypted message. The length of this sequence is calculated based on the length of the received information and the length of the key K_h. First, the size of the original image (assuming $m \times m$ for simplicity) is calculated according to the formula (1): $m = \sqrt{(n_p/8)}$, where n_p is the length of the encrypted placeholder. The number of bits corresponding to the stored reference pixels is calculated according to formula (2): $R = 8 \cdot (2m - 1)$. Next, the number of bits of the prediction models map is calculated (3): $P = (m/b)^2 \cdot 5$, where b is the block size chosen during the preprocessing.

The number of sign bits generated during sign-magnitude encoding is equal to m^2. In this way, the bit number from which the compressed error map begins in the AI sequence is calculated as the sum of bits from reference pixels, bits of the prediction models map, and bits generated during sign-magnitude encoding, increased by 1.

The length of the data hiding key K_h is equal to the length of the compressed error map. Starting from the initial bit of the compressed error map, a sequence of l_{kh} bits is extracted, where l_{kh} is the length of the key K_h. In this way, the encrypted AI sequence is obtained, which is then decrypted using the encryption key K_e. The remaining part in the encrypted message contains information embedded by the data hider, which is extracted and decrypted using the key K_h.

In order to recover the carrier, the obtained error map is decompressed and reshaped into a matrix using sign bits. This matrix, representing the reconstructed image, is initiated with reference pixel values placed initially in the first row and then in the first column. The order of bits representing reference pixel values in the AI sequence corresponds to the order of placing them in the reconstructed image. The process of image reconstruction occurs pixel by pixel, starting from the pixel in the second row and second column. Using the prediction models map extracted from the AI sequence, the pixel value is predicted, and then added to its corresponding element from the error map matrix. This operation is performed sequentially (pixel by pixel).

3 Results

This section presents tests of the implemented RDHEI algorithm, conducted in MAT-LAB R2023b on Windows 11 with a 12th Gen Intel® Core™ i7–12700 processor. Five standard 512 × 512 greyscale images (Baboon, Lake, Plane, Peppers, and Boat) were used [12], along with images from the BOSSbase 1.01 [13] and BOWS2 [14] datasets.

3.1 Security and Image Quality Analysis

To assess the effectiveness of the encryption, commonly used metrics were employed: analyses of histograms of carrier and encrypted images, the Number of Pixel Change Rate (NPCR) indicator, and the Unified Average Changing Intensity (UACI) [10]. Histograms of encrypted images differ from those of original images and do not exhibit any significant features characteristic of carriers (Fig. 1). The NPCR coefficient in all cases reached nearly 100%, which implies that nearly all pixels undergo changes compared to the original image. The UACI result averages 15% for both compression methods, which indicates a satisfactory robustness of the encrypted data against differential attacks.

After decryption and data extraction, reconstructed images were evaluated using Peak Signal-to-Noise Ratio (PSNR) and Structural Similarity Index Measure (SSIM). All images achieved infinite PSNR and SSIM of 1, confirming lossless reconstruction.

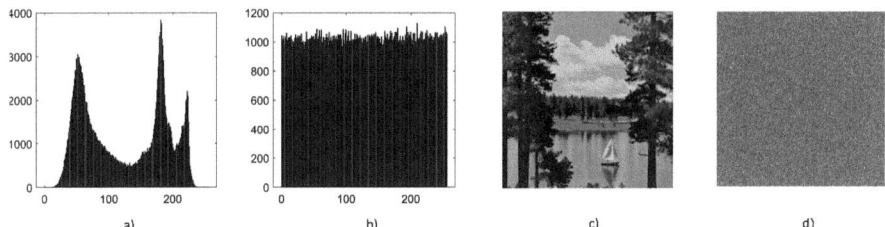

Fig. 1. Histograms of image pixel values distribution: (a) before and (b) after encryption, as well as (c) original and (d) encrypted Lake image.

3.2 Performance Analysis

The discussed algorithm was examined for its performance. In particular, the impact of the applied block size b and compression method on embedding efficiency was investigated. Moreover, the best average L_{fix} value for ERLE compression was identified.

ERLE Performance Test. The impact of the L_{fix} value (ranging from 4 to 20) on ERLE compression efficiency was tested for two block sizes (Fig. 2). For BOSSbase and BOWS2, the optimal value averaged around 8, as it was selected for approximately 2000 images depending on dataset and block size. For the standard images, with varying block sizes, the optimal values of L_{fix} were essentially the same for every tested block size, but larger than those for the majority of images from the datasets.

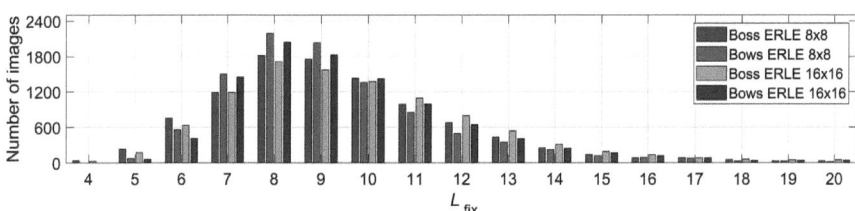

Fig. 2. Optimal L_{fix} values for BOSSbase and BOWS2 images and two fixed block sizes.

Effectiveness of Data Embedding. Embedding performance varied with compression method and block size. Huffman coding provided better average embedding rates (by 0.36 bpp for 8 × 8, and by 0.34 bpp for 16 × 16) across both datasets. However, ERLE achieved higher maximum embedding rates, outperforming Huffman in 574 images (difference > 0.1 bpp), and reaching improvements of over 0.54 bpp in 93 cases. Compared to existing methods (Table 2), ERLE compression provided better embedding results. Although slightly below [8] in average embedding rate, ERLE achieved higher maximum rates (by 13.75% for BOSSbase and 9.24% for BOWS2). Prediction models 1, 4, 7, 16, 21, and 22 were the most frequently used, demonstrating consistent performance across the datasets. Notably, two of the prediction models introduced in this study were among the top performers.

Table 2. Comparison of the embedding rate (bpp) for the discussed RDHEI algorithm with different block sizes and compression methods against selected state-of-the-art methods.

Images	This work $b: 8 \times 8$ ERLE	Huff	This work $b: 16 \times 16$ ERLE	Huff	Method [5]	Method [6]	Method [7]	Method [8]	Method [9]	Method [11]
Baboon	1.079	1.748	1.193	1.828	-	-	0.641	1.204	1.04	-
Lake	1.821	2.433	1.946	2.543	-	-	1.468	-	-	1.944
Airplane	2.781	3.257	2.861	3.321	-	2.219	2.281	3.067	2.75	2.340
Peppers	2.347	3.002	2.397	3.049	-	-	1.798	-	-	1.879
Boat	2.381	2.793	2.455	2.839	-	-	1.519	-	-	-
BOSS (max)	6.705	5.804	6.765	5.864	-	-	-	5.921	-	-
BOSS (aver.)	3.113	3.444	3.216	3.523	2.732	2.026	-	3.389	-	2.435
BOWS2 (max)	6.138	5.519	6.197	5.575	-	-	4.881	5.646	6.11	-
BOWS2 (aver.)	2.961	3.324	3.082	3.419	2.547	-	2.425	3.282	2.9	-

4 Conclusions

In this study, enhancements to an RDHEI algorithm based on block-wise pixel prediction were presented, focusing on increasing embedding capacity through fine-tuned ERLE compression. Results indicate that, on average, ERLE slightly underperforms compared to Huffman coding but excels for images containing repetitive pixel values. Huffman coding remains preferable for images with diverse pixel distributions, such as those in BOSSbase and BOWS2.

Therefore, the choice of compression should reflect image characteristics, with ERLE being likely advantageous for medical (DICOM) images exhibiting large uniform regions. Future research will explore ERLE's application in medical imaging to further optimize the embedding process.

Acknowledgments. This study was funded by the Faculty of Electronics, Telecommunications, and Informatics of Gdansk University of Technology, and by the research subsidy from the Polish Ministry of Science and Higher Education. The authors would like to express their gratitude to Mateusz Myszk for his conducted research, which laid the foundation for the discussed approach.

Disclosure of Interests. The authors declare no conflict of interest.

References

1. Ni, Z., Shi, Y.Q., Ansari, N., Su, W.: Reversible data hiding. IEEE Trans. Circuits Syst. Video Technol. **16**(3), 354–362 (2006)
2. Li, X., Li, B., Yang, B., Zeng, T.: General framework to histogram shifting-based reversible data hiding. IEEE Trans. Image Process. **22**(6), 2181–2191 (2013)

3. Tian, J.: Reversible watermarking by difference expansion. In: Proceedings Workshop Multimedia Security: Authentication, Secrecy, and Steganalysis, pp. 19–22. (2002)
4. Li, X., Li, J., Li, B., Yang, B.: High-fidelity reversible data hiding scheme based on pixel-value-ordering and prediction-error expansion. Signal Process. **93**(1), 198–205 (2013)
5. Dzwonkowski, M., Czaplewski, B.: Reversible data hiding in encrypted DICOM images using sorted binary sequences of pixels. Signal Process. **199**, 1–14 (2022)
6. Yi, S., Zhou, Y.: Separable and reversible data hiding in encrypted images using parametric binary tree labeling. IEEE Trans. Multimedia **21**(1), 51–64 (2019)
7. Tang, Z., Xu, S., Yao, H., Qin, C., Zhang, X.: Reversible data hiding with differential compression in encrypted image. Multimed. Tools Appl. **78**, 9691–9715 (2019)
8. Yin, Z., Xiang, Y., Zhang, X.: Reversible data hiding in encrypted images based on multi MSB prediction and Huffman coding. IEEE Trans. Multimedia **22**(4), 874–884 (2020)
9. Mohammadi, A., Nakhkash, M., Akhaee, M.A.: A high-capacity reversible data hiding in encrypted images employing local difference predictor. IEEE Trans. Circuits Syst. Video Technol. **30**(8), 2366–2376 (2020)
10. Zhang, H., Li, L., Li, Q.: Reversible data hiding in encrypted images based on block-wise multi-predictor. IEEE Access **9**, 61943–61954 (2021)
11. Chen, K., Chang, C.C.: High-capacity reversible data hiding in encrypted images based on extended run-length coding and block-based MSB plane rearrangement. J. Vis. Commun. Image Represent. **58**, 334–344 (2019)
12. Standard images. http://www.dip.ee.uct.ac.za/imageproc/stdimages/greyscale/, https://ccia.ugr.es/cvg/CG/base.htm. Accessed 08 Apr 2025
13. BOSSbase 1.01 dataset, Binghamton's University DDE download section, https://dde.binghamton.edu/download/, last accessed 2025/04/08
14. BOWS2 dataset. https://web.archive.org/web/20221129163351/http://bows2.ec-lille.fr/. Accessed 08 Apr 2025

Information Flow Between Neighboring Housing Markets: A Case from the Seoul Metropolitan Area

Leehyun Jung[1,2], Minhyuk Jeong[1,2], Yena Song[3], and Kwangwon Ahn[1,2(✉)]

[1] Department of Industrial Engineering, Yonsei University, 50 Yonsei-ro, Seodaemun-gu, Seoul 03722, South Korea
{aidan6,wjd9496,k.ahn}@yonsei.ac.kr

[2] Center for Finance and Technology, Yonsei University, 50 Yonsei-ro, Seodaemun-gu, Seoul 03722, South Korea

[3] Department of Geography, Chonnam National University, 77 Yongbong-ro, Buk-gu, Gwangju 61186, South Korea
y.song@chonnam.ac.kr

Abstract. This study explores the directional flow of price information within the housing market network, focusing on the Seoul metropolitan area from 2013 to 2022. Utilizing a network constructed through the multivariate Granger causality test, centrality metrics identify key industrial cities (e.g., Suwon and Hwaseong) as central nodes. In contrast, Seoul, the capital city, plays a marginal role. The findings reveal a shift in influence within the housing market network, with traditionally dominant cities like Seoul ceding prominence to key industrial hubs. Our findings suggest that spatial and industrial dynamics may now outweigh administrative hierarchy in shaping housing market centrality. Future studies should explore how external shocks have driven this transformation over time.

Keywords: Housing market network · Multivariate Granger causality test · Centrality metrics

1 Introduction

The housing market, a key driver of economic growth, requires close monitoring to promote sustainable development and prevent market distortions. A rapid housing price increase in one city can influence neighboring markets, potentially triggering housing bubbles [1]. This interdependence underscores the importance of understanding how price signals propagate across regions, particularly in densely interconnected metropolitan areas. Therefore, analyzing the directional influence of price movements, hereafter referred to as "information flow," among adjacent housing markets is essential.

Prior studies have examined the flow of information between housing markets in regions such as the Netherlands, China, and South Korea [2–4]. In these cases, their capital cities—Amsterdam, Beijing, and Seoul, respectively—were identified as the most influential nodes in their national housing networks, serving as primary sources of price signals and market information. These findings underscore the pivotal role capital cities play in disseminating information across national housing markets.

Despite these valuable insights, previous research has focused primarily on major cities, often overlooking the roles of satellite cities and suburban areas in housing markets. This limitation may arise from restricted data availability for peripheral areas, challenges in modeling complex multilateral dependencies, or a historical emphasis on dominant markets like Seoul. Such gaps are increasingly relevant considering the ongoing trend of suburbanization, characterized by the expansion of business complexes in metropolitan regions of developed countries [5–7]. Therefore, analyzing the structure and hierarchy of urban housing markets requires incorporating these peripheral regions to fully capture the spatial dynamics of information flow.

To address this issue, our study examines a network of 30 cities in the Seoul metropolitan area, encompassing both central urban and suburban regions. By including these underexplored regions, our research offers a comprehensive view of inter-regional dynamics between neighboring housing markets. This broader approach allows us to investigate how spatial proximity and economic interdependencies between central and peripheral cities shape the structure of housing market networks. In moving beyond the capital-centric perspective of previous studies, this research provides a more detailed understanding of price signal propagation within metropolitan areas.

2 Data and Methodology

2.1 Data

This study focuses on 30 cities within the Seoul metropolitan area, including Seoul, Incheon, and 28 cities in Gyeonggi Province. We excluded three rural counties in Gyeonggi from our analysis because of their illiquid housing markets and limited apartment presence [8].

Figure 1 illustrates the study sites where apartment price indices are analyzed, as apartments represent the predominant housing type in South Korea [9]. The dataset, sourced from the Korea Real Estate Board, covers the period from January 7, 2013, to May 9, 2022, and includes 488 weekly observations per city.

2.2 Multivariate Granger Causality Test

The Granger causality (GC) test [10] evaluates whether one time series can predict another by determining if the past values of one time series improve the other's predictions. However, the pairwise nature of this approach may exclude relevant variables, potentially leading to spurious results.

The multivariate GC (MVGC) test [11] addresses this limitation by incorporating multiple variables. This allows for a more robust analysis in multivariate systems. Unlike the traditional GC method, MVGC examines how one set of variables influences another while accounting for complex interactions among all variables. This capability makes it especially effective for studying interconnected systems with multiple pathways. In the MVGC framework, the vector autoregressive model expands to include multiple time

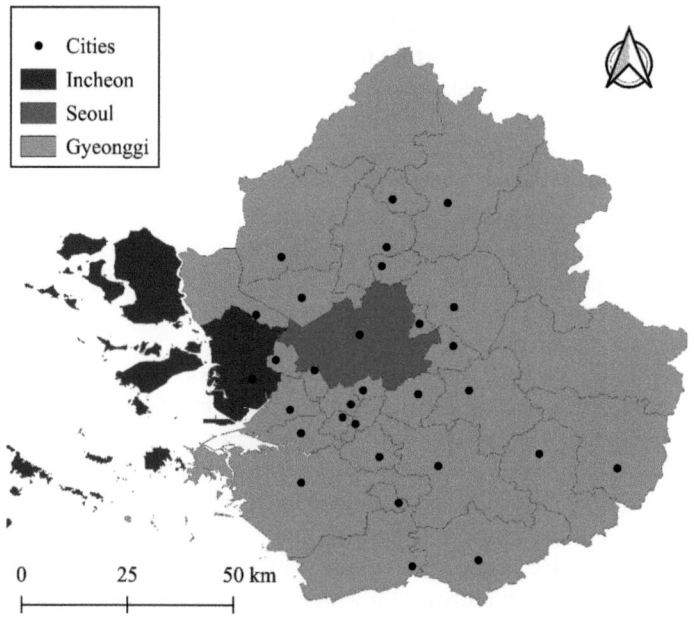

Fig. 1. Seoul metropolitan area. Bounded regions marked with black dots indicate the 30 cities included in this study.

series, represented as $X_t = (x_{1,t}, x_{2,t}, x_{3,t}, \ldots, x_{n,t})'$. Here, n represents the number of different time series data considered. The multivariate form of the model can be expressed as follows:

$$X_t = \sum_{i=1}^{p} B_i X_{t-i} + \epsilon_t. \qquad (1)$$

Here, B_i is the $n \times n$ coefficient matrix for the lag i vector, and ϵ_t is the $n \times 1$ error vector. The MVGC test assesses whether the past values of one series (e.g., $x_{2,t}$) enhance another (e.g., $x_{1,t}$) series' predictions while considering the influence of other series in the system (e.g., $x_{3,t}, x_{4,t}, \ldots, x_{n,t}$). The lag order ($p$) is determined using the Akaike information criterion [12].

Conditioned on $x_{3,t}, x_{4,t}, x_{5,t}, \ldots, x_{n,t}$, the null hypothesis of the MVGC test posits that the past values of $x_{2,t}$ provide no additional predictive power for $x_{1,t}$. We defined the F-statistic as follows:

$$F_{x_{2,t} \to x_{1,t} | \{x_{3,t}, x_{4,t}, \ldots, x_{n,t}\}} = \frac{m - p(n+1)}{p n_y} \cdot \frac{\Sigma'_{xx}}{\Sigma_{xx}}. \qquad (2)$$

Here, m is the number of observations, and n_y is the dimension of $x_{2,t}$. Σ_{xx} is the residual covariance matrix from the unrestricted model, and Σ'_{xx} is from the restricted model. A larger F-statistic exhibits a greater reduction in prediction error when $x_{2,t}$ is included, and this reflects a stronger causal relationship [11]. Therefore, we used the F-statistics from the MVGC tests to determine the node weights within the information network of the Seoul metropolitan housing market.

2.3 Centrality Metrics

Centrality metrics quantify a node's influence within a network [13–16]. This study employs several key metrics: degree centrality [17, 19], hub and authority scores [18], and eigenvector centrality [19, 20].

Degree Centrality. In a graph $G(V, E)$, with vertex set V and edge set E, degree centrality quantifies a node's connectivity [15, 16]. In weighted networks, degree centrality is calculated as the sum of edge weights instead of the number of connections. For directed graphs, degree centrality is divided into out-degree (activity) and in-degree (popularity). The out-degree centrality of node i is as follows:

$$D_{out}(i) = \sum_{j} w_{ij} m_{out}(i, j). \quad (3)$$

Here, w_{ij} is the edge weight from node i to j, and $m_{out}(i, j)$ equals 1 if such an edge exists and 0 otherwise. Similarly, in-degree centrality is as follows:

$$D_{in}(i) = \sum_{j} w_{ji} m_{in}(i, j). \quad (4)$$

Here, $m_{in}(i, j)$ is 1 if an edge exists from node j to i and 0 otherwise.

Hub and Authority Scores. Derived from eigenvector centrality, hub and authority scores capture the mutually reinforcing relationship between nodes. Authority scores assess a node's importance, while hub scores measure its connections to authority nodes. Hub nodes link to numerous authority nodes, and authority nodes are heavily linked by many hubs. Both of these scores are iteratively computed, with numerical weights updated for each node [18].

Eigenvector Centrality. Eigenvector centrality measures a node's influence by considering both its connections and the importance of its neighbors [20, 21]. This is defined as follows:

$$E(i) = \frac{1}{\lambda} \sum_{j=1}^{n} A_{ij} E(j). \quad (5)$$

Here, λ is the eigenvalue scaling the centrality scores, and A_{ij} is the weighted adjacency matrix for a network with n nodes. Nodes with high eigenvector centrality are linked to other highly central nodes.

3 Results and Discussion

Table 1 presents the centrality metrics for cities in the Seoul metropolitan area, emphasizing the three cities with the highest values for each metric. The findings underscore the prominent roles of populous cities in the Seoul metropolitan housing market network. Among Gyeonggi Province's cities, Suwon and Hwaseong emerge as key players. Suwon ranks highly in in-degree, authority, and eigenvector centrality, reflecting its position as a major recipient of housing price signals. By contrast, Hwaseong ranks highest in out-degree and hub scores, indicating its role as a key source of influence.

Hence, Suwon functions as a central information receiver and connector. Conversely, Hwaseong is the network's strongest information transmitter. This distinction may be attributed to the presence of Samsung Electronics' headquarters and factories [22, 23]. Similarly, Yongin, where SK Hynix leads the development of a semiconductor cluster, serves as another influential node transmitting information across the network. These industrial hubs, anchored by national industries, experience high residential demand [24], reinforcing their critical roles in the housing market network.

Table 1. Centrality metrics of cities in the metropolitan Seoul area. The bracketed values represent centrality metrics (rounded to two decimal places) that are calculated based on edge weights obtained from the F-statistics of the MVGC test.

Rank	1	2	3
In-degree	Suwon [0.40]	Yangju [0.21]	Paju [0.20]
Out-degree	Hwaseong [0.26]	Yongin [0.24]	Dongducheon [0.20]
Authority	Suwon [1]	Pyeongtaek [0.27]	Gunpo [0.23]
Hub	Hwaseong [1]	Yongin [0.62]	Incheon [0.33]
Eigenvector	Suwon [1]	Gimpo [0.76]	Yangju [0.75]

Among the two most populous cities, Incheon and Seoul, contrasting roles are evident. Incheon, bolstered by aviation and biotechnology industries [25], exhibits a high hub score, reflecting industrial activities and national economic development initiatives. Conversely, Seoul—despite being the region's largest city and its most economically dominant (contributing 22.6% of South Korea's GDP in 2013 [26])—plays a surprisingly marginal role within the network. Specifically, Seoul ranks 20th for in-degree centrality, 19th for out-degree centrality, 22nd for authority, 26th for hub score, and 14th for eigenvector centrality. Similarly, first-phase new towns in Gyeonggi Province, including Goyang, Seongnam, and Bucheon [27, 28], demonstrate limited influence. These findings suggest a shift in the roles of traditionally dominant cities in the housing market network during the study period. This shift is likely influenced by structural changes in the market caused by external shocks, such as administrative regulations [29] and economic crises [30].

4 Conclusion

This study employed the MVGC test and centrality metrics to analyze the information flow among housing markets in 30 cities within the Seoul metropolitan area. Our results indicate a diminished role for Seoul and other historically dominant cities, underscoring a shift in centrality toward key industrial cities such as Suwon and Hwaseong. These findings are counterintuitive, given existing literature that emphasizes the leading role of capital cities in housing market networks. The results suggest that external shocks, such as policy changes or economic crises, may have altered the information dynamics within the Seoul metropolitan housing market network, warranting further exploration in future research.

Future studies should explore why the findings of this research diverge from previous literature, particularly regarding temporal dynamics. Period-specific analyses could help identify whether external events (e.g., political shifts or the COVID-19 pandemic [31]) have reshaped the structure of information flow. These insights could further inform policy discussions on how government interventions should be tailored to address the evolving dynamics of regional housing markets.

Acknowledgments. This work was supported by the National Research Foundation of Korea (NRF) grant funded by the Korea government (MSIT) (No. 2022R1A2C1004258, Kwangwon Ahn).

Disclosure of Interests. The authors have no competing interests to declare that are relevant to the content of this article.

References

1. Meen, G.: Regional house prices and the ripple effect: a new interpretation. Housing Stud. **14**, 733–753 (1999)
2. Teye, A.L., Knoppel, M., de Haan, J., Elsinga, M.G.: Amsterdam house price ripple effects in the Netherlands. J. Eur. Real Estate Res. **10**, 331–345 (2017)
3. Lee, H.S., Lee, W.S.: Cross-regional connectedness in the Korean housing market. J. Housing Econ. **46**, 101654 (2019)
4. Tsai, I.C.: The connectedness between Hong Kong and China real estate markets: spillover effect and information transmission. Empir. Econ. **63**, 287–311 (2022)
5. Mills, E.S., Price, R.: Metropolitan suburbanization and central city problems. J. Urban Econ. **15**, 1–17 (1984)
6. Mieszkowski, P., Mills, E.S.: The causes of metropolitan suburbanization. J. Econ. Perspect. **7**, 135–147 (1993)
7. Muller, P.O.: The suburban transformation of the globalizing American city. Ann. Am. Acad. Polit. Soc. Sci. **551**, 44–58 (1997)
8. South Korean Ministry of Land, Infrastructure and Transport: Administrative districts' boundaries by cities, counties, and districts.https://www.vworld.kr/dtmk/dtmk_ntads_s002.do?dsId=30604. Accessed 20 Aug 2024
9. Ahn, K., Jang, H., Song, Y.: Economic impacts of being close to subway networks: a case study of Korean metropolitan areas. Res. Transp. Econ. **83**, 100900 (2020)
10. Granger, C.W.: Investigating causal relations by econometric models and cross-spectral methods. Econometrica **37**, 424–438 (1969)
11. Barnett, L., Seth, A.K.: The MVGC multivariate Granger causality toolbox: a new approach to Granger-causal inference. J. Neurosci. Methods **223**, 50–68 (2014)
12. Akaike, H.: Fitting autoregressive models for prediction. Ann. Inst. Stat. Math. **21**, 243–247 (1969)
13. Freeman, L.C.: Centrality in social networks conceptual clarification. Soc. Netw. **1**, 215–239 (1978)
14. Opsahl, T., Agneessens, F., Skvoretz, J.: Node centrality in weighted networks: generalizing degree and shortest paths. Soc. Netw. **32**, 245–251 (2010)
15. Kim, H., Ha, C.Y., Ahn, K.: Preference heterogeneity in Bitcoin and its forks' network. Chaos Solitons Fractals **164**, 112719 (2022)

16. Jeong, M., Joo, K., Kim, J., Kim, J., Kim, J., Ahn, K.: The impact of trading environments on commodity futures: evidence from biofuel feedstocks' network. In: Proceedings of International Conference on Mathematical Modeling in Physical Sciences, pp. 367–374. Springer (2023)
17. Kleinberg, J.M.: Authoritative sources in a hyperlinked environment. J. ACM **46**, 604–632 (1999)
18. Bonacich, P.: Factoring and weighting approaches to status scores and clique identification. J. Math. Sociol. **2**, 113–120 (1972)
19. Maharani, W., Adiwijaya, Gozali, A.A.: Degree centrality and eigenvector centrality in Twitter. In: Proceedings of 2014 8th International Conference on Telecommunication Systems, Services, and Applications, pp. 1–5 (2014)
20. Solá, L., Romance, M., Criado, R., Flores, J., García del Amo, A., Boccaletti, S.: Eigenvector centrality of nodes in multiplex networks. Chaos **23**, 033131 (2013)
21. Newman, M.E.: The mathematics of networks. New Palgrave Encycl. Econ. **2**, 1–12 (2008)
22. Moon, H.C., Parc, J., Yim, S.H., Yin, W.: Enhancing performability through domestic and international clustering: a case study of Samsung Electronics Corporation (SEC). Int. J. Performability Eng. **9**, 75–84 (2013)
23. Jung, U., Chung, B.D.: Lessons from the history of Samsung's SCM innovations: focus on the TQM perspective. Total Qual. Manag. Bus. Excell. **27**, 751–760 (2016)
24. Dalavong, P., Choi, C.G.: Commuting paradox of the technopole newtown: a case study in the Seoul metropolitan area. PLoS ONE **19**, e0306304 (2024)
25. Jo, K., Yoon, S.: State industrial policy and local industrial coalition: evidence from the pharmaceutical industry in Incheon, South Korea (Working Paper No. 19). https://www.geschkult.fu-berlin.de/e/oas/korea-studien/Files/2023/Korea-Focus-Working-Paper-No-19.pdf. Accessed 24 Aug 2024
26. Statistics Korea: Gross regional domestic product. https://www.index.go.kr/unity/potal/main/EachDtlPageDetail.do?idx_cd=1008. Accessed 24 Aug 2024
27. Kim, S.J., Ahn, K.H.: Changes and polarization of social classes in new and old city after new town developments in Seong-Nam city. J. Urban Design Inst. Korea Urban Design **14**, 53–66 (2013)
28. Kim, J.H., Kwon, O.S., Ra, J.H.: Urban type classification and characteristic analysis through time-series environmental changes for land use management for 31 satellite cities around Seoul. South Korea. Land **10**, 799 (2021)
29. Quigley, J.M., Raphael, S.: Regulation and the high cost of housing in California. Am. Econ. Rev. **95**, 323–328 (2005)
30. Gokmen, G., Nannicini, T., Gaetano, O.M., Papageorgiou, C.: Policies in hard times: assessing the impact of financial crises on structural reforms. Econ. J. **131**, 2529–2552 (2021)
31. An, S., Kim, J., Choi, G., Jang, H., Ahn, K.: The effect of rare events on information-leading role: Evidence from real estate investment trusts and overall stock markets. Humanit. Soc. Sci. Commun. **11**, 1628 (2024)

A Bi-Stage Framework for Automatic Development of Pixel-Based Planar Antenna Structures

Khadijeh Askaripour[1], Adrian Bekasiewicz[1(✉)], and Slawomir Koziel[1,2]

[1] Faculty of Electronics, Telecommunications and Informatics, Gdansk University of Technology, Narutowicza 11/12, 80-233 Gdansk, Poland
adrian.bekasiewicz@pg.edu.pl

[2] Department of Engineering, Reykjavik University, Menntavegur 1, 102, Reykjavik, Iceland

Abstract. Development of modern antennas is a cognitive process that intertwines experience-driven determination of topology and tuning of its parameters to fulfill the performance specifications. Alternatively, the task can be formulated as an optimization problem so as to reduce reliance of geometry selection on engineering insight. In this work, a bi-stage framework for automatic generation of antennas is considered. The method determines free-form topology through optimization of interconnections between components (so-called pixels) that constitute the radiator. Here, the process involves global optimization of connections between pixels followed by fine-tuning of the resulting topology using a surrogate-assisted local-search algorithm to fulfill the design requirements. The approach has been demonstrated based on two case studies concerning development of broadband and dual-band monopole antennas.

Keywords: bi-stage design · multiport simulation · pixel-based antennas · surrogate-based optimization · topology development

1 Introduction

Antenna design is a challenging task that involves development of topology followed by its tuning to obtain the desired performance specifications. Given lack of empirical, or analytical solutions that could streamline the process, modern radiators are generated and tuned based on responses obtained from electromagnetic (EM) simulations of their models [1]. EM-driven evolution of topology w.r.t. application-specific requirements is considered a pivotal part of the design process [2]. However, it is also subject to engineering bias which might hinder identification of high performance solutions with unintuitive geometries [1, 3, 4]. From this perspective, automatic design of antennas is an interesting alternative to cognition-driven procedures.

Algorithmic design methods involve EM-driven development of free-form radiators represented using either a set of characteristic points (interconnected using line sections, or splines), or a distribution of primitives (e.g., rectangles) specified in the form of a binary matrix [4–7]. However, a large number of parameters required to ensure flexibility of radiator shape makes both representations impractical for automatic antenna design [5, 7]. This is because exploration of large search spaces associated with free-form geometries involves the use of global-search methods (e.g., population-based meta-heuristics). The latter ones are numerically prohibitive for multi-dimensional problems, due to excessive number of EM simulations (hundreds or even thousands) required to converge [4–7].

For structures represented using primitives, the problem of unacceptable design cost can be mitigated using internal multi-port method (IMPM) [8, 9]. The approach boils down to representation of the radiator as a lattice of dummy components (so-called pixels) interconnected through internal ports [8]. Identification of IMPM model involves extraction of an impedance matrix pertinent to pixel-based antenna from a single, multi-port EM simulation [9]. Electrical performance of the radiator is then approximated through adjustment of impedance between the internal ports. The approach represents a significant advancement compared to conventional models where distribution of primitives (e.g., metallic rectangles) is specified within an associated matrix and structure responses are derived from EM simulations (one at a time) [7]. Instead, IMPM enables adjustment of connections between pixels (and antenna performance characteristics) at a negligible cost during post-processing (simple calculations based on circuit theory). Hence, method can be used to reliably handle global optimization of complex topologies [8]. Regardless of numerical efficiency, IMPM does not support fine-tuning of responses based on adjustment of pixels dimensions.

In this work, a bi-step framework for automatic development of pixel-based antennas is considered. The method involves global optimization of IMPM-based antenna model to approximate the desired solution. The obtained design is then used as a starting point for surrogate-assisted fine-tuning. The approach has been demonstrated using two test cases concerning development of broadband and dual-band antennas based on monopole topology. For the considered numerical experiments, the computational cost of antenna design does not exceed 36 EM model simulations.

2 Design Framework

2.1 Problem Formulation

Let $R(x, y)$ be the EM simulation response of a generic pixel-based antenna model obtained over a frequency range of interest f for the vector x that represents floating-point dimensions and binary vector y that denotes interconnections between the specific pixels (for visualization, see Fig. 1). The design problem can be defined as the optimization task of the form:

$$x^*, y^* = \arg\min_{x,y}(U(R(x, y))) \tag{1}$$

where U is a scalar objective function; x^* and y^* represent the optimum design parameters and pixel connections to be found. Direct minimization of (1) is computationally impractical as it would necessitate hundreds, or thousands of $R(x, y)$ simulations to complete. Instead, the design can be realized as a bi-stage process, where appropriate configuration of y is first obtained as a result of global-search executed on the IMPM representation of the antenna $R_{sy}(x, y)$ with constant x by solving:

$$y^* = \arg\min_{y}\left(U\left(R_{sy}(x, y)\right)\right) \quad (2)$$

Next, the specific dimensions of the radiator x—with constant y^* obtained from (2)—can be subject to fine-tuning using a local-search algorithm.

2.2 IMPM Representation of Pixel-Based Antenna

Consider an EM-based representation of an N_x row and N_y column pixel-based radiator in monopole configuration $R(x, y)$ and a schematic view of its IMPM model $R_{sy}(y) = R_{sy}(x, y)$, both shown in Fig. 1. The vector of antenna floating-point dimensions is given as $x = [l\ d\ \alpha\ \gamma]^T$. The relative variables are: $l_s = lN_y + d(N_y - 1) + 2o$, $l_m = w_g + \gamma$, $l_g = l_s - \alpha$, and $w_g = \beta l_m$, whereas $w_m = 3$, $\beta = 0.4$, and $o = 3$ remain constant. The unit for all dimensions (except unit-less α, β, γ) is mm. Parameter $M = 2N_xN_y - (N_x + N_y)$ specifies the number of unique internal (auxiliary) ports between the pixels, whereas $y = [z_{1.2} \ldots z_{1.N}]^T$ is an M-element bi-value vector of load impedances for individual ports [9]. Its component associated with mth port ($m = 2, \ldots, N$) can be set to either 0, or ∞ when pixels are to be connected (closed port), or disconnected (open port). The IMPM model can be outlined as follows. Let $Z = Z(x, f)$ be an $N \times N$ matrix [8]:

$$Z = \begin{bmatrix} Z_A & Z_B \\ Z_C & Z_D \end{bmatrix} \quad (3)$$

where $Z_A = z_{1.1}$, is pertinent to external port impedance; $Z_B = [z_{1.2} \ldots z_{1.N}]$, $Z_C = [z_{2.1} \ldots z_{N.1}]^T$ represent vectors of external-to-auxiliary ports and Z_D is a matrix of auxiliary ports impedances (cf. Figure 1(b)). The components of (3) are calculated only once for the given configuration of pixel antenna EM-based model [8]. Note that configuration is understood here as the number of pixels and ports M between them, but also as a specific vector of floating point parameters x and the frequency sweep f. Once (3) is extracted, the impedance at external input port can be obtained by solving [8]:

$$Z_{in}(y) = Z_A - Z_B(Y(y) + Z_D)^{-1}Z_C \quad (4)$$

Here, $Y(y)$ is an $N \times N$ diagonal matrix comprising the elements of y vector. Finally, IMPM model response $R_{sy}(y)$ can be derived from (4) as [8]:

$$R_{sy}(y) = \left(\frac{Z_{in}(y)}{z_0} + I\right)^{-1}\left(\frac{Z_{in}(y)}{z_0} - I\right) \quad (5)$$

where z_0 is a normalized impedance (here, $z_0 = 50\ \Omega$) and I denotes an $N \times N$ identity matrix. It should be reiterated that, for the IMPM model, the contents of x only affect the composition of Z in (3), which is obtained as a result of a single multi-port EM simulation. Once Z matrix is extracted, the responses of (5) are only affected by the composition of y. It is worth noting that the cost of deriving (3) is higher compared to evaluation of conventional antenna EM models [8]. Computational overhead is associated with the need to extract Z matrix for multiple ports rather than only one (antennas are predominantly single-port devices) [1, 4, 5]. For more comprehensive discussion on IMPM representation of pixel-based structures, see [8, 9].

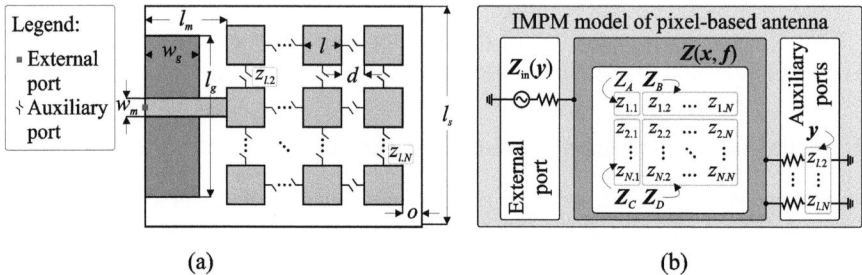

Fig. 1. Pixel-based antenna development: (a) visualization of a generic EM model with highlight on dimensions and ports (along with impedances associated with them) and (b) representation of the structure input impedance using IMPM model.

2.3 Optimization Engine for Local Tuning of Pixel Antenna Geometry

Despite low cost, IMPM model of Sect. 2.2 is suitable only for adjustment of interconnections between the individual pixels. The process can be performed, e.g., using population-based metaheuristics, or exhaustive search [4, 5]. However, fine-tuning of the topology involves optimization of floating-point parameters x (e.g., dimensions of individual pixels, separation between them, or configuration of the fed line). Conventional algorithms require dozens of EM simulations for convergence, even when x is relatively low-dimensional [5]. Here, the problem is mitigated using a gradient-based method embedded within a trust-region (TR) loop. Let $R_f(x,y^*)$ be the equivalent of a pixel-based EM model which substitutes open/closed ports specified in y^* by metal/etched connections between pixels (cf. Figure 2(a)). TR framework generates a series of approximations ($i = 0, 1, 2, \ldots$) to the final design x^* by solving [4, 10]:

$$x^{(i+1)} = \arg \min_{\|x-x^{(i)}\| \leq \delta} \left(U\left(R_{sx}^{(i)}(x)\right) \right) \tag{6}$$

where $R_{sx}^{(i)}(x) = R_f(x^{(i)}, y^*) + J_f(x^{(i)}, y^*)(x - x^{(i)})$ is a first-order Taylor model and J_f is a Jacobian based on large-step finite-differences around $R_f(x^{(i)}, y^*)$ [10]. The trust-region radius δ, i.e., the range around $x^{(i)}$ for which the model $R_{sx}^{(i)}$ is considered acceptably accurate, is controlled based on a gain ratio given as [10]:

$$\rho = \frac{U\left(R_f(x^{(i+1)}, y^*)\right) - U\left(R_f(x^{(i)}, y^*)\right)}{U\left(R_{sx}(x^{(i+1)})\right) - U\left(R_{sx}(x^{(i)})\right)} \tag{7}$$

At first iteration ($i = 0$), the radius is set to $\delta = 1$. Then, it is adjusted based on standard rules, i.e., $\delta = 2\delta$, when $\rho > 0.75$ and $\delta = \delta/3$, when $\rho < 0.25$. The candidate design obtained from (6) is accepted when $\rho > 0$; otherwise it is rejected. The algorithm is terminated either when $\delta < \varepsilon$, or $\|x^{(i+1)} - x^{(i)}\| < \varepsilon$ (here, $\varepsilon = 10^{-2}$ is used). The computational cost of the algorithm is just $D + 1$ EM simulations per successful iteration (D denotes dimensionality of x). Additional EM evaluation is required for each unsuccessful step. Upon termination of the algorithm (6), $x^* = x^{(i)}$ is set. For more comprehensive discussion on TR-based optimization of antennas, see [4, 10, 11].

2.4 Feature-Assisted Representation of Antenna Responses

Antenna can be considered as a transformer between the transmission line (e.g., microstrip) and wireless propagation medium. Consequently, operational frequencies of the radiator are proportional to its physical dimensions [11]. The problem outlined in Sect. 2.1 involves adjustment of y for the given x (that determines size of the radiator) followed by fine tuning of the latter. Due to sequential nature, the procedure might be cumbersome when determination of target operational frequencies is associated with tuning of x. The problem is important for multi-band antennas [4]. It can be mitigated by execution of local-search within the domain of features [12]. The latter represent antenna responses in the form of carefully selected points.

Let $F(x) = [\omega\ L] = P(R_f(x))$ be the feature-based response of a pixel-based antenna extracted from EM simulation using a function P [4, 12]. Here, $\omega = [\omega_1 \ldots \omega_q]^T$ and $L = [L_1 \ldots L_q]^T$ ($q = 1, \ldots, Q$) denote the features defined w.r.t. frequencies (typically in GHz) and levels (in dB) of interest for the design problem. The pair of points can represent frequency at the specific response level (e.g., local minimum pertinent to resonance of narrowband antenna), or level at specific target frequency (e.g., at the expected edge of the operational bandwidth) [12]. When compared against $R_f(x)$, changes of response features are a much less non-linear function of x. Therefore, coordinate points can be used for reliable shifting of resonances over a wide frequency range while maintaining acceptable cost of the process. Representation of antenna responses using feature points is illustrated in Fig. 2. For more comprehensive discussion on the concept, see [4, 12].

It should be emphasized that, when used with optimization engine of Sect. 2.3, the R_{sx} model is not only constructed based on $F(x)$ responses—instead of $R_f(x)$—but also evaluated using different design objectives. The same applies to evaluation of gain ratio.

2.5 Summary of the Bi-Stage Optimization Method

Unsupervised generation and tuning of the pixel-based topology enables streamlined development of antennas according to the desired design specifications. The considered bi-stage optimization method can be summarized as follows.

1. Set N_x, N_y, x, calculate M and generate antenna EM model;
2. Generate $Z(x, f)$ and construct $R_{sy}(y)$ as explained in Sect. 2.2;
3. Find y^* by solving (2) using objective U and global-search method of choice;
4. For frequency-based design, go to Step 6; otherwise go to Step 5;
5. Define features extraction function P and substitute $R_f(x)$ with $F(x)$;

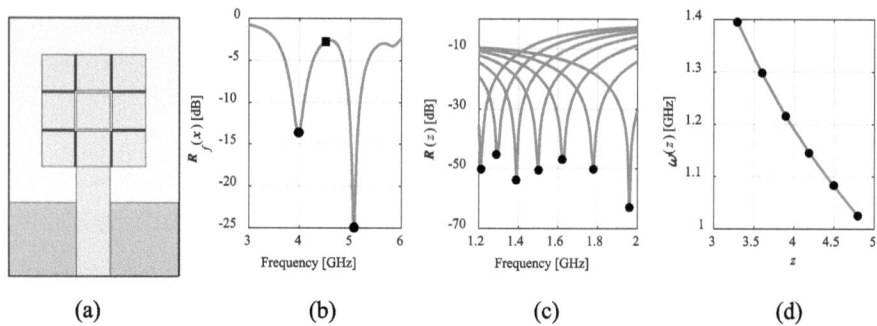

Fig. 2. Pixel-based antenna: (a) example EM model with highlight on metal/etched (red/white) connections between pixels, as well as (b) feature-based representation ($Q = 3$) of $R_f(x)$ response (—) in terms of levels (●) and frequencies of interest (■), (c) frequency sweep along given z, and (d) feature-based sweep. Changes of $\omega(z)$ are much less non-linear compared to $R(z)$ [4].

6. Find x^* by minimization of (6) using selected objective function and END.

Antenna EM model is generated without user inference using a custom script that determines the lattice of pixels constituting the radiator, as well as the ports that enable controlling connections between them. It should be noted that computational cost of the outlined algorithm is low as identification of x^* and y^* involves only a handful of EM simulations.

3 Results

The considered framework is demonstrated based on monopole topology constituted by a lattice of pixels with $N_x = 3$ rows and $N_y = 3$ columns, as well as $M = 12$ auxiliary ports (cf. Section 2.2). The EM models incorporate an FR 4 substrate ($h = 1.6$ mm, $\varepsilon_r = 4.3$, $\tan\delta = 0.025$). The initial parameters are $x^{(0)} = [3\ 0.2\ 0\ 3]^T$. Two test cases that involve development of broadband and dual-band radiators are considered. EM models are generated using CST Studio. Given relatively low number of internal ports and negligible cost of R_{sy}, minimization of (2) is performed using exhaustive search. Lower and upper bounds on x are given as $l_x = [3\ 0.2\ 0\ 2.4]^T$ and $u_x = [5\ 0.6\ 4\ 5]^T$.

3.1 Broadband Antenna Design

The first case concerns development of wideband structure dedicated to work within $f_L = 3.8$ GHz to $f_H = 10$ GHz range. Upon generation of the antenna EM model, the impedance matrix Z is obtained for the design $x^{(0)}$ as explained in Sect. 2.2. The optimization is governed by a min-max function $U = \max\{\max(R - R_{thd}), 0\}_{f_L \leq f \leq fH}$, where $R_{thd} = -10$ dB is selected as a threshold for an acceptable in-band reflection; $f \in f$ (cf. Section 2.1). The pixels configuration $y^* = [\infty\ \infty\ 0\ 0\ 0\ 0\ 0\ \infty\ \infty\ 0\ \infty\ 0]^T$ is found through minimization of (2) using the above objective. Next, the identified geometry is fine-tuned using algorithm of Sect. 2.3. The final solution $x^* = [3.55\ 0.2\ 1.78\ 2.8]^T$ is found after just 6 iterations of (6), which corresponds to a total of 31 $R_f(x)$

simulations. Overall, the computational cost of antenna development amounts to 33.3 R_f simulations (~0.56 h of CPU-time on a dual AMD EPYC 7282 system). Figure 3(a) shows a geometry of the optimized antenna and a comparison of its responses at each stage of the design process. It is worth noting that $R_f(x^{(0)}, y^*)$ slightly violates the imposed design requirements. However, the performance is improved as a result of the second stage of design process. Overall, $R_f(x^*, y^*)$ is characterized by the reflection below the -10 dB level within 3.74 GHz to 10 GHz range.

3.2 Multi-band Antenna Design

The second case involves design of a dual-band antenna dedicated to work at $f_1 = 3$ GHz and $f_2 = 6$ GHz. Given that $x^{(0)}$ might be inadequate for obtaining resonances at target frequencies, the problem has been reformulated to identify the geometry for which the second frequency is shifted w.r.t. the first resonance by the predefined scaling factor K (here, $K = f_2/f_1 = 2$). For such a problem, identification of suitable topology can be performed by minimization of the following objective function $U = \max\{R - R_{\text{thd}}\}_{fr.1, fr.2 \in f}$, where $f_{r.1}$ is the first resonance frequency featuring the reflection below $R_{\text{thd}} = -15$ dB level and $f_{r.2} = Kf_{r.1}$. Similarly as in Sect. 3.1, the optimum solution $y^* = [0 \infty 0 0 \infty \infty 0 0 \infty \infty 0 0]^T$ is found as a result of exhaustive search. It is worth noting that the search space defined by U is subject to discontinuities. Nonetheless, it does not poses a challenge given the nature of selected search mechanism. As shown in Fig. 3(b), the center frequencies for the design $R_f(x^{(0)}, y^*)$ are at 3.74 GHz and 9.1 GHz, which represents substantial shift w.r.t. the target values. Having that in mind, the structure response is converted to features $F(x)$ similar as the ones shown in Fig. 2(b), i.e., frequencies/levels at resonances. The feature-based objective function is $U_F = \beta_1 \max(L - R_{\text{tF}}) + \|\omega - [f_1 f_2]^T\|$, where $\beta_1 = 10$ is used to balance the frequency- and level-related requirements. The design goal involves shifting the resonances to target frequencies, while maintaining the reflection at ω below the R_{thd} level. The optimized design $x^* = [4.88\ 0.6\ 0.79\ 3.41]^T$ is found after 8 TR iterations—minimization of (6) is realized based on $F(x)$ responses—which corresponds to 33 $R_f(x, y^*)$ evaluations. As shown in Fig. 3(b), the optimized design is characterized by reflection below -10 dB level within 2.66 GHz to 4.06 GHz and 4.44 GHz to 6.67 GHz ranges, respectively. The overall computational cost of the design process corresponds to 35.3 R_f simulations (~0.6 h of CPU-time) which is low, given identification of y^* based on the exhaustive global-search.

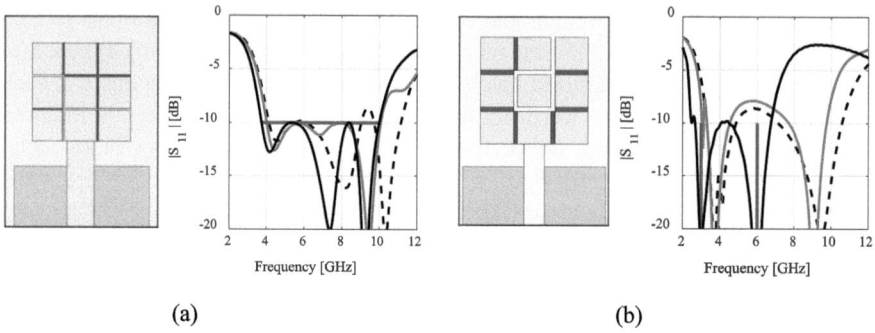

Fig. 3. Pixel antennas – optimized geometries and responses for (gray) IMPM $R_{sy}(x^{(0)}, y^*)$, as well as (black) EM models $R_f(x^{(0)}, y^*)$ (– –), and $R_f(x^* y^*)$ (—): (a) first and (b) second case.

4 Conclusion

In this work, a bi-stage framework for cost-efficient development of pixel-based antennas has been considered. The method involves identification of the desirable antenna topology through global optimization of the cost-efficient IMPM model. The resulting geometry is then subject to fine-tuning through adjustment of floating-point dimensions (e.g., size, and separation of pixels). The method has been demonstrated based on two examples concerning automatic design of broadband and dual-band antennas. For the first case, fine-tuning has been executed using frequency responses. For the second case, structure geometry has been adjusted using the feature-based responses to enable shifting the off-target resonances obtained as a result of global-search. For the considered test cases, the aggregated cost of antenna design does not exceed 36 EM simulations. Future work will focus on optimization of pixels with non-uniform dimensions.

Acknowledgments. This work was supported in part by the National Science Center of Poland Grant 2021/43/B/ST7/01856.

Disclosure of Interests. The authors declare no conflicts of interest.

References

1. Koziel, S., Bekasiewicz, A.: Comprehensive comparison of compact UWB antenna performance by means of multiobjective optimization. IEEE Trans. Ant. Prop. **65**(7), 3427–3436 (2017)
2. Haq, T., Koziel, S.: Novel complementary multiple concentric split ring resonator for reliable characterization of dielectric substrates with high sensitivity. IEEE Sensors J. **24**(10), 16233–16241 (2024)
3. Mroczka, J.: The cognitive process in metrology. Measurement **46**, 2896–2907 (2013)
4. Bekasiewicz, A., et al.: Strategies for feature-assisted development of topology agnostic planar antennas using variable-fidelity models". J. Comp. Sci. **85**, 102521 (2024)
5. Whiting, E.B., et al.: Meta-atom library generation via an efficient multi-objective shape optimization method. IEEE Open J. Ant. Prop. **28**, 24229–24242 (2020)

6. Jacobs, J.P.: Accurate modeling by convolutional neural-network regression of resonant frequencies of dual-band pixelated microstrip antenna. IEEE Ant. Wireless Prop. Lett. **20**(12), 2417–2421 (2021)
7. Alnas, J., Giddings, G., Jeong, N.: Bandwidth improvement of an inverted-F antenna using dynamic hybrid binary particle swarm optimization. Appl. Sci. **11**, 2559 (2021)
8. Zhang, J., et al.: Inverse design of SIW devices based on the internal multiport method. IEEE Microwave Wireless Tech. Lett. **34**(2), 171–174 (2024)
9. Soltani, S., Lotfi, P., Murch, R.D.: Design and optimization of multiport pixel antennas. IEEE Trans. Ant. Prop. **66**(4), 2049–2054 (2018)
10. Bekasiewicz, A.: Optimization of the hardware layer for IoT systems using a trust-region method with adaptive forward finite differences. IEEE IoT J. **10**(11), 9498–9512 (2023)
11. Balanis, C.A.: Antenna Theory Analysis and Design, 3rd edn. John Wiley & Sons, Hoboken (2005)
12. Pietrenko-Dabrowska, A., Koziel, S.: Response Feature Technology for High-Frequency Electronics. Optimization, Modeling, and Design Automation. Springer, New York (2023)

Investigation of CUDA Graphs Performance for Selected Parallel Applications

Oksana Diakun and Paweł Czarnul(✉)

Faculty of Electronics, Telecommunications and Informatics,
Gdansk University of Technology, Narutowicza 11/12, 80-233 Gdansk, Poland
oksdiaku@pg.edu.pl, pczarnul@eti.pg.edu.pl

Abstract. In this paper, we contribute by providing a direct comparison of performance of NAS Parallel Benchmarks implemented with standard CUDA against our extended implementation with CUDA Graphs. The evaluation was conducted for four hardware platforms, using two desktop GPUs—NVIDIA GeForce RTX 2080 and NVIDIA GeForce RTX 4070 Ti—and two server-class GPUs—NVIDIA Quadro 8000 and NVIDIA A100 80GB. The primary focus of the comparison was the execution time of the benchmarks, analyzed for various problem sizes (classes S, A, B, C and D). Two applications exhibited noticeable performance gains from the implementation of CUDA Graphs. The CG code demonstrated the most consistent improvements across all cases, achieving an average relative speedup of 3.3%. The highest result of 4.13% (Class C) for this algorithm was achieved for the NVIDIA GeForce RTX 4070 Ti card. The LU code showed gains primarily on newer generation GPUs, with an average speedup of 2% on the RTX 4070 Ti and 7%, on the A100, with a maximum gain of 11.87% (Class B). In contrast, visible negative performance was observed for MG and some instances of IS, but we attribute that to the relatively small absolute running time in which case additional overheads cannot be mitigated by the new mechanism.

Keywords: GPGPU · CUDA Graphs · Parallel Computing · High-performance computing · GPU Acceleration · Performance evaluation

1 Introduction

NVIDIA CUDA (Compute Unified Device Architecture) technology, for many years already, has allowed programmers to parallelize code on GPU devices [5]. CUDA versions have progressively introduced new features and optimizations. CUDA Graphs [7], investigated in this paper, were designed to tackle the overhead associated with CPU-GPU interactions, especially when handling a large number of small and/or interconnected tasks. CUDA Graphs allows to bundle a series of CUDA operations into a single graph entity that can be executed all at once. This reduces the frequency and cost of CPU-GPU communication, including kernel launch overhead, as visualized in [4].

This paper aims to assess the performance of CUDA Graphs across a diverse set of parallel applications. The main goal of this research is to compare the performance of applications that use CUDA Graphs with traditional CUDA versions without CUDA Graphs. This comparison will help identify situations where CUDA Graphs offer performance improvements and where their use may be less beneficial. Section 2 presents existing examples of applying CUDA Graphs to various applications. Section 2 describes motivations and contribution of this work. Section 3 specifies the set of selected benchmarks and its updating with CUDA Graphs. Section 4 describes the conducted experiments, presents and discusses the results. Section 5 presents a summary and potential future work.

2 Related Work and Our Contribution

CUDA Graphs, introduced in CUDA 10, allow to express more complex execution scenarios in a form of a graph. Nodes of the latter can refer to kernel launches, memory copies, and synchronization events. Edges between nodes define the dependencies. A particular node begins execution only after all its predecessor nodes have completed. This effectively removes the overhead of CPU-side stream synchronization. CUDA Graphs can effectively reduce the overhead associated with launching individual kernels and improve overall application performance [4]. One way of adopting the code to use CUDA Graphs is to encapsulate and capture existing operations put into a CUDA stream, into a graph. This can be done by calling cudaStreamBeginCapture() and cudaStreamEndCapture() API calls. Alternatively, a graph can be created explicitly using cudaGraph Create() and appropriate functions for definition of nodes and dependencies.

The two CUDA Graphs creation modes can be combined for effective code in a similar scenario. In article [11], the author proposes code with 3 kernel invocations where the first two are captured in a stream with static parameters while a third kernel is being added manually with dynamic parameters to the graph extracted with a call to cudaStreamGetCaptureInfo_v2(). In terms of performance, three versions were compared: without CUDA Graphs, with CUDA Graphs with the recapture-then-update approach, using CUDA Graphs with the aforementioned combined approach, the latter two giving speed-ups over the first one: 1.22 and 1.63 respectively. Obviously, potential gains depend on the ratio of the time spent on computations to communication as well.

Article [9] discusses combining CUDA Graphs with an Image Processing Domain-Specific Language (DSL) and Hipacc, a source-to-source compiler. Benchmarks were conducted across ten different image processing applications on two different NVIDIA GPUs: the RTX2080 and the GTX680. The proposed approach allowed to achieve a geometric mean speedup of 1.30 over Hipacc without CUDA graph, 1.11 over CUDA graph without Hipacc.

Paper [12] discusses a novel compiler transformation technique that converts OpenMP code into CUDA Graphs to enhance NVIDIA device programmability.

The approach combines high-level OpenMP's high-level programmability with CUDA's performance benefits. The performance evaluation involved two benchmarks: Saxpy, a structured benchmark with 1024 tasks, and Cholesky decomposition, an unstructured benchmark with 1540 tasks. The evaluation showed that, for the Saxpy benchmark, the CUDA graph implementation reduced execution time by approximately 78%, while the Cholesky decomposition saw a reduction in execution time by approximately 93% when compared to their OpenMP offloading counterparts. It was noted, though, that CUDA Graphs significantly increase the complexity and the number of lines of code needed, especially for Cholesky decomposition. In work [3] authors conducted a study that involved modifying a Breadth-First Search (BFS) application from a benchmark suite to utilize CUDA Graphs, and comparing its performance to the original non-graph version. The testbed system was equipped with an NVIDIA RTX 2060 GPU and a Ryzen 5 3600x CPU. The BFS application from the Rodinia benchmark suite was selected, duplicated, and modified to implement code with CUDA Graphs. The results, for runs repeated 100 times, showed that for input sizes of 524,288 nodes, the CUDA Graph implementation allowed to obtain a 14% speedup compared to the non-graph counterpart, with a running time reduction from 1836 μs to 1610 μs. On the other hand, for smaller input sizes, such as 8,192 nodes, the CUDA Graph version actually exhibited worse performance, with the execution time 306 μs – 7% slower than 285 μs for the standard non-graph version. The study indicates that while CUDA Graphs can offer performance benefits, particularly for medium-sized workloads, the overhead associated with graph instantiation can negate these benefits for smaller workloads.

Article [6] discusses the integration of CUDA Graphs in PyTorch, which is aimed at accelerating PyTorch with advanced CUDA features. Using CUDA Graphs resulted in up to a 50% reduction in CPU usage and an increase in GPU utilization, which led to a significant speedup of up to 30% in the Mask R-CNN model's overall execution time. The article also illustrates how employing CUDA graphs to capture the model leads to the elimination of CPU overhead and synchronization issues resulting in a performance boost of 1.12 times for a large-scale BERT model.

In the context of existing works, we aim at thorough assessment of CUDA Graphs versus the traditional CUDA model. We do that by extending the well established NAS Parallel Benchmarks (NPB) [2] (available at GMAP/NPB-GPU: NAS Parallel Benchmarks for GPU) [1] with CUDA Graphs. We then performed comprehensive comparison on four different GPUs. We discuss relative speed-up, which refers to the difference in execution time between the approach without and with implemented CUDA Graphs in relation to the time obtained for the approach without CUDA Graphs.

3 Testbed Workloads and Our Implementations with CUDA Graphs

The collection of evaluated benchmarks consisted of algorithms Conjugate Gradient (CG), Embarrassingly Parallel (EP), Integer Sort (IS), Multi-Grid (MG),

CFD-related tasks (LU, SP) and one CFS-related task (BT), without Fourier Transform (FT). For the latter, we did encounter unsuccessful verification of the original code on one of the cards.

For the CUDA Graphs implementations, we decided to use the stream capture mode. Details regarding implementation are available in the github repository[1]. The implementation began with the creation of a stream to which kernel functions would be assigned, so that they could then be captured. After that, variables that will describe our graph were created. A bool variable ensures that the graph will be created only one time, when set to false, and true flag allow executing an already existing graph. To capture a stream it was necessary to create a new, non default one, and assign desirable functions to it. After that, with a call to the `cudaStreamBeginCapture(stream, mode)` function it was possible to start capturing. The capturing of the graph could be easily terminated with a `cudaStreamEndCapture(stream, &graph1)` call. To connect captured graph with its execution name, function `cudaGraphInstantiate(graphExec_1, graph_1, NULL, NULL, 0)` was used. Having that it was possible to finally call an instance of created graph by `cudaGraphLaunch(graphExec_1, stream)` [8].

Unfortunately, the presented procedure would not work in every case. It would not be optimal in situations where kernel functions calls are interspersed with operations performed on the CPU, where, for example, the value of a variable used in the kernel had to be changed. Such situations occurred in several places in our implementation and we decided to use other CUDA functionalities so that the mentioned values could always be available in the graph. One option was to use CUDA Graphs with Dynamic Parameters as outlined in [11]. We eventually chose the task-related variable which is allocated on the device and stores the value calculated on the host, transferred via `cudaMemcpy()`.

Another problem is a situation where the function we want to capture contains a set of various other functions calling kernel functions. This situation is problematic, because: firstly – it is possible to encounter a case similar to the one presented previously, and secondly – it is possible that in certain functions it will be necessary to copy data from device to host, e.g., using function `cudaMemcpy()`. The solution to this is to replace the synchronous operation with an asynchronous one. This also effectively requires to prepare an analogous function and add the stream used for capturing within the function parameters. Finally, we verified correctness of results by checking an appropriate flag already implemented in the NAS NPB software.

4 Experiments

The foregoing benchmarks were evaluated on GPUs of various classes/architectures: desktop/Turing NVIDIA GeForce RTX 2080, desktop/Ada Lovelace GeForce RTX 4070 Ti, workstation/Turing Quadro RTX 8000 and server/Ampere A100 80GB PCIe. System configurations for the 4 nodes are as follows: 2 x Intel(R) Xeon(R) Gold 6130, NVIDIA GeForce RTX 2080, 256 GB RAM; 13th Gen

---

[1] https://github.com/odiakun/NPB-GPU-CUDA-Graphs/tree/master.

Intel(R) Core(TM) i7-13700K, NVIDIA GeForce RTX 4070 Ti, 32 GB RAM; Intel(R) Xeon(R) Gold 6248R CPU, 2 x Quadro RTX 8000 + Quadro RTX 5000, 192 GB RAM; 2 x Intel(R) Xeon(R) Silver 4316 CPU, NVIDIA RTX A4500 + NVIDIA A100 80GB PCIe, 256 GB RAM.

We have obtained results on every card for certain algorithm. We conducted 10 iterations for each application across different classes. The average value was calculated alongside with the standard deviation. Detailed execution times are presented for NVIDIA GeForce RTX 4070 Ti and A100 in Fig. 1. Tables with more detailed information are available under the link[2].

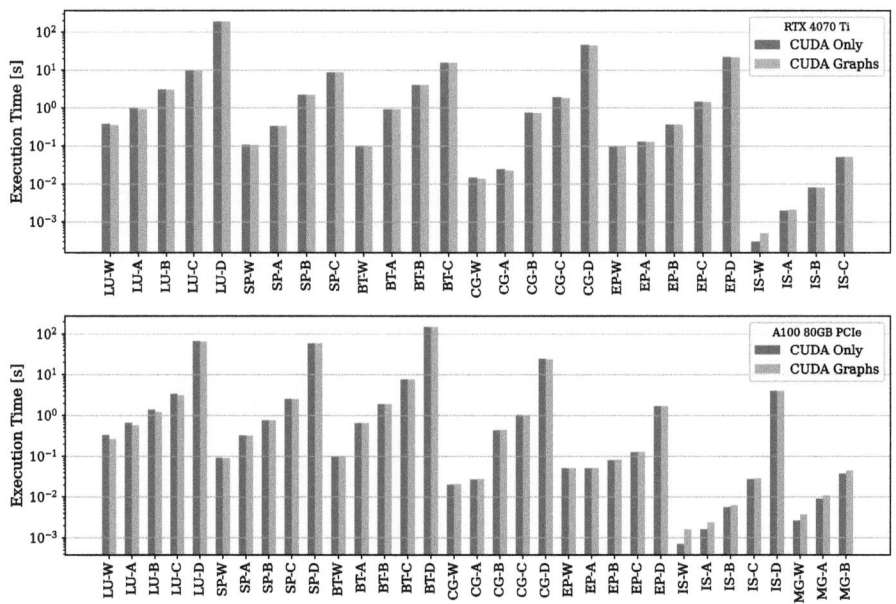

Fig. 1. Execution times per NAS Benchmarks and Class (for two cards).

For each of the cases a relative speed-up was calculated. The values of the execution time obtained for the variant that does not take into account the implementation of CUDA Graphs, were considered as the relative value.

In the following analysis, we focus on the comparison mainly of the larger classes of the benchmarks and assume that gains/losses within $[-1, +1]\%$ range are considered non-conclusive. This is because, for these classes, for the vast majority of cases, standard deviation in mean (percentage) is also within the $[0, 1]\%$ range, although very often significantly below 1%. In very rare cases it exceeds 1% but it is for relatively short execution times, for these classes. Instance time is the execution time of a single kernel instance, and instance

---

[2] https://cdn.files.pg.edu.pl/eti/KASK/CUDA_Graphs_additional/CUDA_Graphs_paper_detailed_data.pdf.

number is the number of kernel instances that have been executed for a given algorithm. These values have been weighted by the proportion of total execution time that each kernel consumes. Data relevant to configurations giving the best results are shown under the link (see Footnote 2). Weighted values are used for axes of the following figures.

Speed-ups for the tested configurations are shown in Fig. 2. It can be observed that gains in application duration were achieved primarily in cases where there were relatively many short-lived kernels.

In the case of the NVIDIA GeForce RTX 2080, there are gains for the CG algorithm for classes B and C with values of 2.97% and 3.41%. As for the other algorithms, LU, SP, IS, BT and EP reach values that are close to zero. On the other hand, one can notice a negative value for the MG (-9.44%) algorithm. In the latter case, however, absolute execution times are very low, in which case additional overheads might not be mitigated by potential benefits.

For the NVIDIA GeForce RTX 4070 Ti, for the LU algorithm, the profit values are: 0.36%, 1.58% and 2.57%, with the value of 0.36% again being too small to make a conclusion. For the CG algorithm, the results are: 2.89%, 3.83% and 4.13%. The values for the IS, SP and BT algorithms are again too small for conclusive statements. The MG algorithm (-5.60%) again shows a negative value, but its execution times are very small in absolute terms.

For the NVIDIA Quadro RTX 8000, there are consistent gains for CG (2.11%, 3.46% and 3.84%). The BT (-1.34% and -4.08%) and MG (-8.56%) algorithms are characterized by slight negative values and in this case the result of the IS algorithm for one of the computing classes (-3.88%) also signals a decrease in the value of the application duration.

For the NVIDIA A100 80GB PCIe, the noticeable difference from the previous GPUs is that the LU algorithm here features significant gains in application time with values of 1.45%, 8.05% and 11.87%. With the values decreasing as the weighted instance time of a single instance increases. For the CG algorithm, gains for two classes amount to 2.78% and 3.56%, similarly to the results on the other GPUs. Negative values, on the other hand, are visible for MG (-19.98%) and IS algorithms for two classes (-2.27% and -10.28%). The remaining cases have values that are too small to be considered significant.

For all the cards, the MG and IS algorithms are characterized by short execution times, even for larger computational classes (B and C). In most cases, the execution times of these applications, in their unmodified forms, do not exceed 1 s. For class D and NVIDIA Quadro RTX 8000, these times are 15.42 s for the IS algorithm and 17.04 s for MG. In comparison, the LU algorithm for class D on the same card has an execution time of 301.76 s, and the CG algorithm 273.48 s. On NVIDIA A100 80GB PCIe, where the MG algorithm achieves a time of 6.17 s and IS 3.96 s. For the LU and CG algorithms, these times are 66.18 s and 24.16 s, respectively. For both of these cards, these times are of different orders of magnitude. With such short execution times, the implementation of CUDA Graphs may introduce some fixed overheads and could slow down the application.

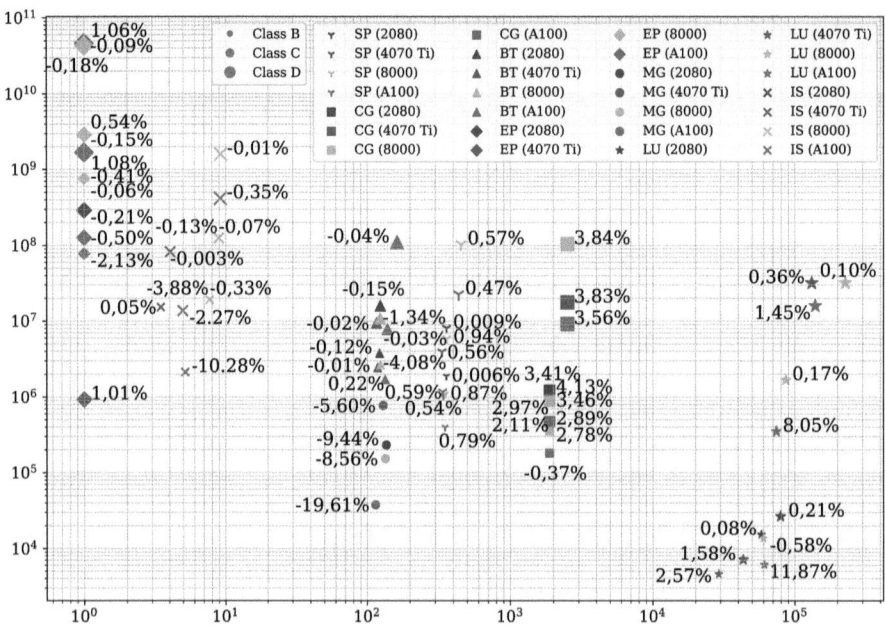

Fig. 2. Speed-ups, Weighted Instance Time in relation to Weighted Instance Number for every GPU.

5 Summary and Future Work

This work is one of the first to explicitly compare CUDA and CUDA Graphs. This paper presents an experimental study based on the GPU implementation of NAS Parallel Benchmark. The research presented in this work includes the evaluation of the aforementioned benchmarks in their unaltered form on NVIDIA GeForce RTX 2080, NVIDIA GeForce RTX 4070 Ti, NVIDIA Quadro RTX 8000, and NVIDIA A100 80GB PCIe high-performance cards. Usage of CUDA Graphs was then incorporated by us into all applications comprising the NAS Parallel Benchmarks, followed by reevaluation on the same hardware, which allowed to compute a relative speed-up. The CG algorithm shows the most consistent gains across various cases. Depending on the card, CUDA Graphs yielded gains of 2 to over 4%. For LU, gains were largest on the A100 and reached even 11.87%. For all of the aforementioned cards, low negative gain values were observed for the MG algorithm (−9.44%, −5.60%, −8.56%, and −19.61%) that we attribute to small absolute running times and impact of fixed overheads of CUDA Graphs. It can be concluded that the CG algorithm consistently yields gains, and the LU algorithm shows improvements in about half of the analyzed cases. Other benchmarks resulted in execution times similar for both cases.

For the future, detailed investigation of the GPU architecture impact could be performed, along with integration of CUDA Graphs into a higher-level DAG processing framework in a cluster [10].

References

1. Araujo, G., Griebler, D., Rockenbach, D.A., Danelutto, M., Fernandes, L.G.: NAS Parallel Benchmarks with CUDA and Beyond. Software: Practice and Experience (2023)
2. Araujo, G.A., Griebler, D., Danelutto, M., Fernandes, L.G.: Efficient NAS parallel benchmark kernels with CUDA. In: 2020 28th Euromicro International Conference on Parallel, Distributed and Network-Based Processing (PDP). IEEE (2020)
3. Demirsu, M., Lervik, A.: Evaluating the performance of CUDA Graphs in common GPGPU programming patterns. School of Electrical Engineering and Computer Science, KTH Royal Institute of Technology (2023). https://www.diva-portal.org/smash/get/diva2:1779194/FULLTEXT01.pdf
4. Gray, A.: Getting Started with CUDA Graphs. NVIDIA Developer, Technical Blog (2019). https://developer.nvidia.com/blog/cuda-graphs/
5. Kirk, D.B., Wen-Mei, W.H.: Programming Massively Parallel Processors: A Hands-on Approach, 3rd edn. Morgan Kaufmann (2016)
6. Nguyen, V., et al.: Accelerating PyTorch with CUDA Graphs (2021). https://pytorch.org/blog/accelerating-pytorch-with-cuda-graphs/
7. NVIDIA Corporation: CUDA Graphs API. NVIDIA Developer Documentation (2019). https://developer.nvidia.com/cuda-graphs
8. NVIDIA Corporation: Cuda C++ best practices guide (2024). https://docs.nvidia.com/cuda/cuda-c-best-practices-guide/
9. Qiao, B., Özkan, M.A., Teich, J., Hannig, F.: The best of both worlds: combining cuda graph with an image processing DSL. In: Proceedings of the 57th ACM/EDAC/IEEE Design Automation Conference, DAC 2020. IEEE Press (2020)
10. Rościszewski, P., Czarnul, P., Lewandowski, R., Schally-Kacprzak, M.: KernelHive: a new workflow-based framework for multilevel high performance computing using clusters and workstations with CPUs and GPUs. Concurr. Comput. Pract. Experience **28**(9), 2586–2607 (2016). https://doi.org/10.1002/cpe.3719
11. Tu, J.: Constructing CUDA Graphs with Dynamic Parameters. NVIDIA Developer, Technical Blog (2022). https://developer.nvidia.com/blog/constructing-cuda-graphs-with-dynamic-parameters/
12. Yu, C., Royuela, S., Quiñones, E.: OpenMP to CUDA graphs: a compiler-based transformation to enhance the programmability of NVIDIA devices. In: Proceedings of the 23th International Workshop on Software and Compilers for Embedded Systems, SCOPES 2020, pp. 42–47. Association for Computing Machinery, New York (2020). https://doi.org/10.1145/3378678.3391881

Instance Selection by Fast Local Set Border Selector

Norbert Jankowski[✉] and Mateusz Skarupski

Department of Informatics, Faculty of Physics, Astronomy and Informatics, Nicolaus Copernicus University, Toruń, Poland
norbert@umk.pl

Abstract. Prototype selection is one of the typical goals of machine learning, which aims to reduce the number of vectors in the training set. The local set border selector (LSBo) algorithm is presented as a Pareto optimal choice between the reduction power of the training set and the classification quality. Its complexity is $O(n^2)$, which means that it is not very advantageous for larger sets. This article presents the Fast LSBo algorithm, which is based on the original idea of the LSBo algorithm. After the applied conceptual changes, the algorithm has achieved a complexity of $O(m \log m)$. Additionally, the analysis of Fast LSBo on several data sets shows that its classification quality and reduction power remain statistically indistinguishable from the original LSBo algorithm.

1 Introduction

Let us assume that we have a learning data set $\mathcal{D} = \{\langle \mathbf{x}_i, y_i \rangle : i = 1, \ldots, m\}$ where $\mathbf{x}_i \in R^n$ are the input vectors and $y_i \in [1, \ldots, c]$ are the class labels. Selection of instances (prototypes) means that we are looking for a subset $S \subseteq \mathcal{D}$ that is enough to build a trustworthy classifier, for example, a k nearest neighbor (kNN) classifier [2]. Vector selection algorithms can be divided into two groups: filtering algorithms and prototype selection algorithms. The first group consists of algorithms whose main goal is to remove erroneous vectors, in other words, vectors that are inconsistent with the rest of the data set. Here, the best examples are the ENN [10] and LSSm algorithms [6]. The second group consists of algorithms that try to select prototypes, i.e. the most important vectors of the training set, i.e. those that carry the basic knowledge about the decision boundaries between classes. Interesting examples of such algorithms include [3,4,9]. It is also worth mentioning here a few review articles devoted to a broad analysis of vector selection algorithms [3,4].

The first time *local sets* was introduced in the algorithm Iterative Case Filtering (ICF) in [1]. The *local set* for a given vector $\mathbf{x}$ from $\mathcal{D}$ is defined as the set of vectors in the largest hypersphere centered in $\mathbf{x}$ that does not contain an instance of the opposite class (an *enemy*): $LS(\mathbf{x}) = \{\mathbf{x}' : ||\mathbf{x}' - \mathbf{x}|| < ||\mathbf{x} - ne(\mathbf{x})||\}$, where $ne(\mathbf{x}) = \arg\min_{\mathbf{x}' \wedge y \neq y'} ||\mathbf{x} - \mathbf{x}'||$ is the nearest enemy. This definition strongly uses the distance to the nearest vector of an alien class as the radius of the hypersphere.

Before presenting the local set border selector (LSBo) algorithm [6], it is necessary to present the local set-based smoother (LSSm) algorithm, which is the necessary first phase of the LSBo algorithm.

The LSSm algorithm is a very good example of a filtering algorithm, i.e., an algorithm that removes potentially inconsistent vectors from the original data set. Therefore, it will be used in the algorithm presented below, however in a modified version of lower complexity, but providing the same quality of selection in terms of its usefulness in classification.

The LSSm uses the idea of local sets to define two properties that are of vital importance for constructing the final algorithm. The first is the *usefulness* of a given instance $\mathbf{x}$, measured as the number of instances which has $\mathbf{x}$ in their local sets:

$$u(\mathbf{x}) = |\{\mathbf{x}' \in \mathcal{D} : \mathbf{x} \in LS(\mathbf{x}')\}| \qquad (1)$$

The second property used in LSSm is the *harmfulness* of $\mathbf{x}$ measured as the number of instances for which the $\mathbf{x}$ instance is the nearest enemy:

$$h(\mathbf{x}) = |\{\mathbf{x}' \in \mathcal{D} : ne(\mathbf{x}') = \mathbf{x}\}| \qquad (2)$$

The difference between those two properties defines the strength of balance between usefulness and harmfulness. The LSSm algorithm removes instances with greater harmfulness than usefulness. The final LSSm algorithm is presented in Algorithm 1. The complexity of the LSSm algorithm is $O(m^2)$.

Now, the LSBo algorithm can be described. The first step of LSBo is to filter the original dataset $\mathcal{D}$ by the LSSm algorithm, thus obtaining a filtered dataset $\mathcal{T}$. The next step is to calculate local sets for $\mathcal{T}$. After that, the vectors in $\mathcal{T}$ are ordered (ascending) by the cardinality of local sets $|LS(\mathbf{x})|$. The initial set of prototypes S is empty and the main step browses the ordered vectors in the previous step, and if neither element of $LS(\mathbf{x})$ is already in S then S is extended by $\mathbf{x}$.

The algorithm's strategy first analyzes what is near to the decision boundaries after filtering out bad vectors. The prototype set is thus built from reliable vectors on the decision sides, and therefore, vectors far from the boundaries do not need to be added to the prototype set. The algorithm is presented in Algorithm 2.

Algorithm 1: LSSm($\mathcal{D}$)	Algorithm 2: LSBo($\mathcal{D}$)
Data: $\mathcal{D}$ — dataset **Result:** S 1 $S = \{\}$ 2 compute local sets($\mathcal{D}$) 3 **foreach** $\mathbf{x} \in \mathcal{D}$ **do** 4 $\quad$ compute $u(\mathbf{x})$ and $h(\mathbf{x})$ 5 $\quad$ **if** $u(\mathbf{x}) \geq h(\mathbf{x})$ **then** 6 $\quad\quad$ $S = S \cup \{\mathbf{x}\}$	**Data:** $\mathcal{D}$ — dataset **Result:** S 1 $S = \{\}$ 2 $\mathcal{T} = LSSm(\mathcal{D})$ 3 compute local sets($\mathcal{T}$) 4 sort $\mathcal{T}$ ascending by the cardinality of LS 5 **foreach** $\mathbf{x} \in \mathcal{T}$ **do** 6 $\quad$ **if** $\mathbf{x}.LS \cap S = \{\}$ **then** 7 $\quad\quad$ $S = S \cup \{\mathbf{x}\}$

2 Fast Local Set Border Selector Algorithm

To construct the Fast LSBo algorithm, it was necessary to reduce the complexity of the LSSm algorithm and the complexity of the central part of the LSBo algorithm. A fast version of the LSSm algorithm was proposed by us in [5]. However, it will be slightly modified here.

The crucial point that the complexity of LSSm and LSBo is $O(m^2)$ is: $\sum_{\mathbf{x} \in \mathcal{D}} LSC(\mathbf{x}) = O(m^2)$. This means that as long as the algorithm uses the LS, the complexity will be $O(m^2)$. For this reason, one of the key changes for both algorithms is to use the firmly local sets (FLS) instead of local sets:

$$FLS(\mathbf{x}) = \begin{cases} \{\mathbf{x}' \,:\, ||\mathbf{x}' - \mathbf{x}|| < ||\mathbf{x} - ne(\mathbf{x})||\}, & \text{if } \exists_{\mathbf{x}' \in N^k(\mathbf{x})}\, y \neq y' \\ N^k(\mathbf{x}), & \text{otherwise,} \end{cases} \quad (3)$$

where $N^k(\mathbf{x})$ is a set of nearest neighbours of $\mathbf{x}$.

Fast LSSm Algorithm: FLS is very crucial for overall complexity because since k is $O(1)$, then the $FLS(\mathbf{x})$ can be computed in average complexity $O(\log m)$. In FastLSSm, we used balanced forests of locality-sensitive hashing to compute nearest neighbours and LS's.

In this version of fast LSSm, the graph-based approximation (HNSW) of the nearest neighbours will be used [7]. This change is dictated by the requirements of the central part of the Fast LSBo algorithm presented in the next section.

The usefulness and harmfulness must be redefined as in [5] by

$$u'(\mathbf{x}) = |\{\mathbf{x}' \in \mathcal{D} \,:\, \mathbf{x} \in FLS(\mathbf{x}')\}|, \quad (4)$$

and by

$$h'(\mathbf{x}) = |\{\mathbf{x}' \in \mathcal{D} \,:\, ne(\mathbf{x}') = \mathbf{x} \wedge \mathbf{x} \in N^k(\mathbf{x}')\}| \quad (5)$$

appropriately.

The main idea behind the LSSm lies in: instance $\mathbf{x}$ remains in set if its usefulness is not smaller than harmfulness. See Algorithm 3.

Hopefully, $|FLS(\mathbf{x})|$ has a more useful property: $\sum_{\mathbf{x} \in \mathcal{D}} |FLS(\mathbf{x})| = O(m)$ because k is fixed (not dependent on m). This proves vital to achieving an overall complexity of $O(m \log m)$, as discussed further in Sect. 3.

Fast LSBo Algorithm: Now, all the elements of the fast LSBo algorithm will be presented. The LSBo algorithm is also based on the FLS definition instead of LS. However, several other modifications must be made to the original algorithm. The first is the use of the fast version of the LSSm algorithm – this is a necessary condition to have a chance of the overall complexity of $O(m \log m)$.

The next step is to use a unique structure to represent the set S, i.e., the set of selected prototypes. The representation of the set S will play an additional role here, apart from storing the selected vectors. The structure of the set S will be used to find the closest vectors—it will be necessary to know which vector in S is closest to a given $\mathbf{x}$.

Algorithm 3: FastLSSm($\mathcal{D}$)

Data: $\mathcal{D}$ — dataset
Result: S

1. $S = \{\}$
2. construct HNSW set on $\mathcal{D}$
3. compute $N^k(\mathbf{x})$ for each $\mathbf{x} \in \mathcal{D}$ using HNSW's
4. **foreach** $\mathbf{x} \in \mathcal{D}$ **do**
5.   find nearest enemy $ne(\mathbf{x})$ in $N^k(\mathbf{x})$
6.   **if** *any enemy* **then**
7. $h'[ne(\mathbf{x})]{+}{+}$

8. **foreach** $\mathbf{x} \in \mathcal{D}$ **do**
9.   **foreach** $\mathbf{x}' \in N^k(\mathbf{x})$ **do**
10.   **if** *no enemy of* $\mathbf{x}$ *in* $N^k(\mathbf{x}) \lor ||\mathbf{x}' - \mathbf{x}|| < ||\mathbf{x} - ne(\mathbf{x})||$ **then**
11. $u'[\mathbf{x}']{+}{+}$

12. **foreach** $\mathbf{x} \in \mathcal{D}$ **do**
13.   **if** $u'(\mathbf{x}) \geq h'(\mathbf{x})$ **then**
14. $S = S \cup \{\mathbf{x}\}$

That is why the graph-based HNSW structure [7] represents S, which allows alternating between adding new elements to the set and querying for nearest neighbors. As we show below, this plays a key role in the most essential point of the LSBo algorithm.

The second essential data structure is the HNSW (J) array of sets. This structure is used twofold: to determine nearest enemy class vectors and to determine the cardinality of the FLS sets. Determining the nearest enemies consists of determining each set's nearest neighbor from the J array of sets except for the own set. On the other hand, building the FLS sets consists of taking the nearest k neighbors from the own class set from the J array, but only those closer than the nearest enemy.

The fast version of the LSBo algorithm also uses the ordering of vectors in $\mathcal{T}$ in the main loop, which determines the prototype vectors S. This time, the ascending sorting occurs here by the cardinality of FLS instead of the cardinality of LS. In the above-specified order, all vectors from $\mathcal{T}$ are then traversed. However, the condition from the line 6 of the LSBo Algorithm 2 had to be significantly reformulated. This condition in LSBo stated whether no vector from $LS(\mathbf{x})$ is in the prototype set of S because if so, $\mathbf{x}$ is necessary for S. However, in the fast version of LSBo, we do not have the set $LS(\mathbf{x})$. Still, the same effect can be obtained by the condition:

$$||\mathbf{x} - \mathbf{x}_n|| > ||\mathbf{x} - \mathbf{x}_e||,$$

in which it is tested whether the nearest neighbor of $\mathbf{x}$ in S ($\mathbf{x}_n$) is more distant than the nearest enemy of $\mathbf{x}$ ($\mathbf{x}_e$). Such a case means no vector from the *real* $LS(\mathbf{x})$ is in the current set S. Consequently, it forces the addition of the vector $\mathbf{x}$ to S.

Algorithm 4: FastLSBo($\mathcal{D}$)

Data: $\mathcal{D}$ — dataset
Result: S — the set of selected prototypes

1. $\mathcal{T} = FastLSSm(\mathcal{D})$
2. S = empty HNSW set
3. J = construct one HNSW per class in $\mathcal{T}$ (one HNSW set for all instances of given class)
4. **foreach** $\mathbf{x} \in \mathcal{T}$ **do**
5.   compute $|FLS(\mathbf{x})|$ using HNSW for class y from J
6. sort $\mathcal{T}$ ascending by $|FLS(\mathbf{x})|$
7. **foreach** $\langle \mathbf{x}, y \rangle \in \mathcal{T}$ **do**
8. $\mathbf{x}_n$ = nearest neighbour in S to $\mathbf{x}$
9. $\mathbf{x}_e$ = nearest enemy of $\mathbf{x}$ in J
10.   **if** $||\mathbf{x} - \mathbf{x}_n|| > ||\mathbf{x} - \mathbf{x}_e||$ **then**
11. $S = S \cup \{\mathbf{x}\}$

3 Complexity and Experimental Results on Benchmark

This section starts with the presentation of the average complexity of Fast LSBo is $O(m \log m)$ and that the accuracy of the fast version of LSBo is so strongly similar (in almost all cases the same) to the original LSBo. The most expensive part of the Fast LSSm algorithm is its first part, i.e. the lines 2 and 3. The average cost of adding m vectors to the HNSW structure is $O(m \log m)$. The cost of determining k nearest neighbors for all m vectors is also $O(m \log m)$. In Fig. 1, a relation is presented between the time of adding vectors and building sets of nearest neighbors using HNSW. On the OX axis we see the values of the number of m vectors in the set $\mathcal{D}$. On the OY axis is the execution time divided by $m \log m$. This means that the presented graph should not grow if building and using the HNSW architecture is to have complexity $O(m \log m)$ as mentioned above. And this is precisely how this graph behaves, which proves the complexity of $O(m \log m)$.

The next three loops of the algorithm have a complexity of $O(m)$ (we assume that k is a constant value).

The Fast LSBo algorithm starts execution with the LSSm algorithm.

In the next part, an empty set S is created, to which a subset of $\mathcal{D}$ will be added. Consequently, this will not worsen the previous complexity. In the next part, a series of HNSW structures (table J) is created, in which the vectors of the appropriate classes will be added (one HNSW structure $J[i]$ for the vectors of i-th class). Then, the sets $FLS(\mathbf{x})$ are determined. The costs of these operations are not greater than the costs of creating an HNSW structure for the entire set $\mathcal{D}$. This means that the complexity of this part is limited by $O(m \log m)$ as well as sooner. The next step is to sort the array of all cardinalities $|FLS(\mathbf{x})|$—once again $O(m \log m)$.

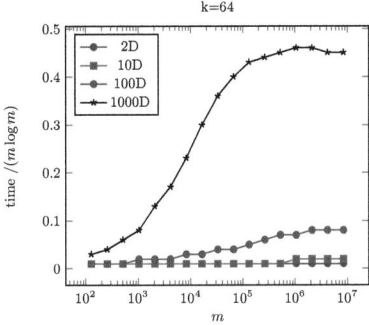

Fig. 1. Time consumption by HNSW.

In the last loop, we have m iterations, and each iteration has complexity $O(\log m)$ because none of the loop elements require more complexity than $O(\log m)$.

This gives the complexity $O(m \log m)$ of the fast LSBo algorithm.

Experimental Results: Since it was shown in [6] that the LSBo algorithm is very effective as a Pareto tradeoff between training set reduction power and classification quality, these tests compare LSBo with state-of-the-art instance selection algorithm and kNN. Hence, here we focus on showing that the classification quality and the quality of prototype selection remain almost the same.

For this purpose, 45 sets were selected from the UCI Machine Learning Repository [8]. Datasets differ in the number of instances, attributes, and balance of classes. In all tests, we used 10-fold stratified cross-validation and all learning machines were trained on the same sets of data partitions. To visualize the performance of all algorithms, we present the average accuracy for each benchmark dataset and each learning machine, Table 1.

In addition, we present the average reduction of the dataset size in separate tables. *Ranks* are calculated for each machine for a given dataset. The ranks are calculated as follows: First, for a given benchmark dataset, the averaged accuracies of all learning machines are sorted in descending order. The machine with the highest average accuracy is ranked 1. Then, the following machines in the accuracy order whose accuracies are not statistically different[1] from the result of the first machine are ranked 1, until a machine with a statistics different result is encountered. That machine starts the next rank group (2, 3, and so on), and an analogous process is repeated on the remaining (yet unranked) machines. Notice that each cell in the main part of is in a form: $acc + std(rank)$, where acc is the average accuracy (for a given data set and given learning machine), std is its standard deviation and $rank$ is the rank described just above. If a given table cell is in bold, it means that this result is the best for the given data set or not worse than the best one (rank 1 = winners).

---

[1] We use the paired t-test to test the significance of statistical differences.

Table 1. Comparison of LSBo and fast LSBo algorithms.

Dataset	Accuracy		Removed %	
	LSBo	LSBoFast	LSBo	LSBoFast
arrhythmia	44.2±20(1)	43.6±19(1)	0.55±0.04	0.55±0.04
autos	60.3±12(1)	59.2±12(1)	0.67±0.02	0.67±0.02
balance-scale	75.9±5.7(1)	75.3±5.3(1)	0.78±0.01	0.84±0.01
blood-transfusion	70±5.1(1)	68.7±5.8(1)	0.75±0.01	0.86±0.01
breast-cancer-diagnostic	92.8±4(1)	92.5±3.8(1)	0.91±0.004	0.91±0.005
breast-cancer-original	94.5±4.5(1)	94.6±3.6(1)	0.97±0.004	0.96±0.004
breast-cancer-prognostic	63.3±13(1)	63.4±13(1)	0.78±0.03	0.78±0.03
breast-tissue	58.7±14(1)	59.5±14(1)	0.73±0.02	0.73±0.02
car-evaluation	87.4±2.2(1)	78.4±2.3(2)	0.52±0.01	0.89±0.005
cardiotocography-1	68.6±3.5(1)	68.5±3.7(1)	0.71±0.004	0.71±0.004
cardiotocography-2	85.9±2.1(1)	86±2.3(1)	0.9±0.003	0.89±0.004
chess-rook-vs-pawn	85.4±1.8(1)	85.5±2(1)	0.8±0.004	0.79±0.004
cmc	42.6±3.8(1)	42.8±3.7(1)	0.61±0.007	0.58±0.008
congressional-voting	89.1±6.5(1)	89.4±6.2(1)	0.88±0.01	0.88±0.01
connectionist-bench-sonar	78.3±9.2(1)	78.2±9.3(1)	0.7±0.01	0.71±0.01
connectionist-bench-vowel	72.1±5.8(1)	72.2±5.7(1)	0.78±0.005	0.78±0.005
cylinder-bands	65.9±8.8(1)	66.3±8.7(1)	0.69±0.01	0.7±0.01
dermatology	87.7±5(1)	87.9±5.4(1)	0.8±0.009	0.81±0.009
ecoli	77.1±6.9(1)	77.5±7.2(1)	0.85±0.01	0.85±0.01
glass	62.5±9.4(1)	62.3±9.2(1)	0.72±0.01	0.73±0.01
habermans-survival	66.6±7.8(1)	67.1±8.1(1)	0.77±0.01	0.78±0.01
hepatitis	80.9±11(1)	81.1±12(1)	0.84±0.03	0.84±0.03
ionosphere	82±6.4(1)	81.8±6.5(1)	0.82±0.01	0.81±0.01
iris	89.7±8.2(1)	88.4±9.2(1)	0.88±0.009	0.89±0.01
libras-movement	68.7±6.7(1)	68.5±7.4(1)	0.68±0.01	0.69±0.01
liver-disorders	60.7±9(1)	59.5±9.3(2)	0.61±0.02	0.62±0.01
lymph	74.4±12(1)	73.4±11(1)	0.74±0.02	0.79±0.02
monks-problems-1	91.8±7(2)	93.3±6.1(1)	0.8±0.001	0.8±0.002
monks-problems-2	54.7±6.4(1)	55.3±5.4(1)	0.64±0.01	0.64±0.02
monks-problems-3	90.6±4.7(1)	91±3.9(1)	0.8±0.003	0.81±0.004
parkinsons	84.7±8.6(1)	84.1±8.9(1)	0.84±0.01	0.84±0.01
pima-indians-diabetes	69.1±5(1)	68.9±5.3(1)	0.75±0.008	0.75±0.008
sonar	78.3±9.2(1)	78.2±9.3(1)	0.7±0.01	0.71±0.01
spambase	85.4±1.9(1)	85.4±1.9(1)	0.88±0.003	0.87±0.005
spect-heart	74±9(1)	72.2±9.4(1)	0.79±0.01	0.91±0.01
spectf-heart	69.6±9(2)	70.7±8.5(1)	0.79±0.02	0.79±0.02
statlog-australian-credit	72.8±6(1)	73.4±5.7(1)	0.77±0.01	0.78±0.01
statlog-german-credit	67±4.6(1)	67±4.4(1)	0.7±0.008	0.71±0.008
statlog-heart	76.1±7.4(1)	75.3±8(1)	0.75±0.01	0.75±0.01
statlog-vehicle	67.2±4.8(1)	66.8±4.8(1)	0.7±0.008	0.71±0.007
teaching-assistant	48.3±14(1)	48.9±13(1)	0.59±0.02	0.59±0.02
thyroid-disease	91±1.4(2)	91.6±1(1)	0.97±0.001	0.96±0.003
vote	91.2±5.6(1)	89.6±6.5(2)	0.86±0.02	0.92±0.01
wine	92.5±6.2(1)	92.4±6.5(1)	0.83±0.01	0.84±0.01
zoo	78.1±10(1)	78.1±11(1)	0.89±0.006	0.89±0.006
Mean	74.8±7.2	74.5±7.2	0.77±0.01	0.78±0.01
Mean Rank	1.07±0.038	1.07±0.038		

As can be seen from the results presented in the table, the differences between the classification quality of the regular version of the LSBo algorithm and the fast version are practically negligible for all data sets. The same is true for the reduction power of the training set—both algorithms work very similarly in this aspect. The goal of creating a fast version of the LSBo algorithm has been achieved.

3.1 Conclusions

The research goal was to create an algorithm that would have lower complexity than the local set border selector algorithm and that would be as good as possible in terms of classification quality on benchmark tests. Thanks to the careful selection of new data structures and appropriate reformulation of the original algorithm, it was possible to create a new algorithm that is just as good but with complexity $O(m \log m)$ instead of $O(m^2)$. Thanks to this, it will be possible to use this algorithm for much more significant problems.

References

1. Brighton, H., Mellish, C.: Advances in instance selection for instance-based learning algorithms. Data Min. Knowl. Disc. **6**(2), 153–172 (2002)
2. Cover, T.M., Hart, P.E.: Nearest neighbor pattern classification. Inst. Electr. Electron. Eng. Trans. Inf. Theory **13**(1), 21–27 (1967)
3. Garcia, S., Derrac, J., Cano, J., Herrera, F.: Prototype selection for nearest neighbor classification: taxonomy and empirical study. IEEE Trans. Pattern Anal. Mach. Intell. **34**(3), 417–435 (2012)
4. Jankowski, N., Grochowski, M.: Comparison of instances selection algorithms: II. Algorithms survey. In: Rutkowski, L., Siekmann, J.H., Tadeusiewicz, R., Zadeh, L.A. (eds.) Artificial Intelligence and Soft Computing, Lecture Notes in Computer Science, vol. 3070, pp. 598–603. Springer, Poland, Zakopane (2004). http://www.is.umk.pl/~norbert/publications/04-zakopane-NJMG.pdf
5. Jankowski, N.: A fast and efficient algorithm for filtering the training dataset. In: Neural Information Processing, vol. 13623, pp. 504–512. Springer, Cham (2022). https://doi.org/10.1007/978-3-031-30105-6_42
6. Leyva, E., González, A., Pérez, R.: Three new instance selection methods based on local sets: a comparative study with several approaches from a bi-objective perspective. Pattern Recogn. **48**(4), 1523–1537 (2015). https://doi.org/10.1016/j.patcog.2014.10.001
7. Malkov, Y.A., Yashunin, D.A.: Efficient and robust approximate nearest neighbor search using hierarchical navigable small world graphs. IEEE Trans. Pattern Anal. Mach. Intell. **42**(4), 824–836 (2020). https://doi.org/10.1109/TPAMI.2018.2889473
8. Merz, C.J., Murphy, P.M.: UCI repository of machine learning databases (1998). http://www.ics.uci.edu/~mlearn/MLRepository.html
9. Wilson, D.R., Martinez, T.R.: Reduction techniques for instance-based learning algorithms. Mach. Learn. **38**(3), 257–286 (2000)
10. Wilson, D.: Asymptotic properties of nearest neighbor rules using edited data. IEEE Trans. Syst. Man Cybern. **2**(3), 408–421 (1972)

Modelling the Transient Evolution of Queues in Plugged-in Electric Vehicles (PEV) Fast Charging Stations

Godlove Suila Kuaban[1](✉)[iD], Tomasz Nycz[2][iD], Monika Nycz[2][iD], Tadeusz Czachórski[1][iD], and Piotr Czekalski[2][iD]

[1] Institute of Theoretical and Applied Informatics, Polish Academy of Sciences,
Baltycka 5, 44-100 Gliwice, Poland
{gskuaban,tadek}@iitis.pl

[2] Silesian University of Technology, Akademicka 2A, 44-100 Gliwice, Poland
{monika.nycz,tomasz.nycz,piotr.czekalski}@polsl.pl

Abstract. The transportation sector is responsible for approximately 23% of global greenhouse gas (GHG) emissions, with road transportation contributing nearly 70% of these emissions. The widespread adoption of electric vehicles (EVs) is transforming this sector by reducing emissions and decreasing reliance on fossil fuels. However, the growing number of EVs presents significant challenges for charging infrastructure, particularly in managing long queues, extended wait times, and limited station capacity. Most existing studies on the performance of electric vehicle charging stations assume Poisson arrivals and exponential charging times, simplifications that often overlook real-world variability. This paper introduces a generalized queueing model that leverages empirical interarrival and charging duration data for more accurate performance evaluation. A transient analysis is conducted, examining two operational optimization strategies aimed at minimizing queue sizes during peak demand: (1) a queue management policy that encourages charging only up to a predefined state-of-charge (SoC) threshold instead of the typical 80–100%, and (2) dynamic control of the number of active charging ports based on demand. The results show that these operational optimization policies improve the efficiency of the charging station and significantly improve the customer experience.

Keywords: Plugged-in Electric Vehicles (PEV) · Fast Charging Stations · diffusion approximation models · Transient analysis · Performance Evaluations

1 Introduction

The transportation sector contributes approximately 23% of global greenhouse gas (GHG) emissions, with road transportation accounting for nearly 70% of these emissions [1]. The rapid adoption of electric vehicles (EVs) is driven by the urgent need to reduce the carbon footprint of transportation, a crucial step in mitigating pollution and other environmental challenges associated with fossil fuel consumption. As a result, plug-in electric vehicles (PEVs) are becoming an integral part of the global transportation system, with their market penetration expected to increase significantly in the coming years.

Municipalities and businesses are accelerating the shift to electric vehicles (EVs) by integrating them into fleets and offering adoption incentives. Interest among individual car owners is also rising globally, driven by the significant benefits of EVs. However, this growing demand requires a corresponding expansion of charging infrastructure and power distribution networks to ensure reliable and efficient EV operation.

Despite their environmental and economic advantages, EVs face challenges such as limited driving range and long charging durations, which can deter widespread adoption [2]. Extended charging times often lead to congestion at charging stations, particularly during peak hours, resulting in delays and reduced user satisfaction [3]. Addressing these challenges requires efficient queue management strategies that account for the stochastic nature of EV arrival patterns and charging durations.

Most existing studies on EV charging station performance assume that EV arrivals follow a Poisson process and that charging times are exponentially distributed. Consequently, the widely used queueing model for charging station analysis is the M/M/c model [3]. However, these Markovian assumptions do not always hold in real-world scenarios, necessitating more generalized models to capture system dynamics accurately. While some studies have incorporated time-varying interarrival rates [4], they remain constrained by Poisson arrival assumptions and exponential service times. An attempt to model a charging station considering non-stationary arrival and charging rates was discussed in [5].

This paper presents a diffusion-based G/G/c/N queueing framework to analyze the performance of EV charging stations under realistic, time-varying arrival rates. Using real-world data, the model captures dynamic fluctuations in customer arrivals and service times, allowing the evaluation of two operational strategies: dynamically limiting the final state of charge (SoC_f) and adjusting the number of active charging ports. Simulation results demonstrate that these strategies can effectively reduce queue lengths during peak hours, offering practical insights for optimising charging infrastructure operations.

2 Time-Dependent Queueing Model for the Fast Charging Station

We model the fast-charging station as a $G/G/c/N$ queue, where $N = c + K$. The notation $G/G/c/N$ indicates that both interarrival and service times follow general distributions. To capture the dynamic evolution of the number of electric vehicles (EVs) in the charging station, we approximate the system using a diffusion process. Specifically, we define a continuous process $\{X(t)\}$, where the probability of having x EVs in the station at time t is given by $\text{Prob}\{X(t) = x\}$ for $x \in [0, N]$. These types of models were started in computer science by a single server G/G/1/N model [6]. The approach to solve diffusion equations in the case of transient states was proposed for the G/G/1/N model in [7]. We use this approach in many applications, e.g. modelling transient behaviour of SDN networks [8] and energy storage systems [9].

The following partial differential equation describes the diffusion process governing the number of EVs in the charging station, e.g. [10]:

$$\frac{\partial f(x,t;x_0)}{\partial t} = \frac{\alpha}{2}\frac{\partial^2 f(x,t;x_0)}{\partial x^2} - \beta\frac{\partial f(x,t;x_0)}{\partial x}, \qquad (1)$$

where βdt and αdt represent the mean and variance of the changes in the diffusion process over an infinitesimal time interval dt. The probability density function (PDF) of the number of EVs in the charging station is given by:

$$f(x,t;x_0) = P[x \leq X(t) < x + dx \mid X(0) = x_0], \qquad (2)$$

where x_0 is the initial state of the process at $t = 0$.

The mean change in the number of EVs in the charging station, β, and its variance, α, depend on the mean arrival rate λ, the squared coefficient of variation of the interarrival time C_A^2, the mean charging rate per port μ, and the squared coefficient of variation of the charging time C_B^2. They also depend on the number of currently occupied charging ports. That means that diffusion parameters depend on the process value. To represent it in an easy form, the diffusion interval $[0, N]$ is partitioned into c sub-intervals: $[0, 1], [1, 2], \ldots, [c-1, N]$. Each sub-interval corresponds to a specific number of occupied charging ports at a given time, from one to c. In the last sub-interval, $[c-1, N]$, all c charging ports are occupied.

The diffusion parameters for each sub-interval are chosen as

$$\begin{aligned}\alpha_i &= \lambda C_A^2 + i\mu C_B^2, \quad \beta_i = \lambda - i\mu \quad \text{for } i-1 < x < i, \quad i = 1, 2, \ldots, c-1\\ \alpha_c &= \lambda C_A^2 + c\mu C_B^2, \quad \beta_c = \lambda - c\mu \quad \text{for } c-1 < x < N\end{aligned} \qquad (3)$$

The state of the diffusion process, x, evolves as EVs arrive at the charging station, begin charging when a port becomes available, and depart upon completing their charge. Probability mass shifts between neighboring sub-intervals whenever the charging ports are not fully occupied and EVs enter or exit the system.

The probability density function (PDF) of the number of EVs in the station is constructed from the diffusion process PDFs across all sub-intervals. These are obtained by solving the diffusion equations for each sub-interval, incorporating the probability flows between adjacent regions as described in [11].

The application used to solve the system is implemented in Python and executed on an HP ProLiant DL580 G7 server, a four-node cc-NUMA system with four Intel Xeon E7-4870 2.4 GHz CPUs (80 logical processors) and 512 GB of DDR3 REG RAM across 32 modules. The system runs SUSE Linux Enterprise Server 15 SP4 with kernel version 5.14.21-150400.24.41.

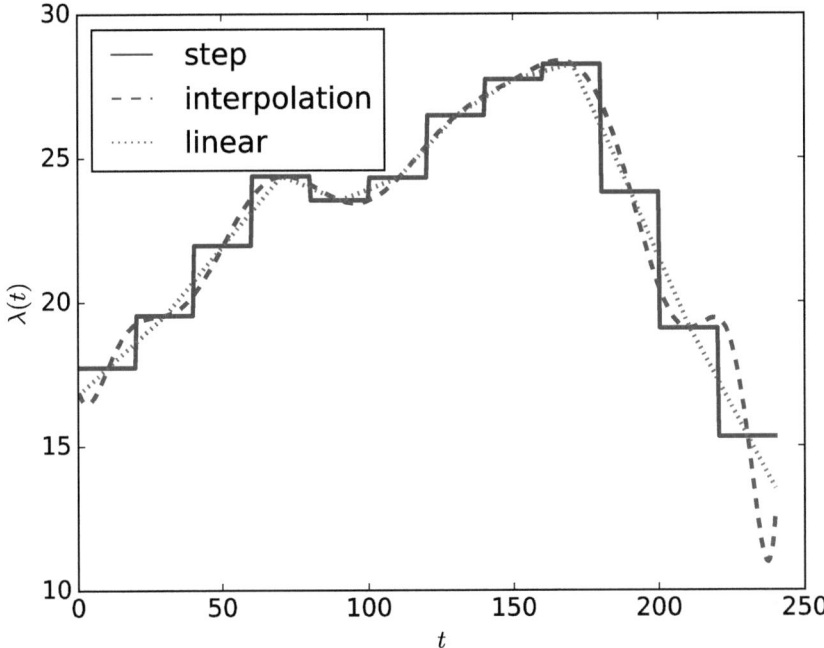

Fig. 1. Dynamic evolution of the mean arrival rate of EVs to the charging station, $\lambda(t)$ from 9:00 to 20:00.

3 Performance Evaluation and Operational Optimisation of the Fast Charging Station

We evaluate the impact of charging station management policies on the performance of the charging station. The performance metrics considered in this study is the mean number of EVs present at the charging at time t denoted as $E[N(t)]$. It is derived using the probability density function of the diffusion process as determined in the previous section

$$E[N(t)] = \int_0^N x f(x,t;\psi) dx. \tag{4}$$

The EV charging time depends on battery capacity, initial and target states of charge (SoC_i and SoC_f), and charging power, P_c (in KW). It is given by:

$$T_c = (SoC_f - SoC_i)\frac{B}{P_c} \tag{5}$$

where $B = Q \cdot V$ is the battery's energy capacity, with Q as nominal capacity (Ah) and V as voltage (V).

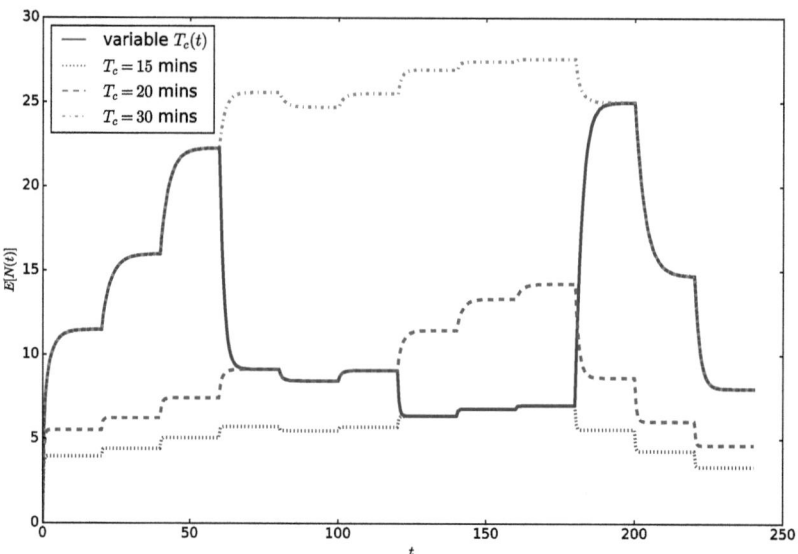

Fig. 2. The influence of dynamic changes of mean charging times $T_c(t)$ (by adjusting the final state of charge at time t denoted as $SoC_f(t)$) on the transient evolution of the mean number of EVs, $E[N(t)]$ from 9:00 to 20:00.

The performance of charging stations mainly depends on the customer arrival rate, λ, and the charging duration, $T_c = 1/\mu$. Arrival rate values used in the simulations are taken from [12]. Figure 1 shows the interpolated hourly mean arrival rate, $\lambda(t)$, from 9:00 to 20:00, based on Tables 1 and 2. All figures use a 6-minute time unit, covering 11 h (220 intervals).

To remain consistent with the dataset in [12], we assume a Poisson arrival process, although the diffusion approximation supports general arrival patterns. Accordingly, the squared coefficients of variation are set to $C_A^2 = C_B^2 = 1$. The system is configured with a maximum parking capacity of $N = 30$, a waiting area for up to $K = 20$ vehicles, and each EV has a battery capacity of $B = 50$ kWh. The charging power is set to $P_c = 50$ kW, identical across all ports, implying uniform charging durations T_c or equivalently, a consistent charging rate μ across all ports.

To evaluate the impact of charging duration and queue management strategies, we vary the final state of charge (SoC_f) of arriving vehicles. Each EV is assumed to arrive with an initial state of charge $SoC_i = 20\%$, and all vehicles have a battery capacity of $B = 50$ kWh. Adjusting SoC_f directly affects the charging duration. For example, charging up to $SoC_f = 70\%$ yields a charging time $T_c = 30$ min, based on Eq. 5.

Table 1. Data for Scenario 1 ($c = 10$, SoC_f in %, and $T_c(t)$ in mins)

t	9:00	10:00	11:00	12:00	13:00	14:00	15:00	16:00	17:00	18:00	19:00	20:00
$\lambda(t)$	17.74	19.55	21.98	24.38	23.54	24.32	26.49	27.73	28.26	23.81	19.09	15.32
SoC_f	70	70	70	53	53	53	45	45	45	70	70	70
$T_c(t)$	30.00	30.00	30.00	20.00	20.00	20.00	15.00	15.00	15.00	30.00	30.00	30.00
$\mu(t)$	2.00	2.00	2.00	3.00	3.00	3.00	4.00	4.00	4.00	2.00	2.00	2.00

Table 2. Data for Scenario 2 ($T_c = 30.00$ min or $\mu = 2.00$)

t	9:00	10:00	11:00	12:00	13:00	14:00	15:00	16:00	17:00	18:00	19:00	20:00
$\lambda(t)$	17.74	19.55	21.98	24.38	23.54	24.32	26.49	27.73	28.26	23.81	19.09	15.32
$c(t)$	10	10	10	13	13	13	15	15	15	10	10	10

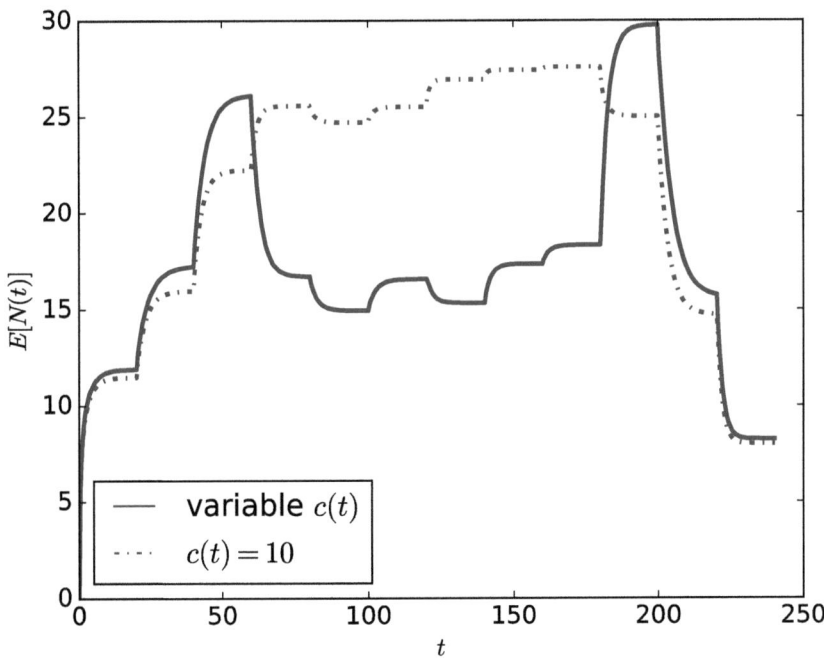

Fig. 3. The influence of dynamic changes of number of charging ports at time t denoted as $c(t)$ on the transient evolution of the mean number of EVs, $E[N(t)]$ from 9:00 to 20:00.

Figures 2 and 3 illustrate two operational optimisation strategies and their effects on the transient queue size at the charging station.

In the first scenario (Fig. 2), T_c or the service rate μ is reduced during peak hours by capping SoC_f, as shown in Table 1. Lowering SoC_f shortens T_c, reducing congestion. We also compare limiting SoC_f only during peak hours with applying it throughout the day. For instance,

$$SoC_f = \{45, 53, 70\}, \quad T_c = \{15, 20, 25, 30\} \text{ min},$$

show how more aggressive restrictions (e.g., $SoC_f = 45\%$, $T_c = 15$ min) significantly reduce queue size but may lower user satisfaction, as most drivers prefer to charge up to at least 80%. Hence, increasing charging power or the number of ports may be more acceptable alternatives.

In the second scenario (Fig. 3), the number of active charging ports is increased during peak demand (12:00–17:00). Once demand drops and grid load increases after 17:00, ports are scaled down. This strategy enhances queue management and aligns with grid efficiency goals.

4 Conclusion

The increasing adoption of electric vehicles (EVs) necessitates the development of efficient charging infrastructure to accommodate rising demand while minimizing congestion and service delays. In this study, we proposed a G/G/c/N diffusion-based queueing model to evaluate the performance of EV charging stations under time-varying arrival rates. Unlike traditional M/M/c models, our approach accounts for general arrival and service time distributions, providing a more realistic representation of charging station dynamics.

Our analysis revealed that queue lengths and waiting times fluctuate based on EV arrival patterns, with peak congestion occurring during high-demand hours. By examining the effects of different charging durations, we demonstrated that imposing a final state of charge (SoC_f) threshold significantly improves station efficiency. Specifically, limiting SoC_f reduces charging times, increases throughput, and lowers the number of lost customers who are unable to charge due to full station capacity. These findings highlight the critical role of queue management strategies in optimizing charging station performance. Future research can extend this work by incorporating more complex traffic patterns, multiple classes of customers and charging ports, and dynamic pricing strategies to further improve the efficiency of EV charging infrastructure.

Funding Acknowledgement. This paper was supported by the ReACTIVE Too project, which has received funding from the European Union's Horizon 2020 Research, Innovation, and Staff Exchange Programme under the Marie Skłodowska-Curie Action (Grant Agreement No. 871163). Scientific work is published as part of an international project co-financed by the program of the Minister of Science and Higher Education entitled "PMW" in the years 2021–2025, contract no. 5169/H2020/2020/2. Additionally, this scientific work is also published as part of an international project, ReACTIVE Too, co-financed by the program of the Minister of Science and Higher Education entitled "PMW" in the years 2024–2025, contract no. 5872/H2020/2024/2.

References

1. Bashmakov, I.A., et al.: Climate change 2022: mitigation of climate change. In: Contribution of Working Group iii to the Sixth Assessment Report of the Intergovernmental Panel on Climate Change, vol. 11 (2022)
2. Zhang, L., Shaffer, B., Brown, T., Samuelsen, G.S.: The optimization of dc fast charging deployment in california. Appl. Energy **157**, 111–122 (2015)
3. Zenginis, I., Vardakas, J.S., Zorba, N., Verikoukis, C.V.: Analysis and quality of service evaluation of a fast charging station for electric vehicles. Energy **112**, 669–678 (2016)
4. Tang, Y., Chau, K., Liu, W.: Charging station placement optimization using queueing model with time-varying arrival rate. In: Proceedings of the 36th International Electric Vehicle Symposium and Exhibition (EVS36), Sacramento, California, USA, June, pp. 11–14 (2023)
5. MacDonald, C.D., Kattan, L., Layzell, D.: Modelling electric vehicle charging network capacity and performance during short-notice evacuations. Int. J. Disast. Risk Reduct. **56**, 102093 (2021)
6. Gelenbe, E.: On approximate computer system models. J. ACM (JACM) **22**(2), 261–269 (1975)
7. Czachórski, T.: A method to solve diffusion equation with instantaneous return processes acting as boundary conditions. Bull. Polish Acad. Sci. Techn. Sci. **41**(4), 417–451 (1993)
8. Czachórski, T., Gelenbe, E., Kuaban, G.S., Marek, D.: Time-dependent performance of a multi-hop software defined network. Appl. Sci. **11**(2469), 1–21 (2021)
9. Kuaban, G.S., Gelenbe, E., Czachórski, T., Czekalski, P., Nkemeni, V.: Energy performance of off-grid green cellular base stations. Perf. Eval., 102426 (2024)
10. Cox, R.P., Miller, H.D.: The Theory of Stochastic Processes. Chapman and Hall, London (1965)
11. Czachórski, T., Kuaban, G.S., Nycz, T.: Multichannel diffusion approximation models for the evaluation of multichannel communication networks. In: Vishnevskiy, V.M., Samouylov, K.E., Kozyrev, D.V. (eds.) DCCN 2019. LNCS, vol. 11965, pp. 43–57. Springer, Cham (2019). https://doi.org/10.1007/978-3-030-36614-8_4
12. Zhu, L., Pu, Y.: Analysis on the operation of a charging station with battery energy storage system. J. Electr. Eng. Technol. **12**(5), 1916–1924 (2017)

Is Heterogeneous Model Soup Tasty? A Multidimensional Evaluation of Diverse Model Soups in Language Model Alignment

Dawid Motyka[1]( ), Paweł Walkowiak[1], Julia Moska[1], Bartosz Żuk[2], Karolina Seweryn[3], and Arkadiusz Janz[1]

[1] Wrocław University of Science and Technology, Wrocław, Poland
{dawid.motyka,arkadiusz.janz}@pwr.edu.pl
[2] Institute of Computer Science, Polish Academy of Sciences, Warsaw, Poland
[3] NASK - National Research Institute, Warsaw, Poland

Abstract. Training and fine-tuning language models is becoming increasingly expensive. "Model soups" offer a promising solution by combining parameters from separately trained models to create a new one with merged capabilities. Our paper explores using heterogeneous model soups to improve LLM alignment by combining models trained with different alignment methods - a novel approach not previously explored in literature. Through empirical evaluation using an "LLM-as-a-judge" approach, we found that mixing different types of models can improve alignment performance, though this requires careful adaptation of interpolation techniques to account for varying alignment objectives. We've shared our model merging source code on GitHub (https://github.com/dawidm/iccs-2025-model-soups).

Keywords: Model Soups · Alignment · Language Models

1 Introduction

In recent years, the dynamic development of language models has significantly improved task-specific performance across a wide range of natural language processing tasks. However, the increasing costs of training and fine-tuning large language models for specific applications pose a significant computational challenge for researchers and practitioners. One promising solution to this problem is a model merging technique known as model soups. Model soups combine the parameters of multiple independently trained models to create a new model with merged capabilities and characteristics. This approach offers several advantages, including reduced computational costs, improved adaptation stability, and the ability to personalize models without the need for retraining. As individual models can often be limited by local minima encountered during optimization, combining multiple models should average their capabilities and mitigate these limitations. This provides a compelling reason for training models multiple times under varying conditions.

Model soups are usually considered in the context of homogeneous models trained with equivalent fine-tuning methods. In contrast, our study explores *heterogeneous* model soups created using diverse alignment methods. We empirically evaluate resultant models formed via DPO, KTO, and ORPO methods, using both linear (LERP) and spherical interpolation (SLERP). Our evaluation focuses on key alignment goals, safety, factual accuracy, linguistic correctness, conciseness, and proactivity. To assess models performance, we established an evaluation framework grounded in a strong language model serving as a judge.

1. We examine the compatibility of DPO-, KTO-, and ORPO-aligned models mixed via LERP and SLERP, assessing their performance across key alignment targets. Heterogeneous mixtures remain underexplored in prior work.
2. Our evaluation utilises a multidimensional framework focused on core alignment goals – safety, factuality, ling. correctness, conciseness, and proactivity.
3. We show, that model mixing techniques such as SLERP, require careful adaptation when combining aligned models.

2 Related Work

Averaging of models' parameters, also called model soups [20], is a widely used approach that showed a range of applications. [7] showed that simply averaging model parameters across its learning trajectory leads to better generalisation. The important concept of *linear mode connectivity* was introduced by [4] who demonstrated that, given shared initialization, two networks tend to converge to the points connected by a line with a relatively flat error rate. [13] showed that the fine-tuned models are similar in the feature and parameter space. [19] and [17] used averaging for an improvement in the out-of-distribution performance. The concept of linear mode connectivity also extends to multiple tasks [12]. [6] demonstrated that *task vectors* can be merged to build a multitask model. [9] also explored the merging of *expert* language models to achieve compositional capabilities.

The less explored field is merging models with different training objectives, which was also shown to be promising. [6] explored merging *task vectors* from both supervised and unsupervised objectives, and [2] used model soups with various loss functions to improve model adversarial robustness.

The concept of model soups has also been employed in the context of the alignment task. [16] investigated RL fine-tuning with interpolation of diverse reward models, while [8] demonstrates the potential to achieve personalized alignment through weighted interpolation of diverse models.

3 Methods

To align the models with human preference data, we used offline alignment, as it allows for direct preference optimization without an explicit reward function, requiring fewer computational resources compared to online methods.

Direct Preference Optimization (DPO) [15] learns implicit reward function by increasing the relative log probability of preferred to non-preferred response, with KL-divergence penalty regularization (reference model).

Odds Ratio Preference Optimization (ORPO) [5] contrastively to DPO, does not incorporate a reference model. It combines the odds ratio between the chosen and rejected responses and a supervised fine-tuning component in its optimization objective.

Kahneman-Tversky Optimization (KTO) [3] – authors proposed an alternative training objective based on Kahneman-Tversky model of human utility, with no need for paired preference data, showing that it exceeds performance of DPO and ORPO in many cases.

In our study, we conducted model alignment using Huggingface TRL[1] implementation of described methods. We used `PLLuM-12B-instruct`[2] base model (further called SFT model) – a 12B parameter model for Polish language, based on `Mistral-NeMo`[3]. All models were trained with AdamW optimizer, learning rate of 2e-6, weight decay of 1e-3 and effective batch size of 64. Regarding method-specific parameters, we used $\beta = 0.1$ for DPO and KTO. $\lambda = 0.2$ for ORPO and $\lambda_D = \lambda_U = 1$ for KTO.

3.1 Model Merging

We utilize two popular model merging techniques–Linear Interpolation (LERP) and Spherical Linear Interpolation (SLERP) to combine models trained with diverse alignment methods. We call both approaches model soups. Merges are conducted in whole model parameter space: MLP, attention layers' transformations, and RMS normalizations (RMSNorm) parameters.

Linear Interpolation (LERP) is a method where the weights of two models are combined linearly, specifically by using linear interpolation between weights of 2 models: $\theta_{LERP} = \lambda \cdot \theta_{m_1} + (1-\lambda) \cdot \theta_{m_2}$, where $\theta_{m_1}, \theta_{m_2}$ are parameters vectors of aligned models and λ is the weighting parameter.

Spherical Linear Interpolation (SLERP) is an alternative interpolation method, which preserves norms of parameters vectors [18]. As the parameters of the aligned models are close to the SFT model, we operate on *alignment vectors*: $\theta_{d_i} = \theta_{m_i} - \theta_{SFT}$. Specifically we calculate:

$$\theta_{SLERP} = \theta_{SFT} + \frac{\sin[(1-\lambda)\Omega]}{\sin \Omega} \cdot \theta_{d_1} + \frac{\sin[\lambda\Omega]}{\sin \Omega} \cdot \theta_{d_2}$$

for corresponding parameter matrices with λ controlling influence of source models. To calculate the angle Ω, two-dimensional matrices are reshaped as vectors.

---

[1] https://github.com/huggingface/trl/tree/v0.13.0.
[2] https://huggingface.co/CYFRAGOVPL/PLLuM-12B-instruct.
[3] https://mistral.ai/news/mistral-nemo.

3.2 Model Evaluation

A high-quality evaluation of aligned large language models is both challenging and time-consuming. We use the "LLM-as-a-Judge" approach [21], which can be used to approximate the evaluation of human preferences in a cheaper and faster way. We choose *pairwise comparison* of the human-written gold answer and evaluate the model response. For the judge model, we use stronger LLM – (Llama3.1-70B[4]).

We employ the **win-tie-rate (WTR)** metric – the percentage of test cases x in which the response z_t from the evaluated model t is either superior to or on par with the corresponding gold-standard answer z_g, with respect to predefined evaluation criteria. The WTR score for a given response evaluation function Q is calculated as follows: $WTR(T, G) = E_x[\mathbb{1}_{Q(z_t|x) >= Q(z_g|x)}]$, where z_t is the response generated by the evaluated model t, T is the set of model responses $z_t \in T$, G is a set of corresponding gold-standard responses $z_g \in G$.

Evaluation Criteria. We selected seven evaluation dimensions that represent typical alignment objectives: safety, factuality, linguistic correctness, conciseness, proactivity, false rejection rate (FAR) and false acceptance rate (FAR). In order to define evaluation function Q for each of them, we used specification of worse answer for every dimension. A precise description of the evaluation guidelines was given to the judge model, along with detailed specifications for each dimension, and a gold answer. We share the prompt with our source code.

3.3 Experimental Protocol

In the experimental part, we create merged models by conducting two model trainings and a single merge operation (we call it a single *run*).

The pipeline of model training and merging consists of model alignment training (ORPO, DPO, KTO), win-tie rate (WTR) measurement for each model, model merging using Linear Interpolation (LERP) and Spherical Linear Interpolation (SLERP), and win-tie rate measurement of the resulting merge.

1. **We investigated linear mode connectivity** between homogeneous LERP merges (ORPO-ORPO, DPO-DPO, KTO-KTO) and heterogeneous LERP merges (ORPO-DPO, ORPO-KTO, DPO-KTO) of two distinct models with $\lambda = 0.5$ (homogenous) and $\lambda \in \{0.25, 0.5, 0.75\}$ (heterogenous). For each model combination, we created 3 merges using diverse models (trained with different shuffling of the dataset). Each source and merged model was evaluated with our protocol (Sect. 3.2). For every evaluation dimension, besides win-tie-rate we used custom metric to assess the performance of merged model $(m_{1,2})$ against average result of source models (m_1 and m_2):

$$WTR_{\text{mean-diff}}(m_1, m_2) = WTR(m_{1,2}) - \frac{1}{2} \cdot (WTR(m_1) + WTR(m_2))$$

---

[4] https://huggingface.co/meta-llama/Llama-3.1-70B

2. SLERP merging led to non-functional models when applied to all parameters. **We investigated models' parameter vectors** to find possible causes and also to better understand the influence of alignment methods on groups of parameters. We calculated the L2 norms for the parameters of the aligned models and the angles Ω for heterogeneous pairs.
3. **We examined SLERP merges** that turned out to be functional after switching to LERP for RMSNorm parameter vectors. We followed the same protocol as for LERP (1.).

4 Datasets

The train dataset[5] for preference alignment consists of more than 20,000 manually annotated preference pairs, including both safety-related and neutral topics. Three distinct annotation methods were used: (1) **rating**, where each response was evaluated according to predefined metrics (informativeness, correctness, safety, fairness, conciseness, reasoning, helpfulness), (2) **ranking**, where responses were ordered according to their quality, and (3) **dialog**, where annotators took part in interactive conversations with models and selected the best responses.

The evaluation dataset contains 181 prompt-response pairs categorized as "safe" or "unsafe". Approximately half were sourced from alignment datasets (AlpacaEval [11], CREAK [14], ECQA [1], QED [10], Toxic DPO v0.2[6], Harmful Behaviors[7], Argilla[8]) and translated into Polish, covering commonsense, explanatory, and hazardous questions. The rest includes human-annotated public affairs examples and auto-generated entries. We believe that this diverse collection of examples ensures comprehensive coverage of alignment scenarios.

5 Results and Discussion

As presented in Fig. 2 and Table 1 we can conclude that despite the dissimilarities of the alignment vectors' directions (measured by the angle between them, Fig. 1), heterogeneous LERP merging results in LLMs that perform well and not worse than homogeneous ones. The key observation is that linear interpolation in parameter space often results in close to linear interpolation in evaluation dimensions (ones that vary the most between alignment methods: conciseness, proactivity, and factuality). This makes LERP an effective technique to obtain a model balanced between the advantages of alignment techniques. Crucially, we did not observe a significant drop in safety metrics in any of the merged models.

---

[5] The dataset used in this study will be publicly released in a future publication.
[6] https://huggingface.co/datasets/unalignment/toxic-dpo-v0.2.
[7] https://huggingface.co/datasets/mlabonne/harmful_behaviors.
[8] https://huggingface.co/argilla.

Table 1. Results of equally-weighted homogenous and heterogenous of **LERP** and **SLERP** soups across evaluation dimensions. Values are $WTR_{mean-diff}$, averaged from 3 runs. LQ – linguistic correctness, Proact. – proactivity.

			Safety	Factuality	LQ	Conciseness	Proact.	FRR	FAR	Avg.
LERP	heterogenous	ORPO-DPO	-0.015	-0.029	**-0.013**	-0.114	-0.198	**-0.003**	-0.025	-0.057
		ORPO-KTO	-0.006	**-0.013**	-0.019	**-0.017**	0.068	-0.006	0.000	**0.001**
		DPO-KTO	**-0.004**	-0.029	-0.022	-0.063	-0.222	**-0.003**	**0.006**	-0.048
		Avg.	-0.008	-0.024	-0.018	-0.065	-0.117	-0.004	-0.006	
	homogenous	ORPO-ORPO	-0.002	**0.031**	-0.007	0.009	0.043	0.005	**0.000**	**0.011**
		DPO-DPO	**0.000**	-0.016	**0.000**	**0.015**	-0.080	**0.008**	-0.006	-0.011
		KTO-KTO	-0.004	-0.019	**0.002**	0.001	**-0.013**	-0.003	-0.006	-0.006
		Avg.	-0.002	-0.001	-0.002	0.008	-0.017	0.003	-0.004	
SLERP	heterogenous	ORPO-DPO	**0.005**	**0.017**	0.004	0.016	0.040	-0.001	**0.019**	0.014
		ORPO-KTO	-0.005	0.016	0.002	-0.031	**0.126**	**0.002**	0.003	**0.016**
		DPO-KTO	0.002	-0.025	**0.005**	**0.085**	-0.117	-0.007	0.003	-0.008
		Avg.	0.001	0.003	0.004	0.023	0.016	-0.002	0.008	
	homogenous	ORPO-ORPO	**0.002**	0.005	-0.009	-0.035	**0.080**	-0.005	**0.019**	0.008
		DPO-DPO	-0.002	0.003	-0.017	0.007	-0.093	-0.005	-0.006	-0.016
		KTO-KTO	-0.002	**0.029**	**-0.004**	**0.014**	0.018	**0.002**	-0.006	0.007
		Avg.	-0.001	0.012	-0.010	-0.004	0.002	-0.003	0.002	

Regarding analysis of models' parameters (Figs. 1), we can observe slight variation between model pairs, with DPO and KTO being the most similar, and lower similarities (such as ORPO and DPO) did not make the models incompatible for merging. Also exceptionally high norm of ORPO language modeling head (LM head) parameter vector did not seem to interfere, although we consider it interesting observation that may be attributed to using prompts from the dataset as additional learning signal (which is unique for ORPO).

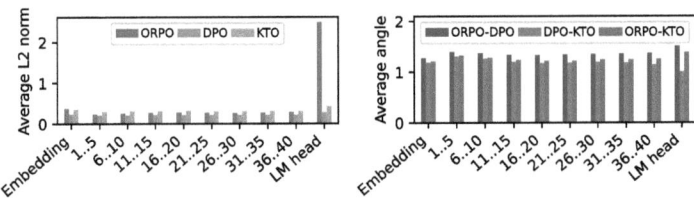

Fig. 1. Mean L2 norms (for aligned models) and angles (between models) for alignment vectors (except RMSNorm) by groups (X-axis, numbers represent transformer layers).

We observed that normalization layer parameters usually remain unchanged during alignment, except for the first transformer layer, where only ~70% of weights match between aligned models (versus nearly 100% in other layers). The SLERP merges were functional only when RMS normalization parameters were merged with LERP, demonstrating that they may need special caution in model merging. The possible reason for this is that the outputs of normalization layers directly affect the residual stream of models as opposed to the outputs of attention or MLP that are always followed by normalization.

We recorded a notable effect of SLERP merges on the evaluation of win-tie rates, specifically the dimensions of factuality, conciseness, and proactivity, while noting no substantial effect on safety metrics. As a result, heterogeneous ORPO-KTO but also homegeneous ORPO-ORPO merges turned out to be overall better compared to the best results of source models, with a main advantage on the proactivity dimension.

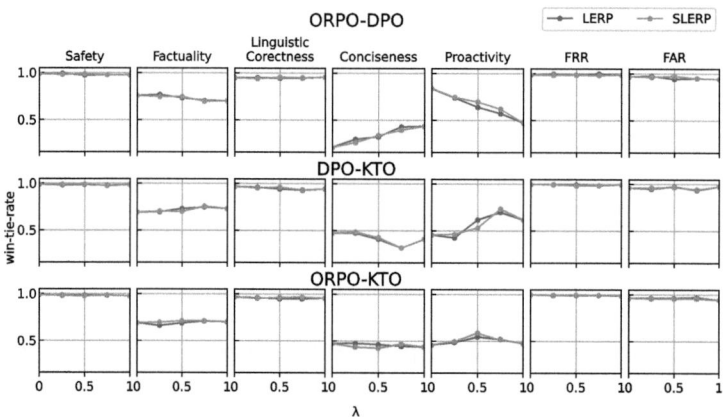

Fig. 2. Evaluation results for LERP and SLERP soups with respect to evaluation dimensions. λ controls influence from the first model, e.g. $\lambda = 0$ ORPO-DPO is a pure DPO model. Each data point is an average from 3 runs.

6 Conclusions

Our study showed that LERP and SLERP merging techniques that operate on whole model's parameter space are compatible between ORPO, DPO and KTO alignment methods utilizing various loss functions. Obtaining a balance between performance in various dimensions of large language model evaluation and models that are better on average was shown to be possible. Considering this and additional insights on alignment vectors' weights, we provided a foundation for further studies on merging aligned LLMs with more advanced techniques, focusing on dissimilarities in the alignment vectors and their connection to various dimensions of evaluation.

Acknowledgments. Financed by: (1) CLARIN ERIC (2024–2026), funded by the Polish Minister of Science (agreement no. 2024/WK/01); (2) CLARIN-PL, the European Regional Development Fund, FENG programme (FENG.02.04-IP.040004/24); (3)

statutory funds of the Department of Artificial Intelligence, Wroclaw Tech; (4) the EU project "DARIAH-PL", under investment A2.4.1 of the National Recovery and Resilience Plan.

Disclosure of Interests. The authors have no competing interests to declare that are relevant to the content of this article.

References

1. Aggarwal, S., Mandowara, D., Agrawal, V., Khandelwal, D., Singla, P., Garg, D.: Explanations for CommonsenseQA: new dataset and models. In: ACL. Association for Computational Linguistics (2021)
2. Croce, F., Rebuffi, S.A., Shelhamer, E., Gowal, S.: Seasoning model soups for robustness to adversarial and natural distribution shifts (2023)
3. Ethayarajh, K., Xu, W., Muennighoff, N., Jurafsky, D., Kiela, D.: KTO: model alignment as prospect theoretic optimization (2024)
4. Frankle, J., Dziugaite, G.K., Roy, D.M., Carbin, M.: Linear mode connectivity and the lottery ticket hypothesis (2019)
5. Hong, J., Lee, N., Thorne, J.: Orpo: monolithic preference optimization without reference model (2024)
6. Ilharco, G., Ribeiro, M.T., Wortsman, M., Gururangan, S., Schmidt, L., Hajishirzi, H., et al.: Editing models with task arithmetic (2022)
7. Izmailov, P., Podoprikhin, D., Garipov, T., Vetrov, D., Wilson, A.G.: Averaging weights leads to wider optima and better generalization (2018)
8. Jang, J., et al.: Personalized soups: personalized large language model alignment via post-hoc parameter merging (2023)
9. Jang, J., Kim, S., Ye, S., Kim, D., Logeswaran, L., Lee, M., et al.: Exploring the benefits of training expert language models over instruction tuning (2023)
10. Lamm, M., Palomaki, J., Alberti, C., Andor, D., Choi, E., Soares, L.B., et al.: Qed: a framework and dataset for explanations in question answering (2020)
11. Li, X., et al.: Alpacaeval: an automatic evaluator of instruction-following models (2023)
12. Mirzadeh, S.I., Farajtabar, M., Gorur, D., Pascanu, R., Ghasemzadeh, H.: Linear mode connectivity in multitask and continual learning (2020)
13. Neyshabur, B., Sedghi, H., Zhang, C.: What is being transferred in transfer learning? In: NeurIPS 2020 (2020)
14. Onoe, Y., Zhang, M.J., Choi, E., Durrett, G.: Creak: a dataset for commonsense reasoning over entity knowledge. OpenReview (2021)
15. Rafailov, R., Sharma, A., Mitchell, E., Ermon, S., Manning, C.D., Finn, C.: Direct preference optimization: your language model is secretly a reward model (2023)
16. Rame, A., Couairon, G., Dancette, C., Gaya, J.B., Shukor, M., Soulier, L., et al.: Rewarded soups: towards pareto-optimal alignment by interpolating weights fine-tuned on diverse rewards. In: NeurIPS (2023)
17. Ramé, A., Ahuja, K., Zhang, J., Cord, M., Bottou, L., Lopez-Paz, D.: Model ratatouille: recycling diverse models for out-of-distribution generalization (2022)
18. Ramé, A., Ferret, J., Vieillard, N., Dadashi, R., Hussenot, L., Cedoz, P.L., et al.: Warp: on the benefits of weight averaged rewarded policies (2024)
19. Ramé, A., Kirchmeyer, M., Rahier, T., Rakotomamonjy, A., Gallinari, P., Cord, M.: Diverse weight averaging for out-of-distribution generalization (2022)

20. Wortsman, M., Ilharco, G., Gadre, S.Y., Roelofs, R., Gontijo-Lopes, R., Morcos, A.S., et al.: Model soups: averaging weights of multiple fine-tuned models improves accuracy without increasing inference time (2022)
21. Zheng, L., Chiang, W., Sheng, Y., Zhuang, S., Wu, Z., Zhuang, Y., et al.: Judging llm-as-a-judge with mt-bench and chatbot arena. In: Oh, A., Naumann, T., Globerson, A., Saenko, K., Hardt, M., Levine, S. (eds.) NeurIPS (2023)

Performance Evaluation of IMS/NGN Network with SDN-Based Transport Stratum

Sylwester Kaczmarek⬤, Magdalena Młynarczuk⬤, and Maciej Sac[✉]⬤

Faculty of Electronics, Telecommunications and Informatics, Gdańsk University of Technology, Narutowicza 11/12, 80-233 Gdańsk, Poland
{kasyl,magdam,Maciej.Sac}@eti.pg.edu.pl

Abstract. The paper continues research on the use of the Software-Defined Networking (SDN) concept in the IP Multimedia Subsystem/Next Generation Network (IMS/NGN) transport stratum to increase flexibility of transport resource control and management mechanisms and make them independent of hardware solutions. The developed simulation model is used to examine the IMS/NGN network performance expressed by standardized Call Processing Performance (CPP) parameters such as mean Call Set-up Delay and mean Call Disengagement Delay. The research is carried out in a multidomain network, taking into account a wide range of service scenarios as well as parameters of network and generated traffic. As a result, the parameters, which are most important from the point of view of the performance are indicated, along with possible further steps to increase the performance.

Keywords: IMS · NGN · SDN · 5G · simulation model · call processing performance

1 Introduction

The IP Multimedia Subsystem (IMS) [1] architecture includes a set of functional units designed to control the process of providing multimedia services to users based on the IP protocol. It is a very important component of the 4G and 5G mobile networks [1]. It also forms the basis for the operation of the Next Generation Network (NGN) [2] service stratum, which is, therefore, also called the IMS/NGN network.

In the NGN architecture, apart from the service stratum (based on IMS), there is also a transport stratum that provides resources for the services requested by users. In the ITU-T standards for NGN [2], there are no assumptions regarding the technologies used in the transport stratum, the only requirement is support for transporting IP packets. Therefore, it is possible to use the Software-Defined Networking (SDN) [3] concept, which provides the ability to control and manage transport resources independently of the technology and hardware manufacturers.

However, the cooperation of the SDN concept with the IMS/NGN network has not yet been standardized and its verification is required. Performed related work review [4] indicated that there are no analytical and simulation models for IMS/NGN network with

SDN-based transport stratum. Moreover, existing practical solutions (testbeds) do not take into account ITU-T standards for NGN networks (primarily in the field of resource control – RACF unit [2]) and do not have the source code available, which causes problems with their further use. In response to these drawbacks, the authors proposed their own testbed [5] which does not have the above-mentioned limitations. The experience gained from the implementation of this testbed (message exchange procedures, data regarding message lengths and processing times, etc.) was used to propose a simulation model and demonstrate that application of the SDN concept in the transport stratum of the IMS/NGN network is possible [4]. The aim of this paper is to use the simulator [4] to examine the performance of such a solution taking into account multiple network domains and a wide set of service scenarios as well as parameters of network and generated traffic. Mean Call Set-up Delay ($E(CSD)$) and mean Call Disengagement Delay Call ($E(CDD)$) are taken as a performance measure, which are a subset of standardized Call Processing Performance (CPP) parameters [6, 7] important for both network users and operators.

The rest of the paper is organized as follows. Section 2 provides a brief description of the performed simulation conditions. Section 3 presents and discusses the obtained research results. Section 4 summarizes the paper and presents further work.

2 Simulation Conditions

The simulation model presented in [4] will be used to conduct research in the IMS/NGN network consisting of two domains managed by two operators. Due to space limitations the details regarding the simulator operation (network structure, service scenarios, software structure, exact course of simulation and methods of obtaining final results) are not provided in this paper and can be found in [4]. To understand the results presented in the next section, only a brief description of the simulator input (Table 1) and output variables (Table 2) is given. Apart from the $E(CSD)$ and $E(CDD)$ times presented in Table 2, associated confidence intervals are also obtained as final simulation results.

In the conducted research, relatively high values of maximum simulation duration time (sim-time-limit = 72000 s) and maximum number of generated calls (call_num_max = 1000000000) were assumed. In this way, the end of each simulation was determined by the values of the confidence intervals for mean CSD and CDD times, which should be less than 5% of these times (conf_interv_max = 5%). The results of the performed experiments are described in the next section.

3 Results

The aim of the research presented in this section was to check how the parameters of IMS/NGN network with SDN-based transport stratum and traffic generated to this network (Table 1) affect its performance, described by CPP parameters (Table 2). The input data sets for the conducted experiments are presented in Table 3. For each set, the value of one input variable is changed in a wide range for all related network elements, while the remaining input variables have the default values given in Table 1.

Table 1. The most vital input variables of the simulator.

Variable name	Description	Default value
sim-time-limit	Maximum simulation duration time [s]	72000
warmup-period	Warm-up period duration time [s]	1250
call_num_max	Maximum number of generated calls	1000000000
meas_per_num	Number of measurement periods	5
conf_level	Confidence level	0.95
conf_interv_max	Threshold value for confidence intervals	5%
delay	Base delay value in [s] defined separately for all modeled network elements. For SUP-FE/SAA-FE servers it applies to all messages. For the remaining elements it concerns a base message, while processing time of other messages is proportional according to the ak variable (more details are available in [4]). The list of base messages for particular network elements is also provided in [4]	0.01 SUP-FE/SAA-FE, 0.001 for programmable switches, 0.0005 for other elements
link_datarate	Link bandwidth in [bps]	50000000
link_length	Link length in [m]	200000
res_info_prob	Probability of controller having information about the resource state in programmable switches	0.7
res_avail_prob	Probability of resource availability in particular programmable switch	1
intrad_call_intensity	Intra-operator call set-up request intensity in [1/s]	50
interd_call_intensity	Inter-operator call set-up request intensity in [1/s]	50
registr_intensity	Registration request intensity in [1/s]	50
multiple_access_areas_ratio	Ratio of intra-operator calls concerning multiple access areas to all generated intra-operator calls	0.5

Table 2. Successful call scenarios and related output simulation variables.

Scenario name and description	Output variables
b1 – successful intra-operator call performed in domain 1 with both UEs connected to the same access areas	$E(CSD)_{b1}, E(CDD)_{b1}$
b2 – successful intra-operator call performed in domain 2 with both UEs connected to the same access areas	$E(CSD)_{b2}, E(CDD)_{b2}$
d1 – successful intra-operator call performed in domain 1 with both UEs connected to different access areas	$E(CSD)_{d1}, E(CDD)_{d1}$
d2 – successful intra-operator call performed in domain 2 with both UEs connected to different access areas	$E(CSD)_{d2}, E(CDD)_{d2}$
f1 – successful inter-operator calls originated in domain 1	$E(CSD)_{f1}, E(CDD)_{f1}$
f2 – successful inter-operator calls originated in domain 2	$E(CSD)_{f2}, E(CDD)_{f2}$

For example, for input data set no. 6, bandwidths of all links in the network are set equally to values ranging from 5 to 200 Mbps, and the values of the remaining

Table 3. Input data sets for the performed investigations.

Set	Input variable	Range	Set	Input variable	Range
1	CSCF_delay	0.1–0.8 [ms]	8	res_info_prob	0–1
2	SUP-FE_SAA-FE_delay	0.1–400 [ms]	9	res_avail_prob	0–1
3	GATEAPP_delay	0.1–1.4 [ms]	10	intrad_call_intensity	1–130 [1/s]
4	CONTROLLER_delay	0.1–1 [ms]	11	interd_call_intensity	1–100 [1/s]
5	SWITCH_delay	0.1–2.1 [ms]	12	registr_intensity	1–450 [1/s]
6	link_datarate	5–200 [Mbps]	13	multiple_access_areas_ratio	0–1
7	link_length	1–1000 [km]			

parameters are consistent with Table 1. The names of the input variables in Table 3 are generally the same as in Table 1. The exception is the "delay" parameter representing the base delay of the network elements (Table 1). In the performed investigations (Table 3, input data sets no. 1–5), it is changed for particular groups of network elements: for all CSCF servers (CSCF_delay), SUP-FE/SAA-FE servers (SUP-FE_SAA-FE_delay), GATEAPP units (GATEAPP_delay), SDN controllers (CONTROLLER_delay) and programmable switches (SWITCH_delay).

In the next part of the section, all obtained research results will be discussed (all input data sets from Table 3). Due to space limitations, only the results of the $E(CSD)_{b1}$, $E(CDD)_{b1}$, $E(CSD)_{f1}$, $E(CDD)_{f1}$ times for selected input data sets will be presented (Figs. 1, 2 and 3). The choice of presented output variables results from identical parameters of elements and traffic in both network domains. Therefore, CPP parameters for the same types of calls generated in domains 1 and 2 are similar and the results for calls generated in domain 2 (scenarios b2, d2, f2; Table 2) are not presented. At the same time, for the examined set of network parameters, the results obtained for scenario d1 are similar to those obtained for scenario b1 and only the latter are presented.

The first analyzed group of IMS/NGN network parameters includes base delays of elements (input data sets no. 1–5 from Table 3). For most network elements (CSCF servers – Fig. 1 (left), GATEAPP units, SDN controllers), increasing their base delay enlarges the number of messages in the queues of their processors. This leads to an increase in all analyzed CPP parameters. At a certain base delay value (dependent on the type of network element, e.g. 0.7 ms for CSCF servers – Fig. 1 (left)), the processor starts having problems with handling incoming messages and the analyzed output variables begin to grow rapidly. Additionally, in Fig. 1 (left) some characteristic relations between mean *CSD* and *CDD* times in IMS/NGN network can be observed. As expected, $E(CSD)$

and $E(CDD)$ times are always higher for inter-operator calls (scenario f1) than for intra-operator calls (scenario b1). Moreover, for particular scenarios, $E(CSD)$ times are always higher than $E(CDD)$ times, which results from more complicated message exchange procedures when establishing a call.

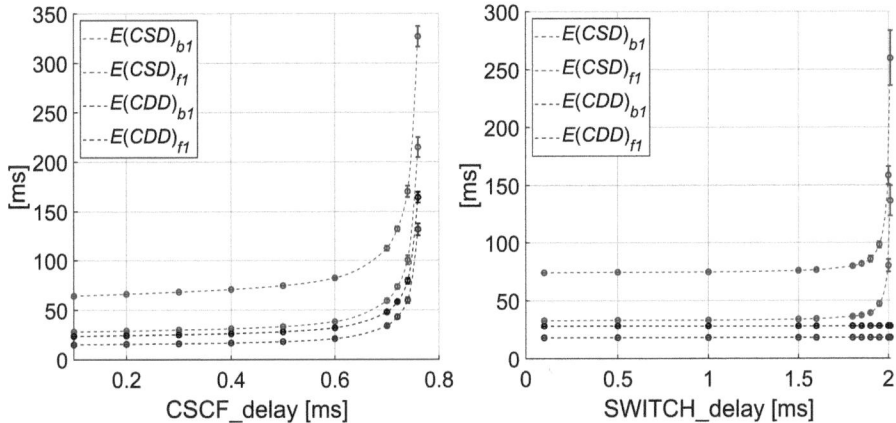

Fig. 1. Results for data set 1 (left) and 5 (right) from Table 3.

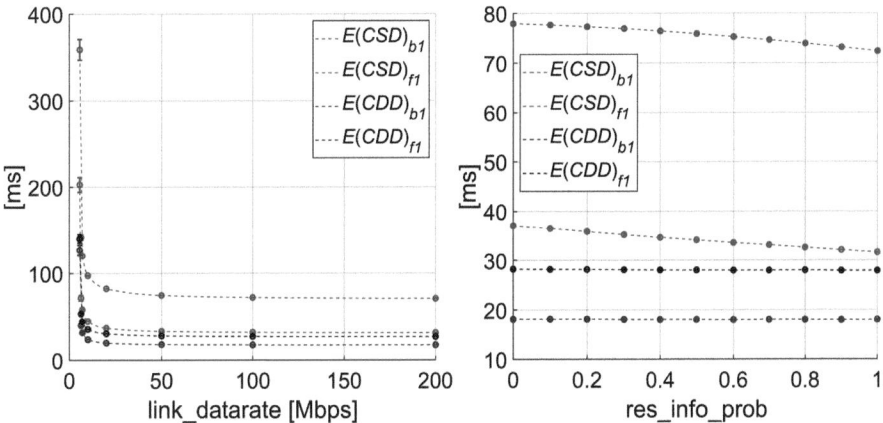

Fig. 2. Results for data set 6 (left) and 8 (right) from Table 3.

For programmable switches (Fig. 1 (right)), a similar increase in $E(CSD)$ times with rising base delay can be observed as in the previously discussed cases (Fig. 1 (left)). However, here there are no changes in $E(CDD)$ times, which results from the fact that during disengaging a call SDN controllers do not wait for the switches' response regarding the release of resources. It is assumed that this operation is always successful [4].

A separate case are the SUP-FE/SAA-FE servers, which do not contain a queue in the simulation model, but respond to each received message after a given base delay.

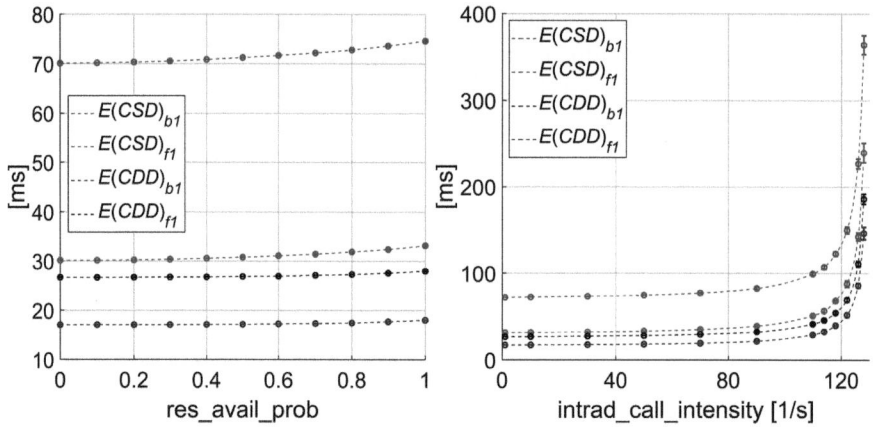

Fig. 3. Results for data set 9 (left) and 10 (right) from Table 3.

These servers are used only when establishing inter-operator calls. This results in a linear increase in the $E(CSD)_{f1}$ time with the base delay of SUP-FE/SAA-FE servers and no changes in the other presented output variables.

Input data sets no. 6 and 7 allow examining how parameters of the links dedicated for signaling messages affect CPP parameters. It is important to ensure appropriate link bandwidths so that they are not overloaded and do not introduce large delays (10 Mbps is sufficient for the tested sets of network parameters, Fig. 2 (left)). Additionally, link lengths, which proportionally increase propagation delay (5μs/km for optical links), can also have a significant impact on mean CSD and CDD times. For the most complex process of establishing an inter-operator call, the values of $E(CSD)_{f1}$ times increase by about 200 ms when link length is changed from 1 km to 1000 km.

The res_info_prob parameter determines the probability that the SDN controller knows the status of resources in programmable switches and has information that transport resources are available. Therefore, there is no need to query the switches about the status of resources and fewer messages are exchanged in the network, which results in shorter mean CSD and CDD times (Fig. 2 (right)). When the controller does not know the status of resources, it queries the programmable switches, which respond positively according to the res_avail_prob probability (Fig. 3 (left)). The lower the value of this probability is, the more call set-up procedures fail, which involve exchange of fewer messages than for the procedures that are successful. This results in a lower load of network elements and lower values of the analyzed output variables. The influence of the res_avail_prob parameter on the analyzed output variables presented in Fig. 3 (left) is relatively small, because for these studies in 70% of cases the controller knows the status of resources and does not have to query the switches about them (res_info_prob = 0.7). For higher number of resource queries sent to switches (lower res_info_prob values) the res_avail_prob parameter will have a stronger impact on the analyzed $E(CSD)$ and $E(CDD)$ times.

Input data sets no. 10–12 show the impact of intra-operator call (Fig. 3 (right)), inter-operator call and terminal registration request intensity on $E(CSD)$ and $E(CDD)$

times. Increasing the above-mentioned intensities results in a larger number of messages in the network and, consequently, causes an increase in the load of network elements. Similarly to the case of base delays, there is a value of each intensity above which the network is overloaded and the analyzed mean *CSD* and *CDD* times increase rapidly. This value is dependent on the request type – the smallest for inter-operator calls (about 90 [1/s]; this is the most complicated message exchange scenario in the network) and the largest for terminal registration (about 400 [1/s]).

For the input data set no. 13 the influence of the multiple_access_areas_ratio parameter on the output variables of the simulator was examined. This input variable turned out to be insignificant – mean *CSD* and *CDD* times increase by about 1–2 ms when the value of the multiple_access_areas_ratio parameter changes from 0 to 1. Such a slight change in the analyzed output variables was caused by the same parameters of network elements and traffic in both domains. When establishing an intra-operator call with several access areas, resources are reserved simultaneously in the access and core networks, but these processes take a similar amount of time. As a result, the P-CSCF server does not wait significantly longer for the result of resource reservation, which would drastically increase mean *CSD* time. Of course, due to additional resource reservation in the core network, slightly more messages are exchanged, but this does not have a critical impact on the analyzed CPP parameters.

To sum up the obtained results, it can be stated that the performance of the IMS/NGN network with the SDN-based transport stratum is the most influenced by:

- base delays of service stratum elements (CSCF servers) and transport stratum elements (GATEAPP, CONTROLLER, SWITCH),
- link parameters (bandwidths and length),
- intensities of generated requests (intra-operator and inter-operator calls, terminal registration).

The large impact of transport stratum elements base delays, which result from their performance, draws attention. Therefore, these elements should have an appropriate processing power. Another approach is to modify the logic of the SDN-based transport stratum to reduce the number of messages exchanged for the resource reservation and release process, which will decrease its load. This will be the subject of our further research.

4 Conclusions

The paper examines the performance of the IMS/NGN network with the SDN-based transport stratum, which is assessed by the $E(CSD)$ and $E(CDD)$ times, a subset of standardized Call Processing Performance parameters important for both network users and operators. The studies use the previously developed and verified simulation model [4]. Several message exchange scenarios (intra-operator and inter-operator calls, terminal registration) are taken into account along with a wide set of parameters of network and generated traffic (including the probability of resource availability, which determines whether a given call request will be successfully handled).

The studies cover 13 input data sets. In each of them, one input variable is modified in a wide range for all related network elements (equally in all domains), and the remaining input variables have default values. As a result of the conducted studies, the parameters that have a significant impact on the performance of the IMS/NGN network with the SDN-based transport stratum are indicated. These are: base delays of selected elements (CSCF servers, GATEAPP, CONTROLLER, SWITCH), link parameters (bandwidth and length), intensities of all types of generated requests. It is demonstrated that with appropriate values of these parameters, the network is able to handle even hundreds of call set-up requests per second.

As already indicated, the efficiency of transport stratum is crucial for performance of the whole IMS/NGN network. In addition to using GATEAPP, CONTROLLER, SWITCH elements with low base delay, another approach is to reduce their load by changing the logic of their operation. One of the solutions that will be tested is the quota-based algorithm [8, 9]. In this approach, resources are allocated with some reserves and most call set-up requests are only mapped in the SDN controller resource database, without communication with programmable switches. It can limit the number of messages exchanged in the network.

Regardless of this, it is planned to extend the presented research in order to include situations in which the parameters of elements and traffic are different in individual network domains. This will allow checking how particular parameters of one domain affect mean *CSD* and *CDD* times for all types of calls generated in all domains.

Acknowledgments. This work was supported by the Gdańsk University of Technology under the grant DEC-25/1/2023/IDUP/I3b/Ag within the Argentum Triggering Research Grants – "Excellence Initiative - Research University" program.

Disclosure of Interests. The authors have no competing interests to declare that are relevant to the content of this article.

References

1. IP Multimedia Subsystem (IMS); Stage 2 (Release 19), 3GPP TS 23.228 v19.1.0, December 2024
2. General overview of NGN, ITU-T Recommendation Y.2001, December 2004
3. Framework of Software-Defined Networking, ITU-T Recommendation Y.3300, June 2014
4. Kaczmarek, S., Sac, M., Adrych, J.: Simulation model for application of the SDN concept in IMS/NGN network transport stratum. In: Franco, L., de Mulatier, C., Paszynski, M., Krzhizhanovskaya, V.V., Dongarra, J.J., Sloot, P.M.A. (eds.) Computational Science – ICCS 2024. Lecture Notes in Computer Science, vol. 14833, pp. 235–250. Springer, Cham (2024)
5. Kaczmarek, S., Sac, M., Bachorski, K.: Implementation of IMS/NGN transport stratum based on the SDN concept. Sensors **23**(12), 5481 (2023)
6. Call processing performance for voice service in hybrid IP networks, ITU-T Recommendation Y.1530, November 2007
7. SIP-based call processing performance, ITU-T Recommendation Y.1531, November 2007
8. Kaczmarek, S., Sac, M.: Performance evaluation of a multidomain IMS/NGN network including service and transport stratum. Appl. Sci. **22**(12), 11643 (2022)
9. Zhang, Z., Duan, Z., Hou, Y.: On scalable design of bandwidth brokers. IEICE Trans. Commun., E84-B, 2011–2025 (2001)

Variable-Resolution Machine Learning for Rapid Multi-Criterial Antenna Design

Anna Pietrenko-Dabrowska[1] and Slawomir Koziel[1,2(✉)]

[1] Faculty of Electronics Telecommunications and Informatics, Gdansk University of Technology, Narutowicza 11/12, 80-233 Gdansk, Poland
anna.dabrowska@pg.edu.pl, koziel@ru.is
[2] Engineering Optimization and Modeling Center, Department of Engineering, Reykjavík University, Menntavegur 1, 102, Reykjavík, Iceland

Abstract. This research introduces a new technique for reduced-cost electromagnetic (EM)-driven multi-objective antenna optimization. Our approach employs artificial neural networks (ANNs) to build a surrogate model of antenna frequency characteristics, acting as a fast predictor providing multiple candidate Pareto-optimal solutions per iteration. The surrogate is refined within a machine-learning framework that leverages accumulated EM simulation data. Computational efficiency is enhanced by incorporating variable-fidelity EM simulations. Verification experiments underscore competitive performance of our method, which requires only two hundred high-resolution EM analyses to complete the MO process. This represents 40% acceleration due to using variable-fidelity models and 90% speedup over traditional single-model surrogate-assisted methods. Our method is also shown competitive concerning design quality.

Keywords: Computer-aided design · antenna engineering · multi-objective optimization · machine learning · EM simulation · variable-fidelity models

1 Introduction

Antennas belong to critical building blocks of wireless communication systems [1, 2]. Satisfying strict performance demands often results in complex structures requiring accurate electromagnetic (EM) analysis for reliable characterization, which is computationally costly. Furthermore, antenna development must balance multiple objectives: practical designs must establish trade-offs between different goals. Identifying these compromise solutions necessitates multi-objective optimization (MO) [3]. Yet, majority of existing procedures are limited to scalar cost functions [4], necessitating objective aggregation to enable multi-objective optimization (MO) [5].

Extensive data concerning trade-off solutions, normally generated as Pareto sets [6], is of high practical value. Predominant MO tools are bio-inspired algorithms that render the complete family of Pareto-optimal solutions in one algorithm execution. Notwithstanding, their applicability to handling EM simulation models is impeded by exceptionally poor cost efficiency. Practical EM-driven MO is often accomplished using surrogate modeling methods [7]. Therein, most computations are delegated to a fast replacement model. Popular modeling techniques include kriging, neural networks, and Gaussian process regression [8, 9]. The surrogate can be constructed beforehand [10] or iteratively during the optimization run, as in the machine learning (ML) frameworks [11]. The candidate designs identified by optimizing the surrogate are validated through EM analysis; the acquired EM data is employed to refine the metamodel. The work [12] provides a review of recent machine learning approaches to MO of antennas. The bottleneck of surrogate-based procedures is building the data-driven model. It is challenging in higher-dimensional spaces or if the spatial extent of the search space is vast. Domain confinement enables addressing dimensionality-related issues. One way is to identify the extreme non-dominated solutions (optimized for individual objectives) and constrain the domain to the smallest interval encapsulating these designs [13]. Constructing the model in the region encompassing high-quality designs, e.g., determined using pre-screening, is another option [14].

This study suggests a novel approach to improved-efficacy antenna MO. Our methodology is an ML algorithm utilizing an artificial neural network (ANN) surrogate. Multiple infill points are rendered in each iteration. The EM data acquired at the candidate designs is used to refine the metamodel. Reduction of the running costs is achieved by utilizing variable-fidelity EM analysis. Extensive verification reveals superior performance of the presented MO over benchmark methodologies. The typical cost of our algorithm corresponds to about 200 high-fidelity EM simulations. It also generates higher-quality Pareto sets compared to the benchmark regarding spatial extent and the Pareto dominance relation. The original contributions of this research include: (i) the development of an ML procedure employing ANN surrogates for high-efficacy MO of antennas, (ii) enhancing the cost efficiency of the search by incorporating multi-resolution EM analysis, (iii) and (iii) the implementation of the entire MO framework and demonstrating its performance using challenging test cases.

2 Multi-criterial Design by Machine Learning

This part of the paper elaborates on the proposed MO algorithm. Problem statement is followed by an outline of the multi-fidelity EM models, the ANN surrogate, and a description of the machine-learning based MO procedure.

2.1 Problem Statement. Variable-Resolution EM Models

Let $F(x) = [F_1(x) F_2(x) \ldots F_{Nobj}(x)]^T$ be a vector of design goals, all to be minimized, where $x = [x_1 \ldots x_n]^T$ represents decision variables. Multi-objective optimization (MO) is understood as finding the Pareto set, a discrete representation of the Pareto front X_P containing all globally non-dominated designs w.r.t. the dominance relation [15]. The designs in X_P are the best available compromises between the objectives of interest.

MO tasks are normally solved using bio-inspired algorithms, which is rarely an option for EM-driven design due to high computational costs. The method proposed in here addresses these issues by incorporating ML, simultaneous rendition of multiple candidate solutions, and variable-resolution EM simulations. Design procedures typically involve a high-fidelity EM model $R_f(x)$ that ensures sufficient reliability in evaluating antenna characteristics. To expedite the process we employ a range of lower-fidelity models $R = (x, L)$, where L is the control parameter governing the discretization density of the antenna under design. The lowest-fidelity model, $R_c(x) = R(x, L_{min})$, is set up to ensure that the EM simulation outputs render all relevant features of the antenna responses, whereas $R_j(x) = R(x, L_{max})$ is set to ensure sufficient reliability (as per designer's requirements).

2.2 Neural Network Surrogates

The primary surrogate model used by the proposed MO procedure is an artificial neural network (ANN). The first model is constructed using N_{init} random samples $x_B^{(j)}$, $j = 1$, ..., N_{init}, allocated by Latin Hypercube Sampling (LHS), and EM simulation outcomes $R(x, L_{min})$. The ANN used is a multi-layer perceptron [16] with two hidden layers (ten neurons each), and a sigmoid activation function. The model is trained using the Levenberg-Marquardt algorithm [16]. The model's inputs are design variables x; the outputs are frequency characteristics (e.g., $|S_{11}|$ or gain vs. frequency), cf. Fig. 1.

2.3 MO by Machine Learning

In this study, the MO process is iterative. In each iteration, the Pareto set is approximated by optimizing the current ANN metamodel with the help of a multi-objective evolutionary algorithm (MOEA) with floating-point representation, fitness sharing with adaptively adjusted niche size, a combination of intermediate and arithmetic crossover, multi-point elitism, and a termination condition based on a sufficient reduction of newly created Pareto-optimal solutions [17]. The population size is set to $N_P = 200$, crossover and mutation probabilities are $p_m = 0.8$ and $p_c = 0.1$.

The candidates are extracted from the current Pareto set generated using MOEA. The EM data is inserted to the training set to refine the surrogate. The infill points $x_I^{(i,j)}$, $j = 1, \ldots, N_{infill}$, are chosen to be possibly close to the target levels $F_j = F_{2.\,min} + (F_{2.\,max} - F_{2.\,min})(j-1)/(N_{infill} - 1)$ of the second objective, where $F_{2.\,min}$ and $F_{2.\,max}$ decide the span of the Pareto front ($N_{infill} = 10$). Enhancing the cost efficiency is realized by employing variable-fidelity EM analysis. The initial metamodel is built using the lowest-fidelity data (L_{min}). During the process, the resolution is enlarged to L_{max} to ensure dependability. The resolution $L^{(i)}$ in the ith iteration is determined as

$$L^{(i)} = \min\left\{L_{max}, L_{min}(L_{max} - L_{min})\frac{i-1}{i - N_{transition}}\right\} \quad (1)$$

According to (1), the model resolution becomes L_{max} after $N_{transition}$, here, set to 5.

The strategy for updating the dataset used to build/refine the ANN surrogate is as follows. If only R_f is employed, all EM data acquired in the MO process is incorporated, including the initial set of samples $x_B^{(j)}$, $j = 1, \ldots, N_{init}$, along with the infill points accumulated up to iteration i inclusive, i.e., $x_I^{(k,j)}$, $k = 1, \ldots, i$, and $j = 1, \ldots, N_{infill}$. For variable-resolution approach, the lowest-fidelity samples are removed in each iteration,

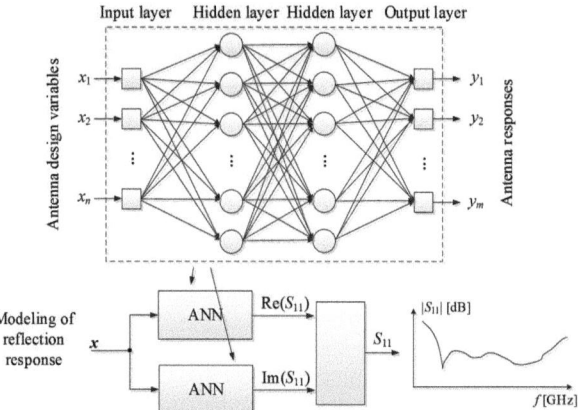

Fig. 1. The architecture of the ANN surrogate model utilized in this study is a multi-layer perceptron. The real and imaginary parts are modeled individually for complex responses. The outputs of the ANN are antenna characteristics at discrete set of frequencies f_1 through f_m.

so that the overall number of data points does not exceed $2N_{init}$; however, if only $\mathbf{R}_f$ points are left, the dataset size may increase beyond $2N_{init}$.

The last component of the algorithm is the termination condition, which is founded on the sufficient resemblance of the Pareto fronts generated in subsequent iterations. The similarity metric is defined as

$$E_i = \left\| F_{nondom}^{(i)} - F_{nondom}^{(i-1)} \right\| \quad (2)$$

The algorithm is terminated if a moving average $E_{a.i} < \varepsilon$ (here, we set $\varepsilon = 1$), where

$$E_{a.i} = \frac{1}{i - \max\{1, i - N_a + 1\} + 1} \sum_{k=\max\{1, i-N_a+1\}}^{i} E_k \quad (3)$$

Using (3) enables smoothing fluctuations of E_i caused by the stochastic components in the optimization routine (specifically, MOEA).

2.4 Complete Framework

The flow diagram of the complete suggested MO procedure is illustrated in Fig. 2. There are several stages involved, which include the initial sampling, building the ANN metamodel, the machine learning optimization loop with iterative generation of multiple infill points, and metamodel refinement, as well as model management process adjusting the EM analysis fidelity during the search operation. The first stages are executed with the lowest-resolution EM analysis (L_{min}), which is gradually increased so that the final dataset only contains high-fidelity samples (L_{max}).

3 Results

This part of the manuscript showcases the operation of the proposed MO technique based on two antennas and juxtaposition to several benchmark algorithms.

3.1 Test Problems

Consider Antennas I and II illustrated in Fig. 3. Antenna I is realized on RF−35 substrate ($\varepsilon_r = 3.5$, $h = 0.762$ mm). The design variables are $x = [L_0\ dR\ R\ r_{rel}\ dL\ dw\ Lg\ L_1\ R_1\ dr\ c_{rel}]^T$, the feed line width is $w_0 = 1.7$ (dimensions in mm). Antenna II is realized on RT6010 substrate ($\varepsilon_r = 10.2$, $h = 0.635$ mm) and its design variables are $x = [s_1\ s_2\ v_1\ v_2\ u_1\ u_2\ u_3\ u_4]^T$; other parameters are $w_1 = w_3 = w_4 = 0.6$, $w_2 = 1.2$, $u_5 = 1.5$, $s_3 = 3.0$ and $v_3 = 17.5$. The EM models are simulated in CST Microwave Studio with the model fidelity controlled using lines-per-wavelength ($L = LPW$). The range of L for Antenna I is from $L_{\min} = 11$ (~210,000 mesh cells, simulation time 42 s) to $L_{\max} = 20$ (~2,300,000 cells, 424 s). For Antenna II, we have $L_{\min} = 17$ (~115,000 cells, 115 s) and $L_{\max}$ (~300,000 cells, 240 s). For Antenna I, the objectives are minimization of the substrate area (F_1) and minimization of the maximum reflection level within the operating band from 3.1 GHz to 10.6 GHz (F_2). For Antenna II, the goals are maximization of the end-fire gain (F_1), and minimization of the maximum $|S_{11}|$ within the operating range from 10 GHz to 11 GHz (F_2).

3.2 Verification Experiment Setup. Benchmark Techniques

The antennas of Fig. 3 are optimized using the suggested algorithm. The outcome is presented as the Pareto set encapsulating non-dominated parameter vectors extracted from the most recent EM dataset $\{x_T^{(i,j)}\}$.

Our technique has been compared to three surrogate-assisted MO frameworks. Algorithms 1 and 2 are one-shot procedures (the metamodel is identified upfront and then optimized by means of MOEA). The difference between these methods is the surrogate modeling approach (kriging for Algorithm 1 and multi-layer perceptron for Algorithm 2). Both methods are run in two versions, different in the training dataset cardinality used to build the surrogate: 400 (version I), and 1600 (version II). Algorithm 3 is a single-objective version of the proposed method, where the optimization is carried out using the high-fidelity EM model. Furthermore, Algorithm 3 employs an accumulative dataset updating scheme: all candidate designs are inserted therein, and no samples are ever eliminated. These methods were specifically implemented for benchmarking.

3.3 Results and Discussion

Figures 4 and 5 show the Pareto sets produced by the suggested procedure and the benchmark techniques for Antenna I and II, respectively. The antenna responses for chosen Pareto optimal solutions can be found in Figs. 6. The design objectives and antenna responses are shown the pictures are evaluated using the respective high-fidelity EM models. The optimization costs have been gathered in Table 1. Note that only the expenses related to EM analysis are included as all other costs (ANN training, surrogate optimization using MOEA) are negligible compared to high-fidelity EM simulation (about four minutes for both Antenna I and II).

The results underscore remarkable performance of the proposed MO procedure regarding reliability and cost-efficiency. As indicated in Figs. 4 and 5, our framework

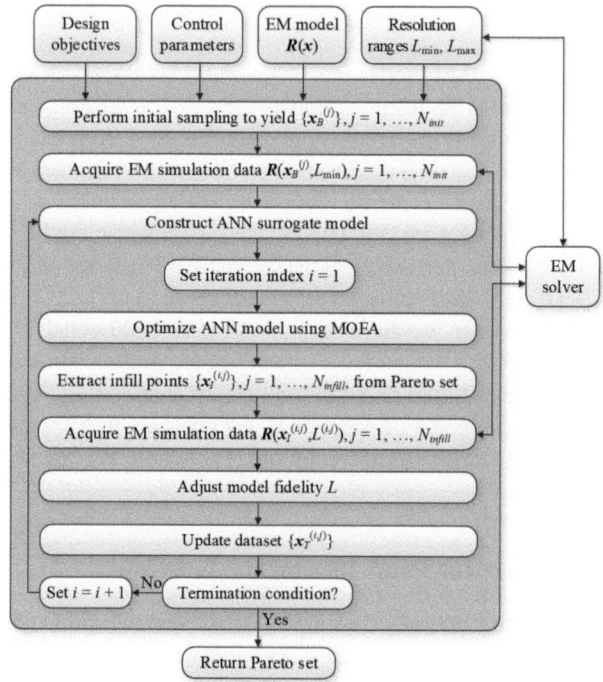

Fig. 2. Flow diagram of the proposed MO algorithm.

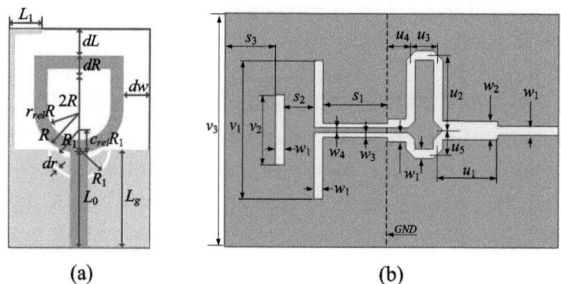

Fig. 3. Verification devices: (a) Antenna I [18], (b) Antenna II [19].

yields significantly better Pareto sets than Algorithms 1 and 2. The reason is the relatively poor accuracy of the surrogate models used by these methods (kriging and ANN). The improvement observed for $N = 1600$ compared to $N = 400$ correlates with enhancing the surrogate's predictive power. For Antenna I, the relative root-mean-squared error (RRMSE) is reduced from around twenty to fifteen percent, whereas for Antenna II, RRMSE falls from eight to five percent. The non-dominated solution sets quality obtained by Algorithm 3 is comparable to the proposed method, demonstrating the advantages of machine learning over one-shot surrogate-assisted procedures.

Concerning cost efficiency (cf. Table 1), the introduced procedure is superior to all benchmark methods. The average CPU cost corresponds to only around 214 high-fidelity EM simulations, which corresponds to 48-percent speedup over Algorithms 1 and 2 (version I), 87-percent speedup over version II of Algorithms 1 and 2, and 38-percent acceleration over Algorithm 3.

Table 1. MO costs: proposed procedure versus benchmark algorithms

Algorithm		Optimization cost[#]	
		Antenna I	Antenna II
This work (variable-fidelity ML with ANN surrogates)		150.4	264.4
Algorithm 1	$N = 400$ (version I)	400	400
	$N = 1600$ (version II)	1600	1600
Algorithm 2	$N = 400$ (version I)	400	400
	$N = 1600$ (version II)	1600	1600
Algorithm 3		320	340

[#] The cost is given in the number of R_f simulations. To compute the expenses for the proposed method, the relationship between the evaluation of time R_f and lower fidelity models is considered.

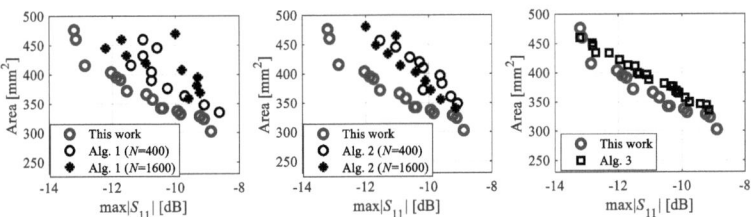

Fig. 4. Pareto sets found for Antenna I: proposed technique versus Algorithms 1, 2, and 3

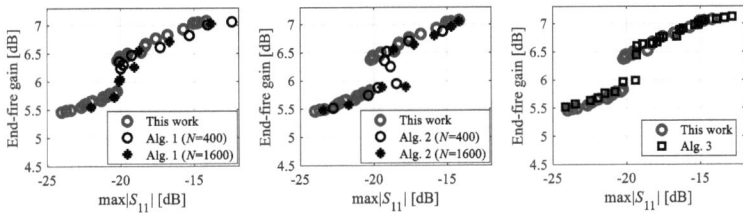

Fig. 5. Pareto sets obtained for Antenna II: proposed algorithm versus Algorithms 1, 2, and 3.

The performance of our methodology makes it an appealing alternative to available MO algorithms. Its reliability and computational efficiency are accompanied by implementation simplicity and straightforward handling. Note these advantages were shown for test problems considerably more challenging than typically found in the MO-related literature.

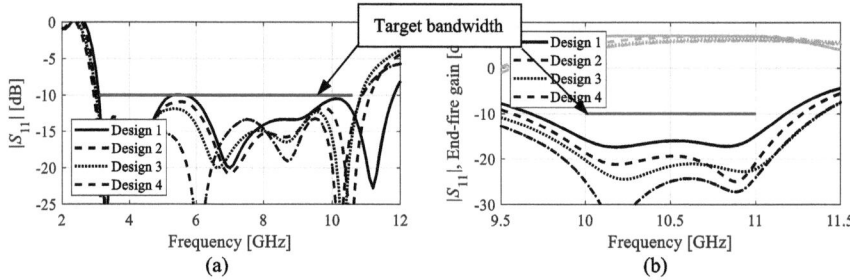

Fig. 6. Characteristics of Antennas I and II at the representative non-dominated designs found with the suggested algorithm: (a) Antenna I: Design 1 ($A = 337$ mm^2), Design 1 ($A = 366$ mm^2), Design 1 ($A = 395$ mm^2), Design 1 ($A = 476$ mm^2); (b) Antenna II: Design 1 (average gain 7.1 dB), Design 2 (average gain 6.5 dB), Design 3 (average gain 5.8 dB), Design 4 (average gain 5.5 dB).

4 Conclusion

This research suggested an alternative methodology for efficient multi-objective optimization (MO) of antennas, involving ANN metamodels and multi-fidelity EM simulations. Extensive validation demonstrates superior performance of the presented strategy regarding the solution quality and low running cost. These features are corroborated through benchmarking against diverse nature-inspired and ML frameworks. Future work will focus on extending our technique's applicability range for more challenging cases including higher-dimensional search spaces and increased numbers of objectives.

Acknowledgment. This work is partially supported by the Icelandic Centre for Research (RANNIS) Grant 2410297 and by the National Science Centre of Poland Grant 2022/47/B/ST7/00072.

References

1. Vinnakota, S.S., et al.: Metasurface-assisted broadband compact dual-polarized dipole antenna for RF energy harvesting. IEEE Ant. Wireless Propag. Lett. **22**, 1912–1916 (2023)
2. Li, M., et al.: Miniaturized slow-wave transmission line-based annular ring antenna with reconfigurable circular polarization. IEEE Ant. Wireless Propag. Lett. **22**, 1766–1770 (2023)
3. Koziel, S., Pietrenko-Dabrowska, A.: Tolerance-aware multi-objective optimization of antennas by means of feature-based regression surrogates. IEEE Trans. Ant. Propag. **70**, 5636–5646 (2022)
4. Wang, J., Yang, X.S., Wang, B.Z.: Efficient gradient-based optimization of pixel antenna with large-scale connections. IET Microwaves Ant. Prop. **12**, 385–389 (2018)
5. Koziel, S., Pietrenko-Dabrowska, A., Mahrokh M.: On decision-making strategies for improved-reliability size reduction of microwave passives: intermittent correction of equality constraints and adaptive handling of inequality constraints. Knowl.-Based Syst., **255**, 109745 (2022)
6. Mirjalili, S., Dong, J.S.: Multi-Objective Optimization using Artificial Intelligence Techniques. Springer Briefs in Applied Sciences and Technology, New York (2019)

7. Easum, J.A., et al.: Efficient multi-objective antenna optimization with tolerance analysis through the use of surrogate models. IEEE Trans. Ant. Prop. **66**, 6706–6715 (2018)
8. de Villiers, D.I.L., et al.: Multi-objective optimization of reflector antennas using kriging and probability of improvement. In: International Symposium on Antennas and Propagation, San Diego, USA, pp. 985–986 (2017)
9. Dong, J., Qin, W., Wang, M.: Fast multi-objective optimization of multi-parameter antenna structures based on improved BPNN surrogate model. IEEE Access **7**, 77692–77701 (2019)
10. De Melo, M.C., Santos, P.B., Faustino, E., Bastos-Filho, C.J.A., Sodré, A.C.: Computational intelligence-based methodology for antenna development. IEEE Trans. Ant. Propag. **10**, 1860–1870 (2022)
11. Nouri, M., Aghdam, S.A., Jafarieh, A., Mallat, N.K., Jamaluddin, M.H., Dor-Emami, M.: An optimized small compact rectangular antenna with meta-material based on fast multi-objective optimization for 5G mobile communication. J. Comp. Electr. **20**, 1532–1540 (2021)
12. Sarker, N., Podder, P., Mondal, M.R.H., Shafin, S.S., Kamruzzaman, J.: Applications of machine learning and deep learning in antenna design, optimization, and selection: a review. IEEE Access **11**, 103890–103915 (2023)
13. Koziel, S., Pietrenko-Dabrowska, A.: Performance-Driven Surrogate Modeling of High-Frequency Structures. Springer, New York (2020)
14. Pietrenko-Dabrowska, A., Koziel, S.: Accelerated multi-objective design of miniaturized microwave components by means of nested kriging surrogates. Int. J. RF & Micr. CAE (2020)
15. Deb, K.: Multi-Objective Optimization Using Evolutionary Alg. Wiley, New York (2001)
16. Vang-Mata, R.: (ed.), Multilayer perceptrons, Nova Science Pub. Inc. (2020)
17. Fonseca, C.M., Fleming, P.J. Multiobjective optimization and multiple constraint handling with evolutionary algorithms–part I: a unified formulation. IEEE Trans. Syst. Man Cybern.: Part A: Syst. Hum. **28**, 26–37 (1998)
18. Alsath, M.G.N., Kanagasabai, M.: Compact UWB monopole antenna for automotive communications. IEEE Trans. Ant. Prop. **63**, 4204–4208 (2015)
19. Kaneda, N., Deal, W.R., Qian, Y., Waterhouse, R., Itoh, T.: A broad-band planar quasi Yagi antenna. IEEE Trans. Antennas Propag. **50**, 1158–1160 (2002)

A Fast and Scalable Genomic Data Compressor for Multicore Clusters

Victoria Sanz[1,2](✉), Adrián Pousa[1], Marcelo Naiouf[1], and Armando De Giusti[1,3]

[1] III-LIDI, School of Computer Sciences, National University of La Plata, La Plata, Argentina
{vsanz,apousa,mnaiouf,degiusti}@lidi.info.unlp.edu.ar
[2] CIC, Buenos Aires, Argentina
[3] CONICET, Buenos Aires, Argentina

Abstract. Genomic data is growing rapidly due to the high demand for precision medicine. Consequently, efficient genome compression algorithms are needed to reduce storage usage in an acceptable response time. This paper introduces HybridHRCM, a hybrid MPI/OpenMP algorithm that harnesses the power of multicore clusters to compress a collection of genomic sequences. We compared our proposal with MtHRCM-opt, its multi-threaded OpenMP counterpart that is suitable for single-node multicore systems. Experimental results demonstrate that HybridHRCM enhances the scalability of MtHRCM-opt for large test collections when using the same number of cores but in a distributed way, while behaves similarly for small test collections. Furthermore, the results reveal that HybridHRCM still achieving good performance when adding more nodes, for all collections.

Keywords: Genomic Data Compression · Multicore Clusters · Hybrid MPI/OpenMP Parallel Programming · Performance · Scalability

1 Introduction

Genomic data is growing rapidly due to the high demand for precision medicine. Consequently, efficient lossless compressors for genomic sequences are needed to reduce storage usage in an acceptable response time [1,2].

In particular, such lossless compression algorithms are classified into two categories, depending on whether they use references during compression or not: (i) reference-free algorithms compress the target sequence using only its internal characteristics; (ii) reference-based algorithms use one or more reference sequences to compress the target sequence, leveraging the high similarity between sequences of the same species [3–5]. Moreover, some reference-based methods allow compressing collections of sequences. Compared to compressing each sequence of the collection individually, batch compression carries out certain steps of the process only once and obtains a higher compression ratio [6,7].

Related to the previously mentioned, HRCM (Hybrid Referential Compression Method) [7] is a compression algorithm for collections of genomes in FASTA format, reference-based and lossless. For each to-be-compressed sequence, the algorithm performs a first matching to find all the segments that are included in the reference sequence. As a result, a compressed sequence is obtained, which contains information about matches and mismatches. Then, first-matching results go through a second matching, where some previously compressed sequences are taken as references (this introduces a data dependency). These second-matching results are appended to a file. After all sequences were processed, the file is compressed with 7-zip.

In order to reduce compression time, the same authors proposed MtHRCM [8], a multi-threaded implementation of HRCM. First, to satisfy the data dependency, the algorithm sequentially solves all the sequences that will be references during the second matching. Then, it uses threads to compress the remaining sequences in parallel. Each compressed output is saved in a separate intermediate file. Finally, the intermediate files are written sequentially (in order) into the final output file, which is then compressed with 7-zip. Although MtHRCM improves the performance of HRCM, it achieves a poor speedup and does not scale well due to its sequential parts and the contention at the I/O system.

Then, we proposed MtHRCM-opt [9], an optimized version of MtHRCM that reduces its sequential component. The experimental results showed that MtHRCM-opt improves the performance of MtHRCM. Also, they revealed that MtHRCM-opt scales well when increasing the number of threads/cores for smaller test collections, but the high amount of simultaneous I/O requests to disk still limits the scalability for larger test collections.

In summary, the aforementioned works focus on compressing genomic sequence collections on a single multicore machine. Single node parallelism is limited by the number of cores available and the concurrent access to shared resources (e.g. memory and disk) that may cause long latencies. In contrast, distributed parallelism allows improving the performance of some applications by leveraging the resources of multiple computers (nodes) in a cluster.

In this paper, we introduce HybridHRCM, a hybrid MPI/OpenMP algorithm based on HRCM that harnesses the power of multicore clusters to compress a collection of genomic sequences. Experimental results demonstrate that HybridHRCM enhances the scalability of MtHRCM-opt for large test collections when using the same number of cores but in a distributed way, while behaves similarly for small test collections (as expected). Furthermore, the results reveal that HybridHRCM still achieving good performance when adding more nodes (resources), for all collections.

The rest of the paper is organized as follows. Section 2 summarizes the HRCM and MtHRCM-opt algorithms. Section 3 describes our proposal. Section 4 shows our experimental results. Finally, Sect. 5 presents the main conclusions and future research.

2 Background

This section describes the HRCM and MtHRCM-opt compression algorithms. Both algorithms compress a collection of FASTA sequences with a lossless reference-based approach.

2.1 HRCM Algorithm

HRCM [7] consists of three stages: *startup*, *matching* and *encoding*.

The *startup* extracts the reference sequence, that is, it keeps all the nucleotides (A, C, G, T) after converting them to uppercase. Then, it constructs a hash table based on the extracted reference sequence, which stores for each possible k-mer (substring of k nucleotides) all its locations in the reference sequence or -1 if it does not exist. This table is used in the matching stage.

Next, the *matching* applies these steps to each to-be-compressed sequence:

- *Extraction:* similar to the extraction step of the startup stage.
- *First-level matching:* this step iterates over the extracted sequence using a sliding window of length k. When the k-mer currently in the window appears in the reference sequence, the position and length of the longest match are recorded in the results. Otherwise, the first base of the k-mer is recorded as a mismatched character. Then, the window is slid. This iteration continues until reaching the end of the sequence.
- *Hash table construction:* only if the sequence will be a reference during the second matching, the results of the previous step along with a hash table built from their entities (taken in pairs) are stored in memory. These data structures are used in the next step. The maximum number of references for the second-level compression is configurable (parameter L).
- *Second-level matching:* the results of the first-level matching are compressed, using a sliding window of length 2 and taking as references the first m *already compressed sequences*, where $m = min(i-1, L)$ and i is the index of the current sequence. From the entities in the window, the longest match among all references is found and the information about the matched segment (sequence id, position, length) is appended to the output file. If a mismatch occurs, the first entity in the window is appended to the output file.

Finally, the *encoding* compresses the output file with 7-zip.

2.2 MtHRCM-opt Algorithm

MtHRCM-opt [9] uses multiple threads to compress the collection of sequences.

First, the main thread executes the *startup* stage. Then, it creates a pool of threads to perform the *matching* stage as follows.

Each thread dynamically picks the next sequence from the collection to be processed. The thread completes the first-level matching, which has no data dependence. Then, it waits for the required data structures before executing the second-level matching (i.e., both the first-level matching results and the associated hash table of the corresponding reference sequences). Specifically, let i be the index of the current sequence, the thread must wait for the data structures of sequences with index between 1 and $min(i-1, L)$ to be ready. Once this condition is met, the thread completes the second-level matching and stores the results in a separate intermediate file.

After all sequences were processed, the main thread *encodes* all the intermediate files with 7-zip, resulting in a single compressed file.

3 HybridHRCM Algorithm

HybridHRCM harnesses the power of multicore clusters to compress a collection of sequences. In particular, it relies on hybrid MPI/OpenMP programming to create a single process per machine/node with multiple threads.

All nodes *collaboratively compress* the collection of genomic sequences in stages. Each node is responsible for processing a subset of sequences that will be second-level references and a subset of the remaining sequences[1]. Let P be the number of nodes, n be the number of sequences in the collection and L be the maximum number of references for the second-level compression, the size of the first subset is $\frac{L}{P}$ and the size of the second subset is $\frac{n-L}{P}$.

First, the main thread on each node performs the *startup* stage, which extracts the reference sequence and constructs the associated hash table. The time of this operation is negligible and its parallelization has no practical sense, for this reason the computation is replicated on each node.

In the first compression stage, intra-node threads perform the first-level matching of the sequences of the first subset (i.e. those that are second-level references). This problem is data parallel, since it only uses the information obtained in the startup. Each thread dynamically picks the next sequence from the subset, extracts it, and then computes and stores the results of the first-level matching and the associated hash table. Then, the thread communicates both data structures to the other nodes through message passing. This information is necessary for the following stages. Message reception is handled by a dedicated receiver thread on each node, which runs concurrently with the worker threads.

The second compression stage will begin once the node has received all the data structures sent by the other nodes. Having such information allows independent processing. At this stage, intra-node threads complete the processing of the sequences of the first subset. Each thread dynamically picks the next sequence from the subset, computes the second-level matching and stores the results in an independent intermediate file.

In the third compression stage, intra-node threads compress the sequences of the second subset. Each thread dynamically picks sequences, extracts each

---

[1] The algorithm assumes that each node can access the files (sequences) of both assigned subsets as well as the reference sequence file. That is, these files can be stored locally on disk or can be obtained from a storage.

one and applies the first and second-level matching, and stores the results in an independent intermediate file.

Finally, the master process executes the *encoding* stage, which compresses all the intermediate files with 7-zip, resulting in a single file. To do this, it must access all the generated intermediate files. Consequently, each node is responsible for leaving the generated intermediate files in a location accessible by the master process (either on its local disk or on a storage).

Notice that communication time does not impact performance since the communicated data are hash tables of limited size ($\sim$8 MB each) and compressed results with high compression ratio.

4 Experimental Results

Our experimental platform is a cluster of multicore nodes. Each node is composed of two Intel Xeon E5-2695 v4 processors (36 cores in total), 128 GB RAM and a SAS disk. Hyper-Threading and Turbo Boost were disabled. All nodes are connected via Ethernet.

Tests considered 1100 commonly used human genomes [9]: 1092 are extracted from the 1000 Genome Project; 5 are the UCSC HG16, HG17, HG18, HG19 and HG38 genomes; 2 are the Korean genomes KOREF_20090131 and KOREF_20090224; and the last is the HuRef genome. We used the UCSC HG13 genome as reference. Human genomes contain 24 chromosomes (identified as 1, 2, .., 22, X, Y) and have a size of $\sim$3000 MB each.

Specifically, we formed 24 test collections in total, one for each chromosome. Each test collection includes the 1100 same-numbered chromosomes from different individuals, and will be compressed against the same-numbered chromosome of HG13. This grouping allows the compressor to leverage the similarity between the to-be-compressed sequences and the reference. It is worth mentioning that: in general, the smaller the chromosome ID, the larger the collection size; each test collection includes sequences of similar size (near to the average); larger test collections are composed of larger sequences, and smaller test collections have smaller sequences. Hence, the aforementioned grouping allows us to evaluate the scalability of the algorithm.

The maximum number of references for the second-level matching (L) was set to 275, which corresponds to 25% of the to-be-compressed collection.

To prove the effectiveness of our proposal, we ran HRCM (sequential code), MtHRCM-opt (multi-threaded code) and HybridHRCM (hybrid MPI/ OpenMP code) on different system configurations. In the former two cases, all the to-be-compressed sequences are stored on the local disk of the single node used. In the case of the hybrid algorithm, each cluster node stores on its local disk the to-be-compressed-sequences of its assigned subsets.

From the experimental results, we first confirmed that the three algorithms achieve the same compression efficiency. That is, for the same to-be compressed collection and the same value of L, all the algorithms obtain the same output (regardless of the system configuration used). This behavior is expected since

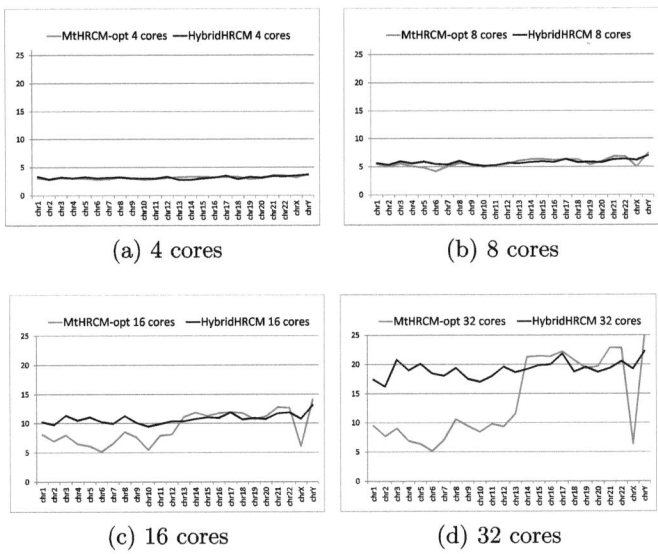

Fig. 1. Speedup comparison between MtHRCM-opt (1 node) and HybridHRCM (2 nodes), for different number of threads/cores.

their compression methodology is identical. In general, all test collections (a total of 3258684 MB or ~3 TB) were compressed to 1318 MB.

Next, we verify the performance behavior of our algorithm, measured in terms of Speedup ($\frac{Time_{seq\_algorithm}}{Time_{par\_algorithm}}$). Figure 1 compares the Speedup of MtHRCM-opt and HybridHRCM, for all chromosomes, with 4, 8, 16 and 32 threads/cores in total. In HybridHRCM threads are distributed equally between two nodes (i.e. 2 nodes × 2 cores, 2 nodes × 4 cores, 2 nodes × 8 cores, 2 nodes × 16 cores). The results show that, for 4 and 8 threads, the Speedup obtained by both algorithms is similar for all chromosomes. However, for 16 and 32 threads, HybridHRCM achieves better Speedup than MtHRCM-opt for chromosomes {1..12, X} (first group), and similar Speedup for the rest of the chromosomes {13..22,Y} (second group).

To explain this behavior, we refer to our previous work [9] where we observed that the main source of overhead in MtHRCM-opt is disk contention, which limits the scalability of large collections (first group). Disk I/O is performed to extract the information of each sequence from file and write the compressed sequences to disk. For a fixed test collection (chromosome), as more threads are involved in compression, more sequences are processed in parallel, therefore there will be more I/O requests that the disk must serve simultaneously. This causes long latencies that affect performance. This overhead is even higher for large collections (first group) since they require more I/O operations. HybridHRCM distributes the to-be-compressed sequences of the collection among nodes. For this reason, the amount of work per node is lower, which implies a lower number of I/O requests to be served simultaneously (regardless of the number of threads used). This results in a performance gain, which is evident for large collections (first group) and a large number of threads.

It should be noted that, for the remaining cases, MtHRCM-opt shows good performance and is not significantly affected by disk contention. Therefore, HybridHRCM obtains similar performance when distributing the work among nodes, using the same number of total cores.

Then, we investigate the scalability of HybridHRCM when increasing the number of threads/cores per node. Figure 2 shows the Speedup of HybridHRCM, for all chromosomes, with 2 and 3 nodes, and 2, 4, 8, 16 and 32 threads/cores per node. The results in both cases reveal that, for all chromosomes, the Speedup improves as the number of cores per node increases. Also, they display that the best performance is obtained when using 32 cores per node.

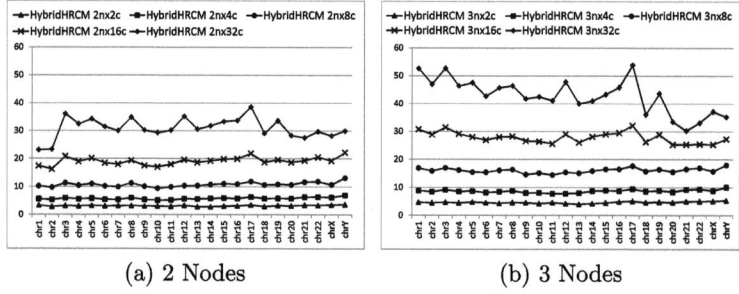

Fig. 2. Speedup of HybridHRCM with 2 and 3 nodes (n) when increasing the number of threads/cores per node (c).

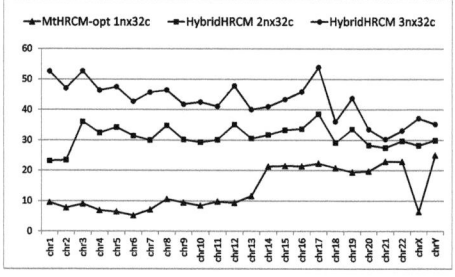

Fig. 3. Speedup of MtHRCM-opt (1 node), HybridHRCM (2 nodes) and HybridHRCM (3 nodes), with 32 threads/cores per node

In addition, we study the scalability of the algorithm when increasing the number of nodes. Figure 3 compares the Speedup of MtHRCM-opt (1 node), HybridHRCM (2 nodes) and HybridHRCM (3 nodes), with 32 threads/cores per node, since this thread configuration gave the best performance, as previously presented. Note that adding a node involves adding 32 processing cores (i.e., the total number of cores with 1, 2 and 3 nodes is 32, 64 and 96 respectively). For all chromosomes, the best performance is achieved with 3 nodes. For smaller test collections (higher ID), the performance does not increase significantly when scaling from 2 to 3 nodes. This is because smaller test collections are smaller in size and thus their compression time is lower, so as more cores are used, the overhead due to parallelism affects more severely the total time (Amdahl's law).

Finally, we analyze the Overall Throughput ($\frac{uncompressed\_data\_size\ (MB)}{compression\_time\ (seconds)}$) of the algorithms to provide a more concrete interpretation of the presented results. Throughput values were calculated considering the total size and the total compression time of all test collections (24 chromosomes), for each algorithm. HRCM (sequential code) achieves a throughput of 16.19 MB/s, MtHRCM-opt (1 node, 32 threads) obtains 144.31 MB/s, while HybridHRCM reaches 480.59 MB/s and 740.63 MB/s (with 2 and 3 nodes, 32 threads per node). As can be derived from these results, on our platform HRCM compresses all data (∼3258684 MB) in ∼56 h, MtHRCM-opt takes ∼6 h 15 m, while HybridHRCM completes the compression in ∼1 h 53 m with 2 nodes and ∼1h 13 m with 3 nodes. In other terms, per human genome (∼3000 MB) HRCM uses about 183 s, MtHRCM-opt uses ∼21 s, while HybridHRCM uses ∼6 s and ∼4 s, with 2 and 3 nodes respectively.

5 Conclusions and Future Work

This paper introduced HybridHRCM, a hybrid MPI/OpenMP genomic data compressor that harnesses the power of multicore clusters.

We compared HybridHRCM with MtHRCM-opt, its OpenMP counterpart for single-node systems, when using the same number of cores but in a distributed way. Experimental results show that the performance of HybridHRCM is more stable than that of MtHRCM-opt when varying the to-be compressed collection. For large test collections and a large number of cores, HybridHRCM outperforms MtHRCM-opt. For the rest of the cases, both algorithms behave similarly (as expected). Hence, HybridHRCM enhances the scalability of MtHRCM-opt.

We further studied the scalability of HybridHRCM when increasing the number of nodes (with 32 threads per node). As a main concrete result, MtHRCM-opt (1 node) compressed the whole 1100 human genomes in ∼6 h 15 m, HybridHRCM (2 nodes) in ∼1 h 53 m and HybridHRCM (3 nodes) in ∼1 h 13 m. Consequently, distributed computing and multicore clusters enable faster compression times for large genomic data.

In future, we plan to compare HybridHRCM with HadoopHRCM (a version of HRCM based on Hadoop MapReduce) and their advantages/disadvantages.

References

1. National Human Genome Research Institute: Genomic Data Science. https://www.genome.gov/about-genomics/fact-sheets/Genomic-Data-Science

2. Stephens, Z.D., et al.: Big data: astronomical or genomical? PLoS Biol. **13**(7), e1002195 (2015)
3. Kredens, K.V., et al.: Vertical lossless genomic data compression tools for assembled genomes: a systematic literature review. PLOS ONE **15**(5), e0232942 (2020)
4. Hosseini, M., et al.: A survey on data compression methods for biological sequences. Information **7**(4), 56 (2016)
5. Wandelt, S., et al.: Trends in genome compression. Curr. Bioinf. **9**(3), 315–326 (2013)
6. Deorowicz, S., et al.: GDC 2: compression of large collections of genomes. Sci. Rep. **5**, 11565 (2015)
7. Yao, H., et al.: HRCM: an efficient hybrid referential compression method for genomic big data. BioMed Res. Int. **2019**, 3108950 (2019)
8. Yao, H., et al.: Parallel compression for large collections of genomes. Concurr. Comput. Pract. Exp. **34**(2), e6339 (2021)
9. Sanz, V., et al.: Fast genomic data compression on multicore machines. In: Naiouf, M., et al. (eds). Cloud Computing, Big Data and Emerging Topics. JCC-BD&ET 2024. Communications in Computer and Information Science, vol. 2189. Springer, Cham (2025). https://doi.org/10.1007/978-3-031-70807-7_1

Anchored Semantics: Augmenting Ontologies via Competency Questions, Self-Attention, and Predictive Graph Learning

Shengqi Li[✉] and Amarnath Gupta[✉]

University of California San Diego, La Jolla, CA 92093, USA
{shl142,a1gupta}@ucsd.edu

Abstract. We propose a framework that enriches ontologies by leveraging competency questions and distant supervision. The process begins by using an LLM to extract domain-relevant entities from the questions, followed by incremental refinement through short definitions anchored to a predefined dictionary. These entities and their hierarchies, along with associated queries, are embedded using a fine-tuned Llama3.2:1b and further processed through a self-attention mechanism to create unified representations. A directed acyclic graph models the dependencies between entities, with additional nodes derived from frequent co-occurrences in queries. A Graph Attention Network (GAT) is used for stable link prediction, discovering latent semantic relationships. These links are then labeled with specific relation types using a fine-tuned RoBERTa module. Evaluations using datasets from HPC training sessions and OpenAlex abstracts show significant improvements in link prediction and ontology enrichment over standard GAT and GraphSage baselines.

Keywords: Ontology Augmentation · Competency Questions · Distant Supervision · Graph Embedding

1 Introduction

Ontologies serve as crucial formal knowledge representations that bridge the gap between human conceptual understanding and machine-processable data structures in modern AI systems. They provide a mathematically rigorous framework for encoding domain expertise, ensuring semantic interoperability across diverse, heterogeneous data by bridging them to a network of related concepts, and answer complex queries that require connecting information across multiple domains. Since they represent domain knowledge, ontologies are often used as a part of a knowledge graph and enables better explainability in AI-based tasks like answering natural language questions.

Many of the early ontologies (e.g., biomedical ontologies such as the Gene Ontology) were constructed by domain experts via human processes, and were regularly maintained and updated as new terminology and new uses emerged [10,16], but such top-down development is slow, non-scalable, and insufficiently agile. Data-driven methods that generate ontologies from text accelerate development but often sacrifice quality and ignore downstream user needs; we therefore require a bottom-up, logically consistent, vocabulary-aligned methodology that can evolve with user demands. Recent advances in LLMs present new opportunities: several studies generate or augment ontologies directly from competency questions—manually or via rule-based techniques [6]—offering deterministic consistency yet limited adaptability, while transformer-based approaches cannot fully capture CQ variability [4]; other work fine-tunes GPT-3 to translate natural language into OWL Functional Syntax [11] or leverages zero- and few-shot learning for ontology alignment [2,8]. Although these enhance expressiveness and matching, our approach differs by integrating extracted CQ topics with distant supervision and deep-learning-based structural prediction to deliver a more automated, scalable ontology augmentation.

We represent user demands via a set of competency questions (CQs), where CQs are natural-language questions that the completed ontology should answer, which have been shown to help resolve ontology defects by introducing entities and relationships the ontology does not capture [3]. We adopt a setting where a consistent but incomplete, task-agnostic ontology exists for some domain, and a CQ bank, obtained from prospective users, is used to computationally extend it while maintaining our design criteria.

In this paper, we investigate how generative models—specifically LLMs, which hold great promise in transfer learning [1] can be effectively used in bottom-up ontology construction using CQs as a guideline (cf. [13]). We propose a framework that leverages an LLM to extract key domain-relevant entities and their relationships from CQs to enhance an existing ontology, systematically addressing both content and structural heterogeneity. Finally, a GAT integrates the explicit links provided by CQs and infers latent semantic relationships between entities, thereby augmenting the expressiveness of the ontology. Specifically, this paper makes the following contributions to CQ-driven ontology augmentation.

- We formalize the problem of ontology augmentation by incorporating competency questions, addressing both content and structural heterogeneity.
- We propose an innovative framework that utilizes LLMs for entity extraction and recursive definition generation, coupled with a multi-head self-attention mechanism to fuse multiple feature modalities.
- We design a comprehensive graph-based model that integrates the question-entity, entity-category, and inferred entity-entity relationships, thus enriching the ontology's expressiveness.
- We empirically demonstrate the effectiveness of our approach on real-world datasets, showing notable improvements in ontology coverage and link prediction accuracy.

2 Our Approach

Our automated ontology construction process has two inputs—an existing ontology O that needs to be augmented and a set of competency questions Q obtained from users. We perform a semantic analysis of each question $q \in Q$, together with $V(O)$, the verbalized version of the ontology. The final ontology is generated in a 3-stage process. In the *Entity Extraction and Definition* stage (Sect. 2.1), semantic entities are extracted from $Q \cup V(O)$ and composed into an initial subClassOfDAG. Next, in the *Embedding Generation and Fusion* stage (Sect. 2.2), each entity's label, hierarchy path, and definition embeddings are combined via self-attention into a holistic representation. Finally, in the *Graph-based Link Prediction and Labeling* stage (Sect. 2.3), a Graph Attention Network identifies new relations and a classifier assigns each predicted edge its relation type (Fig. 1).

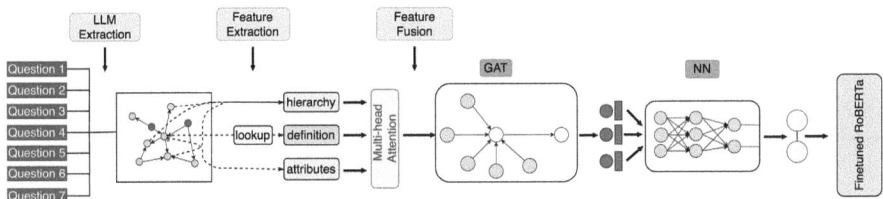

Fig. 1. The primary architecture of the system

2.1 Entity Extraction and Definition

Our process begins with two inputs—an existing ontology O and a set of competency questions Q. We first verbalize O into natural language using an ontology verbalizer [15] and concatenate that text with each $q \in Q$. An LLM, prompted via few-shot examples, parses this combined text to extract candidate entities (noun phrases) along with their hierarchical category paths up to the root node Entity. For each candidate:

- We consult a distant-supervision dictionary of known definitions. If found, the entity is marked "grounded."
- Otherwise, the LLM is prompted to generate a concise, one-sentence definition (cf. [13]). Any new terms in that definition are recursively extracted and resolved until all entities map to dictionary entries.
- As post-processing, we remove cycles, duplicate definitions, and acronyms to ensure the result is a clean subClassOf DAG where vertices are entities and edges capture dependency relationships from the definition process.

2.2 Embedding Generation and Fusion

Each entity e is represented by three facets—its label n, its hierarchy path $\mathcal{C}(e)$, and its definition $d(e)$. We fine-tune a Llama3.2-1b model via LoRA [9] to encode each facet into vectors $\mathbf{v}_n$, $\mathbf{v}_C$, and $\mathbf{v}_d$, drawing training pairs from co-occurrence of entities in OpenAlex abstracts, with positive pairs share a CQ; negative pairs have their nearest common ancestor at least three levels above. Rather than using these vectors independently, we stack them into a matrix and apply a single self-attention layer, whose parameters are trained using the same distant-supervision data mentioned in the earlier encoding stages.

Although there are well-known graph-only embedding approaches, including RDF2Vec [14], OWL2Vec* [5], TransE/DistMult variants. They rely exclusively on triple walks or translation objectives and ignore the rich textual and hierarchical signals in our augmented ontology. In contrast, we obtain embeddings that capture intrinsic semantics, taxonomic context, and relational differences by fine-tuning Llama3.2:1b on our domain corpus and competency questions.

2.3 Graph-Based Link Prediction and Labeling

We construct an augmented graph $G' = (V, E \cup E_r)$ by adding co-occurrence edges

$$E_r = \{(e_i, e_j) \mid \mathcal{Q}(e_i) \cap \mathcal{Q}(e_j) \neq \emptyset\}.$$

A Graph Attention Network (GAT) then propagates each $\mathbf{v}_e$ over its neighborhood:

$$\alpha_{ij} = \text{softmax}_j\big(\text{LeakyReLU}(a^\top [W\mathbf{v}_{e_i} \| W\mathbf{v}_{e_j}])\big), \quad \mathbf{v}'_{e_i} = \sigma\Big(\sum_{e_j \in \mathcal{N}(e_i)} \alpha_{ij} W\mathbf{v}_{e_j}\Big).$$

An MLP over $[\mathbf{v}'_{e_i} \| \mathbf{v}'_{e_j}]$ then scores link existence. For each predicted link, we form an input by concatenating the text of e_i, e_j, and their shared question q_{ij}, and classify it into one of 21 relation types, combining existing HPC labels with a few new predicates via a lightweight fine-tuned classifier.

3 Experiments

3.1 Experiment Design

Datasets. Our primary dataset for ontology augmentation consists of 1,127 competency questions collected from SDSC HPC training sessions. These questions are structured according to the HPC-fair ontology to model relationships within the HPC domain. After extracting the data using GPT-4o, we applied rigorous post-processing steps refining the dataset into a well-structured graph representation.

To fine-tune the LLM for embedding generation, we supplemented our dataset with additional data from a publication resource, OpenAlex. The data collection process began by extracting pairs of entities from the graph constructed using competency questions. We classified entity pairs as relevant if they originated from the same competency question and non-relevant if they shared a common ancestor at a hierarchical distance of three steps.

For each entity pair, we searched for paper abstracts where both keywords co-occurred. In practice, we randomly selected 200 entities from the graph, collecting k_1 (e.g., 3) relevant and an equal number of non-relevant associated words. For each pair of words, we retrieved n (10 in our experiments) abstracts of articles from OpenAlex (Table 1).

Table 1. Dataset Statistics

Competency Questions Dataset	
Questions	1,127
Entities	9,590
Categories	4,954
Question–Entity edges	28,521
Entity–Category edges	17,492
Category–Category edges	9,133
OpenAlex Paper Abstracts	
Word pairs	1,200
Paper abstracts	12,000

3.2 Link Prediction

Which pair of entities is related? To evaluate the effectiveness of our ontology in capturing semantic relationships between entities, we compare our results with pure GAT and GraphSage [7] without self-attention embedding fusion.

The GAT we are using here is trained on the same data as the distant supervision employed in the previous steps. The training labels are derived from the following logic:

– Two entities are *related* if they share at least one competency question,
– Two entities are *not-related* if their closest common ancestor is at least three steps above either node in the hierarchy.

Instead of random sampling part of the graph as training set and the rest as validation and testing set, we consider the case of incremental ontology augmentation. The original graph is used as the training set, while we split a subset of competency questions and corresponding entities as the testing set. Positive samples are selected using the same logic as training labels, and an equal number of negative samples are selected using only the new data. The training-test ratio is set to 8:2.

Table 2. Link Prediction Performance Comparison

Method	Accuracy	F1	AUC
Logistic Regression	0.8665	0.8712	0.8735
GAT	0.9232	0.9235	0.9657
GraphSage	0.9295	0.9294	0.9681
Ours	0.9653	0.9659	0.9830

The result in Table 2 indicates that our method outperforms both conventional GAT and GraphSage models. By incorporating a self-attention layer, our model can selectively integrate and weigh the diverse information provided by the augmented embeddings, capturing subtle semantic nuances and relationships essential for link prediction. It is worth noting that even a Logistic Regressor can reach 0.8665 accuracy, because the embeddings already capture semantic and structural information and separate the dataset, making it easier for simple models to perform well. We explore this further in the *Ablation Study* section.

3.3 Label Prediction

Which relationship does this edge belong to? In this subsection, we solved this problem by finetuning a RoBERTa model. We prepare a dataset comprising every entity pair identified by the GAT module, each record containing the pair, their shared competency questions, and the target relationship. Using few-shot GPT-4o prompting, we generate up to three candidate labels per pair (3,098 labels over 16,455 pairs), embed them with our fine-tuned Llama3.2:1b, cluster by cosine similarity, and manually refine to 27 final labels.

After finetuning, the model achieved an accuracy of 0.7356, precision of 0.7222, recall of 0.7356, and F1 score of 0.7207.

The performance of relationship label classifiers is suboptimal as it shows patterns of errors. The errors arise from contextual ambiguity, where multiple labels may apply, and confusion among semantically similar relations (e.g., causation vs. usage). To enhance accuracy, it is crucial to create a dataset that includes additional contextual details and competency questions to offer a deeper understanding of connections between entities.

4 Discussion

4.1 Embedding Component Contribution

The results clearly show that all these factors, intrinsic, hierarchy, and definition, have a positive impact on total performance, and integration of these provides the best results. In Table 3, we observe that when each of these factors is taken in isolation, intrinsic, hierarchy, and definition give the same results. But combining three components with a self-attention mechanism resulted in better results. For example, hierarchy and definition as a pair provide a higher degree of precision and F1 scores than other pairs of two factors, implying that these factors support each other well.

4.2 Ablation Study

Embeddings play an important role in our ontology augmentation framework. As shown in Table 2, a simple logistic regressor with finetuned llama3.2:1b can already reach 0.8665 accuracy. In this ablation study, we employed three different encodings to evaluate their performance using our model. The result of link prediction is shown in Fig 2. From this figure:

- The finetuned Llama3.2:1b performs better than vanilla Llama3.2:1b, indicating the finetune process can capture domain-specific differences between entities.

Table 3. Performance Metrics for Different Component Combinations

Component Combination	Accuracy	F1 Score	AUC
Intrinsic	0.9295	0.9294	0.9681
Hierarchy	0.9317	0.9318	0.9761
Definition	0.9281	0.9293	0.9765
Intrinsic + Hierarchy	0.9511	0.9521	0.9785
Intrinsic + Definition	0.9461	0.9475	0.9767
Hierarchy + Definition	0.9598	0.9603	0.9813
All Three (Intrinsic, Hierarchy, Definition)	0.9653	0.9659	0.9830

- Finetuned and vanilla Llama3.2:1b outperforms traditional embedding models like word2vec [12], indicating LLM has a deeper understanding of the underlying structure in our augmented ontology.

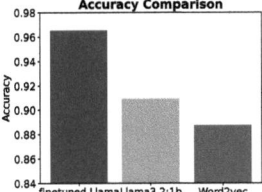 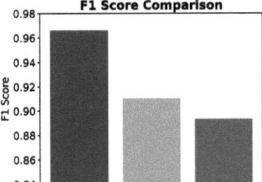 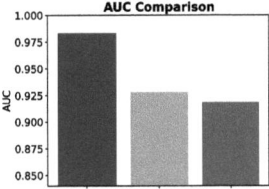

Fig. 2. Performance of various embedding models

5 Conclusion

This paper proposes a new paradigm for competency question based ontology enrichment using large language models, followed by high-level embedding fusion using a self-attention mechanism and graph attention network for robust link prediction. Our experiments demonstrate that the proposed method outperforms conventional methods such as GAT and GraphSage with improved accuracy and link prediction results. Despite some challenges in relationship label prediction, largely due to inherent vagueness and fine-grained semantic overlap, promising results indicate the potential to integrate deep learning and graph-based approaches to enhance the semantic density and structural coherence of ontologies. Future work will focus on enhancing the label prediction process by leveraging more contextual information, with the ultimate goal of further advancing the state-of-the-art in ontology enrichment.

Acknowledgments. This study was was partially funded by USDA Grant 2024-68015-41700.

Disclosure of Interests. Neither author has any competing interest at this time.

References

1. Achiam, J., et al.: Gpt-4 technical report. arXiv preprint arXiv:2303.08774 (2023)
2. Babaei Giglou, H., DSouza, J., Auer, S.: Llms4ol: large language models for ontology learning. In: International Semantic Web Conference, pp. 408–427. Springer, Heidelberg (2023)
3. Bezerra, C., Freitas, F., Santana, F.: Evaluating ontologies with competency questions. In: 2013 IEEE/WIC/ACM International Joint Conferences on Web Intelligence (WI) and Intelligent Agent Technologies (IAT), vol. 3, pp. 284–285. IEEE (2013)
4. Bosselut, A., Rashkin, H., Sap, M., Malaviya, C., Celikyilmaz, A., Choi, Y.: Comet: commonsense transformers for automatic knowledge graph construction. arXiv preprint arXiv:1906.05317 (2019)
5. Chen, J., Hu, P., Jimenez-Ruiz, E., Holter, O.M., Antonyrajah, D., Horrocks, I.: Owl2vec*: embedding of owl ontologies. Mach. Learn. **110**(7), 1813–1845 (2021)
6. Gangemi, A., Lippolis, A.S., Lodi, G., Nuzzolese, A.G.: Automatically drafting ontologies from competency questions with frodo. In: Towards a Knowledge-Aware AI, pp. 107–121. IOS Press (2022)
7. Hamilton, W., Ying, Z., Leskovec, J.: Inductive representation learning on large graphs. Adv. Neural Inf. Process. Syst. **30** (2017)
8. Hertling, S., Paulheim, H.: Olala: ontology matching with large language models. In: Proceedings of the 12th Knowledge Capture Conference 2023, pp. 131–139 (2023)
9. Hu, E.J., Shen, Y., Wallis, P., Allen-Zhu, Z., Li, Y., Wang, S., Wang, L., Chen, W., et al.: Lora: low-rank adaptation of large language models. ICLR **1**(2), 3 (2022)
10. Keet, C., Mahlaza, Z., Antia, M.: Claro: a data-driven cnl for specifying competency questions. ArXiv arxiv:1907.07378 (2019)

11. Mateiu, P., Groza, A.: Ontology engineering with large language models. In: 2023 25th International Symposium on Symbolic and Numeric Algorithms for Scientific Computing (SYNASC), pp. 226–229 (2023). https://doi.org/10.1109/SYNASC61333.2023.00038
12. Mikolov, T., Chen, K., Corrado, G., Dean, J.: Efficient estimation of word representations in vector space. arXiv preprint arXiv:1301.3781 (2013)
13. Petroni, F., et al.: Language models as knowledge bases? arXiv preprint arXiv:1909.01066 (2019)
14. Ristoski, P., Paulheim, H.: RDF2Vec: RDF graph embeddings for data mining. In: Groth, P., et al. (eds.) ISWC 2016. LNCS, vol. 9981, pp. 498–514. Springer, Cham (2016). https://doi.org/10.1007/978-3-319-46523-4_30
15. Zaitoun, A., Sagi, T., Peleg, M.: Generating ontology-learning training-data through verbalization. In: Proceedings of the AAAI Symposium Series, vol. 4, pp. 233–241 (2024)
16. Zhao, Y., Vetter, N., Aryan, K.: Using large language models for ontoclean-based ontology refinement. ArXiv arxiv:2403.15864 (2024). https://doi.org/10.48550/arXiv.2403.15864

Modeling Firm Birth and Death Dynamics Using Survival Fractions and Age Distributions

Yipei Guo[✉][iD], Huynh Hoai Nguyen[iD], and Feng Ling[iD]

Institute of High Performance Computing (IHPC), Agency for Science, Technology and Research (A*STAR), 1 Fusionopolis Way, Connexis (North Tower) #16-16, Singapore 138632, Republic of Singapore
guoyp@ihpc.a-star.edu.sg

Abstract. The birth and death of firms govern firm population dynamics, and understanding these processes can guide urban planning and policy. Longitudinal data with full entry and exit records allow direct analysis of birth and death rates and how they depend on external factors and firm-level properties. However, real-world data are often incomplete, with missing records of extinct firms or exit dates. In such cases, it is unclear if and how we can extract information about the birth and death processes. By modeling how these processes shape firms' age distributions and survival fractions, we show how one can gain insights even from incomplete data. While age distributions are insufficient for inferring both processes, survival fractions reveal how death rates depend on firm age and sector size. Applying our approach to 14 major sectors in Singapore, we find that death rates decline with age and rise with sector size, with a multiplicative interaction between both effects. Assuming sigmoidal dependence on both factors, we infer sector-specific death models that accurately reproduce the data and enable reconstruction of key system features, e.g., sector size trajectories and birth-death rate correlations.

Keywords: birth-death dynamics · age distributions · survival fractions

1 Introduction

The number of firms in a city evolves as new firms emerge and others close. These birth and death rates vary over time and across firm sectors [2,3], and predicting sector size changes can inform urban planning and policy decisions such as space allocation across sectors or targeted incentives.

Understanding firm birth and death dynamics typically requires longitudinal data that track both entry and exit dates. Such data enable the extraction of annual births, closures, sector sizes, and the ages of surviving and exiting firms (Fig. 1a). This allows direct analysis of how births depend on market conditions (e.g., sector size) and how survival rates depend on firm-level properties (e.g., age, sector) and external factors. However, full entry and exit records are often unavailable—a limitation noted in prior studies [5,6,8]. For example, cross-sectional age distribution analyses often rely only on currently active firms [5,6]. In Singapore, the Accounting and Corporate Regulatory Authority (ACRA) provides public data on all firms, including their sector, registration date, and current status (if alive) [1], but not exit dates of inactive firms.

When key information is missing, one potential approach to inferring birth and death dynamics is to use statistical quantities such as age distributions and survival fractions. Age distributions can be derived from registration dates of existing firms alone, while survival fractions for firms born in a given year can be obtained from registration dates and current status, without needing exit dates (Fig. 1a). While these quantities have been empirically studied in various countries [5,6,10,13], the relationship between them and the underlying birth and death processes has been less explored. Hence, it is unclear to what extent they can be used to infer firm population dynamics.

By modeling how age and sector size dependencies in birth and death rates shape age distributions and survival fractions, we show how to infer firm population dynamics under two incomplete-data scenarios (Fig. 1a): (1) extinct firms are not recorded and (2) exit dates are missing. While the age distribution alone—available in both scenarios—is insufficient to recover birth and death dynamics, the survival fractions available in scenario 2 enable inference of how death rates vary with age and sector size. Applying our approach to 14 major sectors in Singapore, we find that death rates decline with age and rise with sector size, with multiplicative interaction between both effects. Assuming a sigmoidal dependence on both factors, we infer sector-specific death models that accurately reproduce empirical data and enable reconstruction of other system properties, e.g., sector size trajectories and correlations between birth and death rates.

2 Results

2.1 The Model

Let $\mu(a, N)$ be the death rate (i.e., probability per unit time) of a firm of age a when N firms (of the same sector) are present. The survival fraction $f(a, T)$ is the probability that a firm survives from the time it was born $t = T - a$ to the current time T:

$$f(a,T) = e^{-\int_{T-a}^{T} \mu(s-(T-a), N(s)) ds}, \quad (1)$$

where $N(t) = \int_0^\infty \rho(a,t) da$, with $\rho(a,t)$ being the density of firms of age a.

Let $B(t)$ be the birth rate of new firms at time t. The density of firms of age a at time T is then the product of their birth rate at time $T - a$ (if $a < T$) or their initial density (if $a \geq T$), and the probability of surviving until T:

$$\rho(a,T) = \begin{cases} B(T-a) f(a,T) & \text{for } a < T \\ \rho(a-T, 0) e^{-\int_0^T \mu(s+a-T, N(s)) ds} & \text{for } a \geq T. \end{cases} \quad (2)$$

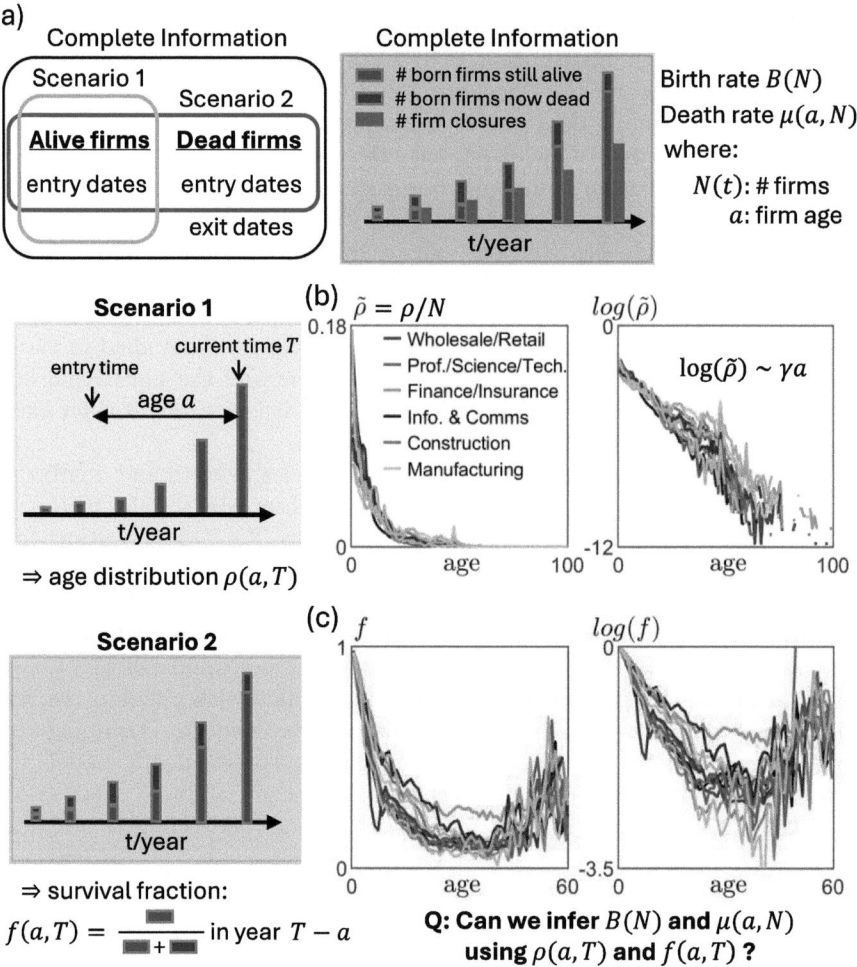

Fig. 1. Gaining insights into birth and death dynamics using incomplete data. (a) With complete information (entry and exit dates of all firms), one can directly extract birth rates $B(N)$ and death rates $\mu(N,a)$, where N is the number of firms and a is firm age. We consider two incomplete-data scenarios. In Scenario 1, only currently existing firms and their entry dates are available, from which the age distribution $\rho(a,T)$ can be computed. In Scenario 2, the dataset includes all firms and their alive status, but lacks exit dates; here, survival fractions $f(a,T)$ can be obtained. We examine whether birth and death models can be inferred in these cases. (b) Normalized age distribution $\tilde{\rho}(a) = \rho(a)/N$ (left) and $\log(\tilde{\rho}(a))$ (right) for each of the 14 largest firm sectors in Singapore. The approximate linear decay in $\log(\tilde{\rho}(a))$ implies that the age distribution approximately follows the exponential distribution. For clarity, only the 6 largest sectors are labeled. (c) The survival fraction $f(a)$ (left) and $\log f(a)$ of each firm sector. A non-linear $\log f(a)$ implies a non-constant death rate. We also ask how the observed non-monotonicity in $f(a)$ can arise.

With the entry dates of all firms (Scenario 2, Fig. 1a), the birth rate trajectory $B(t)$ can be directly extracted. However, this is impossible with only the list of alive firms (Scenario 1, Fig. 1a). In this case, we model the birth dynamics by assuming that time variation in $B(t)$ arises solely from its dependence on $N(t)$: $B(t) = \beta(N(t))N(t)$, where $\beta(N)$ is the birth rate coefficient.

2.2 Scenario 1: Only Age Distributions Are Available

Prior studies in many countries have shown that firm age distributions often follow an exponential form [5,6]. Using registration data for all active firms in Singapore (as of end-2022) and their sector classifications [1], we reconstruct $\rho(a)$ for the 14 largest sectors and find a similarly good approximation to the exponential distribution (Fig. 1b). This raises the question of how exponential age distributions arise. Proposed mechanisms include exponentially growing births with no deaths, or constant birth and death rates [5,6]. A constant death rate with birth rates fluctuating around a constant mean can also yield exponential $\rho(a)$ [5]. However, these mechanisms may not be exhaustive. Notably, while deviations from exponential $\rho(a)$ was found to occur with age-dependent death rates under constant average birth rates [5], the effect of sector size-dependent death rates $\mu(N)$ remains unexplored. Here, we investigate N-dependent birth and death models (Eqs. 2 and 1) and examine their resulting $\rho(a)$.

Constant birth rate. If the birth rate is constant $\left(\beta(N) = \frac{\beta_0}{N} \implies B(t) = \beta_0\right)$, and the death probability is age-independent ($\mu(a, N) = \mu(N)$), then from Eqs. 1–2, $\frac{d \log \rho(a,T)}{da} = \frac{d \log f(a,T)}{da} = -\mu(N(T - a))$. Thus, if $\mu(N(T - a)) = \mu_0$, $\rho(a, T) \sim e^{-\mu_0 a}$, which is independent of β_0. This also implies that if births are approximately constant and $\rho(a)$ is exponential, then the death rate must also be approximately constant over time.

Constant birth rate coefficient. If $\beta(N) = \beta_0$, such that the overall birth rate $B(N) = \beta_0 N$, and $\mu(a, N) = \mu(N)$, then from Eqs. 1–2, $\frac{d \log \rho(a,T)}{da} = -\frac{N'(T-a)}{N(T-a)} - \mu(N(T - a)) = -\beta_0$, where $N'(t) = \frac{dN(t)}{dt} = (\beta_0 - \mu(N))N$ is the dynamics of $N(t)$. Thus, $\rho(a, T) \sim e^{-\beta_0 a}$, regardless of whether $\mu(N)$ is constant or increases with the number of competitor firms.

These results show that while $\rho(a)$ constrains possible birth and death dynamics, multiple models can yield the same $\rho(a)$ shape. Distinguishing between them requires additional information. In particular, since both processes influence firm numbers, access to $N(t)$—e.g., via proxies like corporate tax records or revenue—can, together with $\rho(a)$, enable inference of both $B(N)$ and $\mu(N)$.

2.3 Scenario 2: Survival Fractions Are Also Available

With all firms' entry dates and alive status, we can compute survival fractions $f(a)$ for each birth cohort, where a is the cohort age (Fig. 1b). Unlike the standard cohort-tracking approach (which gives survival fractions over time $f(t)$) [10,13], our method does not need exit dates, which are missing in our dataset.

Across sectors, the $f(a)$ curves are convex and notably non-monotonic (Fig. 1c). While convexity has been observed in $f(t)$ [6,10,13], non-monotonicity is unique to our age-based $f(a)$, as $f(t)$ is always decreasing (by definition). This convexity and non-monotonicity in $f(a)$ are absent in $\rho(a)$ (Fig. 1b), and the shape difference between them implies that $B(t)$ is not a constant (Eq. 2).

Non-Monotonic $f(a)$ Suggests Multiplicative Interactions Between the Effects of Age and Sector Size on Death Rate. Since $f(a)$ depends solely on the death process, its shape can be used to infer properties of the death rate $\mu(a, N)$ (Eq. 1). In particular, a non-exponential $f(a)$, i.e., a non-linear $\log(f(a))$ curve (Fig. 1c, right), indicates that the death rate is not a constant.

To understand how a non-monotonic $f(a)$ can arise, we consider different forms of $\mu(a, N)$ and ask how they affect the sign of $\frac{d \log f(a,t)}{da}$ (using Eq. 1):

(i) **Effects of a and N are additive separable:** $\mu(a, N) = \mu_\alpha(a) + \mu_n(N)$.
In this case, $\frac{d}{da}\mu(a, N) = \mu'_\alpha(a)$, and thus $\frac{d \log f(a,t)}{da} = -\mu_n(N(t-a)) - \mu_\alpha(a) < 0$ for all values of a, implying that $f(a)$ will always be monotonic. This includes cases where the death probability depends only on age ($\mu_n(N) = 0$) or sector size ($\mu_\alpha(a) = 0$). Hence, the non-monotonicity in the observed $f(a,t)$ implies an age- and N-dependent death probability.

(ii) **Effects of a and N are multiplicative separable:** $\mu(a, N) = \mu_\alpha(a)\mu_n(N)$.
In this case, $\frac{d}{da}\mu(a, N) = \mu'_\alpha(a)\mu_n(N)$. Since $f(a,t)$ is non-monotonic in a if $\left.\frac{d \log f(a,t)}{da}\right|_{a \to t} > 0$, the condition for non-monotonicity is found to be: $\int_0^t \mu_\alpha(s)\mu'_n(N(s))N'(s)ds > \mu_\alpha(t)\mu_n(N(t))$. Since the number of firms grows over time ($N'(t) > 0$) and the death probability typically increases with competition ($\mu'_n(N) > 0$), this condition can be satisfied.

These results suggest a multiplicative interaction between age and sector size in determining firm death rates. We thus assume $\mu(a, N) = \mu_\alpha(a)\mu_n(N)$ and assess whether this model reproduces the observed survival and age distributions.

Hill Functions for $\mu_\alpha(a)$ and $\mu_n(N)$ can Reproduce Data. We assume that death rates increase with sector size ($\mu'_n(N) > 0$) and decrease with age ($\mu'_\alpha(a) < 0$), consistent with higher exit rates for young firms [7,11,13]. To keep death probabilities bounded between 0 and 1, we use sigmoidal Hill functions:

$$\mu_\alpha(a) = \mu_{\alpha,ub} - (\mu_{\alpha,ub} - \mu_{\alpha,lb})\frac{a^{m_\alpha}}{a^{m_\alpha} + K_\alpha^{m_\alpha}} \tag{3}$$

$$\mu_n(N) = \mu_{n,lb} + (\mu_{n,ub} - \mu_{n,lb})\frac{N^{m_n}}{N^{m_n} + K_n^{m_n}}, \tag{4}$$

with $\mu_{\alpha,lb}$ ($\mu_{n,lb}$)/$\mu_{\alpha,ub}$ ($\mu_{n,ub}$) being the lower/upper bounds; K_α (K_n) sets the midpoint; m_α (m_n) is the Hill coefficient controlling the transition steepness.

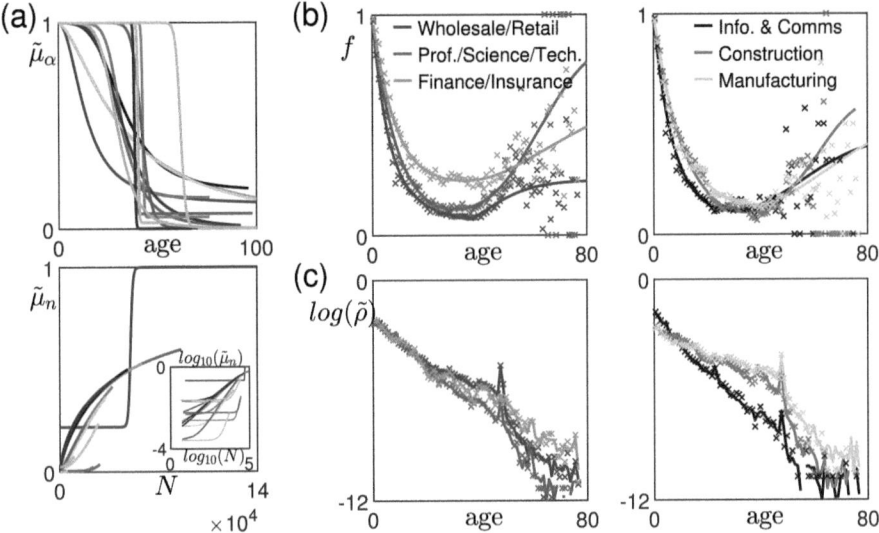

Fig. 2. Inferred death models accurately capture empirical data. (a) Inferred Hill functions for how death rates depend on age $\tilde{\mu}_\alpha(a)$ (Eq. 3) and sector size $\tilde{\mu}_n(N)$ (Eq. 4). (b) Inferred models (solid lines) give rise to $f(a)$ that agree well with the data ('x' markers). (c) Inferred models can recover observed normalized age distributions $\tilde{\rho}(a)$.

Since only the product $\mu_\alpha(a)\mu_n(N)$ affects dynamics, there are 7 effective parameters: $\mu_{\alpha,ub}\mu_{n,ub}$, $\tilde{\mu}_\alpha = \mu_{\alpha,lb}/\mu_{\alpha,ub}$, $\tilde{\mu}_n = \mu_{n,lb}/\mu_{n,ub}$, K_α, m_α, K_n, m_n. Given any parameter values and birth rate data, we solve for $N(t)$ and $f(a,T)$ using Eqs. 1-2. Starting from an initial guess, we iteratively adjust the model parameters to minimize the difference between predicted and observed survival fractions. Through inferring the best-fit parameters, we find that Hill functions for a decreasing $\mu_\alpha(a)$ and an increasing $\mu_n(N)$, combined multiplicative, can provide a good model for the death process (Fig. 2). By combining the predicted survival fractions with birth rate data, we also accurately recover the observed age distributions (Fig. 2c).

Inferred Death Models Provide Insights Into Other System Properties. With the inferred death models (Fig. 2a), we estimated the death rate of a newly born firm today, $\mu(0, N_f)$, where N_f is the current sector size. $\mu(0, N_f)$ falls between 0.08 and 0.15 and does not correlate with N_f (Fig. 3a). Such variation in survival rates across sectors could inform policies, e.g., subsidy allocation.

With both inferred models and birth data, we reconstructed sector sizes $N(t)$, revealing sharp growth transitions and shifts in the economy— while 'Wholesale/Retail Trade' sector has always been largest, 'Professional/Scientific/Technical Activities' only emerged as the second largest in the past 20 years (Fig. 3b). These $N(t)$ also allow model validation if other data sources become available.

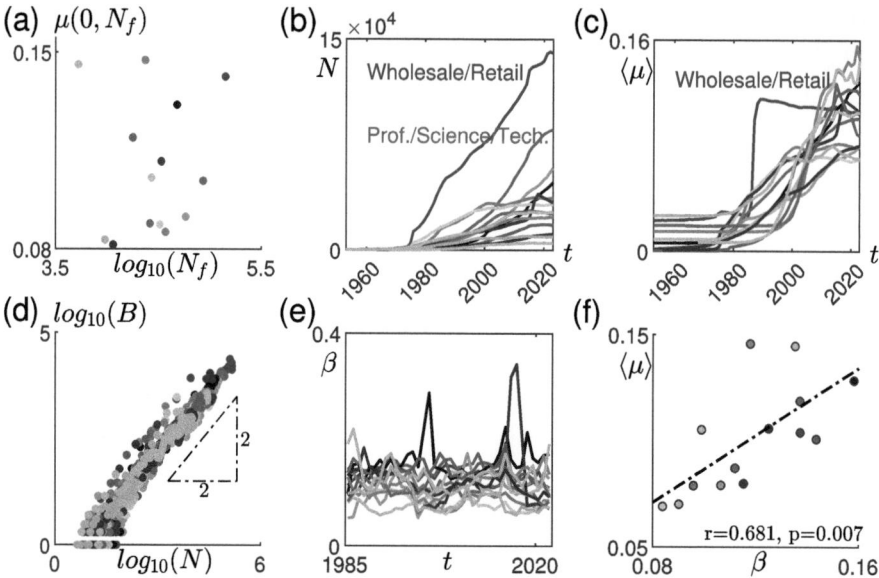

Fig. 3. Inferred death models reveal broader system features. (a) The current death rate of new firms, $\mu(0, N_f)$, shows no correlation with sector size N_f. (b) Reconstructed sector sizes $N(t)$ show rapid growth around 1980 and structural shifts— 'Wholesale/Retail Trade' stayed largest, while 'Professional/Scientific/Technical Activities' rose to second only in the last 20 years. (c) Average death rates $\langle\mu\rangle$ generally rose over time, except in 'Wholesale/Retail Trade', where rates stayed stable for 40 years. (d) Birth rates B scale approximately linearly with N across all sectors. (e) Birth rate coefficients $\beta = B/N$ are approximately constant in the last 40 years. (f) Sectors with higher average death rates tend to have higher β (Pearson correlation coefficient $r = 0.681$, p-value $p = 0.007$).

We also recovered past age distributions $\rho(a, t)$ and used them to compute average death rates $\langle\mu(t)\rangle$. While $\langle\mu(t)\rangle$ rises over time for most sectors, the 'Wholesale/Retail Trade' sector has a stable death rate for many decades (Fig. 3c).

Using reconstructed $N(t)$ and birth data, we find an approximately linear birth model $B(N)$ across all sectors (Fig. 3d,e). Comparing the average birth coefficients $\beta = B/N$ with current death rates $\langle\mu\rangle$ reveals a positive correlation between them (Fig. 3f), consistent with studies in other countries [3,4].

3 Discussion

Incomplete data is a common issue, with some detailed historical information lost for good. In such cases, inferring system dynamics from population-level statistics becomes invaluable. We show how firm age distributions and survival fractions can reveal underlying birth and death dynamics, even without data on extinct firms or exit dates. Though we focused on Singapore, our approach may apply to other countries and population types (e.g., ecological populations).

Age distributions alone constrain but do not fully determine underlying dynamics. In contrast, the convex, non-monotonic survival fraction curves suggest that death rates drop with age and rise with sector size, and both effects combine multiplicatively. Prior studies report differing age-dependence: death rates are constant in Japan but higher for young firms in the U.S. and many developing countries [9,11]. Our findings offer a useful comparison and can guide incentive policies towards younger firms. Unlike age effects, sector size's influence on survival is less studied, and our results may prompt further exploration.

Other firm-level traits—e.g., size or profitability—can also affect survival [11]. While not explicitly modeled, some are implicitly captured (e.g., firm size may correlate with age). Our sector-level approach models average firm dynamics without tracking every micro-level detail. Nonetheless, these factors can lead to heterogeneity in death rates and are critical for predicting individual firm survival. If such granular predictions are desired and the relevant data are available, our framework could be extended to include further dependencies and dynamics.

While we treated sectors independently and found that a common family of models captures their dynamics, cross-sector interactions exist [12]. Also, firm interactions depend on spatial proximity, with birth and death rates influenced by local competition or complementary businesses [14]. Incorporating such effects offers exciting directions for future work.

Acknowledgments. This work is supported by A*STAR Scholars' Development Fund (Y.G.) and Cities of Tomorrow Grant (H.H.N, F.L., project number CoT- H1-2025-3).

Disclosure of Interests. All authors declare that they have no competing interests.

References

1. List of firms in Singapore. https://www.acra.gov.sg/about-bizfile/updates-and-announcements/acra-s-open-data-initiative. Accessed 29 Feb 25
2. Bartelsman, E., Scarpetta, S., Schivardi, F.: Comparative analysis of firm demographics and survival: evidence from micro-level sources in oecd countries. Ind. Corp. Chang. **14**(3), 365–391 (2005)
3. Brixy, U., Grotz, R.: Regional patterns and determinants of birth and survival of new firms in western Germany. Entrep. Reg. Dev. **19**(4), 293–312 (2007)
4. Brown, J.P., Lambert, D.M., Florax, R.J.: The birth, death, and persistence of firms: creative destruction and the spatial distribution of us manufacturing establishments, 2000–2006. Econ. Geogr. **89**(3), 203–226 (2013)
5. Calvino, F., Giachini, D., Guerini, M.: The age distribution of business firms. J. Evol. Econ. **32**(1), 205–245 (2022)
6. Coad, A.: Investigating the exponential age distribution of firms. Economics **4**(1), 20100017 (2010)
7. Coad, A.: Firm age: a survey. J. Evol. Econ. **28**, 13–43 (2018)
8. Haviland, A., Savych, B.: A description and analysis of evolving data resources on small business (2007)

9. Ishikawa, A., Fujimoto, S., Mizuno, T., Watanabe, T.: The relation between firm age distributions and the decay rate of firm activities in the united states and japan. In: 2015 IEEE International Conference on Big Data (Big Data), pp. 2726–2731. IEEE (2015)
10. Mata, J., Portugal, P.: Life duration of new firms. J. Ind. Econ., 227–245 (1994)
11. McKenzie, D., Paffhausen, A.L.: Small firm death in developing countries. Rev. Econ. Stat. **101**(4), 645–657 (2019)
12. O'Leary, D., Doran, J., Power, B.: The role of relatedness in firm interrelationships. J. Econ. Stud. **51**(9), 36–58 (2023)
13. Persson, H.: The survival and growth of new establishments in Sweden, 1987–1995. Small Bus. Econ. **23**, 423–440 (2004)
14. Rosenthal, S.S., Strange, W.C.: Evidence on the nature and sources of agglomeration economies. In: Handbook of Regional and Urban Economics, vol. 4, pp. 2119–2171. Elsevier (2004)

Enhancing Sentiment Analysis Through Multimodal Fusion: A BERT-DINOv2 Approach

Taoxu Zhao[1], Meisi Li[1], Kehao Chen[1], Liye Wang[1], Xucheng Zhou[1], Kunal Chaturvedi[2], Mukesh Prasad[2], Ali Anaissi[1,3], and Ali Braytee[2(✉)]

[1] School of Computer Science, The University of Sydney, Camperdown, Australia
[2] School of Computer Science, University of Technology Sydney, Ultimo, Australia
ali.braytee@uts.edu.au
[3] TD School, University of Technology Sydney, Ultimo, Australia

Abstract. This paper proposes a multimodal sentiment analysis architecture that integrates text and image data to provide a more comprehensive understanding of sentiments. For text feature extraction, we utilize BERT, a natural language processing model. For image feature extraction, we employ DINOv2, a vision-transformer-based model. The textual and visual latent features are integrated using proposed fusion techniques, namely the Basic Fusion Model, Self-Attention Fusion Model, and Dual-Attention Fusion Model. Experiments on three datasets, the Memotion 7k dataset, MVSA-single dataset, and MVSA-multi dataset, demonstrate the viability and practicality of the proposed multimodal architecture.

Keywords: Sentiment analysis · Fusion models · Multimodal learning · Self-Attention mechanism

1 Introduction

The enhancement of unimodal statistical models, which can serve as components within multimodal frameworks, is progressing through the adoption of innovative methodologies. Notably, the emergence of Bidirectional Encoder Representations from Transformers (BERT) [6] and its variants has demonstrated superior performance over preceding models, such as recurrent neural networks (RNNs) and LSTMs in text-based sentiment classification due to its refined understanding of textual syntax and semantics. In parallel, the introduction of Vision Transformers [7], represents a significant extension of the transformer model architecture, originally conceived for NLP applications, into computer vision. These models leverage the strengths of large pre-trained datasets, showing enhanced efficacy than CNNs in many image-related tasks. However, we find that vision transformer models may be able to better deal with the visual features than CNNs.

A good latent representation greatly influences the sentiment analysis results and provides an excellent foundation for the following fusion phase. Modal fusion is one of the main challenges in multimodal constructions. A single concatenation method [16] can effectively integrate the multimodal information. Huang et al. [9] proposed Deep Multimodal Attentive Fusion, which uses the internal correlation between visual and textual features for sentiment analysis. They also pointed out the drawbacks of the early fusion methods and emphasized the great performance of late fusion methods as they cannot always unlock the full potential of each modal data. Yu et al. [19] used the attention technique on the BERT model to deal with target-oriented tasks. Tsai et al. [15] proposed a cross-attention technique to combine the different latent representations effectively, enhancing the performance of each single modality. Recently, Lee et al. [11] explored multimodal learning by leveraging BERT and DINOv2 for modeling social interactions. However, their focus is on aligning language-visual cues for referent tracking and speaker identification, whereas our work aims at sentiment analysis and proposes novel attention-based fusion techniques tailored for this specific task. In this paper, we use two advanced unimodal models: BERT [6] and DINOv2 [13] and then propose three fusion methods to create a multimodal sentiment analysis model combining the extracted features from text and image modalities, named Basic Fusion, Self-Attention Fusion, and Dual-Attention Fusion. The main contributions are as follows:

- Combined extracted features from multimodal data using different fusion methods, such as the Basic Fusion, Self-Attention Fusion, and Dual-Attention Fusion.
- Extensive experiments to compare the effectiveness of the methods using three multimodal sentiment analysis datasets.

2 Method

First, we explain the unimodal information extraction process for both textual and visual context information, respectively. Next, we detail the various fusion methods that combine latent features from both text and image data to predict sentiment analysis. The overall framework is shown in Fig. 1.

We utilize a BERT layer [6] for initial text processing. This involves feeding a sequence of input tokens $X_t = \{x_1, x_2, ..., x_n\}$ into the BERT model, which produces a sequence of output embeddings $H = \{h_1, h_2, ..., h_n\}$. Each embedding h_i is a 768-dimensional vector that captures the contextual information of the corresponding token within the entire sequence. To adapt these embeddings for our multimodal sentiment analysis, we apply a linear transformation directly to the entire sequence of output embeddings from BERT, reducing the dimensionality of each vector from 768 to 256.

For the final textual latent representation t, we use the transformed CLS embedding t_{CLS}, which represents a condensed view of the entire input sequence.

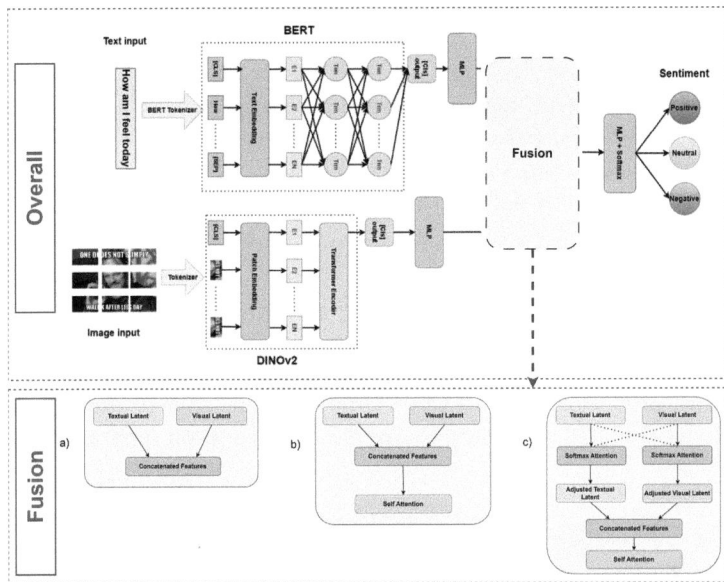

Fig. 1. The overall architecture of the proposed framework (Above). The fusion methodology of the framework (Below): a) Basic Fusion Model; b) Self-Attention Fusion Model; c) Dual-Attention Fusion Model

This approach leverages the CLS token's embedding directly after merging all the necessary information for sentiment analysis following its transformation. The representation is then used as part of the fusion process.

2.1 Visual Features

In the architecture, DINOv2 [13] serves as the foundational layer for extracting complex features directly from image inputs. Each input image is first divided into a sequence of patches, which are then linearly embedded into tokens and processed through the DINOv2 transformer network.

Upon processing images through the transformer, the proposed model employs a custom linear transformation layer, which maps the transformer's output from a 384-dimensional space to a 256-dimensional space for each patch embedding.

For the final visual latent representation v, we use the transformed global feature embedding v_{CLS}, which represents a condensed view of the entire image.

This approach leverages the global feature embedding directly. It is then used as part of the fusion process with textual features extracted from the BERT model in our sentiment analysis framework.

2.2 Attentional Feature Fusion

We propose three fusion methods. A Basic Fusion Model that concatenates the textual and visual latent representation (Fig. 1a). Let t be the textual latent representation obtained from the BERT model, and v be the visual latent representation obtained from the DINOv2 model. Both t and v are vectors. The simple concatenation c is defined in Eq. 1

$$c = [t, v] \qquad (1)$$

where $[t, v]$ represents the concatenation of vectors t and v. The dimension of c will be the sum of the dimensions of t and v. To capture the mutual information of the concatenated latent representation, we introduce a self-attention layer after the concatenation layer as the second method, namely the Self-Attention Fusion Model (Fig. 1b). The output s of the self-attention mechanism is defined in Eq. 2 as,

$$s = \mathrm{Softmax}\left(\frac{cW_Q(cW_K)^T}{\sqrt{d_k}}\right)cW_V \qquad (2)$$

where c is the concatenated vector given from Eq. 1. W_Q, W_K, and W_V are the weight matrices for the query, key, and value, respectively, which are applied to the vector c. d_k is the dimension of the key.

The third method is the Dual-Attention Fusion Model (Fig. 1c). Based on the second fusion method, it additionally uses information from another modality to adjust the latent representation vector of each modality, employing a cross-modal attention mechanism. First, queries, keys, and values for both modalities are computed using Eq. 3,

$$\begin{aligned} Q_t = tW_{Q_t}, \quad K_t = tW_{K_t}, \quad V_t = tW_{V_t}, \\ Q_v = vW_{Q_v}, \quad K_v = vW_{K_v}, \quad V_v = vW_{V_v}. \end{aligned} \qquad (3)$$

Next, we apply softmax attention for cross-modal adjustments as shown in Eq. 4,

$$t' = \mathrm{Softmax}\left(\frac{Q_t K_v^T}{\sqrt{d_k}}\right)V_v \quad v' = \mathrm{Softmax}\left(\frac{Q_v K_t^T}{\sqrt{d_k}}\right)V_t \qquad (4)$$

Finally, we apply concatenation and self-attention as defined in Eq. 5,

$$s' = \mathrm{Softmax}\left(\frac{c'W_Q(c'W_K)^T}{\sqrt{d_k}}\right)c'W_V \qquad (5)$$

where c' is the concatenation of t' and v'.

3 Experiments

3.1 Datasets

Memotion 7k Dataset [14] consists of 6992 samples, each paired with an image and the corresponding caption, representing a complete 'meme'. The dataset is part of a challenge that includes three subtasks: analyzing memes for sentiment, which can be positive, negative, or neutral. MVSA Datasets [12] collected from X, are designed for sentiment analysis on multi-view social data. The MVSA-Single dataset contains 4,869 pairs of images and texts labelled by a single annotator. The MVSA-Multi dataset consists of 19,600 text-image pairs, each labeled by three annotators, ensuring a richer and more robust sentiment analysis. Similarly to the Memotion 7k Dataset, MVSA datasets have three classes: positive, neutral, and negative.

3.2 Experiment Settings

For the Memotion dataset, we employed macro F1 as our evaluation metric. For MVSA-single and MVSA-multi, we used accuracy and F1-score. We utilized focal loss as the loss function and adjusted the γ parameter (γ=2, 3, 4) to ensure the model adequately focuses on the minority classes. To improve the performance of our model, we tuned a set of hyperparameters to facilitate model convergence. Across all datasets, we experimented with different learning rates: 0.01, 0.001, 0.0001, and 0.00001. Adam optimizer is used with dropout rates set to 0, 0.2, and 0.5. Cross-entropy loss function has been used in the experiments.

4 Results and Discussion

MVSA Datasets: The comparative analysis presented in Table 1 showcases the performance of various models on the MVSA-single and MVSA-multi datasets. Notably, the integration of BERT and DINOv2 models, through concatenation, achieves the best performance on the MVSA-single dataset, with an Accuracy of 0.73 and an F1 score of 0.71, surpassing all other models, including MultiSentiNet's attention-based approach (MultiSentiNet-Att) [18], which leads on the MVSA-multi dataset with an accuracy of 0.68 and an F1 score of 0.68.

This approach outperforms the previously established benchmarks, including SentiBank [2], CNN-Multi [4], DNN-LR models [20], and HSAN [17], and even the advanced MultiSentiNet [18] and Dual-Pipeline [3] models. The BERT and DINOv2 model with additional self-attention mechanisms also shows strong performance, underlining the potential of attention mechanisms in multimodal sentiment analysis. This comparison underscores the advances in multimodal sentiment analysis, demonstrating that the fusion of high-performing models like BERT and DINOv2, especially when combined with sophisticated techniques such as self-attention, can lead to substantial improvements. It is worth noting that our model falls short of the state-of-the-art model in the MVSA-multi dataset, which warrants further discussion in the subsequent section. Overall, the performance of our model architectures remains strong, robust, and adaptable.

Memotion 7k Dataset: As shown in Table 2, the proposed models are compared to the state-of-the-art methods [1,5,8,10,14]. Dual-Attention model achieves the highest Macro F1 score of 0.3552, surpassing both the competition

Table 1. Results on MVSA-Single and MVSA-Multi datasets

Model	MVSA-Single		MVSA-Multi	
	Acc	F1	Acc	F1
SentiBank (image only)	0.45	0.43	0.55	0.51
SentiStrength (text only)	0.49	0.48	0.50	0.55
SentiBank + SentiStrength	0.52	0.50	0.65	0.55
HSAN	–	0.66	–	0.67
DNN-LR	0.61	0.61	0.67	0.66
CNN-Multi	0.61	0.58	0.66	0.64
MultiSentiNet-Avg	0.66	0.66	0.67	0.66
MultiSentiNet-Att	0.69	0.69	0.68	0.68
Dual-Pipeline	0.57	0.56	**0.73**	0.69
Ours (Basic Fusion)	**0.73**	**0.71**	0.68	0.67
Ours (Self-Attention)	0.72	0.70	0.68	0.67
Ours (Dual-Attention)	0.72	**0.71**	0.67	0.66

baseline of 0.2176 and several state-of-the-art methods. Notably, it outperforms strong multimodal approaches such as Vkeswani IITK [10], which has a Macro F1 of 0.3546, Guoym [8] with 0.3519, and Aihaihara [14], which has a Macro F1 of 0.3501, all of which incorporate sophisticated combinations of textual and visual features using advanced transformers or ensemble strategies. Compared to our Basic Fusion and Self-Attention variants, the Dual-Attention framework delivers a marked improvement, demonstrating the benefit of simultaneously modeling both cross-modal interactions and intra-modal salience. Furthermore, although several top-performing systems employ powerful feature extractors such as BERT, ResNet, or VGG-16, their relatively close performance suggests diminishing returns from architecture complexity alone. Our results highlight that refined fusion strategies, rather than just deeper encoders, can drive meaningful performance gains in multimodal sentiment classification.

To highlight the limitations of our proposed method, we selected two misclassified images from the testing set. In both of the memes, the individuals are smiling. This likely made the model perceive them as expressing positive emotions (Fig. 2). However, another possible reason for the model's incorrect prediction could be issues with the dataset's annotation. As shown in Example 1 in Fig. 2, although the model has classified a 'neutral' label as 'positive', when we consider what he is saying, from a human perspective, this should be a 'negative' class because his smile is dry. Similarly, as shown in Example 2, the individual is giving a thumbs up with a wide smile, which can easily be interpreted as a display of positive emotion. However, the humor and sarcasm embedded in the

Table 2. Results for the compared methods on Memotion 7k

Model	Macro F1
Competition Baseline [14]	0.2176
Vkeswani IITK [10]	0.3546
Guoym [8]	0.3519
Aihaihara [14]	0.3501
Sourya Diptadas [5]	0.3488
MemoSYS [1]	0.3475
Ours (Basic Fusion)	0.3237
Ours (Self-Attention)	0.3436
Ours (Dual-Attention)	**0.3552**

Example 1

Example 2

Fig. 2. Misclassified memes

memes text Liam Approves! may suggest a satirical or ironic context rather than a genuinely positive emotional state. This highlights the limitations of models that do not integrate textual cues effectively. By using image embeddings that include text, a more nuanced representation of the visual and textual data can be captured, which will help improve the accuracy of emotion classification. This approach will enhance the model's ability to discern more subtle emotional expressions and contextual factors that influence perceived emotions, leading to more accurate predictions.

5 Conclusion

Our proposed multimodal sentiment analysis framework is built on robust unimodal encoders, BERT for text and DINOv2 for images, followed by fusion through three hierarchical strategies: Basic Fusion, Self-Attention Fusion, and Dual-Attention Fusion. Each fusion mechanism is designed to progressively enhance the model's capacity to capture intra- and inter-modal relationships. The results indicated that our fusion methods are highly adept at integrating mutual information across multiple modalities. In the future, we will explore dynamic fusion strategies such as gating mechanisms, transformer-based cross-modal attention, and graph-based modality alignment.

Acknowledgment. We acknowledge the contributions of Yufan Lin to this project.

References

1. Bejan, I.: Memosys at semeval-2020 task 8: multimodal emotion analysis in memes. In: Proceedings of the Fourteenth Workshop on Semantic Evaluation, pp. 1172–1178 (2020)

2. Borth, D., Ji, R., Chen, T., Breuel, T., Chang, S.: Large-scale visual sentiment ontology and detectors using adjective noun pairs, pp. 223–232 (2013). https://doi.org/10.1145/2502081.2502282
3. Braytee, A., Yang, A.S.C., Anaissi, A., Chaturvedi, K., Prasad, M.: A novel dual-pipeline based attention mechanism for multimodal social sentiment analysis. In: Companion Proceedings of the ACM on Web Conference 2024, WWW '24, p. 18161822. Association for Computing Machinery, New York (2024). https://doi.org/10.1145/3589335.3651967
4. Cai, G., Xia, B.: Convolutional neural networks for multimedia sentiment analysis. In: Li, J., Ji, H., Zhao, D., Feng, Y. (eds.) NLPCC -2015. LNCS (LNAI), vol. 9362, pp. 159–167. Springer, Cham (2015). https://doi.org/10.1007/978-3-319-25207-0_14
5. Das, S.D., Mandal, S.: Team neuro at semeval-2020 task 8: multi-modal fine grain emotion classification of memes using multitask learning. arXiv preprint arXiv:2005.10915 (2020)
6. Devlin, J., Chang, M.W., Lee, K., Toutanova, K.: Bert: pre-training of deep bidirectional transformers for language understanding (2019)
7. Dosovitskiy, A., et al.: An image is worth 16×16 words: transformers for image recognition at scale (2021)
8. Guo, Y., Huang, J., Dong, Y., Xu, M.: Guoym at semeval-2020 task 8: ensemble-based classification of visuo-lingual metaphor in memes. In: Proceedings of the Fourteenth Workshop on Semantic Evaluation, pp. 1120–1125 (2020)
9. Huang, F., Zhang, X., Zhao, Z., Xu, J., Li, Z.: Image-text sentiment analysis via deep multimodal attentive fusion. Knowl.-Based Syst. **167** (2019). https://doi.org/10.1016/j.knosys.2019.01.019
10. Keswani, V., Singh, S., Agarwal, S., Modi, A.: Iitk at semeval-2020 task 8: unimodal and bimodal sentiment analysis of internet memes. arXiv preprint arXiv:2007.10822 (2020)
11. Lee, S., Lai, B., Ryan, F., Boote, B., Rehg, J.M.: Modeling multimodal social interactions: new challenges and baselines with densely aligned representations. In: Proceedings of the IEEE/CVF Conference on Computer Vision and Pattern Recognition, pp. 14585–14595 (2024)
12. Niu, T., Zhu, S., Pang, L., El-Saddik, A.: Sentiment analysis on multi-view social data. In: MultiMedia Modeling, p. 1527 (2016)
13. Oquab, M., et al.: Dinov2: learning robust visual features without supervision (2023)
14. Sharma, C., et al.: Task report: memotion analysis 1.0 @SemEval 2020: the visuo-lingual metaphor! In: Proceedings of the 14th International Workshop on Semantic Evaluation (SemEval-2020). Association for Computational Linguistics, Barcelona (2020)
15. Tsai, Y.H.H., Bai, S., Liang, P.P., Kolter, J.Z., Morency, L.P., Salakhutdinov, R.: Multimodal transformer for unaligned multimodal language sequences (2019)
16. Vo, N., et al.: Composing text and image for image retrieval-an empirical odyssey. In: CVPR (2019)
17. Xu, N.: Analyzing multimodal public sentiment based on hierarchical semantic attentional network. In: 2017 IEEE International Conference on Intelligence and Security Informatics (ISI), pp. 152–154 (2017).https://doi.org/10.1109/ISI.2017.8004895
18. Xu, N., Mao, W.: Multisentinet: a deep semantic network for multimodal sentiment analysis, pp. 2399–2402 (2017). https://doi.org/10.1145/3132847.3133142

19. Yu, J., Jiang, J.: Adapting bert for target-oriented multimodal sentiment classification. In: International Joint Conference on Artificial Intelligence (2019). https://api.semanticscholar.org/CorpusID:199465957
20. Yu, Y., Lin, H., Meng, J., Zhao, Z.: Visual and textual sentiment analysis of a microblog using deep convolutional neural networks. Algorithms **9**, 41 (2016). https://doi.org/10.3390/a9020041

Modeling Parallel AI Applications for Performance Analysis on Cloud Environments

Miquel Albert[1]($\boxtimes$), Alvaro Wong[2], Betzabeth Leon[2], Dolores Rexachs[2], and Emilio Luque[2]

[1] Escoles Universitaries Gimbernat (EUG), Computer Science School, Universitat Autonoma de Barcelona, 08174 Sant Cugat del Valles, Spain
miquel.albert@eug.es
[2] Department of Computer Architecture and Operating Systems, Universitat Autonoma de Barcelona, 08193 Bellaterra, Spain

Abstract. In high-performance computing (HPC) environments, efficient execution of AI applications is critical for optimal performance and resource utilization. In this work, we extend the PAS2P methodology to AI applications through message passing on HPC Cloud systems, defining the AI Application Model to describe their performance behavior. This extension identifies phases within AI applications, enabling analysis to focus on these phases instead of the entire application. By concentrating on them, we can better evaluate AI application efficiency, providing insights into system performance and guiding future optimizations for large-scale AI tasks on HPC infrastructure.

Keywords: Performance in HPC · Cloud Computing · AI Applications

1 Introduction

In recent years, the growing demand for AI applications has increased the need for HPC systems to efficiently run large-scale AI workloads [9]. These applications, especially in deep learning and distributed training, require significant computational resources and optimized strategies. However, as AI models grow in complexity, assessing their efficient resource usage has become a major challenge. The problem lies not only in performing a comprehensive performance analysis but also in the uncertainty of whether applications use resources efficiently.

PAS2P [8] (Parallel Application Signature for Performance Analysis and Prediction) is a methodology originally designed to analyze and predict the performance of parallel scientific applications. PAS2P instruments the application to collect performance data from application processes, which are then analyzed to generate an abstract model of the behavior of the application. This model identifies different phases in the execution of the application, each representing a specific segment of parallel code. By analyzing these phases, PAS2P predicts the performance of the application by executing an application signature.

In this work, we extend the Parallel Application Signature for Performance Analysis and Prediction (PAS2P) methodology, originally designed for scientific parallel applications, to AI applications running on HPC systems, including those in cloud computing environments. We propose extending the PAS2P methodology to model the performance of AI applications, enabling the identification of phases within AI workloads. By isolating the phases, we can better analyze the resource efficiency of AI applications, focusing on areas where optimization could have the most significant impact. Quickly evaluating the efficiency of an AI application through its signature helps select cloud resources matched to its performance needs, enabling better resource allocation and optimizing performance and cost.

We have analyzed the MNIST dataset [5], the ResNet-50 model [2], using Horovod [7], k-means [4], a machine learning algorithm for unsupervised learning, DeepGalaxy and pinn-mpi, Physics-informed neural networks (PINNs) are widely used to solve forward and inverse problems in fluid mechanics. The paper is organized as follows: Sect. 2 provides the related works. Section 3 describes the methodology proposed for AI programs. Section 4 presents the model validation followed by a conclusion and future work in Sect. 5.

2 Related Works

In the realm of HPC for AI applications, several approaches have been developed to evaluate the performance of distributed CPU-based workloads. The growing complexity of AI applications, particularly in the areas of deep learning and distributed training, requires efficient techniques to analyze and optimize their performance. Awan et al. [1] analyzed the communication and computational performance of distributed AI applications, focusing on the efficiency of CPU communication in multinode clusters using MPI. Their approach is centered on detailed profiling of the communication workloads using high-performance interconnects like InfiniBand. However, their method involves intensive, time-consuming profiling processes that can be computationally expensive. In contrast, the application signature methodology provides a much faster analysis by focusing on identifying phases and their recurrence, without requiring such detailed profiling.

In application performance benchmarking, several frameworks target AI applications. Tartan [6] uses a multi-GPU benchmark suite, requiring selection of a benchmark that approximates application behavior. Similarly, HPC AI500 [3] evaluates HPC systems for AI workloads with benchmarks for deep learning. In contrast, the signature models application behavior directly, predicting execution time accurately without external benchmarks. It runs in a bounded time frame, making it more efficient and precise. Unlike Tartan and HPC AI500, which rely on selected benchmarks, the application signature focuses on the application's own performance characteristics, making it more adaptive and accurate.

3 IA Paralel Application Model

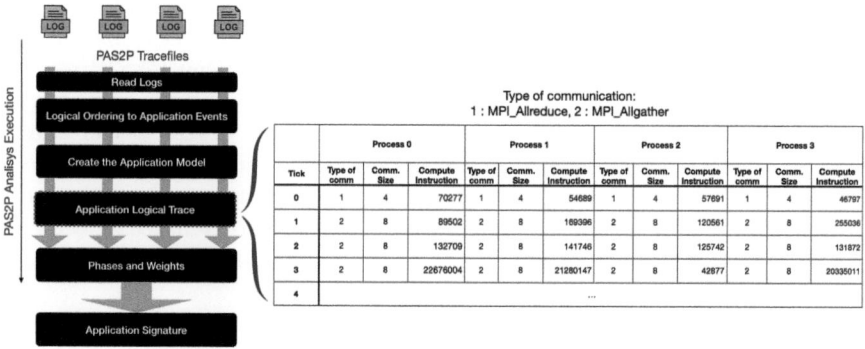

Fig. 1. Logical Trace: Abstract view enabling phase identification.

In distributed AI applications, workload is designed to be uniform as possible, but data partitioning and dynamic training cause computational load variations between processes, leading to differences in executed instructions and challenges in identifying similar phases. We present a methodology to adapt the PAS2P model for detecting execution phases that exhibit similar computational characteristics, despite the inherent differences in workloads across iterations, where each iteration may present slightly different workloads due to the dynamic data distribution. However, the current PAS2P methodology is designed for more deterministic HPC applications. If one process executes a different number of instructions in each iteration, this results in finding multiple phases, as each slight variation is treated as a distinct pattern.

To address this issue, we modeled the PAS2P methodology by designing a matching mechanism to detect phases with similar computational characteristics, despite the differences in workload distribution. This allows us to group phases that share common computational characteristics, even if the number of instructions and CPU times differ slightly between processes. The result of this adaptation is that the phases identified by our model can be used to efficiently analyze performance without being influenced by an excessive number of phases. This approach allows us to identify significant phases that impact performance.

The PAS2P methodology starts by instrumenting the application to extract performance data, saved in trace files. After collecting traces, we generate a logical trace that captures the application's behavior, consisting of communication and computation events. Events like MPI_Allgather and MPI_Allreduce are logged with details such as event ID, data size, timestamp, computational time, and instruction count. These events reveal how the application uses CPU and communication resources across processes.

After collecting performance traces with PAS2P instrumentation, the next step is to generate a logical trace that captures the application's behavior. As

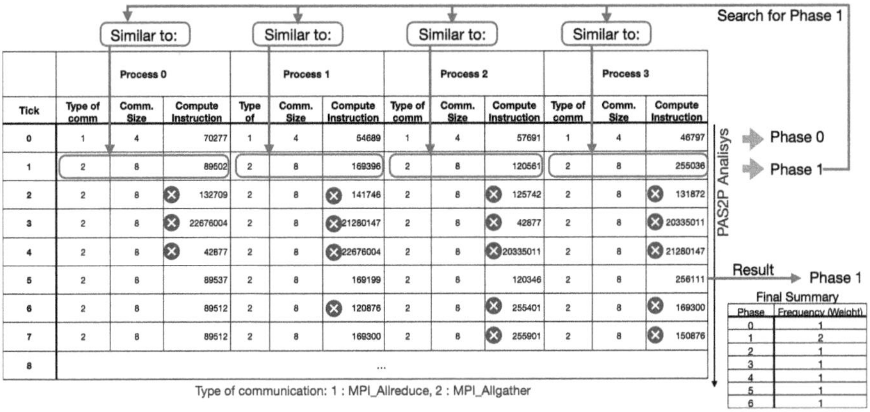

Fig. 2. More processes increase identified phases due to instruction count variations.

shown in Fig. 1, this trace organizes MPI events by logical time, including key operations like MPI_Allgather and MPI_Allreduce. It records details such as communication size and instructions executed per event. Ordering events by logical time creates an abstract performance model, facilitating the identification of patterns or phases during analysis.

The challenge in analyzing the application lies in the variability of instructions executed by each process across iterations, due to factors like data partitioning and workload distribution. As processes receive different data amounts, they execute varying instruction counts, making phase identification difficult. Each process may behave differently depending on its data, leading to many phases—a common issue in AI and deep learning models, where data and workload distribution often fluctuate between iterations.

Current version of PAS2P compares the communication type, communication size, and the number of instructions executed by different processes to identify phases. However, due to the variability in instruction counts across iterations, the conventional approach of using a distance metric to compare instruction counts can result in the identification of numerous phases that are not recognized as similar.

As the number of processes increases, the complexity of identifying phases also increases, as shown in Fig. 2. Each phase is represented by a segment of events across all processes, and this distribution of events makes it harder to identify similar phases. In this example, as more processes are added, PAS2P struggles to group similar phases, especially when instruction counts vary. This variability leads to the identification of more phases, resulting in a higher number of phases being recognized, even when the phases may be functionally similar.

Once the problem was identified, as shown in Table 1, we use a segment of the logical trace from the DeepGalaxy IA application with Horovod and eight processes. The table focuses on the number of instructions executed by each process at each tick. The challenge in identifying similar phases lies in how

Table 1. DeepGalaxy Logical Trace segment with frequencies and SIMO values.

Tick	Processes 0 1 2 3 4 5 6 7	Frequency	SIMO	Tick	Processes 0 1 2 3 4 5 6 7	Frequency	SIMO
0	6 7 6 7 4 6 7 7 0		50	20	6 6 6 5 6 5 7 7 0		48
1	7 6 7 6 6 5 7 7 0		51	21	6 6 5 6 5 6 5 7 0		47
2	6 5 6 6 6 5 7 6 0		47	22	6 5 6 6 6 6 5 7 0		48
3	6 6 6 6 4 5 7 6 0		46	23	7 5 6 6 6 6 7 7 0		50
4	6 6 6 5 4 5 7 6 0		45	24	6 6 5 5 6 6 7 6 0		47
5	6 6 6 5 5 6 7 6 0		47	25	6 6 5 5 6 6 7 6 0		48
6	6 6 6 5 5 5 7 6 0		46	26	6 6 5 5 6 6 5 7 0		46
7	7 7 6 7 5 5 7 7 0		51	27	6 6 5 5 6 5 6 6 0		46
8	7 6 6 7 5 5 7 7 0		50	28	6 6 5 5 5 5 6 6 0		44
9	5 5 6 6 8 8 8 8 0		54	29	7 5 7 5 6 5 7 7 0		49
10	7 8 8 8 8 8 8 8 0		63	30	7 5 6 6 5 6 7 7 0		48
11	5 5 5 5 6 7 6 6 0		45	31	7 5 7 6 7 6 7 7 0		52
12	5 6 5 6 6 7 6 6 0		47	32	7 5 7 5 7 5 7 7 0		50
13	5 6 5 5 5 6 7 6 0		46	33	7 5 7 5 7 6 7 7 0		51
14	5 6 5 6 5 6 7 6 0		46	34	8 8 8 7 8 7 8 8 0		62
15	6 6 5 5 6 7 6 6 0		47	35	8 7 8 7 6 5 6 6 0		53
16	5 6 5 6 6 6 5 5 0		44	36	8 5 8 8 8 7 8 8 0		60
17	6 7 6 6 5 6 6 7 0		49	...	...		
18	6 7 6 6 7 7 6 6 0		51				
19	7 6 6 5 6 7 7 7 0		50				

PAS2P models computation. To address this, we model instruction counts by their order of magnitude, considering segments similar if they share the same order, even with slight variations. After analyzing the trace, we add a column to specify repetitions (weights/frequencies) for each pattern. However, even using the order of magnitude, we did not find similarities between phases, as each tick represents a phase.

As we know, applications aim to balance the workload as evenly as possible across processes, though this depends on the data being processed. In some cases, certain processes must handle more data than others, and this balance can change unpredictably with each iteration. However, we also know that the total workload across all processes remains fixed. That is, the sum of the workload across each process gives the total computational load, as shown by the following equation (where $Compute_i$ represents the workload of each process i, and N is the number of processes):

$$\text{Total Workload} = \sum_{i=1}^{N} Compute_i$$

Tick	SIMO	Group By SIMO Phase ID	Group By SIMO Tolerance +-1 Phase ID	Frequency	Tick	SIMO	Group By SIMO Phase ID	Group By SIMO Tolerance +-1 Phase ID	Frequency
0	50	1	1	5	20	48	10	2	4
1	51	2		4	21	47	3		6
2	47	3	2	6	22	48	10		4
3	46	4		6	23	50	1	1	5
4	45	5	3	2	24	47	3		6
5	47	3	2	6	25	48	10	2	4
6	46	4		6	26	46	4		6
7	51	2	1	4	27	46			6
8	50	1		5	28	44	8	3	2
9	54	6	4	1	29	49	9	1	2
10	63	7	5	1	30	48	10	2	4
11	45	5	3	2	31	52	11	6	1
12	47	3		6	32	50	1	1	5
13	46	4	2	6	33	51	2		4
14	46			6	34	62	12	5	1
15	47	3		6	35	53	14	4	1
16	44	8	3	2	36	60	15	7	1
17	49	9		2	...			...	
18	51	2	1	4					
19	50	1		5					

Final Summary

Phase	Frequency (Weight)
1	11
2	14
3	4
4	2
5	6
6	1
7	1

Fig. 3. SIMO calculation for a DeepGalaxy trace. Grouping SIMO values into phases and applying ±1 tolerance reduces phases from 15 to 7.

Building on this, what we observed is that by summing the instructions of a specific phase across all processes, represented by SIMO (Sum of Instructions in Magnitude Order), we can find the total computational load for each phase. As shown in Table 1, the value of SIMO in the table begins to show similarity at times, which helps to analyze the total computational load for the application. The sum of instructions for each phase can give us a better understanding of the workload distribution and its variability between different processes.

Once we calculate the SIMO, as shown in Fig. 3, the next step is to process the results for phase identification. We first remove the columns related to patterns from the previous figure, focusing on grouping patterns by instruction count similarity. In the Group By SIMO Phase ID column, we assign a unique Phase ID to each set of equal SIMO values. Patterns belong to the same phase if their SIMO values match across processes. In this DeepGalaxy trace, we identified up to 16 distinct phases, showing patterns with equal computational behavior.

However, our goal in performance analysis is not only to find equal patterns but to identify phases meaningful in computational load. To refine this, we introduce a ±1 tolerance for SIMO values in the Group By SIMO Tolerance ±1 Phase ID column. This tolerance groups patterns with slight instruction count variations but similar computational load. Applying this reduces the number of identified phases from 15 to 7, as shown in the Summary in Fig. 3. Furthermore, PAS2P performs a classification of the identified phases, discarding those that are considered insignificant in terms of performance: phases whose execution time considering their weight constitutes less than 1% of whole execution.

Table 2. Phases obtained for DeepGalaxy 4 Processes

Phase	CPU Time(Sec.)	Frequency	CPU Time * Frequency (Sec.)
1	0.000023	2,935	0.067495
2	0.000048	470,550	22.545653
3	0.000037	137,632	5.057869
4	0.000034	202,724	6.828968
5	0.000027	6,550	0.179555
6	0.000634	27,371	17.365881
7	0.000629	152	0.095631
8	0.001395	35	0.048816
Computational Total Time			52.189868

4 Experimental Results

To validate our proposed methodology, we selected a set of AI applications that utilize the Horovod framework. Among them is a ResNet50 model [2], trained for image classification. Another application involves training a model to recognize handwritten digits [5]. We also included the DeepGalaxy application, which simulates galaxy clusters, and trained using 4 to 16 processes. Additionally, we run KMeans clustering, implemented with library mpi4py. These programs were run on two AWS instance types: c7a.4xlarge, with 16 vCPUs, 32 GiB memory, up to 12.5 Gbps network, and AMD EPYC 9R14 processor and c5.9xlarge, with 36 vCPUs, 72 GiB memory, 12 Gbps network, and Intel Xeon 8124M CPU.

We ran the application as shown in Table 2, presenting the results obtained after applying the proposed model to DeepGalaxy with 4 processes. Each table shows: the phase identifier, the CPU time per phase, and the frequency or weight, indicating how many times each phase is executed. By multiplying the CPU time by its weight, we can extrapolate the total computation time for each phase and predict the application's overall execution time. The final row displays the Computational Total Time, summing all phase times multiplied by their weights. This predictive model condenses the application into a few key phases, enhancing the efficiency of performance analysis.

To validate our methodology, we applied the same analysis to all applications in this study. As shown in Table 3, we ran the applications with 4 to 16 processes across 1 and 4 nodes. The table compares the number of phases detected by the previous PAS2P version ("Previous Number of Phases") and the proposed model ("Current Number of Phases"). This reduction will allow for a quicker execution of the signature as it will consist of a smaller set of phases.

Additionally, we performed an analysis on the reduced set of phases. Using the phase information, we calculated the percentage of the total time that applications spent in communications, represented in the column "Percentage of AET

Table 3. Phase Detection and Execution Time Comparison with the PAS2P Model.

Application	Number of processes	Previous number of phases	Current number of phases	Application Execution Time (AET) (Sec.)	Percentage of AET in Comm.	Percentage of AET in Compute Stage
Resnet 50	8	7,079	16	653.55	97.39	2.61
Resnet 50	16	23,993	35	854.83	97.30	2.70
MNIST	4	419	9	555.50	99.42	0.58
MNIST	8	5,335	15	307.73	99.62	0.38
DeepGalaxy	4	285	8	844.24	94.61	5.39
DeepGalaxy	8	4,929	16	581.54	95.33	4.67
K-Means	8	175	16	751.02	0.20	99.80
K-Means	16	526	28	535.23	0.40	99.60
pinn-mpi	16	7	4	642.435	2.81	97.19
Resnet 50	32	80,948	48	455.48	94.12	5.88
MNIST	32	49,855	37	302.79	94.01	5.99
DeepGalaxy	16	18,627	22	201.33	95.70	4.30
K-Means	64	1,258	28	1503.71	2.64	97.36
pinn-mpi	32	6	4	633.671	4.47	95.53
Resnet 50	64	43,558	37	285.349	96.47	3.53
MNIST	64	47,192	49	403.82	98.72	1.28
K-Means	128	1,575	30	1074.31	7.86	92.14
pinn-mpi	64	7	6	627.75	2.51	97.49

in Comm. Stage", and the percentage of time spent in computation, shown in the "Percentage of AET in Compute Stage" column. The results reveal that, for applications using Horovod, most of the time is dedicated to communications, while the time spent on computation is significantly lower. For k-means clustering, the compute stage dominates, accounting for nearly 99% of execution time. This shows k-means is highly compute-intensive with minimal inter-process communication. Thus, cloud instance optimization should prioritize computational power over communication bandwidth, unlike AI training workloads. This insight aids resource management by guiding instance selection based on application reliance on communication or CPU.

5 Conclusions and Future Works

In this work, we proposed extending the PAS2P methodology to model AI applications in HPC Cloud environments. By identifying phases, we enable performance analysis without evaluating the entire application, helping understand resource use and predict execution times. Our results show the methodology effectively models application behavior, reducing phases to improve analysis efficiency and identify inefficiencies.

By analyzing the computational load and communication patterns of each phase, we can identify resource-intensive phases to guide cloud instance selection. For AI applications with Horovod, the communication stage consumes a large part of execution time. The next step is to keep applying this methodology to understand resource usage and explain why applications spend more time in communication or computation, aiding resource allocation and improving performance. As part of future work, we will analyze more AI applications and explore the extension of this methodology to other communication libraries, such as NVIDIA NCCL (NVIDIA Collective Communications Library). This will further improve the methodology's applicability to various AI workloads.

Acknowledgment. This research has been supported by the Agencia Estatal de Investigación (AEI), Spain and the Fondo Europeo de Desarrollo Regional (FEDER) UE, under contracts PID2020-112496GB-I00 and PID2023-146978OB-I00.

References

1. Awan, A.A., Jain, A., Chu, C.H., Subramoni, H., Panda, D.K.: Communication profiling and characterization of deep learning workloads on clusters with high-performance interconnects. In: IEEE Symposium on High-Performance Interconnects (HOTI), pp. 49–53. IEEE (2019)
2. He, K., Zhang, X., Ren, S., Sun, J.: Deep residual learning for image recognition. In: Proceedings of the IEEE Conference on Computer Vision and Pattern Recognition, pp. 770–778 (2016)
3. Jiang, Z., et al.: HPC AI500: a benchmark suite for HPC AI systems. In: Zheng, C., Zhan, J. (eds.) Bench 2018. LNCS, vol. 11459, pp. 10–22. Springer, Cham (2019). https://doi.org/10.1007/978-3-030-32813-9_2
4. Jin, X., Han, J.: K-Means Clustering, pp. 563–564. Springer, Boston (2010). https://doi.org/10.1007/978-0-387-30164-8_425
5. LeCun, Y., Bottou, L., Bengio, Y., Haffner, P.: Gradient-based learning applied to document recognition. Proc. IEEE **86**(11), 2278–2324 (1998). https://doi.org/10.1109/5.726791
6. Li, A., Song, S.L., Chen, J., Liu, X., Tallent, N., Barker, K.: Tartan: evaluating modern cpu interconnect via a multi-cpu benchmark suite. IEEE Trans. Parallel Distrib. Syst. **30**(7), 1358–1370 (2018). https://doi.org/10.1109/TPDS.2018.2805956
7. Sergeev, A., Del Balso, M.: Horovod: fast and easy distributed deep learning in tensorflow. arXiv preprint arXiv:1802.05799 (2018)
8. Wong, A., Rexachs, D., Luque, E.: Parallel application signature for performance analysis and prediction. IEEE Trans. Parallel Distrib. Syst. **26**(7), 2009–2019 (2015)
9. Yi, G., Loia, V.: High-performance computing systems and applications for AI. J. Supercomput. **75**(8), 4248–4251 (2019). https://doi.org/10.1007/s11227-019-02937-z

Simplified Swarm Learning Framework for Robust and Scalable Diagnostic Services in Cancer Histopathology

Yanjie Wu[1], Yuhao Ji[1], Saiho Lee[1], Juniad Akram[1,2,3(✉)], Ali Braytee[2], and Ali Anaissi[1,2]

[1] The University of Sydney, Camperdown, NSW 2008, Australia
{yawu2780,yuji6835,slee6156}@uni.sydney.edu.au,
ali.anaissi@sydney.edu.au, ali.anaissi@uts.edu.au
[2] University of Technology Sydney, Ultimo, NSW 2007, Australia
ali.braytee@uts.edu.au
[3] Australian Catholic University, North Sydney, NSW 2060, Australia
junaid.akram@sydney.edu.au, junaid.akram@uts.edu.au,
junaid.akram@acu.edu.au

Abstract. The complexities of healthcare data, including privacy concerns, imbalanced datasets, and interoperability issues, necessitate innovative machine learning solutions. Swarm Learning (SL), a decentralized alternative to Federated Learning, offers privacy-preserving distributed training, but its reliance on blockchain technology hinders accessibility and scalability. This paper introduces a *Simplified Peer-to-Peer Swarm Learning (P2P-SL) Framework* tailored for resource-constrained environments. By eliminating blockchain dependencies and adopting lightweight peer-to-peer communication, the proposed framework ensures robust model synchronization while maintaining data privacy. Applied to cancer histopathology, the framework integrates optimized pre-trained models, such as TorchXRayVision, enhanced with DenseNet decoders, to improve diagnostic accuracy. Extensive experiments demonstrate the framework's efficacy in handling imbalanced and biased datasets, achieving comparable performance to centralized models while preserving privacy. This study paves the way for democratizing advanced machine learning in healthcare, offering a scalable, accessible, and efficient solution for privacy-sensitive diagnostic applications.

Keywords: Single-cell Sequencing Integration · Multi-Omics · Dimensionality Reduction · Normalization

1 Introduction

The exponential growth in healthcare data, coupled with advancements in machine learning, has catalyzed significant progress in medical diagnostics [2,5,8]. However, challenges such as data privacy, imbalanced datasets, and the lack of interoperable frameworks continue to hinder the effective adoption of artificial intelligence (AI) in healthcare. Swarm learning (SL), a decentralized form of federated learning, has emerged as a promising solution by allowing distributed model training without the need for centralized data aggregation [1,3]. Unlike traditional methods, SL emphasizes privacy preservation by ensuring that sensitive patient data remains on-site while enabling collaborative model development across institutions. These capabilities make SL particularly attractive for domains like cancer histopathology, where data privacy and algorithmic efficiency are paramount [15]. Despite its potential, the adoption of SL frameworks has been constrained by their reliance on complex infrastructures, such as blockchain technology, and a lack of adaptability to resource-constrained environments [13]. Existing frameworks, such as HPE's Swarm Learning, depend on blockchain for model aggregation and consensus, which poses significant barriers for non-technical users and smaller organizations. This complexity restricts the accessibility and scalability of SL in real-world applications, particularly in healthcare [7].

To address these challenges, this study introduces a *simplified peer-to-peer swarm learning (P2P-SL) framework* designed to overcome the limitations of existing SL architectures. By eliminating blockchain dependencies and incorporating lightweight communication protocols, the proposed framework offers a more accessible and scalable alternative. The framework leverages dynamic networks to facilitate model synchronization and aggregation, ensuring robust performance even in imbalanced and biased data scenarios. Furthermore, it integrates state-of-the-art pre-trained models, such as TorchXRayVision, with domain-specific optimizations, demonstrating its efficacy in cancer histopathology tasks. This study is grounded in the premise that simplifying the SL framework while maintaining its privacy-preserving and decentralized nature can unlock new possibilities for healthcare diagnostics. By focusing on histopathology, a domain characterized by high data sensitivity and diagnostic variability, this work highlights the transformative potential of distributed learning frameworks in improving diagnostic outcomes and operational efficiency. The major contributions are summarized as follows:

- *Introduction of a simplified P2P-SL framework* that eliminates blockchain dependencies, making swarm learning more accessible and scalable for non-technical users and resource-constrained environments.
- *Development of dynamic networking mechanisms* to enable seamless model synchronization and aggregation across nodes, ensuring robust performance even in the presence of imbalanced and biased datasets.
- *Integration of optimized pre-trained models* with domain-specific enhancements, demonstrating the framework's applicability and efficacy in cancer histopathology tasks.
- *Comprehensive experimental evaluation* comparing the proposed framework with centralized and standalone models, showcasing its superiority in privacy preservation, scalability, and diagnostic accuracy.

2 Literature Review

Swarm Learning (SL), a decentralized extension of Federated Learning (FL), enables distributed model training across multiple nodes without a central server by employing blockchain-based consensus (e.g., Ethereum) or peer-to-peer synchronization, thereby enhancing privacy, scalability, and fault tolerance [4,12,14]. The core of SL is the model merging algorithm, which aggregates local parameter vectors into a unified representation; while simple arithmetic averaging requires consistent parameter shapes, it often degrades performance when some nodes possess limited data, a shortcoming addressed by the Federated Averaging (FedAvg) algorithm through weighted averaging based on local dataset sizes[9,11], yet FedAvg does not fully consider the geometric implications of parameter aggregation. Model training seeks to solve optimization problems where stochastic gradient descent iteratively updates model weights (points on the loss hypersurface), and averaging such weights corresponds to locating intermediate points whose impact on performance depends on the curvature of the hypersurface. To overcome these limitations, advanced merging strategies such as FedApprox incorporate auxiliary information from participating nodes, and statistical methods leveraging Fisher information and gradient matching refine aggregation by predicting weight trajectories [6]. Alternative aggregation paradigms include ensemble learning, which combines predictions via voting to enhance reliability at the expense of scalability, and transfer learning approaches that merge local models through shared pre-trained representations [10]; notwithstanding these contributions, the development of efficient, scalable, and mathematically principled model merging algorithms remains a pivotal challenge for robust SL frameworks.

3 Proposed Method

The proposed framework emphasizes dynamic networking capabilities, allowing nodes to dynamically discover and register within the swarm network. Each node operates independently, exchanging model updates directly without a central server. To ensure data integrity and security during communication, mechanisms such as TLS/SSL encryption and robust authentication protocols are integrated. Monitoring and logging tools are employed to track training progress, model transformations, and overall system performance across nodes. These tools also aid in identifying and mitigating challenges related to latency, network stability, and resource allocation, enabling robust performance under real-world conditions.

3.1 Dynamic Networking and Decentralized Training

The decentralized training process in P2P-SL follows a structured methodology:

1. Nodes dynamically join the network using discovery protocols, establishing secure communication channels.

2. Each node performs localized training using its available data and updates its model parameters.
3. Model updates are shared directly between nodes at predefined intervals, bypassing the need for a centralized server.
4. Model aggregation is performed locally at each node using mechanisms such as weighted averaging, ensuring adaptive synchronization with other peer nodes.

This dynamic approach enhances scalability, fault tolerance, and adaptability, particularly in scenarios with heterogeneous data distributions or imbalanced datasets.

3.2 Peer-To-Peer Swarm Learning Framework

Unlike traditional SL methods that rely on consensus algorithms built on blockchain technology, the P2P-SL framework focuses on improving local models through peer-based interactions. Model weights are updated and aggregated dynamically across nodes without the need for a global model. This ensures that edge nodes retain greater autonomy while benefiting from collaborative training insights.

For machine learning tasks, the proposed framework incorporates advanced modeling architectures to optimize performance. After evaluating various frameworks, TorchXRayVision—a pre-trained model based on chest X-ray datasets from Harvard University—was selected for its robust feature extraction capabilities. The model was further optimized by integrating a DenseNet decoder module, enhancing its ability to process complex medical imaging data.

3.3 Model Architecture

The training architecture is illustrated in Fig. 1. Input images of size 224×224 are processed through four encoder modules, each consisting of four layers, reducing dimensionality to 1024. A fully connected layer with batch normalization and ReLU activation then transforms these features from 1152 to 512 dimensions. Finally, classification is performed using a fully connected layer that maps features from 512 to three dimensions, employing batch normalization and a sigmoid activation function. This architecture ensures efficient and accurate feature extraction and classification, tailored for histopathology tasks.

4 Results and Discussion

4.1 Experimental Setup

To systematically assess the performance of the peer-to-peer swarm learning (P2P-SL) framework, we deployed four compute nodes, each provisioned with identical hardware (NVIDIA RTX 4090 GPU, 24 GB VRAM; AMD EPYC 7443P CPU; 128 GB RAM) under Ubuntu 22.04. The base dataset comprised

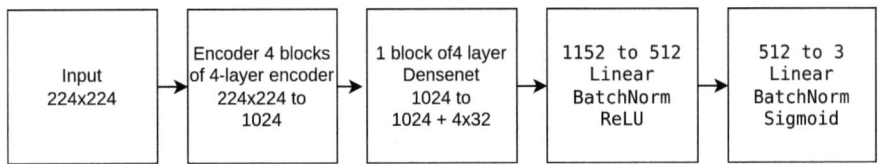

Fig. 1. TorchXRayVision Pre-trained Model Architecture

10,000 annotated histopathology images, pre-processed via Macenko stain normalization, random rotations ($\pm 15°$), horizontal flips, and color jitter (± 0.1). For federated-average unbalanced experiments, Node 0 received 10% (1,000 images), Nodes 1 and 2 each received 30% (3,000 images), and Node 3 received 30% (3,000 images). To probe extreme scarcity, Node 2 was further down-sampled to 25% and Node 3 to 5% in separate trials. Models were fine-tuned for 20 epochs with a batch size of 32, using the AdamW optimizer (weight decay 1×10^{-4}) and a cosine-annealing learning-rate schedule (initial LR 1×10^{-4}), with early-stopping patience of five epochs. Peer exchanges of LoRA-adapter weights occurred every three epochs via gRPC over TLS. Each experiment was repeated five times with different random seeds, and performance metrics (AUC, sensitivity, specificity, F1-score) were computed on a held-out test set of 2,000 images.

4.2 Performance Vs. Centralized and Local Baselines

To quantify the efficacy of P2P-SL relative to centralized and isolated paradigms, we compared swarm-trained models against a simulated centralized "full-data" baseline and fully independent local learners. In the federated-average unbalanced configuration, Node 0 held only 10% of the data while Nodes 1–3 each held 30%. Performance was evaluated on a 2,000-image test set using AUC, sensitivity, specificity, and F1-score. The centralized baseline achieved an AUC of 0.7156, sensitivity of 0.82, specificity of 0.84, and F1-score of 0.78. The standalone Node 0 model achieved an AUC of 0.6192 ± 0.0057, demonstrating severe degradation under limited data. Incorporating P2P-SL with synchronous weight exchanges every three epochs improved Node 0s AUC to 0.6397 ± 0.0036, sensitivity to 0.75, specificity to 0.77, and F1-score to 0.72 (Fig. 2). Node 3s swarm model recovered over 80% of centralized performance, achieving AUC of 0.6892 ± 0.0063, sensitivity of 0.79, specificity of 0.81, and F1-score of 0.73 (Fig. 3).

4.3 Handling Imbalance and Scarcity

We evaluated resilience to skewed sample distributions by simulating nodes with biased data allocations. In one scenario, Node 2s share was down-sampled to 25% of its nominal portion while other nodes retained 30%. The standalone Node 2 model achieved an AUC of 0.6259 ± 0.0028, sensitivity of 0.70, specificity of 0.72, and F1-score of 0.68. After integrating into P2P-SL, Node 2s AUC increased

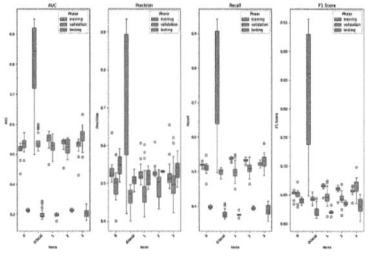
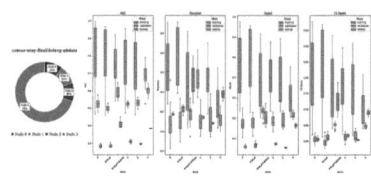

Fig. 2. Federated-average unbalanced: swarm vs. local on Node 0.

Fig. 3. Centralized full-data vs. swarm on Node 3.

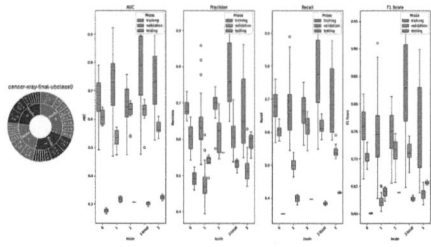

Fig. 4. Testing AUC for Node 2 with 25% data: swarm vs. local.

to 0.6387, sensitivity to 0.74, specificity to 0.76, and F1-score to 0.72 (Fig. 4). To analyze representation learning, we applied t-SNE to penultimate-layer activations: swarm-trained embeddings exhibited 15% lower Davies-Bouldin Index and tighter intra-class clustering compared to local models. Minority-class recall improved by up to 4.5%, demonstrating that collaborative gradient fusion preferentially enhances underrepresented categories.

4.4 Robustness and Overfitting Mitigation

We analyzed training and validation trajectories to elucidate the framework's impact on generalization. In isolated local training, models rapidly attained training AUC ¿ 0.95 within five epochs but validation AUC plateaued at 0.82 before declining, indicative of overfitting. Conversely, swarm-trained models displayed a more gradual ascent, reaching training AUC 0.93 by epoch 10 and maintaining validation AUC of 0.86 ± 0.01 through epoch 20. The generalization gap (training minus validation AUC) at epoch 20 was 0.14 for local versus 0.07 for swarm models—a 50% reduction—while cross-validation variance of AUC decreased by 35% under P2P-SL. Precision-recall analysis further showed an 8% reduction in both false positives and false negatives, confirming that periodic peer aggregation serves as an implicit regularizer, yielding models that generalize more reliably to unseen histopathology scans.

4.5 Discussion

The proposed Peer-to-Peer Swarm Learning (P2P-SL) framework addresses key limitations of existing decentralized learning systems, offering simplified deployment without reliance on blockchain infrastructure, making it accessible to non-expert users. It enhances scalability and fault tolerance through dynamic networking and peer-based aggregation mechanisms while preserving privacy with localized training and secure communication protocols. Compared to HPE Swarm Learning, which depends on a blockchain-based centralized coordinator and predefined aggregation methods, the P2P-SL framework employs a fully decentralized peer-to-peer network. Each device acts independently, sharing model updates every three epochs, and uses a validation-based threshold of 80% to accept model aggregation. This adaptive setup enables flexibility, ease of integration, and improved diagnostic accuracy by leveraging pre-trained models and domain-specific optimizations. Unlike HPE, the P2P-SL framework requires manual intervention to terminate training, allowing for greater customization. Furthermore, swarm learning models in P2P-SL demonstrate superior mitigation of overfitting, as evident in steady improvements in metrics such as Micro-AUC, Precision, Recall, and F1-Score. These advantages make the P2P-SL framework a practical and efficient solution for resource-constrained environments, particularly in healthcare diagnostics, where privacy preservation, adaptability, and reliable model performance are critical.

5 Conclusion

The proposed Peer-to-Peer Swarm Learning (P2P-SL) framework represents a transformative approach to decentralized machine learning, addressing critical challenges of scalability, privacy, and accessibility in distributed environments. By eliminating the complexities of blockchain-based architectures and introducing a dynamic peer-to-peer networking model, the framework achieves robust model performance with minimal resource constraints. Experimental results highlight its effectiveness in mitigating overfitting, handling imbalanced datasets, and ensuring consistent improvements in diagnostic accuracy through adaptive model aggregation. The integration of pre-trained models further enhances its applicability, particularly in sensitive domains like healthcare. Compared to traditional frameworks, P2P-SL offers unparalleled flexibility, allowing customized deployments tailored to specific needs while maintaining strict data privacy. These advancements establish P2P-SL as a practical and scalable solution for real-world applications, paving the way for broader adoption of swarm learning across diverse fields and contributing to the democratization of advanced machine learning technologies.

References

1. Aamir, M., Raut, R., Jhaveri, R.H., Akram, A.: Ai-generated content-as-a-service in iomt-based smart homes: personalizing patient care with human digital twins. IEEE Trans. Consum. Electron. (2024)

2. Akram, A., Akram, J., Alabdultif, A., Anaissi, A., Jhaveri, R.H.: Secure and interoperable iomt-based smart homes. IEEE Consum. Electron. Maga. (2025)
3. Alsharif, M.H., Kannadasan, R., Wei, W., Nisar, K.S., Abdel-Aty, A.H.: A contemporary survey of recent advances in federated learning: taxonomies, applications, and challenges. Internet Things, 101251 (2024)
4. Anaissi, A., Braytee, A., Akram, J.: Fine-tuning llms for reliable medical question-answering services. In: 2024 IEEE International Conference on Data Mining Workshops (ICDMW). IEEE (2024)
5. Asif, S., et al.: Advancements and prospects of machine learning in medical diagnostics: unveiling the future of diagnostic precision. In: Archives of Computational Methods in Engineering, pp. 1–31 (2024)
6. Daheim, N., Mllenhoff, T., Ponti, E.M., Gurevych, I., Khan, M.E.: Model merging by uncertainty-based gradient matching (2023). https://arxiv.org/abs/2310.12808v1
7. Gao, Z., Wu, F., Gao, W., Zhuang, X.: A new framework of swarm learning consolidating knowledge from multi-center non-iid data for medical image segmentation. IEEE Trans. Med. Imaging **42**(7), 2118–2129 (2022)
8. Khan, M., Saad, M.M., Tariq, M.A., Kim, D.: Spice-it: Smart covid-19 pandemic controlled eradication over ndn-iot. Inf. Fusion **74**, 50–64 (2021)
9. McMahan, H.B., Moore, E., Ramage, D., Hampson, S., Arcas, B.A.Y.: Communication-efficient learning of deep networks from decentralized data (2017). https://arxiv.org/abs/1602.05629v4
10. Papernot, N., Abadi, M., Goodfellow, I., Talwar, K.: Semi-supervised knowledge transfer for deep learning from private training data (2016). https://doi.org/10.48550/arXiv.1610.05755. http://arxiv.org/abs/1610.05755
11. Qian, C., Shi, X., Yao, S., Liu, Y., Zhou, F., Zhang, Z.: Optimized biomedical question-answering services with llm and multi-bert integration. In: 2024 IEEE International Conference on Data Mining Workshops (ICDMW). IEEE (2024)
12. Sah, M.P., Singh, A.: Aggregation techniques in federated learning: Comprehensive survey, challenges and opportunities. In: 2022 2nd International Conference on Advance Computing and Innovative Technologies in Engineering (ICACITE), pp. 1962–1967 (2022)
13. Saldanha, O.L., et al.: Swarm learning for decentralized artificial intelligence in cancer histopathology. Nat. Med. **28**(6), 1232–1239 (2022)
14. Sung, Y.L., Li, L., Lin, K., Gan, Z., Bansal, M., Wang, L.: An empirical study of multimodal model mergin (2023). https://doi.org/10.48550/arXiv.2304.14933. http://arxiv.org/abs/2304.14933
15. Warnat-Herresthal, S., et al.: Swarm learning for decentralized and confidential clinical machine learning. Nature **594**(7862), 265–270 (2021)

A Fast MPI-Based Distributed Hash-Table as Surrogate Model for HPC Applications

Max Lübke[1(✉)], Marco De Lucia[2], Stefan Petri[3], and Bettina Schnor[1]

[1] Institute of Computer Science, University of Potsdam, An der Bahn 2, 14476 Potsdam, Germany
{mluebke,schnor}@uni-potsdam.de
[2] GFZ Helmholtz Centre for Geosciences, Telegrafenberg, 14473 Potsdam, Germany
delucia@gfz.de
[3] Potsdam Institute for Climate Impact Research, Member of the Leibniz Association, Telegrafenberg, 14473 Potsdam, Germany
petri@pik-potsdam.de

Abstract. Surrogate models can enhance performance in HPC applications. Cache-based surrogates use pre-calculated simulation results for interpolation or extrapolation. However, this is only effective if retrieval is faster than simulation. This paper proposes an MPI-based distributed architecture where parallel processes share memory to build a distributed hash table. Three DHT approaches for HPC are presented. A lock-free design outperforms both coarse-grained and fine-grained locking DHTs, demonstrating good scaling for read and write performance. The lock-free DHT improved the runtime of a reactive transport simulation up to 42%.

Keywords: Distributed Hash Table · Key-Value Store · Surrogate Model · RDMA

1 Introduction

During the last years, several research groups have demonstrated the benefits of surrogate models for accelerating parallel multi-physics simulations. While some propose physics-informed neural networks, others use fast caches within their surrogate approach [2,3,8,15]. For example, Reaktoro employs On-Demand Machine Learning (ODML) to speed up chemical calculations by extrapolating new chemical states [8]. POET utilizes a distributed hash table (DHT) as fast storage for simulation results, which are reused to approximate subsequent results [2]. However, the efficiency of such cache-based surrogates hinges on query and retrieval times being significantly faster than full physics simulations. Additionally, using a surrogate model results in a trade-off between runtime and accuracy.

The performance of distributed key-value stores has been a focus of much research, including *addressing*, *data consistency*, and *collision handling*. Another key advancement has been the adoption of RDMA-capable networks to accelerate communication [1,2,4–7,12,14]. While most key-value stores are server-based [4,6,7,9,14], this architecture is suboptimal for HPC due to additional hardware and setup requirements. Server-based approaches offer data consistency, but distributed approaches can enable direct data access from remote storage using RDMA, at the cost of requiring additional synchronization protocols. Due to space restrictions, we refer to a deeper disucssion of server-based key-value stores and related work in [10].

MPI, the de facto standard in HPC, offers an RMA API [13, pp. 547], making it a natural choice for implementing key-value stores to leverage HPC advancements [5,12]. An MPI-based DHT was previously integrated into the POET simulator [2].

This paper explores fully distributed key-value store architectures based on MPI for seamless integration into scientific applications. The contributions of this work are:

- Presenting three distributed hash-table architectures using the MPI one-sided communication API: two synchronized and one lock-free approach.
- Evaluating the three approaches, demonstrating the performance advantages of the lock-free approach by up to 1,400 times in a synthetic benchmark.
- The lock-free DHT's excellent scaling for read and write requests, achieving 16 million read and 15 million write operations per second with 640 processes.
- Integrating the lock-free DHT as a fast data cache into a reactive transport simulation, results in runtime improvements of up to 42%.

2 Improving Synchronization Methods for Distributed Hash Tables with MPI

The MPI-DHT API [2] consists of four operations: `DHT_create`, `DHT_read`, `DHT_write`, and `DHT_free`. `DHT_read` and `DHT_write` use MPI's one-sided communication operations `MPI_Put` and `MPI_Get` to access remote memory. This section presents three approaches to solve the *data consistency* problem:

- *Coarse-grained locking*: The original MPI-DHT, employing MPI's passive target synchronization.
- *Fine-grained locking*: An MPI-DHT variant utilizing MPI's atomic operations for explicit synchronization.
- *Lock-free*: MPI-DHT with optimistic concurrency control, using checksums for conflict detection.

The MPI library was chosen for its wide use in HPC, providing a "plug-and-play" solution for MPI-parallelized applications.

2.1 Coarse-Grained Lock-Based MPI-DHT

The original implementation uses MPI's passive target synchronization with MPI_Win_lock and MPI_Win_unlock [2]. Each process allocates memory, which is divided into buckets containing a key-value pair and metadata, and which is shared with the other processes using a MPI_Window. A bucket's address is a process rank and an index of the bucket within the memory window. A 64-bit hash sum of the key determines the target rank, and a set of bucket indices is derived from the hash sum.

For write operations, the target rank and possible bucket indices are calculated. The first bucket is checked; if unoccupied, data is written. If occupied, the next bucket is checked until an empty bucket or a matching key is found. Read operations similarly traverse buckets to find a matching key.

The MPI-DHT implements a Readers&Writers semantic. DHT_read or DHT_write operations lock the *entire* target rank's memory window in shared or exclusive mode, respectively.

2.2 Fine-Grained Lock-Based MPI-DHT

Here, adressing and collision handling are identical to the coarse-grained approach. However, instead of locking the entire memory window, the fine-grained mechanism locks only the bucket being accessed. To avoid excessive memory overhead from using MPI windows for each bucket, the implementation uses self-implemented locking on 8-byte integers with MPI_Compare_and_swap and MPI_Fetch_and_op operations, inspired by Open MPI's passive-target synchronization.

A lock value of $0x10000000$ indicates an active writer (exclusive lock). Write processes atomically set the lock to this value if it is currently zero. Readers increment the lock atomically until the value is less than $0x10000000$. If the value is greater or equal to $0x10000000$, the read request is revoked, and the process tries to acquire the lock again. This allows concurrent reads but exclusive writes. Locks are released by decrementing the lock value. An 8-byte lock is included in each bucket, requiring up to 7 bytes of padding, resulting in a maximum overhead of 15 bytes per bucket compared to the coarse-grained approach.

All windows are pre-locked with MPI_Win_lock_all during setup to ensure that the MPI standard is not violated by performing RMA operations outside an epoch [13, p. 588].

2.3 Lock-Free MPI-DHT

This approach uses the same addressing and collision handling but replaces locking with checksum calculation. Each bucket includes a 32-bit value for storing a checksum. The approach is inspired by the work of Pilaf [14].

For DHT_write, the origin process calculates a checksum of the key-value pair and appends it to the bucket data. A reading process retrieves the bucket data, recalculates the checksum, and returns the key-value pair if both checksums match. Mismatches trigger a retry, and persistent mismatches invalidate the bucket. Write operations can overwrite invalid buckets. This checksum-based approach is a lock-free mechanism, avoiding atomic operations or locks. It adds only 4 bytes of memory overhead per bucket for the checksum compared to the coarse-grained approach.

Similar to the fine-grained approach, all windows are locked by all processes with `MPI_Win_lock_all` prior to any RMA operation.

3 Evaluation

3.1 Test Bed

The benchmarks were conducted on the cluster of the Potsdam Institute for Climate Impact Research (PIK). Each node has two AMD EPYC 9554 CPUs with 64 cores each and a base clock speed of 3.1 GHz. All 128 cores per node were used. The nodes share 768 GB of DDR5 memory and are interconnected via NVIDIA Mellanox ConnectX-7 NDR Infiniband (400 Gbps per port). The GNU Compiler Collection 14.1 and OpenMPI 5.0.6 with UCX 1.17.0 were used for the experiments. The code was compiled with `-O3 -DNDEBUG`. Open MPI was configured to use single atomic remote memory operations[1].

3.2 Synthetic Benchmarks

- **First experiment:** This benchmark evaluates maximum read/write throughput. It generates a random number to derive an 80-byte key, with a value size of 104 bytes (modeling POET's key-value pair size) [2]. Uniform and Zipfian distributions were used for random number generation. The Zipfian distribution (skew of 99, range 1 to 712,500) models POET's access requests. The benchmark writes and then reads 500,000 key-value pairs.
- **Second experiment:** This benchmark evaluates a more read-intensive mixed workload (95% reads, 5% writes), also reflecting POET's access pattern [2].

The experiments used 1 to 5 nodes (up to 640 processes), with each process providing 1 GB of memory for the DHT. The benchmark was replicated five times, and median throughput values, calculated as operations per second, with standard deviations are shown in Figs. 1a and 1b (uniform distribution) and Figs. 2a and 2b (Zipfian).

In the first experiment, the lock-free DHT outperformed both synchronized approaches in read-only and write-only benchmarks with both distributions. For example, with 640 processes, the lock-free DHT achieved over 16 million read operations per second, about 3 times higher than fine-grained locking and 2 times higher than coarse-grained locking. Write throughput was lower than read throughput for all approaches, as expected.

Table 1 shows write-only performance for 640 processes, demonstrating the overhead of locking. The lock-free DHT performed significantly better than the other approaches.

---

[1] See `osc_ucx_acc_single_intrinsic` of Open MPI's MCA configuration: https://docs.open-mpi.org/en/main/mca.html.

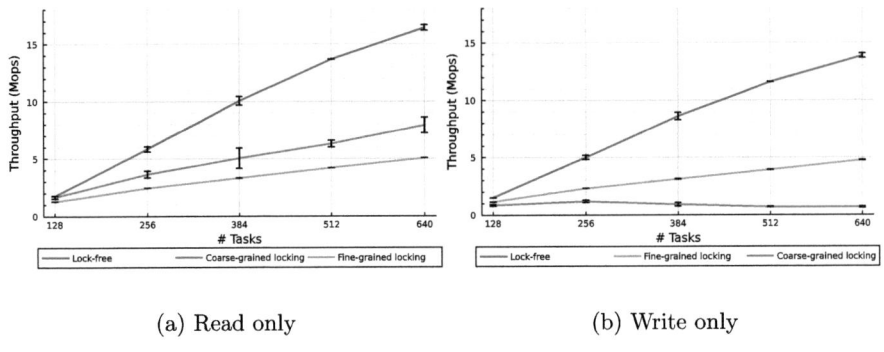

(a) Read only (b) Write only

Fig. 1. Throughput of read and write operations with *uniform* distributed keys.

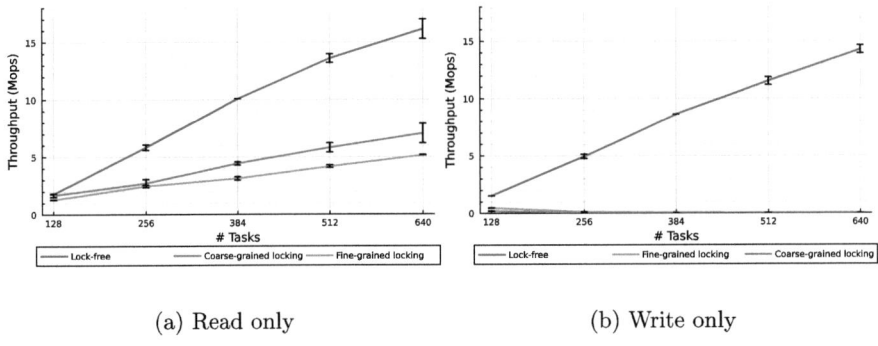

(a) Read only (b) Write only

Fig. 2. Throughput of read and write operations with *zipfian* distributed keys.

For the second mixed workload benchmark (Fig. 3a and 3b), the lock-free DHT's throughput was close to its read-only performance. Fine-grained locking showed some improvement over coarse-grained locking with the uniform distribution, but both synchronized approaches were challenged by the Zipfian distribution. Table 2 shows checksum mismatches in the lock-free DHT. Mismatches occurred only with the Zipfian distribution, indicating concurrent writes, but with negligible results.

Table 1. First experiment: Write-only Performance for 640 Processes in Million Operations per Second.

Workload	Coarse-Grained	Fine-Grained	Lock-Free
uniform	0.67	4.75	13.9
zipfian	0.01	0.03	14.3

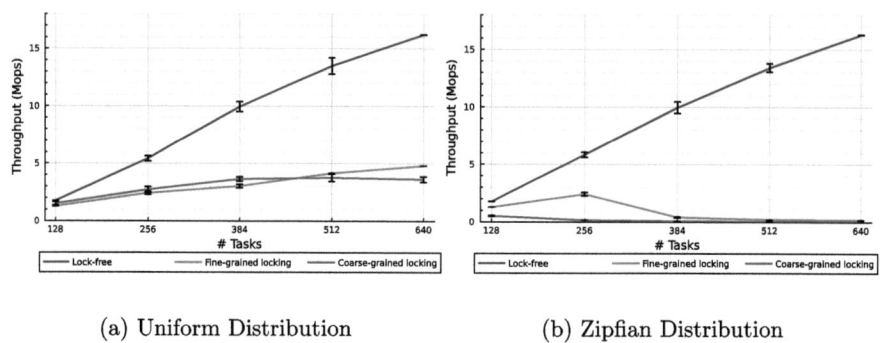

(a) Uniform Distribution (b) Zipfian Distribution

Fig. 3. Throughput of mixed workload with a 95% Read/5% Write ratio for uniform and zipfian distributed keys.

Table 2. Second experiment: Checksum mismatches for the lock-free DHT.

Workload	# of Tasks	# of Mismatches	Percentage [%]
mixed - Zipfian	128	13	$1.1 \cdot 10^{-5}$
mixed - Zipfian	256	16	$6.5 \cdot 10^{-6}$
mixed - Zipfian	384	25	$6.8 \cdot 10^{-6}$
mixed - Zipfian	512	31	$6.3 \cdot 10^{-6}$
mixed - Zipfian	640	64	$1.1 \cdot 10^{-5}$
Others	Any	0	0

3.3 HPC Use-Case: POET

POET is an MPI-parallelized reactive transport simulator that combines solute flow and transport in porous media with geochemical reactions. The simulation used an explicit upwind advection scheme on a 500×1500 grid with homogeneous species concentrations and the same chemistry setup as in previous work [2]. Due to advective transport, a sharp reaction front occurs, allowing caching of previously simulated results in the DHT. Input parameters for the geochemical simulation are rounded to serve as keys for the DHT, and stored values consist of the exact results. The simulation ran for 500 time steps, simulating flux, transport, and geochemical reaction.[2]

The simulation was scaled from 1 to 5 nodes, with one MPI task per CPU core. Experiments were repeated three times for each DHT approach and a reference without DHT. Median runtimes are shown in Fig. 4. The average hit rate over all DHT runs was 91.8%.

The lock-free DHT was the only approach that improved simulation runtime. The best result was a 41.9% reduction in simulation time with 128 processes. Table 3 shows the checksum mismatches observed during simulations with

---
[2] The code is available: [11].

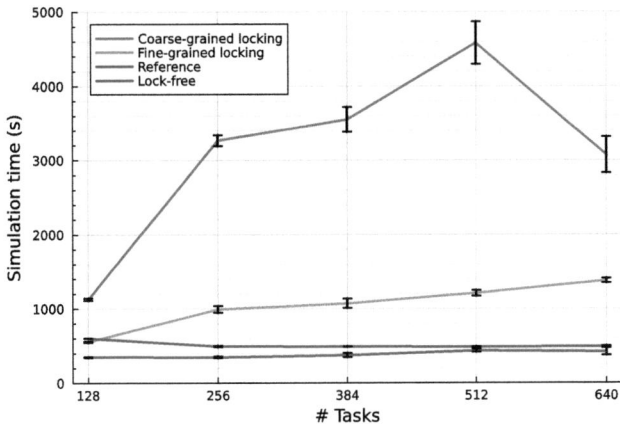

Fig. 4. Runtime of the chemical simulation of POET w/ and w/o DHT.

the lock-free DHT, which were minimal. Additionally, the relative performance gain is shown.

Table 3. Checksum mismatches and performance gain of the POET simulation with lock-free MPI-DHT compared to the Reference Run without DHT.

# of Tasks	# of Mismatches	Percentage [%]	Performance Gain [%]
128	1507	$4.4 \cdot 10^{-4}$	41.9%
256	3049	$8.9 \cdot 10^{-4}$	29.5%
384	4315	$1.3 \cdot 10^{-3}$	23.3%
512	2884	$8.4 \cdot 10^{-4}$	10.1%
640	4421	$1.3 \cdot 10^{-3}$	14.1%

4 Conclusion and Future Work

This paper compared MPI-based DHT implementations with coarse-grained locking, fine-grained locking, and lock-free approaches. The lock-free DHT outperformed the synchronized approaches in synthetic benchmarks, with read throughput of 16 million operations per second and write throughput of 15 million operations per second. In the POET simulation, the lock-free DHT improved runtime, reducing it up to 42%.

Acknowledgments. The authors gratefully acknowledge the Ministry of Research, Science and Culture (MWFK) of Land Brandenburg for supporting this project by providing resources on the high performance computer system at the Potsdam Institute for Climate Impact Research.

Disclosure of Interests. The authors have no competing interests to declare that are relevant to the content of this article.

References

1. Cassell, B., Szepesi, T., Wong, B., Brecht, T., Ma, J., Liu, X.: Nessie: a decoupled, client-driven key-value store using RDMA. IEEE Trans. Parallel Distrib. Syst. **28**(12), 3537–3552 (2017). https://doi.org/10.1109/TPDS.2017.2729545
2. De Lucia, M., Kühn, M., Lindemann, A., Lübke, M., Schnor, B.: POET (v0.1): speedup of many-cores parallel reactive transport simulations with fast DHT-Lookups. Geosci. Model Dev. **14**(12), 7391–7409 (2021). https://doi.org/10.5194/gmd-14-7391-2021
3. De Lucia, M., Kühn, M.: DecTree v1.0 – chemistry speedup in reactive transport simulations: Purely data-driven and physics-based surrogates. Geosci. Model Dev. **14**(7), 4713–4730 (2021). https://doi.org/10.5194/gmd-14-4713-2021
4. Dragojević, A., Narayanan, D., Castro, M., Hodson, O.: FaRM: fast remote memory. In: 11th USENIX Symposium on Networked Systems Design and Implementation (NSDI 14), pp. 401–414 (2014)
5. Gerstenberger, R., Besta, M., Hoefler, T.: Enabling highly-scalable remote memory access programming with MPI-3 One Sided. Sci. Program. **22**(2), 75–91 (2014). https://doi.org/10.3233/SPR-140383
6. Jose, J., et al.: Memcached design on high performance RDMA capable interconnects. In: 2011 International Conference on Parallel Processing, pp. 743–752 (2011). https://doi.org/10.1109/ICPP.2011.37
7. Kalia, A., Kaminsky, M., Andersen, D.G.: Using RDMA efficiently for key-value services. In: Proceedings of the 2014 ACM Conference on SIGCOMM, Chicago, Illinois, USA, pp. 295–306 (2014). https://doi.org/10.1145/2619239.2626299
8. Leal, A., Kyas, S., Kulik, D.A., Saar, M.O.: Accelerating reactive transport modeling: on-demand machine learning algorithm for chemical equilibrium calculations. Transp. Porous Media, 1–44 (2020). https://doi.org/10.1007/s11242-020-01412-1
9. Liang, Z., Lombardi, J., Chaarawi, M., Hennecke, M.: DAOS: a scale-out high performance storage stack for storage class memory. In: Panda, D.K. (ed.) SCFA 2020. LNCS, vol. 12082, pp. 40–54. Springer, Cham (2020). https://doi.org/10.1007/978-3-030-48842-0_3
10. Lübke, M., De Lucia, M., Petri, S., Schnor, B.: A fast MPI-based Distributed Hash-Table as Surrogate Model demonstrated in a coupled reactive transport HPC simulation (2025). https://doi.org/10.48550/arXiv.2504.14374
11. Lübke, M., De Lucia, M., Petri, S., Schnor, B.: MPI-DHT source code and benchmarks (ICCS25) [Data set] (2025). https://doi.org/10.5281/zenodo.15235555
12. Maynard, C.M.: Comparing UPC and one-sided MPI: a distributed hash table for GAP. Technical report, Cray User Group (GUG) (2012). https://cug.org/proceedings/attendee_program_cug2012/includes/files/pap195.pdf
13. Message Passing Interface Forum: MPI: A Message-Passing Interface Standard Vers. 4.1 (2023). https://www.mpi-forum.org/docs/mpi-4.1/mpi41-report.pdf
14. Mitchell, C., Geng, Y., Li, J.: Using one-sided RDMA reads to build a fast, CPU-efficient key-value store. In: Proceedings of the 2013 USENIX Conference on Annual Technical Conference, USENIX ATC'13, USA, pp. 103–114 (2013)
15. Raissi, M., Perdikaris, P., Karniadakis, G.: Physics-informed neural networks: a deep learning framework for solving forward and inverse problems involving nonlinear partial differential equations. J. Comput. Phys. **378**, 686–707 (2019). https://doi.org/10.1016/j.jcp.2018.10.045

Rockburst Forecasting Using Composite Modelling for Seismic Sensors Data

Ilia Revin, Vadim A. Potemkin(✉), and Nikolay O. Nikitin

ITMO University, Saint Petersburg, Russia
{ierevin,vadim_potemkin}@itmo.ru
https://en.itmo.ru/

Abstract. Seismic monitoring is used to ensure the safety of workers in the rock massif. The main security threat is a rockburst, which can be predicted based on the sequence of seismic events. An important task is to develop a mining forecasting model that can take into account the structural heterogeneity of the mountain range and select the necessary forecast horizon depending on monitoring data. In the paper, we propose a flexible approach that combines multiple machine learning models designed to solve various tasks (clustering, time series forecasting) as parts of one composite model. This approach allows for adjustment of the forecast horizon of the model, which enables it to flexibly adapt to rock massifs with different geological structures and seismic monitoring stations. Also, the use of clustering models allows us to take into account the physical and mechanical features of the rockburst formation process. According to experimental results, the resulting composite model showed more accurate results for specific forecast horizons, compared with classical "hierarchical" models and machine learning models. At the same time, the obtained model allows us to interpret the results from the rock mechanics point of view.

Keywords: Rockburst forecasting · Geomonitoring · Clustering · Time series forecasting · Machine learning · Data-driven modeling

1 Introduction

Rockbursts are a highly complex dynamic phenomenon. The formation of a rockburst is influenced by many factors, such as the physical and mechanical properties of the rock mass, stress state, geological structure, and engineering position [1]. The classical method of rockburst forecasting is the use of various statistical criteria. However, those criteria ignore rock massif's physical and mechanical parameters, and it leads to poor quality of forecast [2].

Many machine learning methods are used in rockburst forecasting [3] and it is an actively developing field of signal processing and machine learning methods for solving rock mechanic problems. However, these models can restore only extremely simple and trivial relationships between seismic events and the probability of a rockburst [4]. Another problem is "class imbalance", which is expressed in the fact that rockburst is an extremely rare event, and their number is significantly less than the number of seismic events. Correcting class imbalance leads to a lack of data in the training dataset and the inability to use deep learning models [5]. The principal scheme of the seismic monitoring station and an example of data visualization is shown in Fig. 1. An example of sensor data from a seismic monitoring system is shown in Table 1, where the input data consists of timestamps and coordinates of seismic events.

Table 1. Example of input seismic sensor data

Timestamp	X	Y	Z	Energy
00:00:00	0.120	0.200	0.032	1200
00:01:00	0.1284	0.201	0.038	2000
00:02:00	0.1089	0.513	0.011	11000

The sequence of seismic events appears during mining processes in the massif and can be represented as a time series that consists of discrete events. Each event is characterized by a time coordinate and a characteristic describing the "degree of destruction" (e.g., the magnitude of the signal's energy). Clustering methods are used to localize the spatial zone of a potential rockburst, which corresponds to each event with its spatial cluster (Fig. 1).

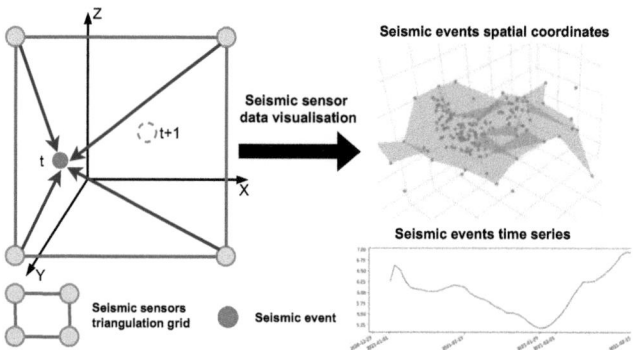

Fig. 1. Principal scheme of the seismic monitoring station, based on the triangulation of seismic events in the rock massif.

However, the geological heterogeneity of the rock massif leads to an uneven distribution of rockburst. It leads to the appearance of clusters with an increased or decreased probability of a rockburst.

In this paper, we propose a data-driven automated hybrid modelling approach, which intends to solve the problems described above. The main idea is to combine temporal and spatial sensor data in one feature space and apply an automated forecasting model design using a graph-based pipeline representation and evolutionary optimization. We use spatial clustering methods based on machine learning models for spatial and temporal data seismic monitoring combinations. Then, we present the data within each cluster as a discrete time series. The proposed approach is fully automated and has been tested on experimental data, where it has shown its effectiveness compared to similar approaches.

2 Related Works

Classical methods of rockburst forecasting are based mainly on a deterministic empirical approach [6]. This approach cannot be adapted to the uncertain conditions of a complex dynamic system, which is the studied rock massif. In the rock mechanic "hierarchical model" theory, the criterion for the rockburst formation is a violation of the stationarity/quasi-stationarity conditions of the modeled process (for example, the Poisson process). Hierarchical models can usually predict only the total number and time of rockbursts at fixed spatial coordinates.

Short-term prediction methods based on monitoring seismic data can simultaneously predict both the time of the occurrence and the expected location of the rockburst [4]. These methods are primarily focusing on signal processing and filtering. Nevertheless, this approach has proven to be more flexible and scalable than the empirical, numerical, and physical models [7].

Deep learning models have long established themselves as an effective method for modeling various processes using both temporal and spatial data types. In several works [7], the use of such models for the problem of rockburst prediction has shown high efficiency. The reason is the ability to model complex non-linear relationships between factors affecting the probability of rockburst occurrences. However, the disadvantages of this approach include the low interpretability of the model results and the tendency to overfit due to the relatively small size of the datasets.

The hybrid approach has also found its application in rockburst forecasting. This approach combines the capabilities of existing rock mechanic models and machine learning models. Classical models provide a representation of geological heterogeneities in the rock massif. ML models reproduce complex nonlinear connections between inhomogeneities, external factors, and the seismic activity of the rock massif. Accordingly, the combination of classical models that take into account the physical laws and machine learning models capable of modeling complex nonlinear dependencies is the most promising trend of development in this area [8]. Another hybrid approach introduced a hybrid model combining KMeansSMOTE oversampling with Random Forest optimization using Optuna (KMSORF). This model demonstrated high accuracy in predicting rockburst levels in real-world mining projects while addressing challenges like imbalanced datasets. Another work integrated Particle Swarm Optimization (PSO) with neural networks (e.g., BPNN) and ensemble methods like XGBoost, achieving prediction accuracies exceeding 0.9 in practical applications [12]. These developments highlight how combining classical geomechanical principles with machine learning techniques can enhance prediction precision and robustness under complex geological conditions.

3 Proposed Approach

Currently, the identification of data-driven models with a complex heterogeneous structure remains an unsolved problem [9]. The desired mathematical model can be developed using a single machine learning model and a hybrid (composite) approach [10]. The set of clustering models includes HDBSCAN, KMeans and Spectral models. For time series forecasting, such models as Singular Spectrum Analysis (SSA), Random Forest, and XGBoost are used.

The aim of the proposed approach is to predict rockbursts in a technologically disturbed rock massif. We can consider our massif as a discrete dynamical system $X_{next} = F(X_{cur})$, where X_{cur} is the current state of the massif. The discrete-time propagator F is given by the flow map:

$$F(X_{cur}) = X_{cur} + \int_{k\Delta t}^{(k+1)\Delta t} f(x(r)) dr \qquad (1)$$

Since one of the stages of the model is spatial clustering of seismic events, X_{cur} can be represented as a matrix $X \in \mathbb{R}^{N \times M}$, where N is the number of clusters, and M is the length of the time series of seismic events.

$$X = \begin{bmatrix} x_1(t_1) & \dots & x_1(t_m) \\ x_2(t_1) & \dots & x_2(t_m) \\ x_n(t_1) & \dots & x_n(t_m) \end{bmatrix} \qquad (2)$$

The future state of the system X_{next} can be expressed in a similar way:

$$X_{next} = \begin{bmatrix} x_1(t_{m+1}) & \dots & x_1(t_{m+k}) \\ x_2(t_{m+1}) & \dots & x_2(t_{m+k}) \\ x_n(t_{m+1}) & \dots & x_n(t_{m+k}) \end{bmatrix} \qquad (3)$$

Here the hyper-parameters of the proposed model are N and K—the length of the forecast.

The task of finding function F is simultaneously a data-driven modeling task and a multi-criteria optimization task [11], shown in Eq. 4. Since both spatial and temporal coordinates describe each seismic event, the proposed model consists of various machine learning models most suitable for a particular data type in a single composite model.

$$\hat{U} = \bigcap_{i=1}^{m} \arg\min_{u \in W} f_i(u) \qquad (4)$$

where:

- f_i—objective criteria that characterizes the modelling quality;
- W—a set of possible solutions (search space);
- $\hat{U}$—a vector of composite model hyperparameters;
- $\bigcap_{i=1}^{m}$—an intersection of the set of solutions for each of the criteria, and m is the number of criteria used during optimization.

To solve the problem of multi-criteria optimization, we proposed the following criteria:

- **Silhouette criterion.** The silhouette shows how the average distance to the objects in its cluster differs from the average distance to the objects of other clusters. This value is in the range $[-1, 1]$. Values close to -1 correspond to poor (scattered) clusterization, and values close to zero indicate that clusters intersect and overlap.
- **FAR - The proportion of false alarms, and MAR - the proportion of missed alarms.** A sliding window is used to calculate this criterion. Both metrics have identical lower and upper bounds - $[0, 1]$. The selection of such metrics is based on its applicability for expert use and in order to take into account class imbalance.

In this paper, three values of the detection window width were taken. The short-, medium-, and long-term forecast horizons correspond to the values of six hours, two days, and seven days before and after the appearance of the rockburst.

4 Experiments

In order to evaluate the effectiveness of the proposed approach, experimental comparisons of the composite model with existing approaches were carried out. Obtained composite model consists of three machine learning models. The HDBSCAN model is used for spatial clusterization of seismic events, KNN model is used to fill in gaps in time series, and SSA model is used to time series forecasting. As a dataset, we used a synthetic dataset that was developed considering key characteristics of real seismic phenomena—seven rockbursts were distributed unevenly in space and time, reproducing the class imbalance problem typical for real monitoring data. The results show that this approach to synthetic data generation allows for effective testing and comparison of various model architectures (hierarchical, LSTM, and the proposed composite model), revealing their strengths and weaknesses across different forecasting horizons. In order to correctly evaluate the work of similar time series prediction models, we carried out preliminary clustering of seismic events using the proposed composite model. Selected cluster is "stable" in time, i.e. they include seismic events during the entire monitoring period.

Figure 3 shows the solutions considered during optimization. FAR/MAR metric value shows the normalized ratio of the sum of false and missed predicted rockbursts to true rockbursts. The closer the value is to 1, the greater the number of false and missed rockbursts.

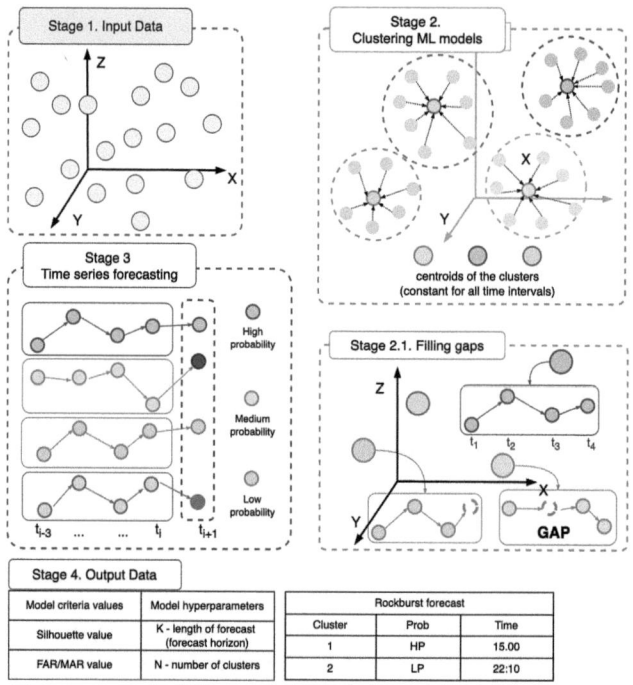

Fig. 2. Proposed approach for rockburst forecasting.

For model comparison, we used the "hierarchical model" as the baseline, the LSTM as the DL approach and our composite model. The results of the experiment are shown in Table 2 (*Italics* indicate the best result for each of the forecast horizons, **bold** indicates the best result among all horizons). Composite model showed superior result among all models with long-term forecasts and the best result among all models when using a medium-term forecast. Such results are related to the fact that the use of spatial data in the composite model allows localizing time-stable clusters of seismic events. The probability of rockburst in such clusters increases over time, which explains the effectiveness of the composite model in the medium and long-term forecast horizon. This also explains the effectiveness of the LSTM model in short-term forecasts, because in the absence of formed clusters, such a model allows better modeling of complex nonlinear dependencies that lead to rockbursts.

5 Conclusion

This paper proposes an approach to the problem of rockburst forecasting, based on seismic monitoring data. The idea of the approach is to automatically combine machine learning models based on temporal and spatial data into a single composite model.

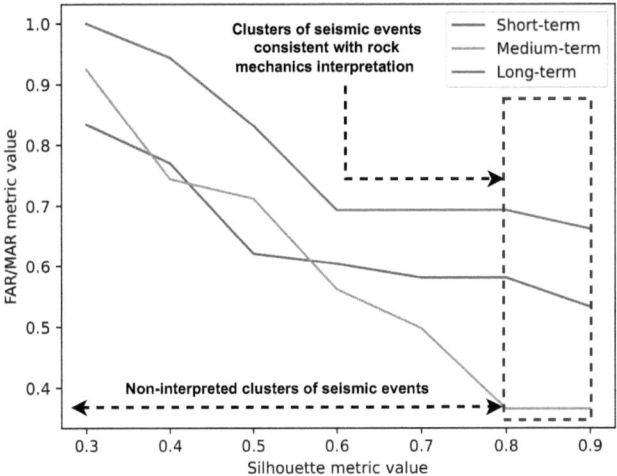

Fig. 3. Comparison of FAR/MAR metrics for three proposed forecast horizons.

Table 2. FAR-MAR criteria comparison

Model	Short-term (6 h)	Mid-term (2 days)	Long-term (7 days)
Hierarchical model	0.993	0.974	0.758
LSTM	*0.417*	0.714	0.688
Composite model	0.533	**0.365**	*0.562*

The multi-criteria optimization of the proposed composite model has shown its effectiveness. Appropriate values of the forecast horizon and the seismic event silhouette criterion make it possible to obtain results that are superior to hierarchical and DL models.

The proposed model implements a data-driven approach. On the one hand, it simultaneously uses spatial and temporal coordinates of seismic events, using the entire amount of information obtained during seismic monitoring. On the other hand, it reproduces rock mechanics phenomena, such as zones of the stress-strain state of the console, making it more interpretable than other models. The use of a synthetic seismic monitoring dataset in our research serves as an experimental demonstration of the importance of creating directionally generated synthetic multidimensional time series for improving the robustness of machine learning models.

We can conclude composite model is a more effective means of predicting rockbursts than classical hierarchical models and is slightly inferior to DL models when using short-term forecasts. The proposed approach based on machine learning and signal processing methods, is an effective forecasting algorithm. However, it can be improved by including more complex forecasting models and new criteria.

The implemented algorithms and examples of their application are available in https://github.com/ITMO-NSS-team/RockBurst.AI repository.

Acknowledgments. This research is financially supported by The Russian Scientific Foundation, Agreement # 24-71-10093, https://rscf.ru/en/project/24-71-10093/

Disclosure of Interests. The authors have no competing interests to declare that are relevant to the content of this article.

References

1. Gao, F., Kang, H., Li, J.: Numerical simulation of fault-slip rockbursts using the distinct element method. Tunn. Undergr. Space Technol. **110** (2021). https://doi.org/10.1016/J.TUST.2020.103805
2. Chao, Z., et al.: Prediction of rockbursts in a typical island working face of a coal mine through microseismic monitoring technology. Tunn. Undergr. Space Technol. (2021). https://doi.org/10.1016/J.TUST.2021.103972
3. Tian, H., Xue, Y.: Prediction of rockburst classification using support vector machine. IOP Conf. Ser. Earth Environ. Sci. **861** (2021). https://doi.org/10.1088/1755-1315/861/6/062088
4. Ma, T.H., Tang, C.A., Tang, L.X., Zhang, W.D., Wang, L.: Rockburst characteristics and microseismic monitoring of deep-buried tunnels for Jinping II Hydropower Station. Tunn. Undergr. Space Technol. **49**, 345–368 (2015)
5. Pu, Y., Apel, D., Liu, W., Mitri, H.: Machine learning methods for rockburst prediction-state-of-the-art review. Int. J. Mining Sci. Technol. **29** (2019). 10.1016/j.ijmst.2019.06.009
6. Ochmann, J., Jurczyk, M., Rusin, K., Rulik, S., Bartela, Ł., Uchman, W.: Solution for post-mining sites: thermo-economic analysis of a large-scale integrated energy storage system. Energies (2024). https://doi.org/10.3390/en17081970
7. Zhou, J., Li, X., Mitri, H.S.: Classification of rockburst in underground projects: comparison of ten supervised learning methods. J. Comput. Civ. Eng. **30**(5), 04016003 (2016)
8. Sarafanov, M., Borisova, Y., Maslyaev, M., Revin, I., Maximov, G., Nikitin, N.: Short-term river flood forecasting using composite models and automated machine learning: the case study of lena river. Water **13**, 3482 (2021). https://doi.org/10.3390/w13243482
9. Nikitin, N., Revin, I., Hvatov, A., Vychuzhanin, P., Kalyuzhnaya, A.: Hybrid and automated machine learning approaches for oil fields development: the case study of Volve field, North Sea. Comput. Geosci. **161**, 105061 (2022). https://doi.org/10.1016/j.cageo.2022.105061
10. Sarafanov, M., Pokrovskii, V., Nikitin, N.: Evolutionary automated machine learning for multi-scale decomposition and forecasting of sensor time series. In: Proceedings of IEEE CEC (2022). https://doi.org/10.1109/CEC55065.2022.9870347
11. Gong, W., et al.: Multi-objective adaptive surrogate modeling-based optimization for parameter estimation of large, complex geophysical models. Water Resour. Res. **52** (2015). https://doi.org/10.1002/2015WR018230
12. Li, S., Lu, P., Liang, W., Chen, Y., Da, Q.: Performance evaluation of hybrid PSO-BPNN-AdaBoost and PSO-BPNN-XGBoost models for rockburst prediction with imbalanced datasets. Appl. Sci. **14** (2024). https://doi.org/10.3390/app142411792

Accelerating LBM with C++ STL Asynchronous Parallel Model

Ziheng Yuan[1] and Takashi Shimokawabe[2](✉)

[1] Department of Electrical Engineering and Information Systems,
Graduate School of Engineering, The University of Tokyo,
Hongo 7-3-1, Bunkyo-ku, Tokyo 113-8656, Japan
Zihengyuan22@g.ecc.u-tokyo.ac.jp
[2] Information Technology Center, The University of Tokyo,
6-2-3 Kashiwanoha, Kashiwa-Shi, Chiba 277-0882, Japan
shimokawabe@cc.u-tokyo.ac.jp

Abstract. Asynchronous computation is an important optimization technique in scientific computation. The upcoming C++26 standard introduces a new asynchronous execution framework, stdexec, enabling the development of high-performance code using only standard C++. This paper explores the parallelization of single-GPU and multi-GPU lattice Boltzmann method computations using stdexec and further optimizes performance through its asynchronous execution model. Experimental results show that asynchronous stdexec achieves approximately 83.5%–105.4% of the performance of C++ stdpar. These results suggest potential for further optimizations in the future, providing additional options for high-performance computing development in pure C++.

Keywords: Parallel Computing · High Performance Computing · GPU · stdexec · lattice Boltzmann method

1 Introduction

The parallel programming libraries can currently be categorized into two approaches. The first approach is hardware-specific libraries designed for particular hardware platforms, characterized by their ability to provide fine-grained control over hardware. Representative examples include CUDA [1], OpenCL [2] and HIP [3]. The second approach libraries provide high-level APIs to abstract hardware details. Typical examples of this category include Kokkos [4], OpenMP [5] and OpenACC [6]. To enable programming using pure C++ and enhance code compatibility, two new features have been or will be introduced into the C++ standard, referred to as C++ standard language parallel model (stdpar) [7] and C++ standard model for asynchronous execution (stdexec) [8], following the second approach. stdpar has already been integrated into C++17. As an extension to the existing algorithm, stdpar enables parallel execution support for specific function in standard algorithm library. stdexec provides the asynchronous execution framework, which is not supported by stdpar. Before stdexec is integrated into C++26 in the future. a prototype of stdexec library provided by NVIDIA is available for testing. This paper focuses on evaluating the feasibility of performing parallel programming using C++ stdexec, analyzing its performance and comparing it with other parallel programming labraries. Lattice Boltzmann method (LBM) [9] is selected as the benchmark problem for evaluation due to its wide applicability in the field of fluid dynamic and its suitability for parallel computation.

2 Background Knowledge

2.1 C++ stdexec

The stdexec sender/receiver model is defined by three critical components: executor, sender/receiver, and scheduler. The role of the executor is to provide a uniform task execution interface by abstracting hardware resources. The sender/receiver is responsible for supplying tasks to the executor. The scheduler serves to provide an abstract interface for the management of hardware resource [10]. It is important to note that one scheduler can only manage one hardware resource associated with it, this hardware could be CPU or GPU. Listing 1.1 and 1.2 illustrates the structure of C++ STL parallel model with an example.

```
auto A = stdexec::just()
 | exec::on(sched, stdexec::bulk(n, GPU_task1))
 | stdexec::then(CPU_task)
 | stdexec::let_value([&]{return stdexec::just()
 | exec::on(sched, stdexec::bulk(N, GPU_task2));});
```
Listing 1.1. C++ stdexec synchronous example

```
auto A = stdexec::when_all(stdexec::just()
 | exec::on(sched_low,   stdexec::bulk(N, GPU_task2)),
stdexec::just()
 | exec::on(sched_high, stdexec::bulk(n, GPU_task1))
 | stdexec::then(CPU_task));
```
Listing 1.2. C++ stdexec asynchronous example

In this scenario, three tasks executed on either the CPU or GPU: CPU_task, GPU_task1 and GPU_task2. Among them, both CPU_task and GPU_task2 are waiting the result from GPU_task1. Listing 1.1 and 1.2 represent different approaches to executing the same tasks. stdexec::bulk() is the stdexec equivalent of a for-loop, utilizing the resources allocated by stdexec::on() to execute tasks [11]. stdexec::when_all() provides synchronization functionality. Listing 1.1 implements basic synchronous execution. GPU_task1, CPU_task and GPU_task2 is executed in serial order. Listing 1.2 implements asynchronous execution by assigning schedulers with different priority levels. The program assigns high priority scheduler `sched_high` to `GPU_task1` and low priority scheduler `sched_low` to `GPU_task2`. While `GPU_task2` is being continuously executed, `GPU_task1` with higher priority, will complete its execution first and subsequently proceed to execute `CPU_task`, overlap the execution time.

2.2 Lattice Boltzmann Method

LBM is a fluid dynamic algorithm that simulates fluid by emulating the streaming and collision of virtual fluid particles in mesoscale. Particles motion is characterized by discrete velocity set according to the Boltzmann equation [9]. The LBM uniformly divides the computational domain into lattice like orthogonal grids [12]. This experiments use D3Q27 model as the simulation model. Necessary physical values include density ρ, velocity u and equilibrium distribution function f_i are computed from Eq. (1)–(3).

$$\rho = \sum_i f_i \qquad u = \frac{1}{\rho} \sum_i c_i f_i \qquad (1)$$

$$f_i(x + c_i \Delta t, t + \Delta t) = f_i(x, t) \qquad (2)$$

$$f_i(x, t + \Delta t) = f_i^*(x, t + \Delta t) + \Omega(x, t) \qquad (3)$$

In these equations, c_i represent for velocity set, Δt for time step, Ω for collision model. Equation (1) shows density and velocity computation method. Equation (2) calculates streaming step. Equation (3) calculates the collision step. In this experiment, we use the classic BGK collision model [13] and bounce-back boundary condition [14] for LBM.

Boundary computation and LBM streaming-collision computation is the two main steps for LBM computation. The boundary code computes the numerical values at the boundaries based on the condition, storing the results for use in the computation of streaming-collision step. For multi-GPU experiments, data exchange between multiple GPUs is based on MPI. The performance benchmark used is Mega Lattice Updates Per Second (MLUPS), which represents the number of LBM lattice updates (computations) per unit time. The specific calculation method is outlined as follows:

$$MLUPS = \frac{lattice\ size \times number\ of\ iterations}{total\ time\ consumption} \qquad (4)$$

3 Experiment and Result

3.1 Experiment Condition

Hardware Resource. The experiments were conducted on a Wisteria-Aquarious HPC platform provided by The University of Tokyo, utilizing A100 40 GB GPUs. Each computation node in Wisteria-Aquarious contains 8 GPUs. The compilation was performed using the nvidia/24.1 compiler with the C++20 standard.

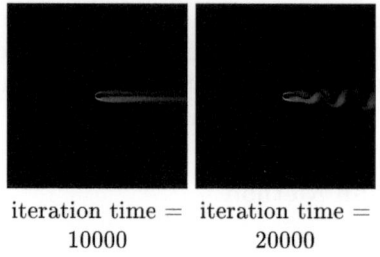

iteration time = 10000   iteration time = 20000

Fig. 1. Visualization of simulation result. The content of these images represents the variation of the velocity norm along the z-axis over time.

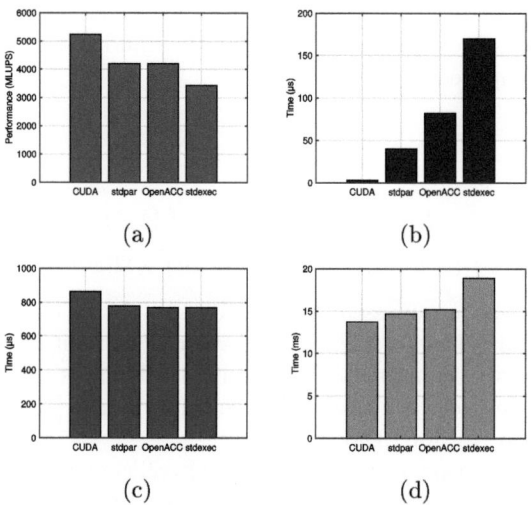

Fig. 2. Performance of single GPU execution and time breakdown comparison of stdexec, CUDA, stdpar and OpenACC in single GPU case. (a): Performance of single GPU execution. (b): Kernel launch time consumption. (c): Boundary computation time consumption. (d): LBM streaming-collision computation time consumption.

LBM Condition. The code presented in this article is applied to the benchmark simulation case of 3D flow around a cylindrical obstacle. A solid cylinder is fixed along the z-axis. A uniform fluid flow enters the computational domain along the x-axis. The computational domain has a size of 512 × 512 × 256 for both single GPU and multi GPU conditions. Reynolds number Re = 1000. Figure 1 illustrates the evolution of the flow field over iterations.

3.2 Single GPU Result

First, the performance of the code is tested under a single GPU scenario. The implementation of stdexec code is similar to Listing 1.1, where GPU_task1 corresponds to boundary computation and GPU_task2 corresponds to streaming-collision computation. However, the code does not include the CPU_task part, as it represents the function required by multi-GPU communication. Figure 2(a) illustrates the performance of different codes when solving the LBM problem under the same condition described in Sect. 3.1. It can be observed that CUDA achieves the best parallel computing performance. OpenACC and stdpar exhibit similar performance, while stdexec shows the lowest performance. The performance of stdexec reaches 65.2% of CUDA and 81.6% of stdpar. This result leads to the conclusion that stdexec does not offer a performance advantage when relying solely on stdexec::bulk() for parallel computation. An analysis of the function's execution time is required to find the reason.

According to Fig. 2(b), Compared to the fastest CUDA implementation, the kernel launch time of stdexec is approximately 60 times longer [15]. As shown in Fig. 2(c), for boundary kernels, which involve small amounts of data, the performance across different implementations is similar. However, Fig. 2(d) reveals a more significant performance gap for large data size LBM streaming collision computation, with stdexec showing a execution time consumption difference of approximately 38.0% compared to CUDA.

3.3 Multi GPU Result

The performance of the code is then tested in a multi-GPU scenario. The implementation of stdexec code is similar to Listing 1.1, with CPU_task corresponds to boundary data exchange function based on MPI, this introduces additional communication cost. Figure 3 presents the performance of different codes when solving the LBM problem under the same condition described in Sect. 3.1.

Similar to the results from the single-GPU tests, the performance of stdexec in the multi-GPU scenario still shows a significant gap compared to CUDA according to Fig. 4, while OpenACC and stdpar exhibit similar performance. In the 2 GPU case, the performance of stdexec is approximately 65.5% of CUDA and 81.0% of stdpar. In the 4 GPU case, the performance of stdexec is approximately 58.7% of CUDA and 73.8% of stdpar. In the 8 GPU case, the performance of stdexec is approximately 62.1% of CUDA and 71.4% of stdpar. In the 16 GPU case, the performance of stdexec is approximately 59.9% of CUDA and 88.1% of stdpar.

3.4 Multi GPU Asynchronous Result

Finally, the performance of asynchronous computation in a multi-GPU setting is evaluated. Figure 4 presents the performance of different implementations when solving the same LBM problem, with experimental conditions remaining the same as before. In this experiment, stdexec async utilize the methods introduced in Listing 1.2. The high-priority GPU kernel for boundary computation completes first and initiates MPI communication, effectively overlapping the MPI communication time with the execution of the low-priority scheduler GPU kernel for streaming-collision. In the 2 GPU case, the performance of stdexec async is approximately 67.8% of CUDA, 83.7% of stdpar and 100.3% of stdexec.

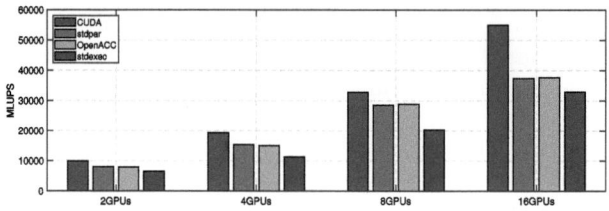

Fig. 3. Performance of multi GPU execution.

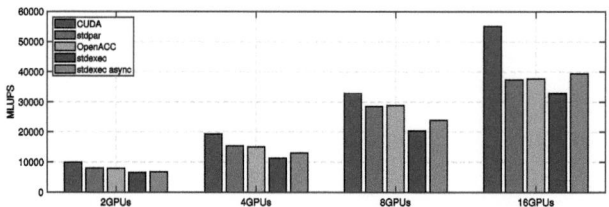

Fig. 4. Performance of multi GPU asynchronous execution. To compare the performance of stdexec async, the results for CUDA, stdpar, OpenACC, and stdexec are the same as those presented in Fig. 3.

In the 4 GPU case, the performance of stdexec async is approximately 67.3% of CUDA, 84.7% of stdpar and 114.8% of stdexec. in the 8 GPU case, the performance of stdexec async is approximately 72.6% of CUDA, 83.5% of stdpar and 117.0% of stdexec. in the 16 GPU case, the performance of stdexec async is approximately 71.6% of CUDA, 105.4% of stdpar and 119.7% of stdexec.

By analyzing the results, it can be observed that compared to stdexec without asynchronous computation, the asynchronous version of stdexec achieves up to a 19.7% performance improvement. To determine the source of performance improvements, it is necessary to analyze the execution time of individual kernels. Figure 5 presents the kernel execution times for a multi-GPU stdexec program without asynchronous computation, as well as for async, when running on 16 GPUs. Figure 5(a) illustrates the boundary computation time consumption per iteration, showing that performance of the two is nearly identical. A similar trend is observed for both LBM streaming collision computation and MPI communication shown in Fig. 5(b) and Fig. 5(c). The only notable difference lies in the overall kernel execution time shown in Fig. 5(d), which is the combination of MPI communication of LBM streaming collision computation time consumption. The performance of 16 GPUs in stdexec async is higher because inter-node communication is hidden, as expected. Since the overall computational domain does not change as GPU number increases, the size allocated to each GPU decreases. At the same time, the size of boundary remains unchanged, resulting in a gradual increase in the proportion of communication, which creates room for optimization. This shows the importance of asynchronous communication and may be explained as the strength of stdexec's ability to introduce asynchronous communication into pure C++ code.

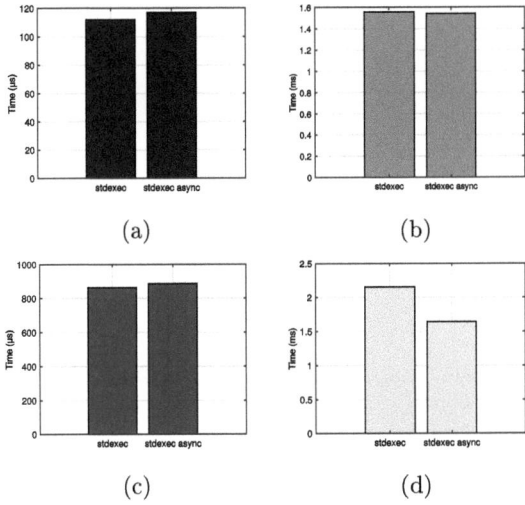

Fig. 5. Time breakdown comparison of sync and async stdexec in 16 GPUs case. (a): Boundary computation time. (b): LBM streaming-collision computation time. (c): MPI communication time. (d): Overall kernel execution time.

Observations indicate that stdexec async achieves 23.2% reduction in execution time compared with stdpar. 95.7% of the communication time was hidden in stdexec async. This reduction is attributed to the asynchronous execution of MPI communication and LBM streaming collision computation. However, due to the high kernel launch overhead at the start of each iteration (approximately 170 ms), part of the performance gains achieved by stdexec async is lost. As a result, the overall reduction in program runtime is smaller than the reduction in kernel execution time.

4 Conclusion

In summary, this paper evaluates the application of the C++26 stdexec prototype in the LBM. The performance of stdexec is compared with other programming languages and implementation approaches, and kernel analysis is conducted to explain the observed performance differences. The test code is based on the prototype provided by NVIDIA and utilizes advanced C++ syntax introduced in C++17 and later versions. The experimental results demonstrate the baseline performance of stdexec, showing that its performance is comparable to stdpar in some cases. This suggests that stdexec provides an alternative option for developing high-performance C++ code. We also encountered software engineering challenges related to compilers and libraries, and shared information with NVIDIA engineers. We will continue to track the progress of the C++26 standard and explore further optimizations for the code in the future.

Acknowledgments. This work was partly supported by JSPS KAKENHI Grant Number JP24K02947. This work was also partly supported by JHPCN projects jh240052 and jh250037.

References

1. NVIDIA CUDA Toolkit. https://developer.nvidia.com/cuda-toolkit. Accessed 10 Feb 2025
2. OpenCL Official guide. https://github.com/KhronosGroup/OpenCL-Guide. Accessed 15 Feb 2025
3. Diederichs, D.: Available now: new HIP SDK helps democratize GPU computing. https://community.amd.com/t5/instinct-accelerators/available-now-new-hip-sdk-helps-democratize-gpu-computing/ba-p/621029. Accessed 10 Feb 2025
4. Trott, C.R., Lebrun-Grandié, D., Arndt, D., Ciesko, J., Dang, V., et al.: Kokkos 3: programming model extensions for the exascale era. IEEE Trans. Parallel Distrib. Syst. **33**(4), 805–817 (2021)
5. Haseeb, M., Ding, N., Deslippe, J., Awan, M.: Evaluating performance and portability of a core bioinformatics kernel on multiple vendor GPUs. In: 2021 International Workshop on Performance. Portability and Productivity in HPC (P3HPC), St. Louis, MO, pp. 68–78. IEEE (2021)
6. Farber, R.: Parallel programming with OpenACC. Newnes, USA (2016)
7. Lopez, G., Olsen, D., Adelstein Lelbach, B.: Accelerating Standard C++ with GPUs Using stdpar. NVIDIA Technical Blog. https://developer.nvidia.com/blog/accelerating-standard-c-with-gpus-using-stdpar/. Accessed 10 Feb 2025
8. Garland, M., et al.: A Unified Executors Proposal for C++ | P0443R14. https://www.open-std.org/jtc1/sc22/wg21/docs/papers/2020/p0443r14.html. Accessed 10 Feb 2025
9. Kruger, T.: The Lattice Boltzmann Method: Principle and Practice (2017). ISBN 978-3-319-44647-9
10. Arutyunyan, R.: P2500R0 C++17 parallel algorithms and P2300 Published Proposal. ISO/IEC JTC1/SC22/WG21 (2022)
11. Dominiak, M., et al.: std::execution Published Proposal. https://www.open-std.org/jtc1/sc22/wg21/docs/papers/2023/p2300r7.html. Accessed 10 Feb 2025
12. Lagrava, D., Malaspinas, O., Latt, J., Chopard, B.: Advances in multi-domain lattice Boltzmann grid refinement. J. Comput. Phys. **231**(14), 4808–4822 (2012). https://doi.org/10.1016/j.jcp.2012.03.015
13. Qian, Y.H., et al.: Lattice BGK models for navier-stokes equation. Europhys. Lett. **17**(6), 479 (1992). https://doi.org/10.1209/0295-5075/17/6/001
14. Ladd, A.J.C.: Numerical simulations of particulate suspensions via a discretized Boltzmann equation. J. Fluid Mech. **271**, 285–309 (1994). https://doi.org/10.1017/S0022112094001783
15. Haseeb, M., Wei, W., Deslippe, J., Cook, B.: That's right – the same C++ STL asynchronous parallel code runs on CPUs and GPUs. In: SC 2023, Denver, USA (2023)

Accelerating Cloud-Based Transcriptomics: Performance Analysis and Optimization of the STAR Aligner Workflow

Piotr Kica[1,2](✉), Sabina Licholai[1,3], Michał Orzechowski[1,2,3], and Maciej Malawski[1,2]

[1] Sano Centre for Computational Medicine, Kraków, Poland
{p.kica,s.licholai,m.orzechowski,
m.malawski}@sanoscience.org
[2] Faculty of Computer Science, AGH University of Kraków, Kraków, Poland
[3] Academic Computer Centre Cyfronet AGH, Kraków, Poland

Abstract. In this work, we explore the Transcriptomics Atlas pipeline adapted for cost-efficient and high-throughput computing in the cloud. We propose a scalable, cloud-native architecture designed for running a resource-intensive aligner – STAR – and processing hundreds of terabytes of RNA-sequencing data. We implement optimization techniques that significantly reduce cost and execution time. The impact of particular optimizations is measured in medium-scale experiments followed by a large-scale experiment that leverages all of them and validates the design. Early stopping optimization allows us to reduce the total alignment time by 23%. For the cloud environment, we identify suitable EC2 instance types and verify the applicability of spot instances usage.

Keywords: Transcriptomics · Optimization · STAR · Alignment · High-Throughput Computing · Cloud-computing · AWS

1 Introduction

The cloud is, in many cases, the infrastructure of choice for large-scale genomic pipelines, as it promises scalability, cost efficiency, and availability of on-demand resources. There are multiple recent examples of pipelines that have been developed for genomic data, built using cloud services [6,13,15]. Such cloud-native pipelines take advantage of the capabilities of cloud services, allowing parallelism, scalability, and elasticity, often with autoscaling, and follow the principles of infrastructure as code for application deployment and environment setup. However, cloud infrastructure is complex, as there are multiple services offered with a wide range of configuration options, so exploiting the full potential of clouds requires combining application-specific expertise with the ability to fine-tune cloud infrastructure configuration. For example, there are multiple compute service options, storage systems, databases, and queuing/messaging systems, etc., so making the right choice becomes a challenge. In this paper, we present a case study of running a Transcriptomics Atlas pipeline in the AWS cloud. It is a data- and compute-intensive pipeline, based on a sequence aligner – STAR [9] – that processes hundreds of terabytes of RNA-seq data. Our aim is to answer research questions, such as: (1) How to take advantage of the intermediate results to reduce time and cost? (2) How to select the optimal level of parallelism within a single node? (3) Which instance types are the most cost-efficient for alignment? (4) How suitable are spot instances for running resource-intensive aligners?

2 Background and Related Work

Transcriptome and NGS Sequencing. The human transcriptome consists of different types of RNA, ranging from messenger RNA (mRNA) encoding proteins to various forms of non-coding RNA with mostly regulatory functions [10]. Transcriptome-releated analyses are essential in modern medical research and typically performed in a comparative context using next-generation sequencing (NGS) data. Lack of comprehensive NGS data under standard conditions significantly increases the cost of experiments [7] and in response to this, we designed the Transcriptomics Atlas, where data from a representative collection of human tissues were processed in a uniform manner. NGS is based on short reads and enables rapid and precise identification of the transcriptome composition within a biological sample. However, due to the limited length of reads, subsequent assembly into a complete transcriptome necessitates specialized bioinformatics algorithms in a process called alignment [3].

RNA-Sequencing in the Cloud. In [17], authors explore different parallel computing strategies and limitations for multiple genomic tools, including architecture-aware and data-storage optimizations. Research on STAR-based workflow as well as cost and throughput analysis for cloud and HPC experiments are carried out in [16]. Pseudoaligners (e.g. Salmon, Kallisto) are recommended by [13] when cost plays a critical role. Research carried out in [8] shows that serverless computing for RNA sequencing is a valid approach when high parallelism is the end goal, with HiSat2 running on AWS Lambda. However, deploying STAR to serverless services is more challenging compared to less resource-intensive aligners. Although possible, it is not recommended for large-scale processing due to decreased cost-efficiency compared to VM-based solutions [12]. Furthermore, in [4] authors moved an HPC workflow to serverless services and identified multiple challenges with efficient data partitioning, transfer and insufficient object storage performance. Our previous work regarding the Transcriptomics Atlas [5] focused on understanding the pipeline requirements for a similar workflow in HPC and the cloud.

3 Pipeline and Cloud Architecture

Pipeline Description. As presented in Fig. 1, the first phase consists of accessing an *SRA* file using `prefetch` and converting it into *FASTQ* with

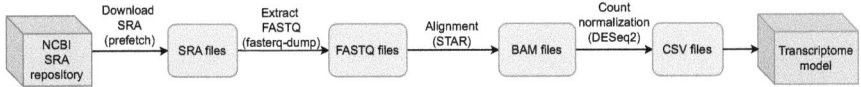

Fig. 1. Transcriptomics Atlas Pipeline.

fasterq-dump. The next and most important, time-consuming step is the alignment with STAR. Finally, the acquired *BAM* file is normalized using DESeq2. Instead of STAR, one can use alternative aligners (HiSat2) or pseudoaligners (Salmon). We are running STAR v2.7.10b with "--quantMode GeneCounts" option. The steps are connected using a Python script, and the current implementation of the Transcriptomics Atlas project is publicly available on GitHub under the MIT license [2].

Input Dataset. Data were obtained from the NCBI Sequence Reads Archive (SRA) [1], focusing on nucleotide sequences from human samples, filtered by tissue type. The query for the SRA database targeted publicly available, human-origin data sequenced with Illumina machines. We downloaded metadata for matching SRA IDs and selected those with compressed sequence sizes between 200 MB and 30 GB. We define the range by taking into account the size of libraries for typical transcriptome sequencing and output from the most commonly used wet lab sequencing protocols. For the Transcriptomics Atlas, we aim for 100–200 (with a good mapping rate) per tissue, selecting up to 400 samples per tissue, resulting in 7216 files totalling 17 TB of SRA data.

Resource Requirements. STAR is a resource-intensive aligner that uses a precomputed genome index in the alignment step, which has to be loaded into the system memory. The generation of such an index is a one-time task. Depending on the type and release version of the genome, the index differs in size. We use the human genome of the "Toplevel" type distributed by Ensembl [14]. Previous work [11] showed that an older release (108) results in an index of 85 GiB in size, and using a newer one (release 111) is much smaller (29.5 GiB) and faster (12 times). STAR requires additional memory for sorting a BAM file, usually 1–2 GiB, but outliers may require even 20.5 GiB. The pipeline requires enough disk space to handle intermediate files such as *FASTQ* files along with *SRA* and *BAM* files. The fasterq-dump tool creates *FASTQ* files, on average, 7.5 times bigger than the original *SRA* file; however, outliers can be even 17 times larger. Moreover, additional space is required during conversion. This use case focuses on files within the 200 MB–30 GB range. Therefore, we can estimate that the required space should not exceed 550 GiB.

Cloud Architecture. for the Transcriptomics Atlas pipeline is presented in Fig. 2. The main processing is performed on a virtual cluster of EC2 spot instances launched from a custom machine image containing all the required software. The initial step for each worker is to connect to an NFS instance and load the STAR index into memory. Subsequently, workers acquire the SRA IDs from the queue and process them using the pipeline (Fig. 1). The transcriptomics

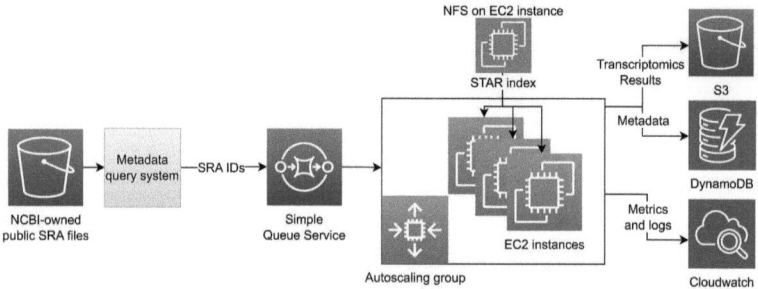

Fig. 2. Cloud architecture for Transcriptomics Atlas Pipeline.

results are stored in a dedicated S3 bucket. Execution metadata are gathered for performance analysis and saved in a Dynamodb table. Metrics are saved using the CloudWatch service, which gives insight into resource utilization. The proposed approach is easily scalable and adaptable for similar workflows. Having extensive control over the underlying compute resources allows us to fine-tune the configuration for given requirements, which improves cost-efficiency.

4 Application-Specific Optimizations

Early Stopping During Alignment. Early stopping is a common method to stop the training of machine learning models when the desired accuracy is achieved. We apply a similar approach to alignment by discarding low-quality or invalid sequences during processing. It is possible, as STAR reports on the intermediate mapping rate at runtime, which allows the identification of such sequences. As shown in [11], utilization of live metrics from *Log.progress.out* file can boost alignment throughput by up to 19.5%. This feature is beneficial when we cannot determine the quality of the FASTQ beforehand.

For the Transcriptomics Atlas project, the mapping rate threshold is set at 30%. However, this is highly dependent on the use case. Pipelines that utilize STAR in a similar scenario and require sequences of the highest quality will greatly benefit from this feature. In Fig. 3, we present an extended analysis based on the experiment carried out in [11]. If we set the threshold at 80%, we would reduce the total compute time by 60%. The minimal number of processed spots was set to 10%.

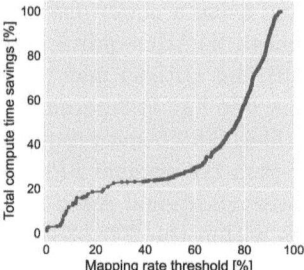

Fig. 3. Impact of early stopping threshold on total alignment time.

Cost-Efficient Allocation of Cores. The recommended approach for STAR aligner is to match the thread count with the number of available cores in the node. Authors of the original STAR publication [9] claim that "STAR exhibits close to linear scaling of the throughput rate with the number of threads". However, the original research lacks detailed performance analysis, which is important to maximize CPU efficiency. We decided to test the scalability of STAR to confirm these claims and find an optimal number of cores per node, which will improve the throughput of the pipeline. The test suite consists of 3 *FASTQ* files of different sizes. Execution times are measured on two different 16-vCPU instance types - with and without Simultaneous Multi-Threading (SMT).

The test results are presented in Fig. 4, and we see the benefit of the increased number of threads. However, there is a noticeable drop in efficiency - for 16 threads and m7a.large instance, we get 84% and 72% efficiency for 16 GiB file and 81 GiB respectively. This is especially visible for the m6a.large instance, which uses SMT and exceeding 8 threads further decreases efficiency. Based on the acquired metrics, we focused on 8-vCPU instances for the best cost-efficiency.

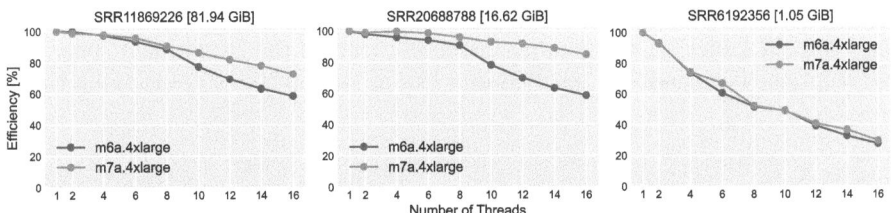

Fig. 4. Efficiency of the STAR aligner.

5 Cloud Infrastructure Optimizations

Instance Type Comparison in AWS. According to the requirements in Sect. 3 and knowing that STAR is a memory-intensive program, we decided to focus on instances with a higher memory per CPU factor with at least 64 GiB of RAM. Using instance types which have more cores but less memory per CPU may require faster block storage and result in increased under-utilization during other, much less CPU-intensive steps (e.g. *prefetch*). Using more cores in a single worker node would also reduce CPU efficiency during alignment as described in Sect. 4. The selected instance types of the current generation that meet these requirements are compared in Table 1. This table also presents the total cost and time for performing STAR alignment on 50 random *FASTQ* files. The results indicate r7a.2xlarge as the fastest and cheapest type. However, when using spot instances, the availability of a given type is also important.

Cost-Efficiency of Spot Instances. Spot instances on AWS offer compute resources at a lower cost (depending on instance type and current market demand). However, such instances can be terminated with a 2-minute notice. For example, an r7a.2xlarge instance can be acquired with 50%–60% discount. Our use case fits this model as the optimized pipeline runs relatively quickly (mean =

Table 1. Cost-efficiency analysis of selected instance types.

Instance type	vCPU	Cores	RAM [GiB]	On-demand price [h]	Total STAR execution time [h]	Total cost
r6a.2xlarge	8	4	64	$0.4536	8.00	3.63 $
r6i.2xlarge	8	4	64	$0.5040	8.04	4.05 $
r7a.2xlarge	8	8	64	$0.6086	5.48	3.33 $
r7i.2xlarge	8	4	64	$0.5292	7.66	4.05 $

8 min). Unfortunately, interruptions result in an additional STAR initialization phase and restarting the computations on a new instance. However, with a good configuration (instance types with a low interruption rate), using spot instances should result in relatively stable computations. In Fig. 5 we present the processing 1000 *SRA* files on r7a.2xlarge instances. During the experiment, only five interruptions occurred and resulted in a loss of <1% of the total running time.

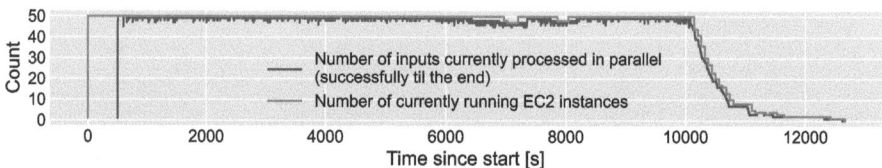

Fig. 5. Spot instances usage experiment timeline.

6 Large-Scale Experiment

The goal is to test the pipeline on a larger scale, measure, and analyze resource usage. We use the input data from Sect. 3 and the configuration:

- EC2 (Spot): 50 r7a.2xlarge instances (EBS: 550 GB, GP3, 500 MiB/s, 3000IOPS)
- Input: 7216 *SRA* files (2.5 GB avg, 17.9 TB total size, max = 29.9 GB)
- Index: Based on Toplevel human genome, release 111, 29.5 GB size.

The experiment used 1102.5 node hours in total and the implemented optimizations gave the expected improvements. The timeline is presented in Fig. 6. We processed 130 TB of *FASTQ* data and acquired an average mapping rate between 57%–87%, depending on the tissue. Early stopping feature reduced the total run time of STAR by about 23%. Using spot instances saved 50% the compute costs, but 138 interruptions occurred, wasting only 2.9% of the total instances' run time. The average CPU utilization across all instances was about

58% for the entire pipeline and 78% for STAR exclusively. 93% of all instances run time the RAM utilization was between 45% and 55%, and only 1.7% required more than 60% of the instance memory, suggesting an area for improvement. STAR accounted for 71% of the total workload time. In Fig. 7, we show aggregated CPU and memory usage during STAR for normalized metrics gathered during alignments longer than 10 min (n = 1091). The estimated cost of the experiment is about $477 - including compute (70%), storage (18.5%) and data transfer costs (11.5%). This is equivalent to about $0.066 per *FASTQ* file.

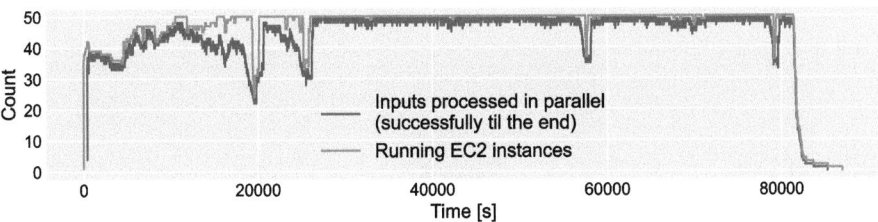

Fig. 6. Large experiment timeline for instances and successfully computed files.

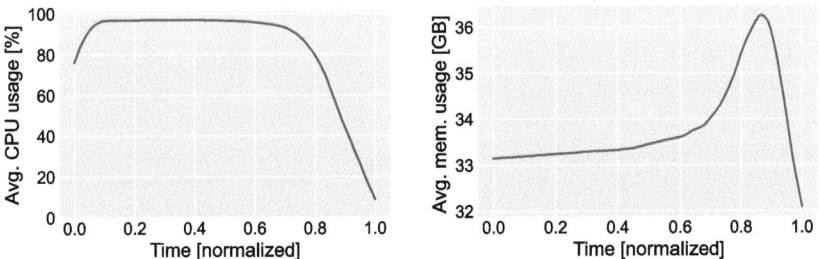

Fig. 7. STAR aggregated and normalized metrics for CPU and memory usage.

7 Conclusions

This work presents the cloud architecture for the Transcriptomics Atlas pipeline with the STAR aligner as its core. The optimizations described here significantly increased performance and throughput. Analysis of a large-scale experiment showed that early stopping saved 23% of the total running time of STAR. For use cases with a high mapping rate threshold, this feature would be even more beneficial. In addition, we observe that the pipeline is a great fit for spot instances and reduces computational costs by 50%–60%. Analysis of STAR's efficiency will help to choose the right configuration in order to maximize the throughput in similar scenarios. We identified one of the best instance types on AWS for alignment in terms of processing time and cost. Many insights in this work are applicable outside the cloud environment, extending the research results for HPC centres and workstations. As concluded in [11], a faster STAR alignment can also improve the time required for a clinician to make a diagnosis.

Acknowledgments. The publication is supported by the Polish Minister of Science and Higher Education, contract number MEiN/2023/DIR/3796; EU Horizon 2020 Teaming grant agreement No 857533; IRAP program of the Foundation for Polish Science MAB PLUS 2019/13; and EU Horizon Europe grant NEARDATA No 101092644.

Disclosure of Interests. The authors have no competing interests to declare that are relevant to the content of this article.

References

1. NCBI NIH: The Sequence Read Archive (2024). https://www.ncbi.nlm.nih.gov/sra
2. Transcriptomics Atlas (2024). https://github.com/SanoScience/NearData
3. Alser, M., et al.: Technology dictates algorithms: recent developments in read alignment. Genome Biol. **22**(1), 249 (2021)
4. Arjona, A., et al.: Scaling a variant calling genomics pipeline with FaaS. In: Proceedings of the 9th International Workshop on Serverless Computing, pp. 59–64 (2023)
5. Bader, J., et al.: Novel approaches toward scalable composable workflows in hyper-heterogeneous computing environments. In: Proceedings of the SC 2023 Workshops of the International Conference on High Performance Computing, Network, Storage, and Analysis, pp. 2097–2108 (2023)
6. Camacho, C., Boratyn, G.M., Joukov, V., Vera Alvarez, R., Madden, T.L.: ElasticBLAST: accelerating sequence search via cloud computing. BMC Bioinform. **24**(1), 117 (2023). https://doi.org/10.1186/s12859-023-05245-9
7. Casamassimi, A., Federico, A., Rienzo, M., Esposito, S., Ciccodicola, A.: Transcriptome profiling in human diseases: new advances and perspectives. Int. J. Mol. Sci. **18**(8), 1652 (2017)
8. Cinaglia, P., Vázquez-Poletti, J.L., Cannataro, M.: Massive parallel alignment of RNA-seq reads in serverless computing. Big Data Cogn. Comput. **7**(2), 98 (2023). https://doi.org/10.3390/bdcc7020098
9. Dobin, A., et al.: STAR: ultrafast universal RNA-seq aligner. Bioinformatics **29**(1), 15–21 (2013). https://doi.org/10.1093/bioinformatics/bts635
10. Gibbs, R.A.: The human genome project changed everything. Nat. Rev. Genet. **21**(10), 575–576 (2020). https://doi.org/10.1038/s41576-020-0275-3
11. Kica, P., Lichołai, S., Orzechowski, M., Malawski, M.: Optimizing star aligner for high throughput computing in the cloud. In: 2024 IEEE International Conference on Cluster Computing Workshops (CLUSTER Workshops), pp. 162–163. IEEE (2024). https://doi.org/10.1109/CLUSTERWorkshops61563.2024.00039
12. Kica, P., Orzechowski, M., Malawski, M.: Serverless approach to running resource-intensive star aligner. arXiv preprint arXiv:2504.05078 (2025)

13. Lachmann, A., Clarke, D.J., Torre, D., Xie, Z., Ma'ayan, A.: Interoperable RNA-Seq analysis in the cloud. Biochimica et Biophysica Acta (BBA)-Gene Regulatory Mechanisms **1863**(6), 194521 (2020)
14. Martin, F.J., et al.: Ensembl 2023. Nucleic Acids Res. **51**(D1), D933–D941 (2023). https://doi.org/10.1093/nar/gkac958
15. Wiewiórka, M., Szmurło, A., Stankiewicz, P., Gambin, T.: Cloud-native distributed genomic pileup operations. Bioinformatics **39**(1), btac804 (2023)
16. Wilks, C., et al.: recount3: summaries and queries for large-scale RNA-seq expression and splicing. Genome Biol. **22**, 1–40 (2021)
17. Zou, Y., Zhu, Y., Li, Y., Wu, F.X., Wang, J.: Parallel computing for genome sequence processing. Briefings Bioinform. **22**(5), bbab070 (2021)

Adaptive Modular Housing Design for Crisis Situations

Anastasiya Pechko, Katarzyna Grzesiak-Kopeć, Barbara Strug(✉), and Grażyna Ślusarczyk

Institute of Applied Computer Science, Jagiellonian University, Łojasiewicza 11, 30-059 Kraków, Poland
anastasiya.pechko@doctoral.uj.edu.pl,
{katarzyna.grzesiak-kopec,barbara.strug,grazyna.slusarczyk}@uj.edu.pl

Abstract. This paper presents an innovative approach to using computational methods in designing modular housing estates with the Wave Function Collapse (WFC) algorithm. Currently, as a result of fast changing humanitarian situations in various parts of the world, like the one caused by the war in Ukraine or climate changes, many people have to leave their homes. It results in complex problem of providing large groups of people with conditions that will allow them to function with dignity for a long time. Therefore the adaptation of the WFC algorithm to design modular settlements is proposed. The applied heuristics allow the solutions to adapt to specific project requirements, generating various modular settlement designs that consider functionality and social aspects. The proposed approach is illustrated by examples of generated arrangements of housing modules for family-type floor plans.

Keywords: modular design · heuristics · WFC algorithm · constraint programming

1 Introduction

Nowadays the growing number of natural and human-made crisis situations requires immediate and effective solutions to rebuild and improve the living conditions of people affected by conflicts or other emergencies. Modular construction is increasingly recognized as an effective, fast, and ecological method for building homes and offers a promising approach to address these needs.

The current armed conflict on the territory of Ukraine has led to significant damage to urban infrastructure, a housing crisis, and a massive refugee migration. Ukrainian architects from the Balbek studio have designed a modular system of rooms with an area of 21 m², accommodating living rooms, kitchens, bathrooms, and meeting rooms. These modules can be used to create both small settlements for hundreds of people and larger towns for thousands [1]. This innovative solution forms the basis for the proposed tool that should support architects in designing process. The created WFC-Ukraine Housing software uses the

Wave Function Collapse (WFC) algorithm [2] to generate solutions in accordance with the above-mentioned project. The WFC algorithm has been modified and enhanced with specific heuristics to meet the design requirements of functionality, aesthetics, and social factors.

2 Related Work

Recently great deal of research has been conducted in the field of computer-aided architectural design particularly in the automatic generation of room arrangements for floor plans [7,9]. In most existing approaches, the graph-based representation of floor layouts, where nodes represent specific modules, is used [8,15]. Many techniques also rely on artificial intelligence and machine learning methods [12,13]. However, most of these methods are not suitable for complex layouts design, because of their high time complexity and a large amount of data needed to attain good accuracy.

Generating plans of modular settlements can be considered as a problem of the procedural content generation in architectural and urban design. Procedural content generation techniques, like Cellular Automata, enable efficient content generation, automating the process and reducing the need to manually create individual elements [3]. Another well-known approach is based on shape grammars [4,5,10]. In [11] an agent system combined with shape grammars was used to support floor layout designs. However, this approach does not give much possibility for designer interaction.

The WFC algorithm can also be used to design architectural projects. It is derived from an algorithm originally known as Model Synthesis [16]. However, its limitations include considering only local dependencies between grid tiles, poor scalability, and a lack of parameterization. As a result, many modifications of the WFC algorithm have been proposed, enabling its effective use as a tool for procedural content generation [6].

3 Generating Modular Housing Estates Using a Modified WFC Algorithm

The modular housing system used as the basis of the research was proposed by architects from the Ukrainian studio Balbek. The proposed system allows for quick organization of space while minimizing costs and ensuring comfortable conditions. The design concept emphasizes the importance of functionality and social factors. Four types of modules (residential, public, kitchen and sanitary) were proposed as the basic construction elements of the estate, from which the final solution is assembled and six different variants of module layouts (Fig. 1). The generated layouts of housing estates are easily modifiable, which allows them to be quickly adapted to various environmental conditions.

The WFC algorithm operates on a grid of basic units, called cells, sequentially determining cell values and propagating changes based on specified rules and

Fig. 1. The RE:Ukraine Housing project: the family arrangement type and 1–3 identified types of standard layouts.

constraints. The algorithm fills the grid cells with the specified tiles. Initially, each cell in the grid can potentially be any tile from the tile set, representing a superposition of all possible states. The entropy measures the uncertainty of a state of a cell (i, j) and is the number of possible tiles for this cell. The cell in the grid is collapsed when its state has been determined, reducing its possible values to a single tile. The neighborhood of a tile (i, j) defines its local surroundings that may affect the evolution of a grid. Adjacency rules define how tiles can be placed next to each other. The adjacency list for a tile specifies the allowed neighboring tiles in the grid and their possible position.

The WFC algorithm generates housing layouts in the following steps:

1. Initialization:
   - The tile set is established and augmented with rotated and mirrored variants to increase diversity.
   - The adjacency rules are defined based on tile compatibility.
   - The grid size and initial cell states are defined.
2. Generation:
   - **Observation**: identify the cell with the lowest entropy, collapse the cell to a single tile based on heuristic selection methods; if the entropy value is equal to zero a contradiction has been reached and there is no information about available moves; if the maximum number of iterations is reached, the algorithm will not be able to find a solution. This may mean that the constraints are poorly defined or there are too many of them.
   - **Propagation**: update the neighboring cells' possible tile sets based on the collapsed cell, ensuring all adjacency rules are maintained.

- **Iteration**: continue the observation and propagation steps until all cells are collapsed, resulting in a complete and coherent layout.
- **Render Observations**: generate the output image.

The modifications and enhancements introduced to WFC algorithm to effectively apply it to the design of modular housing estates are as follows: (1) heuristic functions, (2) enhanced tile encoding, (3) establishing excluded areas, (4) adjacency rules and constraints and (5) customizable parameters. These enhancements ensure that the algorithm not only generates functional and aesthetically pleasing layouts but also adheres to the specific requirements of modular construction and the unique needs of housing projects for displaced people.

Several *heuristic functions* were introduced, namely the bath placement, the symmetry patterns, the door and window placement and the residential standard one. They incorporate domain-specific knowledge and design principles to influence the selection and placement of tiles. The elaborated *enhanced tile encoding* includes additional metadata to provide more detailed information about each tile, such as: functional attributes (space type), structural attributes (doors and windows) and service attributes. Thanks to the use of *establishing excluded areas* enhancement, in which designated paths or areas remain free from any building modules, the algorithm gains greater flexibility and adaptability to various user requirements, existing infrastructure and local urban conditions.

The *adjacency rules* have been extended to incorporate complex relationships between tiles, which respects functional and aesthetic criteria. For instance, kitchens should be adjacent to dining areas, and bathrooms should not open directly into living rooms. *Constraints* related to the structural integrity and safety of the modular units have been integrated, ensuring that generated layouts are compliant with building codes and standards. *Customizable parameters* specify the density of modules needed to accommodate different population sizes, and set priorities for different functional areas (e.g., more space for communal areas or more bedrooms) based on the specific needs of the project.

4 Modular Housing Generation System Implementation

The implementation of a modular housing generation system was carried out using Visual Studio Code. Key libraries and languages used in the project include JavaScript, p5.js and jQuery.

The Re:Ukraine Housing modular housing system assumes four types of modules: residential, public, kitchen, and sanitary marked on the design with different colors (Fig. 1) [1]. It also includes six different variants of module layouts and the first of them (Type I a family-type section) was taken as the starting point in our project. Based on the different possible arrangement of modules, three types of standard layouts were distinguished:

1. Standard 1: includes one double bedroom, a kitchen and a bathroom.

2. Standard 2: includes one double bedroom, one triple bedroom, a kitchen and a bathroom.
3. Standard 3: includes two double bedrooms, a kitchen and a bathroom.

Bitmaps representing the appropriate standards were divided into regular tiles of square shape (Fig. 2) and subsequently, the set of tiles was extended by using rotation and reflection transformations. Each tile represents a specific functional or structural component of the housing units, such as living spaces, kitchens, and bathrooms and is encoded with metadata that provides detailed information about its attributes and relationships with other tiles. This metadata includes:

1. Functional attributes: descriptions of the tile's purpose and usage, ensuring that the generated layout meets the functional requirements.
2. Structural attributes: information about walls, doors, windows, and other structural elements, ensuring the structural integrity of the housing units.
3. Service attributes: indications of the presence of plumbing, electrical wiring, and other utilities, ensuring that the necessary infrastructure is in place.

Tile encoding should include information describing possible relations between them. Symbols of a set $C = \{h, f, m, g, v, p, l, o, e, c, s, q, t, n, j, u, r, z\}$ represent connections of tiles, while four special symbols represent an outer wall (a), door (x), window (w), and empty space (d). The primary relations defined between tiles are: connector relation: $\{\{i_C, i_C\} : i_C \in C\}$, window relation: $\{\{w, d\}\}$, door relation: $\{\{x, d\}\}$, outer wall relation: $\{\{a, d\}, \{a, a\}\}$ and empty relation: $\{\{d, i_S\} : i_S \in S\}$. Having relations between tiles defined explicitly, the algorithm can efficiently determine the compatibility of adjacent tiles. In Fig. 2 a set of tiles with the assigned codes for all three standard layouts is presented.

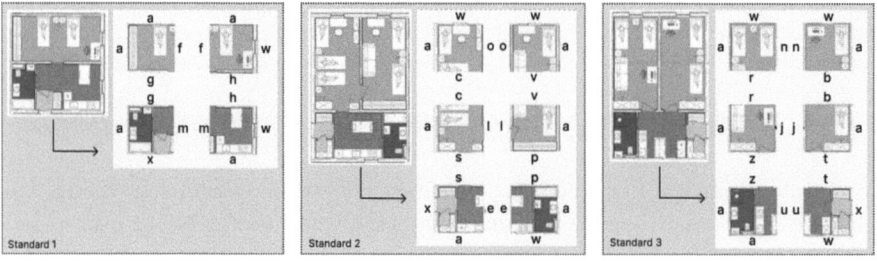

Fig. 2. Set of tiles with the assigned codes.

Heuristics are essential for guiding the WFC algorithm to produce coherent, functional, and aesthetically pleasing layouts. **Residential Standard Heuristic** ensures that the generated layouts meet the predefined residential standards and functional requirements. It involves assigning weights to tiles based on the desired number of occurrences of each standard. **Bath Placement Heuristic** ensures that bathrooms are placed logically within the housing layout. It considers grouping bathrooms with kitchens to optimize plumbing infrastructure, and ensures that bathrooms are placed away from communal areas. **Symmetry Patterns Heuristic** is crucial for ensuring that the generated housing layouts maintain a visually pleasing and structurally balanced design. Balanced layouts can reduce the complexity of plumbing, electrical, and HVAC systems, as these systems can be designed more efficiently in symmetrical spaces. **Establishing Excluded Areas Heuristic** allows specific paths or regions within the layout to be designated as non-buildable zones. This is particularly useful for maintaining open spaces, pathways, or respecting existing urban infrastructure. The algorithm accepts user-defined input for excluded areas. **Accessibility Heuristic** ensures that the layout is free from illogical connections and maintains logical flow. For example, the algorithm can generate layouts of modular housing estates where doors lead to inaccessible or closed areas. In order to eliminate the above anomalies, an additional step was introduced into the algorithm, which analyzes the available doors and checks whether there is at least one coherent path connecting them.

5 Experimental Results

The experiments were conducted on a system running macOS Ventura 13.2.1, equipped with 16 GB of RAM, a 12-core CPU, and a 19-core GPU, with a 512 GB SSD. Over 100 iterations were performed to ensure the stability and consistency of the results. Different grid sizes (e.g., $9 \times 9, 10 \times 10, 11 \times 11, 12 \times 12$) were tested to evaluate the scalability of the algorithm. Various heuristic functions, including Symmetry V, Symmetry H, Bath Placement, and their combinations, were applied to assess their impact on the layout generation process. Four predefined initial configurations of excluded areas were specified to simulate real-world constraints and test the algorithm's flexibility and adaptability.

The **residential standard heuristic** was verified by defining two levels of comfort and generating layouts corresponding to the expected density of people in a given area. The **bath placement heuristic** reduced the chaotic nature of generated designs and minimized the distance between bathroom modules (Fig. 3A). Two **symmetry patterns heuristic**, symmetry H and symmetry V, were tested to maintain visual appeal and structural balance. Obtained symmetrical designs were consistently rated higher in terms of aesthetic quality without compromising functionality (Fig. 3B). The algorithm effectively handled excluded areas by the **excluded area heuristic** without compromising the functionality of the housing layouts. Also, this heuristics with the one taking into account appropriate symmetry can improve the quality of the generated modular estate layouts and increase the number of successful generations (Fig. 4).

The algorithm demonstrated efficiency in generating layouts across different grid sizes and heuristic applications. The average generation time increased with grid size, which was expected due to the larger search space and complexity. The conducted experiments systematically compared various configurations by

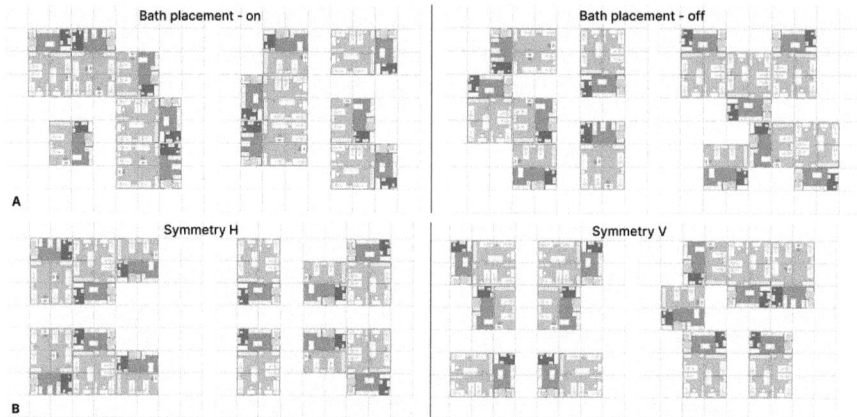

Fig. 3. The bath placement (A) and symmetry (B) heuristics.

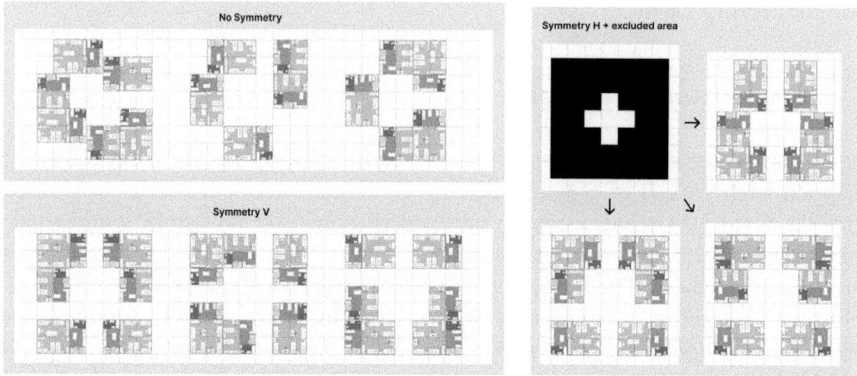

Fig. 4. The excluded area heuristic along with the symmetry patterns heuristic.

modifying parametric variables. The results were analyzed both qualitatively and quantitatively, providing a comprehensive understanding of the impact of different variables on the efficiency of generating modular layouts. The modified WFC algorithm offers a high degree of *customizability*, allowing users to define specific standards, comfort levels, and environmental constraints.

6 Conclusion

This study explored the feasibility and effectiveness of using the WFC algorithm in designing modular housing estates. The implementation of the generation tool demonstrated the algorithm's ability to generate modular housing layouts that are functional, scalable, and aesthetically pleasing. The heuristic functions significantly enhanced the quality of the layouts, with the Bath Placement and Symmetry Patterns heuristics proving particularly effective. The algorithm maintained performance and quality across different grid sizes.

Future research directions include exploring additional heuristic functions, integrating real-world data, further developing hierarchical approaches, and extending the algorithm's application to other domains such as urban planning and emergency housing.

References

1. RE:UKRAINE – balbek bureau. https://www.balbek.com/reukraine-eng. Accessed 23 July 2024
2. Javadi, H.: Wave function collapse and CPH Theory (2021)
3. Temuçin, M.B., Kocabaş, I., Oğuz, K.: Using cellular automata as a basis for procedural generation of organic cities. Eur. J. Eng. Technol. Res. **5**(12), 116–120 (2020)
4. Radies, C.: Procedural random generation of building models based Geobasis data and of the urban development with the software CityEngine. Bernburg, Germany **175**, 184 (2013)
5. Hansmeyer, M.: Michael Hansmeyer - L-Systems in Architecture. https://www.michael-hansmeyer.com/l-systems. Accessed 02 July 2024
6. Kleineberg, M.: Infinite procedurally generated city with the Wave Function Collapse algorithm. https://marian42.de/article/wfc/. Accessed 28 June 2024
7. Upasani, N., Shekhawat, K., Sachdeva, G.: Automated generation of dimensioned rectangular floorplans. Autom. Constr. **113**, 103–149 (2020)
8. Ślusarczyk, G.: Graph-based representation of design properties in creating building floorplans. Comput. Des. **95**, 24–39 (2018)
9. Strug, B., Grabska, E., Ślusarczyk, G.: Supporting the design process with hypergraph genetic operators. Adv. Eng. Inform. **28**, 11–27 (2014)
10. Grzesiak-Kopeć, K., Ogorzałek, M.J.: Intelligent 3D layout design with shape grammars. In: Proceedings of 6th International Conference on Human System Interactions (HSI), Sopot, Poland, 6–8 June 2013, pp. 265–270 (2013)
11. Grabska, E., Grzesiak-Kopeć, K., Ślusarczyk, G.: Designing floor-layouts with the assistance of curious agents. In: Alexandrov, V.N., van Albada, G.D., Sloot, P., Dongarra, J. (eds.) ICCS 2006. LNCS, vol. 3993, pp. 883–886. Springer, Heidelberg (2006). https://doi.org/10.1007/11758532_115
12. Wu, W., Fu, X.M., Tang, R., Wang, Y., Qi, Y.H., Liu, L.: Data-driven interior plan generation for residential buildings. ACM Trans. Graph. **38**, 1–12 (2019)
13. Hu, R., Huang, Z., Tang, Y., van Kaick, O., Zhang, H., Huang, H.: Graph2Plan: learning floorplan generation from layout graphs. ACM Trans. Graph. **39**, 1–14 (2020)
14. Flemming, U., Coyone, R., Gavin, T., Rychter, M.: A generative expert system for the design of building layouts–version 2. In: Topping, B. (ed.) Artificial Intelligence in Engineering Design; Computational Mechanics: UK, pp. 445–464 (1999)

15. Grzesiak-Kopeć, K., Strug, B., Ślusarczyk, G.: Evolutionary methods in house floor plan design. Appl. Sci. **11**, 8229 (2021)
16. Merrell, P.: Model Synthesis. Ph.D. Dissertation, University of North Carolina at Chapel Hill (2009)

Evaluating Parameter-Based Training Performance of Neural Networks and Variational Quantum Circuits

Michael Kölle[✉], Alexander Feist, Jonas Stein, Sebastian Wölckert, and Claudia Linnhoff-Popien

LMU Munich, Oettingenstraße 67, Munich, Germany
michael.koelle@ifi.lmu.de

Abstract. In recent years, NNs have driven significant advances in machine learning. However, as tasks grow more complex, NNs often require large numbers of trainable parameters, which increases computational and energy demands. VQCs offer a promising alternative: they leverage quantum mechanics to capture intricate relationships and typically need fewer parameters. In this work, we evaluate NNs and VQCs on simple supervised and reinforcement learning tasks, examining models with different parameter sizes. We simulate VQCs and execute selected parts of the training process on real quantum hardware to approximate actual training times. Our results show that VQCs can match NNs in performance while using significantly fewer parameters, despite longer training durations. As quantum technology and algorithms advance, and VQC architectures improve, we posit that VQCs could become advantageous for certain machine learning tasks.

Keywords: Variational Quantum Circuits · Parameter Efficiency · Quantum Supervised Learning · Quantum Reinforcement Learning

1 Introduction

Machine learning has advanced rapidly in recent years, with neural networks (NNs) playing a pivotal role in this progress [1]. As tasks become more complex, NNs often require a large number of trainable parameters, increasing computational and energy demands [4,25]. Variational quantum circuits (VQCs) are a promising alternative to classical NNs [5,9]. They harness quantum mechanics to model intricate relationships and usually need fewer parameters [17,22].

Quantum computing shows considerable promise in supervised learning (SL) and reinforcement learning (RL). Schuld et al. [22] introduced a scalable VQC architecture and showed that it achieves strong SL performance with fewer parameters than NNs. Their design inspires the architecture used in our work. Chen et al. [6] illustrated that VQCs can perform well in RL by approximating the action-value function through Q-learning in simple environments. Inspired by this work, we use their custom Frozen Lake environment and Q-learning approach to evaluate NNs and VQCs with varying parameter counts. Kruse et al. [15] explored architectural factors in VQCs for the Pendulum and LunarLander tasks [3], revealing that design choices like input encoding, layering, and qubit count strongly affect outcomes. Despite using about 96% fewer parameters than NNs, VQCs earned lower rewards and struggled with scalability and robustness.

In this work, we evaluate the potential of VQCs relative to NNs on simple SL and RL tasks. To our knowledge, no existing work thoroughly compares NNs and VQCs for machine learning tasks with a detailed focus on model architectures, parameter counts, and training times. We carry out most VQC experiments on a simulator but approximate real hardware training times by running selected circuits on an actual quantum device. Our findings align with prior work, showing that VQCs can achieve performance similar to NNs while using fewer parameters. Although training VQCs takes longer, we suggest that continued advances in quantum technology, improvements in VQC architectures, and algorithmic optimizations may make VQCs appealing for certain applications. All code for the experiments is available here[1].

2 Approach

For each machine learning task, we evaluate NNs and VQCs with varying parameter counts to identify one of each with comparable performance. To ensure fairness, both models function as black-box components in the same classical learning algorithm. Although we conduct the VQC experiments primarily with a quantum simulator, we estimate real-hardware training times by running selected circuits—collected from simulator-based training—on an actual device.

2.1 Classical Neural Network Architecture

We employ a fully connected feedforward NN. The input layer has as many nodes as the input size, followed by one or more hidden layers whose quantities and sizes are hyperparameters. Each hidden layer uses element-wise ReLU activation [14,18]. The output layer has as many nodes as the number of possible outputs and uses a softmax activation to produce a probability distribution [14].

2.2 Variational Quantum Circuit Architecture

The proposed VQC follows a circuit centric design [22] and involves three stages: state preparation, variational layers, and measurement.

State Preparation. We use angle embedding or amplitude embedding, chosen as a hyperparameter. Angle embedding maps each input feature to a rotation angle, using at least as many qubits as the input dimension. Amplitude embedding directly maps input values to the amplitudes of an n-qubit state, which requires $\lceil \log_2(D) \rceil$ qubits to represent D-dimensional data [22,23].

---

[1] https://github.com/alexander-feist/nn-vqc-params.

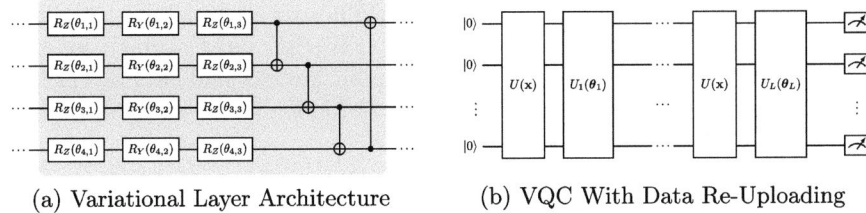

(a) Variational Layer Architecture (b) VQC With Data Re-Uploading

Fig. 1. (a) The l-th variational layer with 4 qubits, where $\boldsymbol{\theta}_l = [\theta_{1,1}, \theta_{1,2}, \ldots, \theta_{4,3}]$ are the trainable parameters for layer l. (b) A VQC with L layers; $U(\mathbf{x})$ encodes the input $\mathbf{x}$, and $U_l(\boldsymbol{\theta}_l)$ represents the trainable operations in layer l [5,22].

Variational Layers. Variational layers are composed of single-qubit rotations, followed by CNOT gates for entanglement (Fig. 1a). The trainable parameters $\boldsymbol{\theta}$, initialized randomly in $[-1, 1]$, are passed through $\varphi(\mathbf{z}) = \pi \cdot \tanh(\mathbf{z})$ to constrain angles to $(-\pi, \pi)$ [12]. We also use data re-uploading [20,24], which embeds the classical input values before every variational layer (Fig. 1b).

Measurement. We measure the expectation value of the Pauli-Z operator on the first K qubits, where K matches the output dimension, and add trainable biases (initialized in $[-0.001, 0.001]$). A softmax function is applied to derive output probabilities. To expand the $[-1, 1]$ Pauli-Z range, we include a trainable scaling parameter (initialized at 1) to enhance the effective output range [24].

3 Experimental Setup

We primarily use PyTorch [19] and PennyLane [2] with the `default.qubit` device for statevector simulation. All experiments run on a Linux cluster with Intel® Core™ i9-9900 processors. To ensure reproducibility, we fix seeds and repeat each experiment ten times (seeds 0–9). For each task, we conduct an exhaustive grid search over model-based hyperparameters. This yields equally sized sets of NNs and VQCs spanning a range of parameter counts, allowing us to identify pairs of models with comparable performance but different complexities.

3.1 Supervised Learning Experiments

We conduct SL classification tasks on the Iris, Wine, and WDBC (Wisconsin Diagnostic Breast Cancer) datasets. We use accuracy as the main performance metric. Features are scaled to $[0, 1]$ and data is split into 75% training and 25% testing, with the test portion evenly divided into validation and test sets.

We train for 50 epochs with cross-entropy loss and Adam [11] at a learning rate of 0.01, using a batch size of 8. Validation performance is checked after each epoch. The checkpoint with the highest validation accuracy is evaluated on the test set. For Iris and Wine, the NN grid search covers {1,2,3} hidden layers ×{3,6,9,12} nodes per layer. For the VQC, we use angle or amplitude embedding and 1 to 6 variational layers. For WDBC (30 features), we limit VQCs to amplitude embedding to avoid 30-qubit circuits and vary layers from 1 to 6. To match, we search for NNs with {1,2} hidden layers ×{3,6,9} nodes.

3.2 Reinforcement Learning Experiments

Our RL experiments use Q-learning with the models (NN or VQC) approximating the action-value function Q. The test reward serves as the main performance metric. Following Chen et al. [6], we use the deterministic (non-slippery) Frozen Lake environment [3] with custom rewards. The 4×4 grid contains safe (frozen) tiles and holes. The agent starts in the top-left and must reach the bottom-right goal. Steps yield −0.01, reaching the goal +1.0, and falling into a hole −0.2.

We train for 500 episodes, each limited to 100 steps. The model observes a 4-dimensional binary-encoded state (one of 16 tiles) and outputs four action values. We use identical policy and target models, updating the target every 20 steps. Actions follow an ϵ-greedy strategy: ϵ starts at 1.0 and decays by 0.99 after each episode until reaching 0.01. We use experience replay [16] with memory size 1000, sampling a batch of 16 transitions per step. The policy model is trained via Adam [11] at a learning rate of 0.01, using MSE loss and a discount factor of 0.95. After training, we evaluate over 50 test episodes without exploration. The grid search for NNs spans {1,2,3} hidden layers ×{3,6,9,12} nodes. For VQCs, it varies embedding (amplitude or angle) and the number of layers from 1 to 6.

3.3 Executing Quantum Circuits on Real Quantum Hardware

Running full training on real hardware is costly, so we log certain circuits (inputs and parameters) during simulator-based training and re-run those circuits on actual quantum processors through IBM's cloud-based Qiskit Runtime. Specifically, we pick circuits from five epochs/episodes under seed 0, unparameterize them with the logged values, and import them into Qiskit [10]. The circuits are executed on `ibm_fez` (version 2) backed by IBM's Heron R2 processor. By comparing Qiskit Runtime's usage metric to the simulator time for the same circuits, we compute an average ratio and apply it to circuit execution times of simulator-based training to estimate real-hardware training times.

4 Results

All metrics are averaged over ten runs (seeds 0–9). We first examine grid-search outcomes to select NNs and VQCs with comparable performance for each task.

4.1 Supervised Learning Results

We focus on models that perform well and choose an NN-VQC pair whose test accuracies and training curves suggest similar performance. Across all datasets, the NNs performed overall better compared to the VQCs. However, for each dataset, we could find an NN-VQC pair where both models are well-performing (more than 96% test accuracy) and comparable in performance. The VQCs have fewer parameters but consistently require substantially longer training times. Figure 2 illustrates training accuracy for our NN–VQC pairs. Although the VQCs converge faster initially, the NNs ultimately achieve slightly higher accuracy.

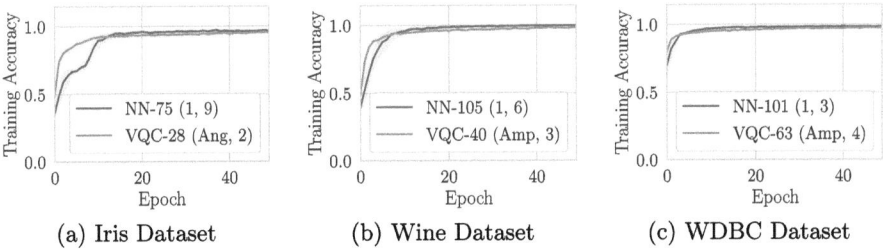

(a) Iris Dataset   (b) Wine Dataset   (c) WDBC Dataset

Fig. 2. Accuracy curves for each chosen NN and VQC. Averaged across ten runs (seeds 0–9); shaded areas are 95% confidence intervals.

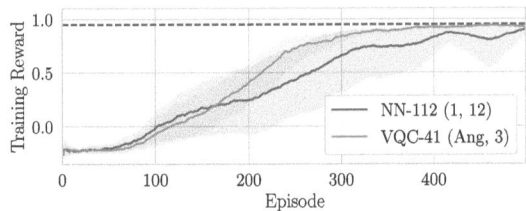

Fig. 3. Training reward (moving average over up to last 50 episodes) on Frozen Lake for the comparable NN and VQC. Mean across ten runs (seeds 0–9); shaded areas are 95% confidence intervals. Dashed red line (0.95) marks solved. (Color figure online)

4.2 Reinforcement Learning Results

We select models that achieve the maximum test reward of 0.95 and choose an NN–VQC pair with similar learning dynamics but relatively low training times. Training-time variance is high because episode length depends on agent behavior. Over all models, VQCs generally outperform NNs here, which may be due to the grid search only considering low-parameter models, favoring VQCs. However, the VQCs take much longer to train. Figure 3 shows the training reward of our most comparable NN and VQC. The VQC converges faster and is more stable.

4.3 Training Times Using Real Quantum Hardware

Estimated real-hardware training times for VQCs are significantly higher than simulator times, as shown in Table 1, based on per-circuit execution durations.

Using angle embedding leads to lower circuit depth compared to amplitude embedding, resulting in smaller simulator runtimes but higher hardware-to-simulator time ratios. For 5-qubit circuits, the hardware ratio decreases, indicating that as qubit counts increase, the simulator grows slower relative to hardware, consistent with existing literature [7,8]. Since overhead and noise are excluded, these estimates likely represent ideal scenarios, but further optimizations (e.g., fewer shots, specialized training environments) could significantly reduce real-hardware training times.

Table 1. Mean execution time per circuit on the simulator vs. real hardware, as well as the hardware-to-simulator time ratio.

Task	VQC	Qubits	Circuit Depth	Simulator (s)	Real Hardware (s)	Ratio
SL: Iris	VQC-28 (Ang, 2)	4	17	0.011	0.314	28.995
SL: Wine	VQC-40 (Amp, 3)	4	100	0.034	0.349	10.295
SL: WDBC	VQC-63 (Amp, 4)	5	261	0.084	0.329	3.932
RL	VQC-41 (Ang, 3)	4	25	0.016	0.322	20.406

Table 2. Mean training times (with 95% confidence intervals) for comparable NNs and VQCs.

Task	Model		Training Time (s)		
	NN	VQC	NN	VQC Simulator	VQC Real Hardware
SL: Iris	NN-75	VQC-28	1.8 ± 0.0	92.6 ± 0.2	1806.1 ± 4.1
SL: Wine	NN-105	VQC-40	1.7 ± 0.0	313.5 ± 1.3	2437.4 ± 11.6
SL: WDBC	NN-101	VQC-63	5.4 ± 0.0	2482.9 ± 12.7	7732.5 ± 37.2
RL	NN-112	VQC-41	95.0 ± 30.7	2511.4 ± 219.3	39330.9 ± 3458.8

4.4 Evaluating Training Performance

Table 2 shows training times of the selected NNs and VQCs. For similar performance, VQCs require 62.7% (Iris SL), 61.9% (Wine SL), 37.6% (WDBC SL), and 63.4% (Frozen Lake RL) fewer parameters, but take much longer to train. For the RL task, equalizing the two training times would require the VQC to be about 414 times faster, which may sound large but could become feasible as quantum hardware matures much faster than classical systems. Architectural and algorithmic improvements—such as specialized VQC optimizers—may also reduce this ratio. Furthermore, although VQCs often converge faster in accuracy or reward, our fixed training schedule does not exploit early convergence.

Even in our small-scale tasks, we see a trend of longer training times for larger models, especially for VQCs. This trend may be more pronounced in complex tasks where standard NNs can have millions of parameters, potentially offering a more substantial advantage to VQCs that require fewer parameters [13,17].

However, it remains unclear whether VQCs can scale effectively to complex tasks and still match NNs [15,21]. Rather than replacing NNs outright, VQCs may find value in scenarios where they offer distinct benefits—especially if quantum hardware, training algorithms, and circuit designs continue to improve.

5 Conclusion

We created a unified environment to compare classical NNs and VQCs as interchangeable models for multiple machine learning tasks. VQCs achieved comparable performance with fewer parameters—especially in RL—but required much longer training times. Because our tasks used at most five qubits, real-hardware execution was slower than simulation. Our findings underscore the simplicity of the tasks, yet suggest that as quantum technology matures, VQC-friendly algorithms improve, and circuit architectures evolve, VQCs may offer advantages for specific applications. Future work could explore more complex tasks and assess quantum device fidelity to better evaluate VQC performance in realistic settings.

Acknowledgements. This work is part of the Munich Quantum Valley, which is supported by the Bavarian state government with funds from the Hightech Agenda Bayern Plus. This paper was partly funded by the German Federal Ministry of Education and Research through the funding program "quantum technologies—from basic research to market" (contract number: 13N16196) and the BWMK (01MQ22008A).

References

1. Alom, M.Z., et al.: A state-of-the-art survey on deep learning theory and architectures. Electronics **8**(3), 292 (2019)
2. Bergholm, V., et al.: Pennylane: automatic differentiation of hybrid quantum-classical computations. arXiv preprint arXiv:1811.04968 (2018)
3. Brockman, G., et al.: OpenAI Gym. arXiv preprint arXiv:1606.01540 (2016)
4. Brown, T., et al.: Language models are few-shot learners. Adv. Neural. Inf. Process. Syst. **33**, 1877–1901 (2020)
5. Cerezo, M., et al.: Variational quantum algorithms. Nat. Rev. Phys. **3**(9), 625–644 (2021)
6. Chen, S., Yang, C., Qi, J., Chen, P.Y., Ma, X., Goan, H.S.: Variational quantum circuits for deep reinforcement learning. IEEE Access **8**, 141007–141024 (2020)
7. Chen, Z.Y., Zhou, Q., Xue, C., Yang, X., Guo, G.C., Guo, G.P.: 64-qubit quantum circuit simulation. Sci. Bull. **63**(15), 964–971 (2018)
8. Cicero, A., Maleki, M.A., Azhar, M.W., Kockum, A.F., Trancoso, P.: Simulation of quantum computers: review and acceleration opportunities. arXiv preprint arXiv:2410.12660 (2024)
9. Du, Y., Hsieh, M.H., Liu, T., Tao, D.: Expressive power of parametrized quantum circuits. Phys. Rev. Res. **2**(3), 033125 (2020)
10. Javadi-Abhari, A., et al.: Quantum computing with qiskit. arXiv preprint arXiv:2405.08810 (2024)

11. Kingma, D.P.: Adam: a method for stochastic optimization. arXiv preprint arXiv:1412.6980 (2014)
12. Kölle, M., et al.: Weight re-mapping for variational quantum algorithms. In: International Conference on Agents and Artificial Intelligence, pp. 286–309. Springer (2023)
13. Kölle, M., Topp, F., Phan, T., Altmann, P., Nüßlein, J., Linnhoff-Popien, C.: Multi-agent quantum reinforcement learning using evolutionary optimization. arXiv preprint arXiv:2311.05546 (2023)
14. Krizhevsky, A., Sutskever, I., Hinton, G.E.: Imagenet classification with deep convolutional neural networks. In: Advances in Neural Information Processing Systems, vol. 25 (2012)
15. Kruse, G., Dragan, T.A., Wille, R., Lorenz, J.M.: Variational quantum circuit design for quantum reinforcement learning on continuous environments. arXiv preprint arXiv:2312.13798 (2023)
16. Lin, L.J.: Self-improving reactive agents based on reinforcement learning, planning and teaching. Mach. Learn. **8**, 293–321 (1992)
17. Lockwood, O., Si, M.: Reinforcement learning with quantum variational circuit. In: Proceedings of the AAAI Conference on Artificial Intelligence and Interactive Digital Entertainment, vol. 16, pp. 245–251 (2020)
18. Nair, V., Hinton, G.E.: Rectified linear units improve restricted Boltzmann machines. In: Proceedings of the 27th International Conference on Machine Learning (ICML 2010), pp. 807–814 (2010)
19. Paszke, A., et al.: Pytorch: an imperative style, high-performance deep learning library. In: Advances in Neural Information Processing Systems, vol. 32 (2019)
20. Pérez-Salinas, A., Cervera-Lierta, A., Gil-Fuster, E., Latorre, J.I.: Data reuploading for a universal quantum classifier. Quantum **4**, 226 (2020)
21. Qian, Y., Wang, X., Du, Y., Wu, X., Tao, D.: The dilemma of quantum neural networks. IEEE Trans. Neural Netw. Learn. Syst. (2022)
22. Schuld, M., Bocharov, A., Svore, K.M., Wiebe, N.: Circuit-centric quantum classifiers. Phys. Rev. A **101**(3), 032308 (2020)
23. Schuld, M., Fingerhuth, M., Petruccione, F.: Implementing a distance-based classifier with a quantum interference circuit. Europhys. Lett. **119**(6), 60002 (2017)
24. Skolik, A., Jerbi, S., Dunjko, V.: Quantum agents in the gym: a variational quantum algorithm for deep q-learning. Quantum **6**, 720 (2022)
25. Strubell, E., Ganesh, A., McCallum, A.: Energy and policy considerations for modern deep learning research. In: Proceedings of the AAAI Conference on Artificial Intelligence, vol. 34, pp. 13693–13696 (2020)

Towards an Open Science—An Academic Recommendation Cloud Platform

Anna Kobusińska[1(✉)], Damian Tabaczyński[1], and Victor Chang[2]

[1] Faculty of Computing and Telecommunications, Poznan University of Technology, Poznań, Poland
{Anna.Kobusinska,Damian.Tabaczynski}@cs.put.poznan.pl
[2] Aston Business School, Aston University, Birmingham, UK
v.chang1@aston.ac.uk

Abstract. This paper provides a cloud-based academic recommender system that integrates multiple heterogeneous big data academic resources to deliver personalized recommendations within the academic networks. The proposed system introduces a hybrid recommendation mechanism combining content-based and collaborative filtering within a graph-based relationship model.

Keywords: scholarly datasets · hybrid recommendation mechanism · cloud computing · serverless

1 Introduction

Nowadays, the number of new publications is reaching millions per year worldwide [4]. The available, overwhelming amount of academic data makes it difficult for researchers to find relevant literature effectively. Existing traditional search engines and databases often provide keyword-based searches without personalization and context awareness, thus not meeting the requirements of researchers. This challenge requires the development of intelligent recommendation systems that utilize big data, cloud computing and advanced filtering techniques. Existing solutions such as Google Scholar, Microsoft Academic Graph and OpenAIRE Research Graph provide partial solutions but do not offer a fully integrated, personalized recommender system. For example, Google Scholar relies mainly on ranking-based citations but does not consider user preferences. The Microsoft Academic Graph provides extensive bibliometric data, but lacks advanced recommendation mechanisms. In turn, the OpenAIRE Research Graph integrates multiple repositories but cannot offer personalized recommendations. Therefore, this paper addresses these gaps by proposing a scalable, adaptive, and user-oriented academic recommendation platform. The solution proposes an approach integrating heterogeneous academic databases to improve recommendation coverage and completeness. In addition, it introduces a recommender algorithm that enables personalized recommendations based on the user's search history while working with the system and the user's scientific relationships. To this end, it leverages content-based and collaborative filtering within a graph-based relationship model. Finally, the proposed system optimizes performance by leveraging cloud-based environments, ensuring scalability and responsiveness. The proposed solution allows for context-aware and highly relevant suggestions, significantly enhancing the research discovery process.

The structure of the paper is as follows: Sect. 2 provides a brief overview of the related works. Section 3 presents a general idea of the proposed solution. The relationship graph model and the recommendation mechanism are proposed in Sect. 4 and Sect. 5, respectively. Finally, the conclusions are discussed in Sect. 6.

2 Related Work

Academic recommendation systems have been widely explored to improve research discovery. Existing solutions can be divided into content-based filtering, collaborative filtering, and hybrid approaches. Among the available platforms and tools to facilitate the discovery of academic literature is Google Scholar, which uses a ranking based on citations but lacks personalized recommendations. Microsoft Academic Graph provides extensive bibliometric metadata, but does not offer real-time recommendations. Semantic Scholar, on the other hand, implements recommendations based on artificial intelligence but relies on NLP techniques without a strong graph-based model. Connected Papers, on the other hand, generates graph-based visualizations but requires manual input of article identifiers. Finally, ArnetMiner uses social network analysis to classify authors but has limited scalability for real-time recommendations. Each system has limitations regarding scalability, integration of heterogeneous data sources, and personalization mechanisms. Our approach addresses these gaps by incorporating graph-based modelling with hybrid recommendation techniques to provide a more comprehensive academic discovery experience.

Among studies dedicated to academic recommender systems, [2] can be mentioned, where the authors investigated citation recommendations based on deep learning in scientific networks. In turn, [1] presents a comprehensive study of article recommendation techniques, including collaborative filtering and content-based methods. In [14] digital libraries were analyzed, demonstrating increased search efficiency. A knowledge graph-based academic recommender system integrating citation networks and semantic analysis has been proposed in [11]. Article [15] proposes a contextual recommender framework for digital research repositories. Much of this research focuses on improving specific aspects of recommender systems, such as natural language processing (NLP) for article recommendations, modelling user behavior, or citation network analysis. However, few works address integrating multiple data sources, personalization, and scalable cloud deployment within a single framework. The proposed solution builds on these elements, offering an improved graph that optimizes the performance and adaptability of discoveries and a cloud-native recommendation engine.

3 The General Idea of the Proposed Solution

This paper proposes a recommendation platform, called *Academic Advisor*, designed to assist researchers by suggesting relevant publications, datasets, organizations, and scientific collaborations. The system leverages a novel recommendation algorithm that establishes relationships among academic entities, enhancing contextual discovery. The solution is user-oriented and designed to require minimal preliminary information input. The platform generates recommendations by comparing the retrieved data with its knowledge base. This eliminates users' need to input detailed preferences manually, streamlining the recommendation process.

To enhance personalization, users can rate recommendations, influencing future results. This feedback loop allows dynamic adaptation of the algorithm, refining the accuracy and relevance of recommendations. Additionally, an Academic Advisor can suggest publications, authors, institutions, and projects, recognizing that academic collaboration extends beyond literature.

The system architecture comprises three main components: clients (who interact with the system through a web application), application servers (which handle clients' requests, retrieve data, and process recommendations), and data source modules (which integrate data from multiple scholarly data sources). The high-level abstraction diagram of the solution with marked cloud parts is shown in Fig. 1. To generate recommendations, the client sends a request to the application server. Before responding to the client, the server retrieves the necessary data from distributed academic sources by sending a data request to these repositories. To avoid sending massive amounts of data, the data is filtered explicitly on the data source side. The returned data set contains raw, filtered data and has to undergo pre-processing, ensuring that only the most relevant and filtered information is used, thus minimizing the data transmission load. After receiving the data, the client executes the recommendation algorithm in the Edge Worker, a component specifically designed for local computing. Edge Worker, which is part of the client's application, refines and classifies the recommendations before presenting the results via a graphical user interface. This approach optimizes performance by reducing backend server load and shortening response times.

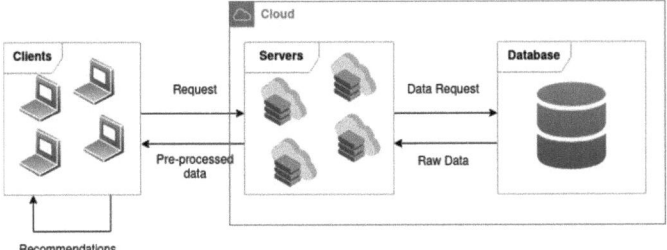

Fig. 1. Solution scheme.

To ensure scalability, the system is deployed in a serverless cloud environment, eliminating the need for manual infrastructure management while providing automated scaling. The use of cloud computing also enhances system availability, security, and performance. Data is stored in replicated cloud databases, ensuring fault tolerance and fast access. Additionally, internal cloud networking reduces latency and improves security by avoiding exposure to external networks. This hybrid cloud-edge approach optimizes computational efficiency while maintaining a seamless user experience.

4 Graph-Based Relationship Model

The Academic Advisor uses a relationship graph to model academic units and their interrelationships. This model represents **Users** (authors, their profiles and interests), **Organizations** (authors' affiliations, e.g. universities or research institutes), **Projects** (understood as funded research initiatives and their results) and **Resources** (publications, data sets and software) as nodes. The following relationships between nodes, represented by edges, are considered: **Authorship** models a bidirectional relation between User and Resource node. This indicates that the given user is the author of the given resource. **Membership** is a bidirectional relation between the Organization and User node, and identifies a specific user as a member of a particular organization. **Origin** represents a bidirectional relation describing the origin of a particular resource. It models a connection between the Organization and Resource node. **Outcome** is a bidirectional Project–Resource relation that connects this specific project to its outcome nodes. Finally, **Like/Dislike** unidirectional relation between a specific user and any other node (even other users) specifies if users like or dislike this node as a recommendation. In the recommendation algorithm, this is the only relationship that does not influence connections.

The combination of three well-known representative academic networks: OpenAire Graph [10], Microsoft Academic Graph [9], and Google Scholar [5] is considered in the proposed solution as a data source. The OpenAIRE Research Graph is an open resource system that aggregates a collection of research data properties (metadata, links) available within the OpenAIRE Open Science infrastructure [8,10,13]. It contains information about organizations, funders, funding streams, projects, communities, data sources, and scientific products, including literature, datasets, software, etc. The Microsoft Academic Graph (MAG) is a heterogeneous graph containing scientific publications, citation relationships between publications, and information regarding authors, institutions, journals, conferences, and fields of study, among others [3,12]. The MAG is marked by its completeness and stability. The third considered solution, Google Scholar network [6,7], was chosen due to the system's and its users' constant updates. It is used as an additional set to extend the knowledge provided by the MAG and the OpenAIRE environments.

An example of a graph with a high degree of abstraction is shown in Fig. 2. Such a graph structure enables effective browsing and filtering of information based on user preferences and relevance metrics derived from the system.

5 Hybrid Recommendation Mechanism

The decision–making process for selecting the best recommendations is carried out by a dedicated edge worker on the client's behalf. As the specific set of data on which the worker operates is determined in advance, the worker is not extensively involved in the pre-processing process. The application server is responsible for selecting and preprocessing the data before transferring them to the final algorithm. In contrast, the data sources are responsible for data filtering. Ultimately, the client side (edge worker) implements the recommendation mechanism and visually presents the algorithm results.

The recommendation engine integrates content-based and collaborative filtering to provide precise and dynamic suggestions. The proposed algorithm is divided into two phases:

1. **Phase 1 Node Usability Function**: each entity in the graph is assigned a usability score that evaluates its relevance based on intrinsic attributes and network influence. The function value is calculated differently, based on the node type:

$$f(u) = \alpha * V_n + \beta * V_r + \gamma * V_l \tag{1}$$

   where u is the node for which calculations are performed, $\alpha, \beta, \gamma \in [0, 1]$ are the factors of specific part of formula. There are three main components to this formula:

   V_n node value—represents node-specific attributes (e.g., citation count for papers, impact score for authors), and indicates the evaluation of the node itself. The value $V_n \in [0, 1]$ is strongly related to the node information, thus it tends to remain relatively constant over time. A node's value is calculated in two ways. First, if a node does not contain citation information, then:

$$V_n = (base_x + C)/2 \tag{2}$$

   where $base_x \in [0, 1]$ is the constant base value for a specific type of node, while $C \in [0, 1]$ is constant compensation for lack of information about citation number. If a node contains citation information, its value is calculated as follows:
   - for User:

$$V_n = (base_u + min(cit_u/T_u, 1))/2 \tag{3}$$

   where cit_u denominates number of citations for user and T_u user citations threshold.
   - for Organization:

$$V_n = (base_o + min(Avg_{cit}/T_o, 1))/2 \tag{4}$$

   where Avg_{cit} stands for an average number of citations of 10% the most cited users in this organization, and T_o is the organization citations threshold.

– for Project:
$$V_n = 1, \qquad (5)$$
A project value is always equal to one because, from a logical standpoint, this type of node aggregates other results. As a result, it should be prioritized and the value should be the maximum, so 1.
– for Resource:
$$V_n = (base_r + min(cit_r/T_r, 1))/2 \qquad (6)$$
where cit_r denominates number of citations for resource and T_r resource citations threshold.

V_r **reputation value**—captures reputation from the academic community. The value $V_r \in [-1, 1]$ represents the average values from Like relationships that end at this vertex, and varies greatly over time. A strong correlation exists between this indicator and the actions taken by the users of the entire system. The formula is presented as follows:

$$V_r = (Avg_{likes}/Max_{like}) * min(likes/T_{likes}, 1) \qquad (7)$$

where Avg_{likes} is an average value of ratings (Like relationships) from all users from the system, Max_{like} is the maximum value of the rating, $likes$ is the number of Like/Dislike relationships and T_{likes} likes count threshold.

V_l **like value**—integrates personalized user preferences (e.g., prior likes/dislikes). The value $V_l \in [-1, 1]$ refers to the user's evaluation of a particular node, and varies considerably over time. Since only one "Like" relation may exist for a given user to a given node, it is a single number value originating from the Like relation. The formula is as follows:

$$V_l = v_{like}/Max_{like} \qquad (8)$$

where v_{like} is a value of the Like relationship and Max_{like} maximum value of rating.

Ultimately, the usability function for a node is the following:

$$f(u) = \alpha * \begin{cases} (base_x + C)/2, \text{ if citations undefined} \\ (base_u + min(cit_u/T_u, 1))/2, \text{ if} \\ \text{node=User} \\ (base_o + min(Avg_{cit}/T_o, 1))/2), \\ \text{if node=Organization+1 or node=Project} \\ (base_r + min(cit_r/T_r, 1))/2, \\ \text{if node=Resource} \end{cases} + \qquad (9)$$

$$+ \beta * (Avg_{likes}/Max_{like}) * min(likes/T_{likes}, 1)$$

$$+ \gamma * (v_{like}/Max_{like})$$

2. **Phase 2 Longest Path Sorting**: the topological sorting algorithm was applied to the directed acyclic graph (DAG) to determine the most influential academic units. The priority of the recommendations is based on the weighted longest path, ensuring the effective disclosure of high-impact research. DAG is a subgraph of the relation graph contained in the data source. Selecting the longest route can begin when all vertices have been sorted according to their topological position and values calculated. For each vertex, the longest path is calculated relative to the starting vertex, which is the user for whom the recommendations are made. Since edges in the graph do not have their values, but nodes do, the edges are assigned the values of the nodes. More precisely, the value of the edge is equal to the vertex value it leads to. *Calculation route* is a path calculated by summing the vertices values of all the vertices on the path. The calculation is performed once and individually for each node. Calculations are performed in topological sort order for optimization purposes, and previous calculations are utilized to prevent redundant calculations. As part of the calculation route, all relationships except for Like relationships are considered. Specifically, this relation indicates only the value for the first phase of the algorithm. During the second phase of the recommendation mechanism, it does not affect any vertices. The result of each calculation route gives the final value of the recommender system for node (item), which can be viewed in the user interface. Figure 3 shows an example of the calculation route.

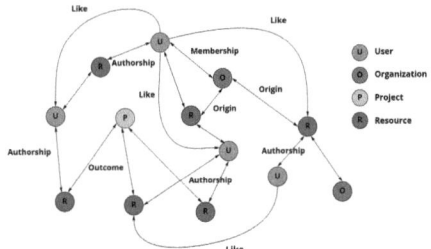

Fig. 2. Exemplary relationship graph.

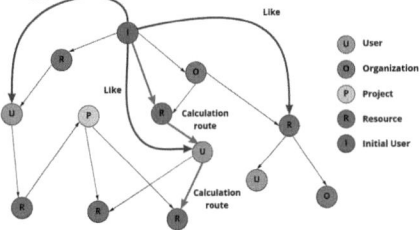

Fig. 3. Exemplary subgraph with parameters and factors for relationship graph in Fig. 2.

6 Conclusion

This paper proposes an academic recommendation system to address challenges in scholarly resource discovery. The idea is based on a hybrid filtering approach that combines collaborative and content-based recommendation techniques within a graph structure. The graph integrates multiple academic databases (OpenAIRE, Microsoft Academic Graph, Google Scholar) for comprehensive recommendations. Furthermore, the serverless cloud infrastructure ensures scalability and performance, optimizing resource utilization. Ongoing research is focused on experimentally validating the system. The tests include a comprehensive performance evaluation in which recommendation accuracy, computational efficiency, and scalability will be assessed under real–world conditions. Future work also plans to explore improved personalization techniques, integration with additional academic data sources, and reinforcement learning-based recommendation strategies. Future work includes refining recommendation algorithms using reinforcement learning, expanding entity types, and improving integration with external repositories. Moreover, the experimental evaluation is planned.

References

1. Beel, J., Gipp, B., Langer, S., Breitinger, C.: Research-paper recommender systems: a literature survey. Int. J. Digit. Libr. **17**(4), 305–338 (2016)
2. Choi, J., Lee, J., Yoon, J., Jang, S., Kim, J., Choi, S.: A two-stage deep learning-based system for patent citation recommendation. Scientometrics **127**(11), 6615–6636 (2022)
3. Färber, M., Ao, L.: The microsoft academic knowledge graph enhanced: author name disambiguation, publication classification, and embeddings. Quant. Sci. Stud. **3**(1), 51–98 (2022)
4. Fire, M., Guestrin, C.: Over-optimization of academic publishing metrics: observing goodhart's law in action. GigaScience **8**, 06 (2019)
5. Google. Google Scholar. https://scholar.google.com/intl/en/scholar/about.html
6. Hacohen, S., Medina, O., Shoval, S.: Autonomous driving: a survey of technological gaps using google scholar and web of science trend analysis. IEEE Trans. Intell. Transp. Syst. **23**(11), 21241–21258 (2022)
7. Kalhor, G., Sarijalou, A.A., Sadr, N.S., Bahrak, B.: A new insight to the analysis of co-authorship in google scholar. Appl. Netw. Sci. **7**(1), 21 (2022)
8. Vichos, K., et al.: A preliminary assessment of the article deduplication algorithm used for the openaire research graph. In: Proceedings of the 18th Italian Research Conference on Digital Libraries, Padua, Italy, 24–25 February 2022 (hybrid event). CEUR Workshop Proceedings, vol. 3160. CEUR-WS.org (2022)
9. Microsoft. Microsoft Academic Graph. https://www.microsoft.com/en-us/research/project/microsoft-academic-graph/
10. OpenAIRE. Openaire research graph. https://graph.openaire.eu/
11. Padmaja, B., Sucharitha, G., Krishna Rao Patro, E.: KGRecSys: knowledge graph-based recommendation systems: a comprehensive overview (2025)
12. Qin, H., Zeng, J., Ma, X.: Trend analysis of research direction in computer science based on microsoft academic graph. In: CONF-CDS 2021: The 2nd International Conference on Computing and Data Science, Stanford, CA, USA, 28–30 January 2021, pp. 18:1–18:4. ACM (2021)
13. Schirrwagen, J., et al.: Data sources and persistent identifiers in the open science research graph of openaire. Int. J. Digit. Curation **15**(1), 1–5 (2020)

14. Troussas, Ch., Krouska, A., Koliarakis, A., Sgouropoulou, C.: Harnessing the power of user-centric artificial intelligence: customized recommendations and personalization in hybrid recommender systems. Computers **12**(5) (2023)
15. Yang, J., Cheng, X., Zhou, W.: Research review and progress on practice of the resource recommendation research based on context-awareness. J. Mod. Inf. **40**(2), 153–159 and 167 (2020)

Algorithm Selection in Short-Range Molecular Dynamics Simulations

Samuel James Newcome[✉], Fabio Alexander Gratl, Manuel Lerchner, Abdulkadir Pazar, Manish Kumar Mishra, and Hans-Joachim Bungartz

Chair of Scientific Computing in Computer Science, Department of Computer Science, Technical University of Munich, Boltzmannstr. 3, 85748 Garching bei Muenchen, Germany
samuel.newcome@tum.de

Abstract. Recent works have highlighted the advantages of algorithm selection to optimise scientific simulations, presenting a range of approaches from classical time-series prediction, to expert-guided, to data-driven. In this work, we present novel variations upon these approaches for Molecular Dynamics simulations, implemented in the algorithm selection particle simulation library AutoPas, and compare them in terms of both performance and practicality. We demonstrate that these approaches can achieve speedups of up to 1.25 compared to an optimal single algorithm without dynamic algorithm selection.

Keywords: Algorithm Selection · Molecular Dynamics · AutoPas

1 Introduction

Molecular Dynamics (MD) simulations are an important technology with applications including thermodynamics [8] and bio-membrane simulation [5]. Molecules are simulated as point bodies moving according to Newton's equations of motion, propagated with some numerical integrator. This results in a large number of timesteps, each typically computationally dominated by inter-particle force calculations, and between them the distribution of particles changes only minimally. Where the forces are short-ranged, a common optimisation is to introduce a cutoff distance beyond which the forces are neglected. Simulations with this optimisation are well studied, resulting in the development of numerous algorithms and parallelisations, however no algorithmic configuration (AC) is optimal in all scenarios [2] and the relative performance of these ACs can vary on different hardware or with different force models [7].

Such findings motivate the development of the algorithm selection and tuning library AutoPas[1], which aims to select the optimal from a large range of ACs [2]. However, prior works [2,7,10] have utilised only naive selection strategies which

---

[1] https://github.com/AutoPas/AutoPas.

involve significant time spent trialling suboptimal ACs, which can perform orders of magnitude worse than the optimal.

In this work, we present and compare three methods that aim to avoid this. The **Predictive** strategy uses past time series data of an AC's performance to predict how it will perform in the future. The **Expert** strategy uses expert-written rules to determine in which scenarios an AC should be trialled. The **Random Forest** strategy uses a random forest trained on pre-collected data, which maps scenario-dependent metrics to optimal ACs. We discuss the practicalities of these methods as well as their performance in three varied simulation scenarios.

2 AutoPas

AutoPas is designed to provide a black-box particle container to the developer of an arbitrary cutoff-based particle simulator. It contains a variety of algorithms and parallelisations which combine to form an AC for efficiently calculating the forces on each particle. AutoPas aims to dynamically select the optimal of these ACs over the course of the simulation, as well as tune them. As the focus of this work is on the selection strategies, we will not discuss the different ACs themselves. For such a discussion, the reader is referred to Gratl et al. [2] and Gratl-Gaßner [3]. Overall, a total of 116 possible ACs will be used in this work.

As AutoPas primarily handles shared memory parallelism, distributed memory parallelism is intended to be implemented by the simulator developer. Seckler et al. [10] proposes decomposing the domain into a number of subdomains and assigning each an MPI rank with its own AutoPas container. The consequence of this is that different subdomains can use different ACs [10].

AutoPas' algorithm selection consists of a series of *tuning phases* that occur in regular intervals. During each tuning phase, all ACs are trialled for a small number of iterations and the average is taken. The AC with the smallest average is then used until the following tuning phase. Naively trialling all ACs is referred to as a Full Search. Selection Strategies, such as those presented in Sect. 5, are intended to reduce the number of ACs trialled [3].

3 Related Work

Armstrong et al. [1] investigated a very similar dynamic algorithm selection problem also applied to short-range particle simulations, using a temporal difference reinforcement learning agent and a linear regression model to predict the optimal algorithm out of two.

Mohammed et al. [6] investigated the dynamic selection of OpenMP load-scheduling algorithms out of a range of twelve algorithms. They considered random search, exhaustive search, and expert search methods. The expert search method takes relevant measurements and applies them to expert-written Fuzzy Logic systems, which can result in changes to the algorithm.

Stylianou and Weiland [11] investigated the selection of one of six different sparse matrix storage formats using a decision tree and random forest. Unlike the previously discussed works, they considered independent matrices rather than a problem that evolves over time and, as such, their selection methods are solely based on features from the matrices.

The first work is similar to the Predictive strategy, and the latter two serve as inspiration for an Expert and a Random Forest strategy; however, the ideas must be non-trivially redeveloped into a form that is applicable to AutoPas. Unlike all of the above works, we will consider a much larger number of algorithms.

4 Live Simulation Statistics

In order to implement expert knowledge or data-driven approaches, as discussed previously, statistics need to be extracted from the simulation, from which decisions on the optimality of ACs can be made. We consider a simple yet computationally efficient scheme in which particles are binned into small cells, and the mean, standard deviation, median, and maximum numbers of particles per cell are collected. Furthermore, we also gather statistics on the number of bins, how many of these are empty, the number of OpenMP threads, and the skin[2].

5 AutoPas' Selection Strategies

In this section, we will describe three novel dynamic algorithm selection methods for AutoPas: the **Predictive**, **Expert**, and **Random Forest** methods. For a deeper discussion of the implementation of the methods, see Gratl-Gaßner [3], Lerchner [4], and Pazar [9] respectively. For details on the exact live simulation statistics used, model parameters, rules, and training data, see the repository mentioned in the Appendix.

5.1 The Predictive Strategy

In the predictive strategy, each AC is trialled during the first two tuning phases. After this, at the start of each following tuning phase, a linear model for each AC fitted to the last two data points for that AC, is used to predict the performance of that AC in that tuning phase. Only the ACs expected to perform within a relative threshold of the expected optimum are trialled.

To avoid ACs that previously performed poorly, never being trialled again, even if the simulation has changed significantly, we retrial ACs after a number of tuning phases where they have not been trialled, even if they are expected to perform worse than the threshold. On the other hand, a blacklist is applied to extremely poor-performing ACs, and they are never retrialled.

The predictive tuning strategy's simplicity is a clear benefit compared to later methods. No data is required outside of the data generated within the simulation itself, and, to a degree, no expert knowledge is required to use it.

---

[2] See Gratl et al. [2].

5.2 The Expert Strategy

A fuzzy-logic-based system was designed to map the simulation statistics using expert-written fuzzy-logic rules to choose an optimal AC. Fuzzy logic is a good choice for such a system, as experts cannot feasibly create exact rules for this.

The clear downside of this method is the requirement for an expert to spend significant effort to develop such rules. Furthermore, Newcome et al. [7] showed that one set of rules might be suboptimal on different hardware or with a different force model. In addition, as AutoPas itself is developed, with new ACs and improvements upon existing ACs, such rules would have to be adapted.

As the above points suggest the infeasibility of a general set of expert rules that could be universally applied, we instead design rules specifically targeted towards the experiments in Sect. 6. The exact design of the rules was determined using data gathered by running the very experiments intended to be optimised. This could be useful, for example, when conducting several experiments with a range of parameters: one experiment is run with slow full searches to collect data to build the rules, and the rest can then use the Expert strategy. We should emphasise that these rules will not perform optimally for all experiments that could be run with AutoPas.

5.3 The Random Forest Strategy

The Random Forest strategy trains a random forest that maps the live simulation statistics to a suggested AC. To train the model, a dataset needs to be collected prior to the simulation being run.

Such a data-driven method is far easier to use than the Expert strategy, requiring only a set of simulation scenarios from which to generate performance data. This avoids much of the difficulties mentioned previously for expert-knowledge methods as, if the set of scenarios is already provided, trialling the ACs and training a model requires only a minor amount of effort.

The key issue comes from generating the set of simulation scenarios: the scenarios should be representative of the experiments intended to be optimised with the strategy, but generating the scenarios should not be so computationally expensive that it negates the benefits of the strategy nor, for the same reason, should expensive, irrelevant scenarios be included. As such, some expert decisions regarding the dataset still need to be made.

To avoid the first of these issues, we generate scenarios using simple random and fixed-grid distributions of particles, which are likely physically unrealistic (molecules too close together) but are representative at the algorithmic level. To address the second issue, these "fake" scenarios were designed to mimic a variety of different real scenarios, ultimately with the experiments in Sect. 6 in mind.

6 Experimental Setup

We chose three scenarios to demonstrate our strategies. Between these scenarios, there is minimal overlap in optimal AC. For brevity, we only discuss the behaviour of the simulations at a high level. The full details of the software versions used and the experiment parameters can be found in the repository mentioned in the Appendix. We tested our selection strategies on HSUper.[3]

Heating Sphere. In the heating sphere experiment, a small sphere of molecules is placed in the centre of a large domain with reflective boundaries. Only one MPI rank is used. The simulation begins at a low temperature, which keeps the molecules together. The temperature is raised slowly, causing the sphere to start losing molecules until the domain fills sparsely. Due to the molecules dispersing, a change in the optimal AC can be seen.

Exploding Liquid. In the exploding liquid experiment used in prior AutoPas works [3], a thick layer of molecules is placed in the centre of an otherwise only sparsely filled tube-like domain with periodic boundaries. The thick layer explodes outwards, along the tube, in two dense "waves" of molecules in either direction. The domain is split into six subdomains along the length of the tube, each with its own MPI rank and AutoPas container. As such, there are different optimal ACs in each subdomain depending on whether it contains the dense wave, the remnants left after the wave, or if the wave has not reached it yet.

Rayleigh-Taylor. In the Rayleigh-Taylor experiment, a layer of larger, lighter molecules is placed under a denser layer of smaller, heavier molecules, with the molecule layers mixing as the larger molecules rise to the top and the smaller molecules sink to the bottom. The simulation was run on 40 MPI ranks with the distribution of the two molecule types, leading to different optimal ACs on each MPI rank.

7 Results and Analysis

We first determined which single AC was optimal for each experiment (i.e. without any algorithm selection) and found that each experiment had a different optimal single AC. These make good targets to reach with our algorithm selection methods. If we achieve a speedup relative to these, we see an advantage of *dynamically* selecting and changing ACs in different regions and at different points in time. However, in the case of the Predictive and Random Forest strategies, getting close to these targets is still a success as a user without prior knowledge of which single AC is optimal will still benefit. But this second criterion of success is not valid for the Expert strategy as the methodology in Sect. 5.2 suggests that this single optimal AC would be known, and so a success for this strategy requires a speedup relative to this AC.

---

[3] Compute nodes featuring 256GB of RAM, 2 Intel Icelake sockets each with an Intel(R) Xeon (R) Platinum 8360Y processor with 36 cores; https://portal.hpc.hsu-hh.de/documentation/hsuper/.

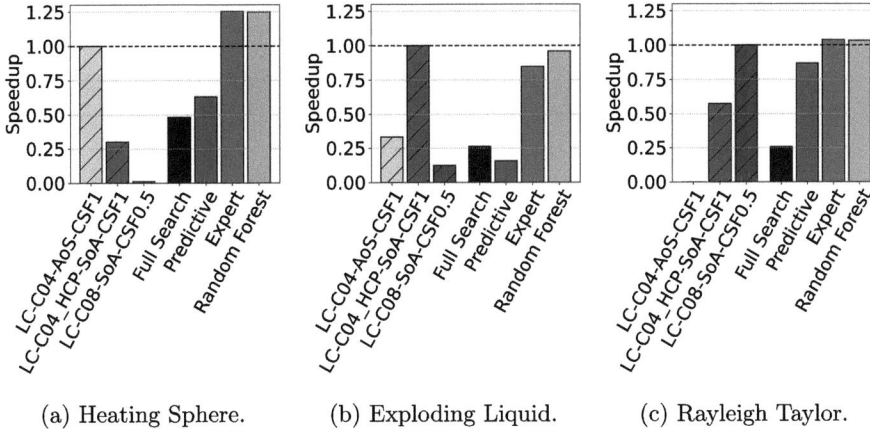

(a) Heating Sphere.   (b) Exploding Liquid.   (c) Rayleigh Taylor.

Fig. 1. Comparison of the accumulated time spent calculating forces for each thread across the entire simulation, for the three algorithmic configurations that are the optimal single configurations for each experiment ((a) `LC-C04-N3L-AoS-CSF1`, (b) `LC-C04_HCP-N3L-SoA-CSF1`, (c) `LC-C08-N3L-SoA-CSF0.5`) and each selection strategy. Results are shown as speedups relative to the optimal single configuration for that experiment. Note that the `LC-C04-N3L-AoS-CSF1` with the Rayleigh-Taylor experiment timed out upon reaching the maximum wall time allowed on HSUper – about 7 times the wall time that `LC-C08-N3L-SoA-CSF0.5` took and only achieving approximately three-quarters of the total iterations.

In Fig. 1, we see the speedup of these three optimal ACs, the naive Full Search strategy, and the three strategies presented in this work relative to each experiment's optimal single AC.

We see that the one experiment's optimal AC could perform significantly worse in other experiments. While, to an expert, such optimal ACs make sense in hindsight, suggesting such an AC with no experience running similar simulations is challenging. Furthermore, in some time periods and subdomains, other ACs perform significantly better than these (and other works have found other optimal ACs [7]) Therefore, we see benefit from dynamic algorithm selection from a wide range of ACs.

Considering now the selection strategies, we see that the Predictive strategy achieves mixed success, achieving near-1.0 speedup only in the Rayleigh-Taylor experiment. That it achieves a worse performance than Full Search in the Exploding Liquid experiment can be explained by the dramatic changes in the computational profile invalidating the linear models.

The Expert and Random Forest strategies perform similarly, with Random Forest performing slightly better in the Exploding Liquid scenario. In the Heating Sphere experiment, both strategies achieve speedups of 1.25, correctly following the switch in the optimal AC. This represents the significant potential of the methods and dynamic algorithm selection more generally. In the Exploding Liquid experiment, however, the methods achieve "speedups" of 0.85 and 0.89, respectively and in the Rayleigh-Taylor experiment, they achieve 1.04 and 1.03, respectively.

In these experiments, the optimal single AC is the AC, which is optimal on the MPI ranks with the most work (dense regions of particles), even if these denser regions move and thus, the ranks with this high workload change. As a dense region moves out of a rank, the optimal AC changes, but as the workload of the rank also reduces, selecting a new AC has a minimal impact overall. We nevertheless should remember that, for the Random Forest approach, a near-1.0 speedup is still successful, and we expect that improved MPI load balancing would lead to a greater potential for above-1.0 speedup in such experiments and the current results suggest the Expert and Random Forest strategies could realise this potential.

The overhead of evaluating all the models and rules was generally found to be negligible.

8 Conclusion and Outlook

Of the three methods presented in this work, the Random Forest strategy appears the most successful – balancing performance with accessibility – and is able to achieve speedups of 1.25 compared to simply picking a single AC. If we instead compare against the prior naive Full Search strategy, we see a speedup of 4.05, demonstrating a significant improvement. Whilst it is expected that the Expert strategy could be improved to match or even beat the Random Forest strategy, this is not a tractable solution to the problem and is unsustainable. The Predictive strategy struggles significantly, particularly in dramatically changing scenarios; however, the further development of such an accessible, data-free approach could be valuable in some situations.

As previously mentioned, a key issue with such data-driven models is the need for good data, the generation of which is itself an expert decision. A more accessible solution could be to combine online learning with an understanding of confidence: in scenarios where the model has not been trained on similar data, the model becomes unconfident in its suggestions, triggering exploration of the search space and thus online learning, and therefore better performance in further similar simulation runs. Thus, the user generates data relevant to the simulations they want to optimise.

Acknowledgements. This work has been financially supported by, as well as computational resources (HSUper) from, the project hpc.bw, which is funded by dtec.bw - Digitalisation and Technology Research Centre of the Bundeswehr. dtec.bw is funded by the European Union — NextGenerationEU.

The authors would like to thank Amartya Das Sharma and Ruben Horn, of Helmut-Schmidt-University, for their invaluable assistance and advice with the Python-C interface on HSUper. The authors would like to express their deep thanks to all contributors to the open-source AutoPas project.

Appendix

All input files, outputs, job scripts for training data and the experiments, models, and fuzzy rule files, as well as the software versions used can be found at https://github.com/SamNewcome/Algorithm-Selection-in-Short-Range-Molecular-Dynamics-Simulations.

References

1. Armstrong, W., et al.: Dynamic algorithm selection using reinforcement learning. In: 2006 International Workshop on Integrating AI and Data Mining, pp. 18–25. IEEE (2006)
2. Gratl, F.A., et al.: N ways to simulate short-range particle systems: automated algorithm selection with the node-level library autopas. Comput. Phys. Commun. **273**, 108262 (2022)
3. Gratl-Gaßner, F.A.: Automated Dynamic Algorithm Selection for HPC Particle Simulations. Doctoral thesis, Technical University of Munich (2025)
4. Lerchner, M.: Exploring fuzzy tuning technique for molecular dynamics simulations in autopas. Bachelor's thesis, Technical University of Munich (2024)
5. Marrink, S.J., et al.: Computational modeling of realistic cell membranes. Chem. Rev. **119**(9), 6184–6226 (2019)
6. Mohammed, A., et al.: Automated scheduling algorithm selection and chunk parameter calculation in openmp. IEEE Trans. Parallel Distrib. Syst. **33**(12), 4383–4394 (2022)
7. Newcome, S.J., et al.: Towards auto-tuning multi-site molecular dynamics simulations with autopas. J. Comput. Appl. Math. **433**, 115278 (2023)
8. Nitzke, I., et al.: ms2: a molecular simulation tool for thermodynamic properties, release 5.0. Comput. Phys. Commun. 109541 (2025)
9. Pazar, A.: Optimizing algorithm selection in autopas with decision trees and random forests. Master's thesis, Technical University of Munich (2024)
10. Seckler, S., et al.: Autopas in ls1 mardyn: massively parallel particle simulations with node-level auto-tuning. J. Comput. Sci. **50**, 101296 (2021)
11. Stylianou, C., Weiland, M.: Optimizing sparse linear algebra through automatic format selection and machine learning. In: 2023 IEEE International Parallel and Distributed Processing Symposium Workshops (IPDPSW), pp. 734–743. IEEE (2023)

SOPMOA*: Unleashing Shared-Open Parallelism for High-Performance Multi-Objective Pathfinding

Long Viet Truong(✉)[iD], Tien Minh Dam[iD], Tuan Anh Nguyen, Linh Thuy Thi Nguyen, and Duong Trung Dinh

Viettel High Technology Industries Corporation, Hanoi, Vietnam
{longtv28,tiendm9,tuanna63,linhntt6,duongdt2}@viettel.com.vn

Abstract. The Multi-Objective Shortest Path (MOSP) problem generalizes the classic shortest path problem by simultaneously optimizing multiple, often conflicting, cost functions. Recent advances in MOSP have yielded algorithms that employ sophisticated heuristic-based techniques and dimensionality reduction to expedite search. However, most existing methods rely on strictly sequential frameworks, leaving parallelized approaches relatively underexplored - especially for high-dimensional objectives. In this paper, we introduce SOPMOA* (Shared-Open Parallelized Multi-Objective A*), an algorithm that addresses this gap by enabling any number of concurrent sub-searchers to cooperate via a shared-memory priority queue. Each sub-searcher independently processes labels, performs dominance checks against locally stored partial Pareto fronts, and contributes to a global frontier of non-dominated solutions. We propose mechanisms for safe and efficient updates to shared data structures, ensuring correctness without excessive locking overhead. Empirical evaluations on benchmark multi-objective road networks demonstrate that SOPMOA* scales favorably with increasing parallelism and consistently outperforms state-of-the-art algorithms such as EMOA*, LTMOA*, and NWMOA* in both speed and robustness. These results underscore the substantial potential of shared-memory parallelization in tackling challenging multi-objective pathfinding tasks.

Keywords: Multi-Objective Shortest Path (MOSP) · Parallel A* Search · Shared-Memory Parallelism · Heuristic Search · Concurrent Algorithms · Multi-Objective Optimization

1 Introduction

1.1 Problem Definition

Consider a directed graph $G = \langle V, E, c \rangle >$, where V is a finite vertex set, E is the set of directed edges (u, v) with $u \neq v$, and $c : E \to \mathbb{R}^d_{\geq 0}$ is a non-negative cost function with d attributes. A path p is a sequence of vertices v_i ($i \in \overline{1..n}$) such that $(v_i, v_{i+1}) \in E$. The total cost of p is $c(p) = \sum_{i=1}^{n-1} c(v_i, v_{i+1})$.

A cost vector α *weakly dominates* β ($\alpha \preceq \beta$) if $\alpha_i \leq \beta_i$ for all i; α *dominates* β ($\alpha \prec \beta$) if $\alpha \preceq \beta$ and $\alpha \neq \beta$. If neither dominates the other, they are *mutually undominated*. The vector α is *lexicographically smaller* than β ($\alpha <_{lex} \beta$), if the first k indices $i \in \overline{1..k}$ satisfy $\alpha_i = \beta_i$, and at index $k+1$, $\alpha_{k+1} < \beta_{k+1}$. The truncated version of $\alpha \in \mathbb{R}^d$, denoted $Tr(\alpha)$, is the $(d-1)$-dimensional vector obtained by removing its first component (i.e., α from the second to last index).

In a search problem with start node s and target t, a heuristic $h: V \to \mathbb{R}^d_{\geq 0}$ estimates the cost from a node to t. In the scope of this paper, h is consistent and admissible in each cost attribute. A path p_{sv} from s to t is represented by a label l with $node(l) = v$, $g(l) = c(p_{sv})$, and $f(l) = g(l) + h(v)$. The Pareto front $\mathcal{P}_v$ at node v is the set of mutually undominated labels among all paths from s to v. The Multi-Objective Shortest Path (MOSP) problem seeks the Pareto front $\mathcal{P}_t$ at the target node, representing all optimal trade-off solutions.

1.2 Algorithmic Background

In recent years, the Multi-Objective Shortest Path (MOSP) problem has attracted increasing attention in network optimization. Heuristic search methods for obtaining the full Pareto set utilize a priority queue of labels: each iteration pops the most promising label, applies a "dominance check" against expanded non-dominated labels, expands and generates successors if qualified. The algorithms terminate when the queue is empty.

The concept of "dimensionality reduction", introduced in NAMOA*$_{dr}$ [8], decreases the size of the non-dominated set, significantly reducing the dominance-check runtime. NAMOA*$_{dr}$'s Eager Check maintains mutually non-dominated labels per node in *OPEN*, guaranteeing that popped labels are already Pareto-optimal, but incurs overhead removing dominated intermediates and re-structuring queue, which reduces queue efficiency. By contrast, LTMOA* [6] and EMOA* [9] employ Lazy Check: *OPEN* may hold dominated labels from the same node, so each popped label must be checked against the Pareto front before expansion. Both share the same framework and utilize dimensionality reduction; EMOA* stores the truncated front in an AVL tree, while LTMOA* uses a linear list or array. NWMOA* [2], originally for negative-weight graphs, also excels on non-negative graphs by using a bucket priority queue, obviating strict lexicographical order, adopting alternative dominance and frontier-update strategies.

Parallelism in MOSP remains under-explored: early parallel algorithms (e.g. [7,10]) are often complex or yield modest gains. BOBA* [3] introduces a bi-objective bidirectional search - one forward from the start, one backward from the target - each prioritizing a different objective. Extending this to d objectives, [1] assigns up to d workers to cyclic permutations of objectives, each exploring a distinct region of the Pareto front, though parallelism remains bounded by d. Other approaches, such as BDA [11] and MDA [4], parallelize subtasks like

Algorithm 1: SOPMOA* High-level

Input: graph $G\langle V, E, c\rangle$, heuristic h, start s, target t, num. workers N
Output: Complete Pareto optimal solutions $\mathcal{P}_t$

1 $OPEN \leftarrow \emptyset$; $Sols \leftarrow \emptyset$; $G_{cl}[v] \leftarrow \emptyset, \forall v \in V$; $active[i] \leftarrow \textbf{true}, \forall i \in \overline{1..N}$;
2 Initialize $start\_label$ of node s; $g(start\_label) \leftarrow \vec{0}$; $f(start\_label) \leftarrow h(s)$;
3 Push $start\_label$ to $OPEN$;
4 **for** $id \leftarrow 1$ **to** N **do in parallel** $SubSeacher(id)$;
5 Remove dominated solutions from $Sols$;
6 **return** $Sols$;

dominance checks and $OPEN$ updates. T-MDA [5] further incorporates BOBA*-style bidirectionality for bi-objective search.

2 Shared-Open Parallelized Multi-Objective A*

This section introduces the Shared-Open Parallelized Multi-Objective A* (SOPMOA*) algorithm. The algorithm begins by initializing shared-memory components:

- Priority queue $OPEN$: A priority queue storing generated labels in lexicographically ascending f values order.
- List of Pareto fronts G_{cl}: Stores cost vectors g of expanded labels belonging to the truncated Pareto fronts of each node. Lexicographical order is maintained on full form of the costs, while dominance checks and frontier update use dimensionality reduction.
- Solution set $Sols$: Stores qualified labels of target node t, used as algorithm's output.
- Array $active$: Tracks worker status using boolean values, enabling workers to update their own status and monitor others.

A start label is inserted into $OPEN$, then SOPMOA* creates N parallel workers with each act as a sub-searcher (Algorithm 1). These sub-searchers share the workload by simultaneously take labels from $OPEN$ and expand the labels to the shared search tree (represented by G_{cl}). A sub-searcher of SOPMOA* follows the baseline of multi-objective search framework, is shown in Algorithm 2.

The notable difference of SOPMOA* sub-searchers from other algorithms is that they utilize memory-locking mechanism for managing shared attributes. When a sub-searcher detects that $OPEN$ is empty, it does not terminate but instead monitors the status of other sub-searchers, anticipating new labels generated by others. The terminate condition is met when all workers are inactive, indicating that no new label will be generated and the $OPEN$ set is truly empty.

In this parallelism setting, labels may not be processed in strict lexicographical order, unlike in synchronized algorithms. To handle potential dimensionality

Algorithm 2: Sub-Searcher

Input: Sub-searcher's index ID

1 **while** $\exists i \in \overline{1..N} : active[i] = $ **true do**
2   **lock** $OPEN$ **do**
3   **if** $OPEN$ is empty **then**
4             $active[ID] \leftarrow$ **false**;
5   **continue**;
6   **else**
7   Pop label x on top of $OPEN$;
8             $active[ID] \leftarrow$ **true**;
9   **if** $FC(node(x), g(x))$ **or** $FC(target, f(x))$ **then continue**;
10    $FUpdate(node(x), g(x))$;
11   **if** $node(x) = target$ **then**
12   **lock** $Sols$ **do** Add label x to $Sols$;
13   **continue**;
14   **for** $succ \in successors(node(x))$ **do**
15   Create new label y of node $succ$;
16         $g(y) \leftarrow g(x) + c(node(x), succ)$; $f(y) \leftarrow g(y) + h(succ)$;
17   **if** $FC(succ, g(y))$ **or** $FC(target, f(y))$ **then continue**;
18   **lock** $OPEN$ **do** Push y to $OPEN$;

reduction dominance check issues, we use specialized frontier check and update procedures:

- **Frontier Check**: Given a node v, label x, or target node t with cost vector α (i.e., $g(x)$ or $f(x)$), the frontier check compares α against the Pareto front $G_{cl}[v]$. A snapshot of $G_{cl}[v]$ is taken under a shared-lock allowing concurrent reads but blocking updates. Vectors lexicographically larger than α are excluded, while dominance is checked on the truncated vectors of those lexicographically smaller than or equal to α, as in the synchronized version.
- **Frontier Update**: If a label is undominated by both frontiers, its $\alpha = g(x)$ is inserted into $G_{cl}[v]$ under an exclusive lock. Before insertion, any vector with a truncated version dominated by α is removed. α is then placed to maintain lexicographical order.

Though true Pareto optimal labels cannot be falsely discarded, false Pareto labels, which do not belong to the Pareto front, can still be retained due to non-monotonic expansion order. These labels are not large in number and are suboptimal because they have to qualify the frontier check beforehand. This abundance does not affect the overall result, as will be proven in Lemma 4.

In the end, previously expanded false Pareto labels are removed from $Sols$. The algorithm returns the complete set of Pareto optimal solutions.

Algorithm 3: Frontier Check (FC)

Input: a node v, a cost vector α
Output: **true** if α is dominated by any of $G_{cl}[v]$, else **false**
1 **shared-lock** G_{cl} at node v **do** $X \leftarrow copy(G_{cl}[v])$;
2 **for** $i_{ub} \leftarrow |X|$ **down to** 0 **do**
3 $\quad$ **if** $i_{ub} > 0$ **and** $X[i_{ub}] \prec_{lex} \alpha$ **then break**;
4 **for** $j \leftarrow 1$ **to** i_{ub} **do** $\qquad\qquad$ // no loop if $i_{ub} = 0$
5 $\quad$ **if** $Tr(X[j]) \preceq Tr(\alpha)$ **then return true**;
6 **return false**;

Algorithm 4: Frontier Update ($FUpdate$)

Input: a node v, a cost vector α
1 **lock** G_{cl} at node v **do**
2 $\quad$ **for** $i \leftarrow 1$ **to** $|G_{cl}[v]|$ **do**
3 $\quad\quad$ **if** $Tr(\alpha) \prec Tr(G_{cl}[v][i])$ **then** Delete at index i of $G_{cl}[v]$;
4 $\quad$ **for** $i_{ins} \leftarrow |G_{cl}[v]|$ **down to** 0 **do**
5 $\quad\quad$ **if** $i_{ins} > 0$ **and** $G_{cl}[v][i_{ins}] \prec_{lex} \alpha$ **then break**;
6 $\quad$ Insert α to $G_{cl}[v]$ at index $(i_{ins} + 1)$;

3 Theoretical Results

Lemma 1. *Let cost vector α be lexicographically smaller than cost vector β. Then, α is not dominated by β.*

Proof. If β dominates α, then $\forall i \in \overline{1..d} : \beta_i \leq \alpha_i$. Since $\alpha <_{lex} \beta$, there exists at least one index j such that $\alpha_j < \beta_j$, implying that α is not dominated by β.

Lemma 2. *Let cost vector α lexicographically smaller than cost vector β, then, α weakly dominates β iff $Tr(\alpha)$ weakly dominates $Tr(\beta)$.*

Proof. $\alpha <_{lex} \beta$ implies that $\alpha_1 \leq \beta_1$. For $\alpha \preceq \beta$, all $\alpha_1 \leq \beta_1$, $\alpha_2 \leq \beta_2$, $\alpha_3 \leq \beta_3$, ..., $\alpha_d \leq \beta_d$ must be satisfied. Since the first inequality is already met, it must hold $\alpha_2 \leq \beta_2$, $\alpha_3 \leq \beta_3$, ..., $\alpha_d \leq \beta_d$, equivalent to $Tr(\alpha) \preceq Tr(\beta)$.

Lemma 3. *The procedure using dimensionality reduction in SOPMOA* frontier check without lexicographical-ordering guarantee, returns accurate result (i.e. equivalent to the frontier check on full-form vectors).*

Proof. Let α be the cost and X the Pareto front, partitioned as $X = X_{\leq \alpha} \cup X_{>\alpha}$, where $X_{\leq \alpha}$ contains labels lexicographically non-larger than α, and $X_{>\alpha}$ the rest. The labels in $X_{>\alpha}$ cannot dominate α from Lemma 1, thus are discarded from frontier check. A label in $X_{\leq \alpha}$ can be equal to α, which truncated form weakly dominates $Tr(\alpha)$. Otherwise, the truncated comparisons are valid when the labels are truly lexicographically smaller than α (proved in Lemma 2).

Lemma 4. *A true Pareto label will not be discarded in frontier check by the false labels in the Pareto front.*

Proof. Let α be the cost and X the Pareto front, partitioned as $X = X_{\leq \alpha} \cup X_{>\alpha}$, where $X_{\leq\alpha}$ contains labels lexicographically non-larger α, and $X_{>\alpha}$ the rest. False labels in $X_{>\alpha}$ are excluded from comparisons with α. Otherwise, considering a false label $\beta \in X_{\leq\alpha}$, it must hold that $Tr(\beta) \not\preceq Tr(\alpha)$. Since α is of the Pareto front while β is not, α either dominates β (ruled out by Lemma 1) or is mutually undominated with β and $\alpha \neq \beta$. In the latter case and $\alpha \neq \beta$, there exists indices i and j that $\alpha_i < \beta_i$ and $\alpha_j > \beta_j$. Since $\beta <_{lex} \alpha$, in the first k ($k \geq 1$) indices, α is no smaller than β, forcing $i > k$ (hence $i > 1$). Thus $Tr(\beta) \not\preceq Tr(\alpha)$, and no false label β can affect α.

Corollary 1. *The pareto optimal labels will always be expanded (will not be discard in dominance check).*

Theorem 1. *SOPMOA* computes cost-unique complete pareto optimal solutions.*

Proof. SOPMOA* employs N parallel sub-searchers following the same multi-objective search framework. The process of frontier check in each sub-searchers is guaranteed to be accurate (Lemma 3) and will not discard true Pareto labels of every node (Lemma 4). Consequently, all true Pareto labels are expanded in frontier updates (sequentially at each node, avoiding conflicts) and generate their successors. This ensures the exploration of all optimal labels in the complete search tree, including the target's Pareto front. However, false Pareto labels are also expanded due to non-monotonic lexicographical order, this only creates abundant branches to the search tree, not affecting the result. Those of the target node will be stripped after the parallelized process, thus SOPMOA* returns complete pareto optimal solutions.

4 Experimental Results

In this section, we evaluate SOPMOA* via two experiments. First, we vary the number of workers to assess how performance scales with parallelism. Second, we compare SOPMOA*'s best and worst configurations against state-of-the-art algorithms EMOA* [9], LTMOA* [6], and NWMOA* [2].

All algorithms were implemented in C++ using standard libraries. SOPMOA* employs OpenMP with `std::mutex` and `std::shared_mutex`, and Intel TBB's `concurrent_priority_queue` for $OPEN$. EMOA*, LTMOA*, and NWMOA* (with bucket PQ) were re-implemented; LTMOA*'s array-G_{cl} version was omitted due to unstable memory handling. Experiments ran on a 24-core Intel Xeon Gold 6242R (3.10 GHz, 32 GB RAM) with a 3600 s timeout.

Benchmarks use the DIMACS New York map with two base objectives (distance, time), extended to four by adding economic cost, random integers in [1,100]. Heuristics derive from d backward Dijkstra one-to-all searches and are excluded from reported runtimes.

Table 1. The average runtimes, the average number of generated labels, the average number of expanded labels on 100 random NY instance with 3 objectives, run by five configurations of SOPMOA* (4, 8, 12, 16, 20 workers)

50 random NY instances (3 objectives) - avg. solutions ≈ 8930			
SOPMOA*	avg. runtime (s)	avg. generated	avg. expanded
4 workers	521.231	11 908 051	9 517 206
8 workers	408.084	11 915 313	9 519 256
12 workers	334.939	11 920 603	9 520 361
16 workers	274.189	11 922 836	9 520 568
20 workers	250.243	11 923 698	9 521 368

Table 2. The number of solved instances and the runtime statistics on 100 instances with 3 objectives and 50 instances with 4 objectives, run by EMOA*, LTMOA*, LazyLTMOA*, NWMOA* and two versions of SOPMOA* (4 and 20 workers)

Algorithms	Solved	Runtime (s)			
		Min	Max	Mean	Median
100 random NY instances (3 objectives) - avg. solutions ≈ 10220					
EMOA*	100/100	**0.209**	1896.45	357.84	58.36
LTMOA*	99/100	0.313	3600.00	823.95	158.38
LazyLTMOA*	100/100	0.273	3528.92	690.43	131.18
NWMOA*	100/100	0.234	3456.15	553.50	102.41
SOPMOA* - 4 workers	100/100	0.311	3024.69	563.13	97.60
SOPMOA* - 20 workers	100/100	0.257	**1730.34**	**248.58**	**51.83**
50 random NY instances (4 objectives) - avg. solutions ≈ 16866					
EMOA*	38/50	**0.507**	3600.00	1344.37	852.71
LazyLTMOA*	33/50	0.896	3600.00	1766.10	1804.97
NWMOA*	37/50	0.579	3600.00	1398.89	980.76
SOPMOA* - 20 workers	**46/50**	0.768	3600.00	**1096.97**	**649.82**

Experiment 1: SOPMOA* was run with 4, 8, 12, 16, and 20 workers on 50 random start-target pairs. Table 1 shows average runtime, generated nodes, and expanded nodes. Runtime drops sharply from 4 to 8 workers (≈113 s), then tapers (only 23 s from 16 to 20). Generated and expanded labels rise slightly, likely due to parallelism-induced extraction disorder. Overall, SOPMOA* benefits substantially from increased parallelism.

Experiment 2: We compared SOPMOA* (4- and 20-worker setups) to EMOA*, LTMOA*, LazyLTMOA*, and NWMOA* on 100 random 3-objective instances. LTMOA* failed one instance; all others solved all 100. The 20-thread SOPMOA* is ≈1.5× faster than EMOA* and more than 3× faster than LTMOA*, while 4-worker SOPMOA* matches NWMOA*. On 50 4-objective instances, 20-thread

SOPMOA* solved 46/50 and maintained superior runtimes, though its minimum runtime is higher due to multithreading overhead (Table 2).

These results confirm that SOPMOA* achieves superior speed and consistency, underscoring the critical role of parallelization in multi-objective search.

5 Conclusions

In conclusion, we introduced SOPMOA*, a novel parallel MOSP algorithm that delivers complete Pareto optimal solutions via multiple simultaneous subsearchers on a shared priority queue. It features unique strategies for expanded Pareto fronts, dominance checking, and frontier updating, and in its early development has shown substantial improvements over state-of-the-art algorithms, highlighting its potential to significantly enhance performance and leverage parallelism in MOSP.

References

1. Ahmadi, S.: Parallelizing multi-objective A* search (extended abstract). In: Proceedings of the International Symposium on Combinatorial Search, vol. 17, pp. 253–254 (2024). https://doi.org/10.1609/socs.v17i1.31567
2. Ahmadi, S., Sturtevant, N.R., Harabor, D., Jalili, M.: Exact multi-objective path finding with negative weights. In: Proceedings of the International Conference on Automated Planning and Scheduling, vol. 34, pp. 11–19 (2024). https://doi.org/10.1609/icaps.v34i1.31455
3. Ahmadi, S., Tack, G., Harabor, D., Kilby, P.: Bi-Objective Search with Bi-Directional A*. Schloss Dagstuhl - Leibniz-Zentrum für Informatik (2021)
4. Maristany de las Casas, P., Sedeno-Noda, A., Borndörfer, R.: An improved multiobjective shortest path algorithm. Comput. Oper. Res. **135**, 105424 (2021). https://doi.org/10.1016/j.cor.2021.105424
5. Maristany de las Casas, P., Kraus, L., Sedeño-Noda, A., Borndörfer, R.: Targeted multiobjective dijkstra algorithm (2021). https://doi.org/10.48550/ARXIV.2110.10978. https://arxiv.org/abs/2110.10978
6. Hernández, C., et al.: Multi-objective search via lazy and efficient dominance checks. In: Proceedings of the Thirty-Second International Joint Conference on Artificial Intelligence, IJCAI 2023, pp. 7223–7230. International Joint Conferences on Artificial Intelligence Organization (2023). https://doi.org/10.24963/ijcai.2023/850
7. Medrano, F.A., Church, R.L.: A parallel computing framework for finding the supported solutions to a biobjective network optimization problem. J. Multi-Criteria Decis. Anal. **22**(5–6), 244–259 (2015). https://doi.org/10.1002/mcda.1541
8. Pulido, F.J., Mandow, L., Pérez-de-la Cruz, J.L.: Dimensionality reduction in multiobjective shortest path search. Comput. Oper. Res. **64**, 60–70 (2015). https://doi.org/10.1016/j.cor.2015.05.007
9. Ren, Z., Zhan, R., Rathinam, S., Likhachev, M., Choset, H.: Enhanced multi-objective A* using balanced binary search trees. In: Proceedings of the International Symposium on Combinatorial Search, vol. 15, no. 1, pp. 162–170 (2022). https://doi.org/10.1609/socs.v15i1.21764

10. Sanders, P., Mandow, L.: Parallel label-setting multi-objective shortest path search. In: 2013 IEEE 27th International Symposium on Parallel and Distributed Processing, pp. 215–224. IEEE (2013). https://doi.org/10.1109/ipdps.2013.89
11. Sedeño-noda, A., Colebrook, M.: A biobjective Dijkstra algorithm. Eur. J. Oper. Res. **276**(1), 106–118 (2019). https://doi.org/10.1016/j.ejor.2019.01.007

From Recursion to Parallelization: Plug & Play Dynamic Programming

Jiang Long[✉]

Duke Kunshan University, Kunshan, China
`jiang.long@dukekunshan.edu.cn`

Abstract. Dynamic Programming (DP) is fundamental to computational science education and application, traditionally taught through tabulation methods that emphasize manual loop construction. This paper introduces a modular, systematic plug-and-play framework that greatly simplifies DP algorithm design and parallelization. Our approach begins with a recursive divide-and-conquer analysis, decomposing DP into reusable components: `Refactored Recursion (RR)`, `OrderSpec`, `TileSpec`, `dp_solve`, `dp_tile_solve`, and `dag_run`. These modules encapsulate recursive structures and facilitate seamless parallelization via dynamic Directed Acyclic Graph (DAG) scheduling. We demonstrate the versatility of this framework using three classical textbook problems: Longest Common Subsequence (LCS) highlights the plug-and-play simplicity, Matrix Chain Multiplication (MCM) employs transitive reduction for dependency clarity, and the Cut-Rod problem illustrates previously obscured tiling optimizations and parallel solutions. This new modular paradigm significantly reduces DP's learning curve, shifting educational focus from code-centric methods to intuitive, reusable patterns, bridging theoretical recursion with practical implementations and greatly enhancing computational science education.

Keywords: Dynamic Programming · Plug-and-Play Framework · Algorithm Design Patterns · Computer Science Education · Parallelization

1 Introduction and Motivation

Dynamic programming (DP) solves complex problems by decomposing them into overlapping subproblems, widely used in optimization and bioinformatics. However, traditional teaching methods, which emphasize recursion and simple tabulation, overlook practical requirements for scalability and parallelism. Our framework addresses these challenges by offering an intuitive, modular approach that seamlessly integrates recursion with efficient parallelization.

Standard DP education begins with recursion, moves to memoization, and ends with loop-based tabulation [3]. Yet, deriving efficient loops from recursive solutions is challenging for beginners and often neglects critical real-world considerations like multi-core parallelism and data-intensive computing.

We introduce a modular framework utilizing OrderSpec for defining traversal order and tileSpec for parallel execution through dependency graphs (DAGs). By converting recursive algorithms into modular components integrated with automated functions like dp_solve and dp_tile_solve, the approach simplifies learning. Demonstrated on problems like LCS, MCM, and an extended rod-cutting example, our approach makes DP accessible and practical.

This paper provides a condensed review of related work (Sect. 2), details our DP framework (Sect. 3), demonstrates examples (Sect. 4), and concludes with teaching impacts and future directions (Sect. 5).

2 Related Work

Dynamic Programming (DP) is a core algorithmic technique for optimization, bioinformatics, and computational science, often requiring parallelization to manage computational costs. Efforts to optimize DP range from compiler-based transformations to task-based frameworks.

OpenMP is widely used for DP parallelization due to its simplicity [4], offering directives like #pragma omp parallel for concurrency. However, manual dependency management can be complex for irregular DP structures. Tools like the Pluto compiler [1] and LLVM Polly [13] automate parallel code generation for affine loops, enhancing locality and parallelism. Task-based frameworks, such as Taskflow [7] and Dask [12], dynamically infer dependencies, offering flexibility but requiring task decomposition. Recent work by Maleki et al. [10] parallelizes DP via rank convergence, enabling concurrent computation of dependent stages for algorithms like Needleman-Wunsch, achieving notable speedups.

Graph-based methods capture DP dependencies effectively. Algebraic Dynamic Programming (ADP) [6] abstracts state traversal and scoring, aiding bioinformatics tasks like sequence alignment. Petri nets model dependencies as token transitions [5], automating scheduling but adding overhead.

Hardware optimizations include GenDP [8] for genome sequencing and DP-HLS [2] for high-level synthesis in bioinformatics. These improve efficiency but demand specialized expertise.

DP is critical in bioinformatics for sequence alignment [11] and RNA structure prediction [14], and physics simulation-based experiments [9], all of which demonstrates the need for efficient and parallel implementations for large-scale problems.

This paper presents a systematic approach to transform recursive DP solutions into parallel implementations via recursion, dependency analysis, tabulation, tiling, and parallelism, making DP a plug & play component once recursion is defined.

3 DP Algorithmic Framework

Dynamic Programming (DP) algorithms systematically solve optimization problems by filling tables using nested loops, exploiting subproblem dependencies.

Classic examples such as CutRod, Matrix Chain Multiplication (MCM), Longest Common Subsequence (LCS), and Optimal Binary Search Tree (BST) showcase diverse traversal patterns (Fig. 1).

DP Problem	Dimension	Start Loc	End Loc	Ordering	Filter(i,j)
CutRod	1D	0	n	row-major	True
MCM	2D	(n,1)	(1,n)	wavefront/row-major	$i \leq j$
LCS	2D	(1,1)	(m,n)	grid/row-major	True
Optimal BST	2D	(n-1,0)	(0,n-1)	wavefront/row-major	$i \leq j$

Fig. 1. Traversal patterns from MIT textbook [3].

Figure 1 summarizes traversal dimensions, start and end locations, traversal ordering, and filtering conditions.

To formalize DP tabulation, our framework in Fig. 2 introduces Python classes: `OrderSpec2D` and `TileSpec2D`. `OrderSpec2D` defines traversal order through starting/ending coordinates, wavefront preference, row-major selection, and cell filtering conditions (e.g., $\lambda i,j : i \leq j$). Its **gen** method generates sequences of cell indices, replacing traditional nested loops.

```
class RRO:
    def prep(self): return
    def fill(self,i,j):
        return
    def result(self): return
class OrderSpec2D:
    def __init__(self,
        start, end,
        waveFront=False,
        rowMajor=True,
        Filter=labmda i,j:True):
        ...
    def gen(): ...
class TileSpec2D:
    def __init__(self,
        start,end,tileHeight,
        tileWidth): ...
    def gen(): ...

def dp_solve(rro, orderSpec):
    rro.prep()
    for i,j in orderSpec.seqGen():
        rro.fill(i,j)
    return rro.result()
def dp_tilie_solve(rro,tileSpec,inTileOrder):
    rro.prep()
    for tileId,(x0,y0),(x1,y1) in \
                        tileSpec.gen():
        inTileOrder.start = (x0,y0)
        inTileOrder.end = (x1,y1)
        for i,j in inTileOrder.gen():
            rro.fill(i,j)
    return rro.result()
def dag_run(rro, tileSpec,
            inTileOrder, tileSchedule):
    rro.prep()
    ... # e.g. setup and run dask
    return rro.result()
```

Fig. 2. DP Algortihmic Framework Library.

The `dp_solve` function encapsulates sequential computation, accepting an `OrderSpec2D` instance and a Refactored Recursion Object (RRO). The RRO modularizes recursion into preparation (`prep`), cell computation (`fill`), and result extraction (`result`). Thus, DP algorithm design becomes a straightforward process: creating RRO and `OrderSpec2D` instances.

For parallel computation, we introduce the `TileSpec2D` class, defining traversal tiles with specific coordinates, tile dimensions, and unique identifiers (`tileID`). The `dp_tile_solve` function iterates over tiles sequentially, applying `OrderSpec2D` within each tile for correctness.

Parallel execution is achieved through directed acyclic graph (DAG) scheduling, explicitly modeling tile-level dependencies beyond loop boundaries, enhancing flexibility over approaches like OpenMP. We utilize existing DAG schedulers such as Dask [12] (Python) or Taskflow [7] (C++), conceptualized in the `dag_run` function. `dag_run` combines tile dependency data (`tileDep`), traversal specifications, and RROs, generating executable task DAGs.

Our framework simplifies DP algorithm design and parallelization into defining modular RROs, `OrderSpec2D`, and `TileSpec2D`. This paper demonstrates these steps through examples: LCS (2D grid), MCM (wavefront with $i \leq j$), and CutRod (1D to 2D lifted parallel traversal).

4 Case Studies

4.1 Longest Common Subsequence (LCS)

The LCS problem is a standard example of 2D dynamic programming (DP). Given two strings X and Y, the goal is to find the length of their longest common subsequence. The recursive formulation is given in Eq. (1), with a corresponding Python implementation shown in Fig. 3. Calling `LCS(X, Y, 0, 0)` returns the correct result.

$$LCS(i,j) = \begin{cases} 1 + LCS(i+1, j+1), & \text{if } X[i] = Y[j] \\ \max(LCS(i, j+1), LCS(i+1, j)), & \text{otherwise} \end{cases} \quad (1)$$

```
def LCS(X, Y, i, j):
    if i >= len(X): return 0
    if j >= len(Y): return 0
    if X[i]==Y[j]:
        return LCS(X,Y,i+1,j+1)
    else:
        return max(LCS(X,Y,i+1,j),
                   LCS(X,Y,i,j+1))
```

Fig. 3. LCS Recursive Formulation.

```
def lcs_dep(X, Y, i, j):
    if i>= len(X): return
    if j>= len(Y): return
    gen_edge((i,j), (i+1,j))
    gen_edge((i,j), (i,j+1))
    gen_edge((i,j), (i+1,j+1))
    lcs_dep(X,Y,i+1,j)
    lcs_dep(X,Y,i,j+1)
    lcs_dep(X,Y,i+1,j+1)
```

Fig. 4. LCS Dependency Graph Generator.

With a small modification (Fig. 4), we generate the LCS dependency graph, where each node corresponds to (i,j) in the DP table, and edges represent dependencies on `LCS(i+1,j+1)`, `LCS(i+1,j)`, and `LCS(i,j+1)`.

Figure 5 shows a dependency graph for len(X) = 5 and len(Y) = 4. This informs the OrderSpec2D definition (Fig. 6) used in our dp_solve framework. The RRO encapsulation appears in Fig. 7, and tiling configuration with dp_tile_solve is shown in Fig. 8.

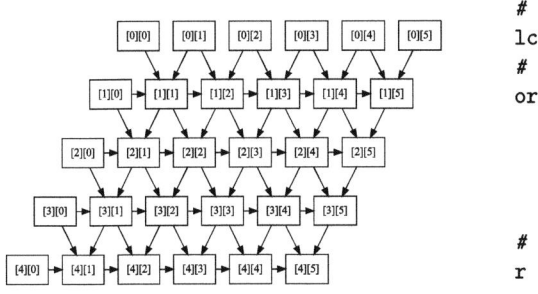

```
# instantiate the LC problem
lcs = LCS_RR("...", "...")
# define ordering
orderSpec = OrderSpec2D(
    start = (lcs.m, lcs.n),
    end = (0,0),
    RowMajor=True,
    WaveFront = False )
# solve
r = dp_solve(lcs, orderSpec)
```

Fig. 5. Dependency Graph.

Fig. 6. LCS OrderSpec2D Usage.

All four configurations of tabulation orders (combinations of wavefront/grid and row-/column-major) are valid and verifiable within the framework. For tiled DP, combining 4 inter-tile and 4 in-tile orders yields 16 configurations. Including forward/backward traversal, a total of 32 valid orderings can compute LCS.

This variety highlights the flexibility of our framework in validating diverse traversal strategies.

```
import numpy as np
class LCS_RR(RRO):
    def __init__(self, X,Y):
        self.X,self.Y = X,Y
        self.m, self.n = len(X),len(Y)
    def prep(self): self.dp = \
        np.zeros((self.m+1, self.n+1))
    def fill(self, i, j):
        if self.X[i] == self.Y[i]:
            self.dp[i][j] = ...
        else:
            self.dp[i][j] = max(...)
    def result(self):
        return self.dp[0][0]
```

```
rr = LCS_RR(...)
tileSpec = TileSpec(
    tileHeight = ...,
    tileWidth = ... ,
    tileOrder = OrderSpec(
        start=(rr.m,rr.n),
        end=(0,0),
        RowMajor=True,
        WaveFront = False) )
    inTileOrder = OrderSpec(
        start=None, end=None,
        RowMajor=False,WaveFront=True)
r = dp_tile_solve(rr,tileSpec,
                        inTileOrder)
```

Fig. 7. LCS RRO Encapsulation.

Fig. 8. LCS Tiling Configuration.

4.2 MCM with Transitive Reduction

Matrix Chain Multiplication (MCM) is a classic dynamic programming optimization problem. Its recursive formulation is:

$$m(i,j) = \begin{cases} 0, & \text{if } i = j \\ \min_{k=i}^{j-1}(m(i,k) + m(k+1,j) + p_{i-1} \cdot p_k \cdot p_j), & \text{if } i \leq j \end{cases} \quad (2)$$

From this, we generate a dependency graph, shown in Fig. 9.

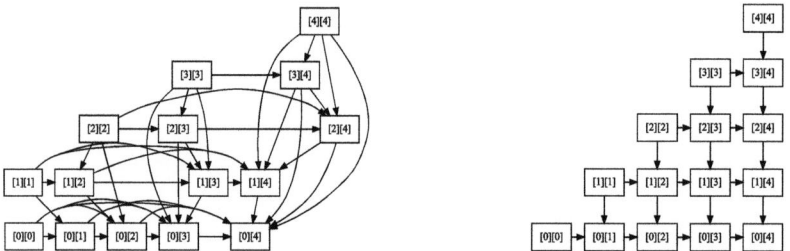

Fig. 9. Raw MCM Dependency Graph.   **Fig. 10.** Reduced Dependency Graph.

The graph contains redundant edges (e.g., $A \to C$ if $A \to B \to C$ exists). Transitive reduction removes these, simplifying Fig. 9 to Fig. 10. This reduced graph clarifies start/end locations per Table 1, enabling an RRO instance for the DP framework, like LCS. We set `OrderSpec2D` with `Filter = lambda i,j: i<=j` to traverse the upper triangular table. Unlike textbook [3] wavefront ordering, all four `OrderSpec2D` configurations are valid, requiring no middle-diagonal start.

4.3 Case Study CutRod

The Rod Cutting problem is a classic dynamic programming optimization task. Its recursive formulation is:

$$R(n) = \begin{cases} 0, & \text{if } n = 0, \\ \max_{1 \leq i \leq n}(p[i] + R(n-i)), & \text{if } n \geq 0. \end{cases} \quad (3)$$

We generate a dependency graph (Fig. 11), simplified via transitive reduction to Fig. 12.

Fig. 11. Raw Dependency Graph.   **Fig. 12.** Transitive Reduced Graph.

A 1D DP implementation is shown in Fig. 13. Viewing `r` as a 2D array, each iteration computes `r[i][j]`, with `r[i]` as `max(r[i][:i])`. This leads to a 2D version in Fig. 14. The 2D version's dependency graph (Fig. 15, n=5) is generated by instrumenting loops, unlike LCS and MCM.

This graph shows non-neighboring dependencies (e.g., (0,0) to (5,0)). Grid or wavefront ordering from (0,0) satisfies these, but irregular dependencies challenge OpenMP. Our DAG-based scheduling framework enables plug-and-play multicore execution, as shown with LCS and MCM.

An additional optimization is that the 2D array can revert to 1D by using Fig. 13's line 5, optimizing memory while preserving correctness, since the max operator is communicative, associative, and monotonic.

```
1  def cutrod(P, n):
2      r = [0] * (n+1)
3      for i in range(n+1):
4          for j in in range(i):
5              r[i] = max(r[i],
6                         r[j]+P[i-j])
7      return r[n]
```

Fig. 13. CutRod DP Algorithm.

```
1  import numpy as np
2  def cutrod_dp_2d(P, n):
3      r = np.zeros((n+1, n+1))
4      for i in range(1, n+1):
5          for j in range(i):
6              r[i][j] = max(
7                  r[j][j], r[j][j]+P[i-j])
8          r[i][i]=max([r[i][:i]])
9      return r[n][n]
```

Fig. 14. CutRod DP Algorithm in 2D.

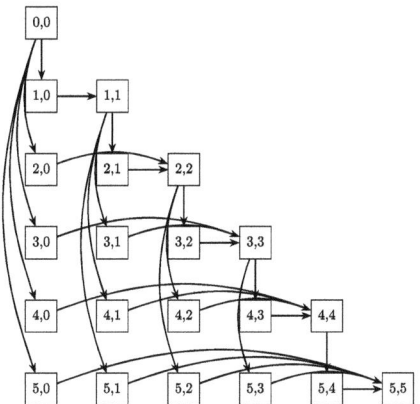

Fig. 15. Dependency Graph for Fig. 14.

5 Conclusion and Future Work

We presented a modular framework for dynamic programming (DP) that begins from recursion, captures dependencies explicitly, and transitions naturally into sequential and parallel implementations. Instead of teaching loop-centric DP from the outset, our approach advocates starting from the recursive formulation, generating the dependency graph, and using that to drive tabulation order and parallelism.

This method reshapes how DP is taught: students gain intuition by visualizing dependencies, then implement sequence generators as modular programming exercises. They no longer struggle to derive nested loops from recurrences—instead, they construct and test different tabulation strategies and plug them into a reusable library.

Beyond education, this framework bridges theory and practice by supporting task-based parallelization through DAG schedulers like Dask or Taskflow. It scales to real-world workloads while remaining accessible for students and researchers.

Future directions include classroom deployment with visualization tools, integrating tiled DP with GPU or distributed execution, and extending the framework to support more complex problems like irregular and higher-dimensional DP, which are common in computational science and engineering.

References

1. Bondhugula, U., Hartono, A., Ramanujam, J., Sadayappan, P.: Pluto: An automatic polyhedral parallelizer and locality optimizer. ACM SIGPLAN Notices **43**(6), 101–113 (2008). https://doi.org/10.1145/1375581.1375595
2. Cao, Y., Gupta, A., Liang, J., Turakhia, Y.: DP-HLS: a high-level synthesis framework for accelerating dynamic programming algorithms in bioinformatics. CoRR, abs/2411.03398 (2024). https://doi.org/10.48550/arXiv.2411.03398, arXiv:2411.03398
3. Cormen, T.H., Leiserson, C.E., Rivest, R.L., Stein, C.: Introduction to Algorithms, 3rd edn. MIT Press (2009)
4. Dagum, L., Menon, R.: OpenMP: an industry standard API for shared-memory programming. IEEE Comput. Sci. Eng. **5**(1), 46–55 (1998). https://doi.org/10.1109/99.660313
5. Esparza, J., Knoop, J.: Petri nets for dynamic programming. In: Application and Theory of Petri Nets, pp. 210–229. Springer (2010). https://doi.org/10.1109/HICSS.2004.1265209
6. Giegerich, R., Meyer, C.: Algebraic dynamic programming. In: Proceedings of the 9th International Conference on Algebraic Methodology and Software Technology, AMAST 2002, pp. 349–364. Springer, Heidelberg (2002)
7. Huang, T.-W., Lin, C.-X., Guo, G., Wong, M.: CPP-taskflow: fast task-based parallel programming using modern C++. In: 2019 IEEE International Parallel and Distributed Processing Symposium (IPDPS), pp. 974–983. IEEE (2019). https://doi.org/10.1109/IPDPS.2019.00105
8. Liang, Y., Wang, H., Wolf, M., Zhou, S.: GenDP: a general-purpose framework for accelerating dynamic programming on gpus. IEEE Trans. Parallel Distrib. Syst. **27**(1), 260–274 (2016). https://doi.org/10.1145/3579371.3589060
9. Loxley, P.N., Cheung, K.-W.: A dynamic programming algorithm for finding an optimal sequence of informative measurements. Entropy **25**(2), 251 (2023). https://doi.org/10.3390/e25020251

10. Maleki, S., Musuvathi, M., Mytkowicz, T.: Parallelizing dynamic programming through rank convergence. In: Proceedings of the 19th ACM SIGPLAN Symposium on Principles and Practice of Parallel Programming, pp. 219–232. ACM (2014). https://doi.org/10.1145/2555243.2555264
11. Needleman, S.B., Wunsch, C.D.: A general method applicable to the search for similarities in the amino acid sequence of two proteins. J. Mol. Biol. **48**(3), 443–453 (1970). https://doi.org/10.1016/0022-2836(70)90057-4
12. Rocklin, M.: Dask: parallel computation with blocked algorithms and task scheduling. In: Proceedings of the 14th Python in Science Conference, pp. 130–136 (2015). https://doi.org/10.25080/Majora-7b98e3ed-013
13. Lengauer, C., Grosser, T., Groesslinger, A.: Polly - performing polyhedral optimizations on a low-level intermediate representation. https://polly.llvm.org
14. Zuker, M., Stiegler, P.: Optimal computer folding of large RNA sequences using thermodynamics and auxiliary information. Nucleic Acids Res. **9**(1), 133–148 (1981). https://doi.org/10.1093/nar/9.1.133

NeoN: A Tool for Automated Detection, Linguistic and LLM-Driven Analysis of Neologisms in Polish

Aleksandra Tomaszewska[✉], Dariusz Czerski, Bartosz Żuk, and Maciej Ogrodniczuk

Institute of Computer Science, Polish Academy of Sciences, Warsaw, Poland
{aleksandra.tomaszewska,dariusz.czerski,
bartosz.zuk,maciej.ogrodniczuk}@ipipan.waw.pl

Abstract. We introduce NeoN, a tool for detecting and analyzing Polish neologisms. Unlike traditional dictionary-based methods requiring extensive manual review, NeoN combines reference corpora, Polish-specific linguistic filters, an LLM-driven precision-boosting filter, and daily RSS monitoring in a multi-layered pipeline. The system uses context-aware lemmatization, frequency analysis, and orthographic normalization to extract candidate neologisms while consolidating inflectional variants. Researchers can verify candidates through an intuitive interface with visualizations and filtering controls. An integrated LLM module automatically generates definitions. Evaluations show NeoN maintains high accuracy while significantly reducing manual effort, providing an accessible solution for tracking lexical innovation in Polish (The prompt templates and a number of NeoN interface screenshots are available at https://zil.ipipan.waw.pl/NeoN-appendix).

Keywords: neologism detection and filtering · new words · lexical innovation · Large Language Models

1 Introduction

Neologism detection in Polish remains challenging due to the reliance on dictionary methods and manual curation. Traditional approaches, while informative, are time-consuming and prone to human bias [7,10,13]. Semi-automatic systems [1,7] provide interactive interfaces for candidate review but their dependence on basic filtering mechanisms limits their ability to capture the full spectrum of emerging lexical phenomena in Polish. In addition, although recent advances in Large Language Models (LLMs) have transformed many areas of natural language processing, no existing tool has yet used LLMs for analysis of new words in Polish. In this paper, we present NeoN, a tool designed for Polish neologism detection, monitoring, and analysis. Rather than relying only on dictionary lookups, NeoN continuously processes RSS feeds through a multi-layered filtering pipeline. In NeoN, we focus on words that have entered usage after 2020 to monitor ongoing changes in the language. The filters tailored for Polish help extract candidate neologisms and consolidate variants through frequency analysis, structural constraints, context-aware lemmatization, and the final LLM-driven precision-boosting filter. Its integrated LLM module automatically generates definitions. By combining corpus-based filtering with LLM-driven analysis, NEON provides a scalable framework for tracking lexical innovation in Polish.

2 Related Work

Earlier studies [13] employed discriminant-based approaches to identify markers flagging potential neologisms. In the Polish context, the dictionary developed by the Language Observatory of the University of Warsaw (UWLO) [10], serves as a recognized but manually curated resource. Semi-automated systems like NEOCRAWLER [7] and NEOVEILLE [1] have also emerged, using dictionaries and rule-based filters to extract new lexical units from sources such as websites, blogs, and press releases. Although such tools reduce manual workload, they still require significant expert oversight and are seldom tailored to Polish. Statistical and machine-learning approaches have also advanced the field, e.g., Falk et al. [5] proposed a framework integrating form-related, morphological, and thematic features to detect neologisms in French newspapers. More recent efforts have incorporated LLMs for e.g., definition generation and translation [11,17] and unsupervised techniques that normalize variant forms via embedding-space mapping [16].

3 Functionalities

Our system processes daily RSS feed data through a unified pipeline integrating candidate extraction, variant grouping, and multi-layered filtering for removing noise candidates. All functionalities are integrated into a web-based interface allowing users to customize filter settings, review candidate lists, and export the resulting data in CSV format.

3.1 Form Filtering

In neologism detection, various filters are applied to determine the likelihood of a word being new or non-standard. The filters in NEON are (1) **Frequency and occurrence:** Document frequency, Term frequency, Unique domain frequency, Domain distribution; (2) **Structural constraints:** Word length constraints, Invalid character check, Presence of digits, Triple repeated characters, (3) **Lexical validation:** Common Polish corpus check, English loanword detection; (4) **Spelling and typographical errors:** Polish word matching, Edit distance with diacritics, Adjacent character swap detection; (5) **Contextual analysis:** English context detection, Name Entity heuristic, Capitalization pattern; (6) **Other:** Compound word detection, Filtering using a few-shot LLM prompt.

The first innovation in our filtering framework is incorporating corpora alongside a dictionary as references for neologism validation. Users can select up to 4 reference corpora based on their research needs, such as general language vs web contexts or temporal relevance. NEON cross-references extracted lexemes against the corpora to exclude existing Polish words, improving detection accuracy. General reference corpora are used by default, with an option to include a web corpus for recent data. The corpora include frequency lists from the National Corpus of Polish [14] (up to 2010), the Corpus of Contemporary Polish [8] (2011–2020), the web corpus NEKST [2] (up to 2020), and the latest Polish Wikipedia dump. The second innovation is employing LLMs as the final filtering stage. We use the `Llama-3.3-70B-Instruct` with a few-shot prompt with 3 positive and 3 negative examples to enhance identification. Recent studies show that LLMs can outperform humans in this task [17].

Experimental Setup. The experiments involved a corpus of 233,538 web documents (873 RSS) from approx. 2 months. The documents underwent a processing pipeline that involved language detection, main content extraction to isolate the primary textual content from web pages, and NLP analysis to process the text using the `Hydra` [9] for NLP. Following filtering with a Polish language dictionary (Available at: https://sjp.pl), we generated a set of 200,696 candidate neologisms. The candidates were then refined through multi-step filtering to distinguish neologisms from noise or established lexemes.

Filtering Pipeline. We implemented an iterative sequence of filters, each designed to eliminate specific types of non-neologisms. For consistency evaluation purposes (neologisms that appeared after 2020), only selected filters were used. (1) **Length constraints:** at least 3 characters and no more than 20 characters; (2) **Numerical content:** must not contain digits; (3) **Frequency:** must appear in more than 5 documents; (4) **Case sensitivity**: must appear in lowercase at least 5 times; (5) **Proper noun exclusion:** must not function as proper nouns in at least 5 occurrences; (6) **Edit distance:** the minimum edit distance to any known word in the Polish dictionary must exceed 0.5; (7) **Spelling:** must not be diacritical variations, result from swapping adjacent letters of existing Polish dictionary words and not contain triple repetitions of the same letter; (8) **English dictionary check:** if a word appears in an English dictionary, it must occur in at least 5 Polish-language contexts; (9) **Exclusion from other dictionaries:** must not be present in the corpora or dictionaries. (10) **LLM filtering:** filtering (we used the `Llama-3.3-70B-Instruct` model) based on a few-shot prompt.

For evaluation, we used data added to the UWLO after 2020 [10]. This timeframe aligns with our primary reference corpora. We preprocessed the dataset by removing neologisms already listed in the Polish dictionary, as these typically reflect existing words with new meanings rather than new lexemes. After preprocessing, our set included 610 neologisms. To ensure the corpus was suitable for evaluation, we expanded it by gathering the top 100 Google search results for each neologism in the training set. For a more precise evaluation, we conducted a manual review of 1,740 neologisms identified during the final filtering stage – before applying the LLM filter – excluding those already in the UWLO. This method enables a more effective evaluation of our tool's precision. The results obtained are presented in Table 1 as an additional section labeled 'Including human-annotated data'. The assessment was conducted by 3 individuals: 2 annotators and 1 adjudicator who resolved conflicting evaluations. The recall for this set is identical to that of the test set, as the manual annotations did not change the number of detected neologisms. This process only validated the existing candidates without adding or removing any, allowing the focus to shift to precision and F1 scores based on these annotations.

Experimental Procedure. The experiment was conducted iteratively, each iteration introducing an additional filter. At each stage, we evaluated filtering performance using precision, recall, and F1 score, comparing filtered candidates to the testing set ground truth. The results are presented in Table 1.

Table 1. Results of incremental filtering in the neologism detection. (*) The recall measure cannot be effectively calculated based only on annotated data.

Conditions	Test set				
	All	Matches	Precision	Recall	F1
No filter	200 696	610	0.003	0.993	0.006
+ Min Token Len	199 977	610	0.003	0.993	0.006
+ Max Token Len	199 289	609	0.003	0.992	0.006
+ No Digits	186 422	609	0.003	0.992	0.007
+ Freq ≥ 5	33 801	607	0.018	0.989	0.035
+ Non-Uppercase Freq ≥ 5	5 116	603	0.118	0.982	0.210
+ Non-NE Freq ≥ 5	4 198	597	0.142	0.972	0.248
+ Min Edit Distance	3 130	552	0.176	0.899	0.295
+ Spelling	2 726	549	0.201	0.894	0.329
+ Non-Eng Freq ≥ 5	2 657	549	0.207	0.894	0.336
+ Not in NKJP	1 784	538	0.302	0.876	0.449
+ Not in KWJP100	1 740	536	0.308	0.873	0.455
+ LLM filtering	1 056	536	0.508	0.873	**0.642**
	Including human-annotated data				
+ Not in KWJP100	1 740	1 385	0.796	(*)	–
+ LLM filtering	1 056	968	**0.917**	0.699	**0.793**

Summary of Results. The detection pipeline, integrating rule-based filters and a LLM, effectively identified new Polish lexemes in a noisy web corpus, achieving F1 scores of 0.642 on the test set and 0.793 on the annotated set. At the final filtering stage (before applying the LLM filter), recall cannot be computed because the process does not retain information about false negatives. In earlier stages, only the candidates that passed the filter are tracked, so any items mistakenly removed (false negatives) are lost, making it impossible to determine recall accurately. Starting with 200,696 candidates, the pipeline reduced this to 1,056 highly probable neologisms, with precision rising to 0.508 and recall settling at 0.873. Rule-based filters drastically reduce non-neologistic candidates with minimal recall loss, while the LLM filter increases precision from 0.308 to 0.508 (test set) and from 0.796 to 0.917 (annotated set) by leveraging contextual and semantic cues. This demonstrates that LLMs today can substantially enhance the neologism detection process. Although some recall was sacrificed (settling at 0.699) for significant precision gains, this trade-off proved valuable for applications where false positives are costly, e.g., in linguistic studies or when tracking emerging terminology. Each filter contributed to a stepwise improvement, making this approach highly effective.

3.2 Form Grouping

NEON detects alternative spellings, inflectional forms, and syntactic variants of neologisms, including multi-word forms (e.g., *tusko-bus*, *tuskobus*). Post-processing groups related forms (e.g., hyphenated, spaced), aggregates frequencies, and lemmatizes variants—a key step for Polish's rich morphology. We evaluated four tools: `Stanza` [15] and `spaCy` [4] (general NLP toolkits), `Hydra` [9] (Polish-specific), LLMs `GPT4o` [12] and `DeepSeek-R1` [3] with custom prompts.

Using a UWLO dataset with 978 neologisms (≥ 3 forms each; 3,659 total), we assessed lemmatization quality. Standard accuracy fails to capture consistency across inflectional groups. We propose 2 group-based metrics over neologism groups G: **Group Accuracy** $A_{gr} = \frac{S}{G}$, where S = groups with all forms mapped to the same lemma. **Strict Group Accuracy** $A_{strict} = \frac{K}{G}$, where K = groups all correctly mapped.

Experimental Setup. We tested lemmatization in 2 setups: isolated words (e.g., *NFTs* → *NFT*) and contextualized sentences (e.g., *They NFTs gained popularity* → *NFT*) (Table 2).

Table 2. Neologism lemmatization results.

Experiment	Model	Accuracy	Group Accuracy	Strict Group Accuracy
Without context	SpaCy	50.18%	14.52%	13.50%
	Stanza	73.41%	**53.58%**	**50.41%**
	Hydra	72.01%	49.08%	46.22%
	GPT4o	72.81%	53.07%	49.90%
	DeepSeek-R1	**75.13%**	51.53%	49.80%
With context	SpaCy	52.94%	16.26%	15.44%
	Stanza	73.35%	51.94%	48.77%
	Hydra	**79.31%**	62.47%	**60.22%**
	GPT4o	78.57%	**62.99%**	59.41%
	DeepSeek-R1	77.51%	57.16%	55.32%

Summary of Results. Experiments on neologism lemmatization show large performance gaps across models. Basic tools like SpaCy perform poorly ($\approx$50%), while Stanza reaches $\approx$73% but surprisingly drops slightly with context. Hydra, optimized for Polish, performs best with 79.31% accuracy and 60.22% strict group accuracy. LLMs like GPT4o and DeepSeek-R1 also perform well, especially without context, and remain competitive with it. These results underline the limits of basic tools, the contextual strength of Hydra, and the promise of LLMs. Future research should focus on fine-tuning a specialized LLM that integrates Hydra's contextual strengths with LLMs' robustness.

3.3 Definition Generation

We conducted experiments to evaluate LLMs' capability to automatically generate neologism definitions, focusing on the most recent lexemes. We only selected neologisms registered in 2024 in UWLO. For each lexeme, we obtained definitions and usage examples from the UWLO website. We filtered out entries with fewer than 5 examples, resulting in a final dataset of 81 neologisms. Our experiments used Llama-3.3-70B-Instruct [6], with a knowledge cutoff date of December 2023, and DeepSeek-R1 [3], no known cutoff date as of February 28, 2025. We chose DeepSeek-R1 to compare newer reasoning-focused models against traditional LLMs like Llama-70B.

Evaluation Protocol. We test 3 prompting setups: (1) the 0-shot setup where we do not provide any examples of neologism usage, (2) 3-shot, and (3) 5-shot where we provide 3 and 5 examples of their usage, respectively. For all the experiments, we sampled the models using the recommended temperature of 0.6 and top-p value of 0.95. We evaluated the generated definitions using the *LLM-as-a-judge* approach [18], which employs LLMs to score, rank, or select from candidate options. For our experiments, we used the GPT4o model [12] (knowledge cutoff: October 2023) as the judge, performing pointwise evaluations - the judging LLM compared the generated definition against a human-made reference, outputting CORRECT or INCORRECT. This setup largely follows [17] and focuses exclusively on the definition correctness. To increase quality, we included all 5 usage examples in the prompt.

Summary of Results. Figure 1 shows the accuracy of models in the pointwise evaluation. Performance improved monotonically with additional usage examples. DeepSeek-R1 outperformed Llama-70B across all setups, achieving the max. 96% accuracy in the 5-shot setup compared to Llama-70B's 88%. Table 3 presents results for each setup.

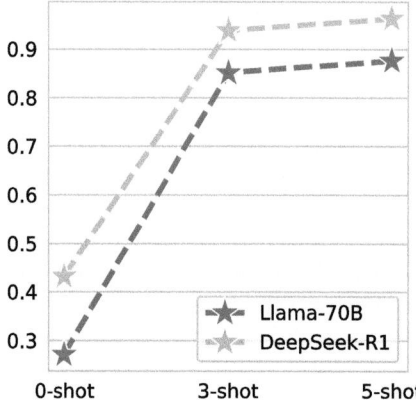

Fig. 1. Accuracy of DeepSeek-R1 and Llama-70B in pointwise evaluation across three prompting setups.

Table 3. Results for pointwise evaluation of DeepSeek-R1 and Llama-70B across 3 prompting setups.

	Verdict	
	Correct	Incorrect
Llama-70B 0-shot	22	59
Llama-70B 3-shot	69	12
Llama-70B 5-shot	71	10
DeepSeek-R1 0-shot	35	46
DeepSeek-R1 3-shot	76	5
DeepSeek-R1 5-shot	78	3

Meta Evaluation. Upon manual inspection, we conducted a meta evaluation verifying GPT4o's effectiveness as a judge. Using 3 human annotators, we evaluate the generated definitions. We focus only on the 5-shot setup as it produces the best results across both models. The evaluation followed our previously described protocol. The results of the meta evaluation are presented in Fig. 2. Human annotators showed high agreement with GPT4o's judgments, consistently rating Llama-70B lower than DeepSeek-R1, which aligns with the results in Fig. 1. Annotators varied in their strictness: Annotators 2 and 3 marked more definitions as incorrect compared to GPT4o, while Annotator 1 was more lenient, marking only 2 Llama-70B definitions as incorrect and none for DeepSeek-R1.

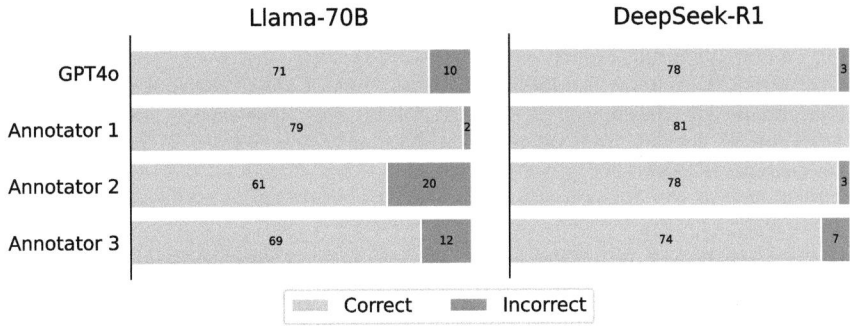

Fig. 2. Results of pointwise meta evaluation shown for 3 human annotators and GPT4o (LLM judge) across 2 judged models: Llama-70B and DeepSeek-R1.

4 Conclusions and Future Work

We presented NEON, a web-based system that integrates corpus-driven filtering, context-aware lemmatization, and LLM-based validation and definition generation to automate Polish neologism detection. Our multi-stage pipeline reduced an initial set of 200 696 candidate tokens to 1 056 high-confidence neologisms, achieving an F1 score of 0.642 on held-out data (0.793 on expert-annotated data) and exceeding 0.90 precision after the LLM filter. In 5-shot prompting, the LLM module produced definitions with up to 96% accuracy, as confirmed by three linguists. This end-to-end framework markedly lowers manual intervention, consolidates inflectional and orthographic variants, and offers visualizations for tracking lexical innovation, all without requiring programming expertise. Future work will enable researchers to upload custom corpora, extending beyond RSS feeds, and will explore fine-tuning open LLMs on manually and semi-automatically annotated neologism datasets, supplemented with synthetic examples to improve base-form detection. We also plan to develop fully LLM-driven detection workflows and to release benchmark datasets and standardized evaluation protocols to foster reproducible research in automated neology and lexical innovation analysis.

References

1. Cartier, E.: Neoveille, a web platform for neologism tracking. In: Proceedings of the Software Demonstrations of the 15th Conference of the European Chapter of the ACL, pp. 95–98. ACL (2017)
2. Czerski, D., Ciesielski, K., Dramiński, M., Kłopotek, M., Łoziński, P., Wierzchoń, S.: What NEKST?—Semantic search engine for polish internet. In: Challenging Problems and Solutions in Intelligent Systems, pp. 335–347. Springer, Cham (2016)
3. DeepSeek-AI, Guo, D., Yang, D., et al.: DeepSeek-R1: incentivizing reasoning capability in LLMs via reinforcement learning (2025)
4. Explosion: spaCy: Industrial-Strength Natural Language Processing in Python (2020)

5. Falk, I., Bernhard, D., Gérard, C.: The Logoscope: a semi-automatic tool for detecting and documenting French new words (2018)
6. Grattafiori, A., Dubey, A., Jauhri, A., et al.: The LLaMA 3 herd of models (2024)
7. Kerremans, D., Stegmayr, S., Schmid, H.J.: The NeoCrawler: identifying and retrieving neologisms from the internet and monitoring ongoing change, pp. 59–96. De Gruyter Mouton, Berlin (2018)
8. Kieraś, W., et al.: Korpus Współczesnego Języka Polskiego. Dekada 2011–2020. Język Polski (2024)
9. Krasnowska-Kieraś, K., Woliński, M.: Parsing headed constituencies. In: Proceedings of the 2024 Joint International Conference on Computational Linguistics, Language Resources and Evaluation, pp. 12633–12643. ELRA and ICCL (2024)
10. K?osi?ska, K.: Mikro- i makrostruktura słownika neologizmów Obserwatorium Językowego Uniwersytetu Warszawskiego. LingVaria **19**(2(38)), 47–61 (2024)
11. Lerner, P., Yvon, F.: Towards the machine translation of scientific neologisms. In: Proceedings of the 31st International Conference on Computational Linguistics, pp. 947–963. ACL (2025)
12. OpenAI, Achiam, J., Adler, S., et al.: GPT-4 Technical report (2024)
13. Paryzek, P.: Comparison of selected methods for the retrieval of neologisms. Invest. linguisticae **16**, 163–181 (2008)
14. Przepiórkowski, A., Bańko, M., Górski, R.L., Lewandowska-Tomaszczyk, B. (eds.): Narodowy Korpus Języka Polskiego. Wydawnictwo Naukowe PWN, Warsaw (2012)
15. Qi, P., Zhang, Y., Zhang, Y., Bolton, J., Manning, C.D.: Stanza: a python natural language processing toolkit for many human languages. In: Proceedings of the 58th Annual Meeting of the ACL: System Demonstrations, pp. 101–108 (2020)
16. Zalmout, N., Thadani, K., Pappu, A.: Unsupervised neologism normalization using embedding space mapping. In: Proceedings of the 5th Workshop on Noisy User-generated Text, pp. 425–430. ACL, Hong Kong (2019)
17. Zheng, J., Ritter, A., Xu, W.: NEO-BENCH: evaluating robustness of large language models with neologisms. In: Proceedings of the 62nd Annual Meeting of the ACL (Volume 1: Long Papers), pp. 13885–13906. ACL (2024)
18. Zheng, L., et al.: Judging LLM-as-a-judge with MT-bench and chatbot arena. In: Proceedings of the 37th International Conference on Neural Information Processing Systems, pp. 46595–46623 (2023)

Surrogate Models for Analyzing Performance Behavior of HPC Applications Using the RAJA Performance Suite

Befikir Bogale[1], Ian Lumsden[1], Dalal Sukkari[1], Dewi Yokelson[2], Stephanie Brink[2], Olga Pearce[2(✉)], and Michela Taufer[1(✉)]

[1] University of Tennessee, Knoxville, TN 37919, USA
mtaufer@utk.edu
[2] Lawrence Livermore National Laboratory, Livermore, CA 94550, USA
olga@llnl.gov

Abstract. Optimizing supercomputer software requires identifying parameter configurations that maximize performance. However, the wide range of parameter values and their varying impact across systems make traditional identification methods insufficient, highlighting the need for new approaches to performance prediction and parameter tuning. We propose Surrogate-Based Modeling (SBM) as an efficient method for characterizing performance across the parameter landscape. Using data from the RAJA Performance Suite's computational kernels (RAJAPerf), we show that SBM outperforms the standard k-Nearest Neighbors (kNN) model, achieving predictions up to 54% more accurate while requiring 33% less data. Thus, SBM emerges as a powerful tool for enhancing performance predictions across diverse parameter combinations.

Keywords: Performance analysis · Surrogate-based modeling · Parameter tuning · Performance modeling · Magma Library

1 Introduction

Parameters such as rank count, memory use, and data distribution strategies influence HPC application performance. These parameters vary widely in value, each affecting performance differently across diverse hardware platforms, from CPUs to GPUs and accelerators. Choosing the optimal parameter configuration is complex due to platform heterogeneity: a setting effective on one platform might fail on another. This requires extensive parameter sampling to map the performance landscape. Traditional methods of exploring configuration spaces involve either exhaustive sampling, which is accurate but computationally expensive, or local search algorithms (LSAs), which, like grid hill climbing or simulated annealing, are more efficient yet unpredictable, often converging to local optima or yielding inconsistent outcomes.

B. Bogale and I. Lumsden—Contributed equally to this work.

Surrogate-Based Modeling (SBM) [3,4] offers a compelling alternative to exhaustive and local searches. By using limited samples, SBM builds predictive models to estimate performance landscapes, achieving near-optimal results by carefully selecting the necessary sampling points. This method can predict optimal configurations not explored during learning, giving it an edge over traditional LSAs, while efficiently balancing exploration costs and prediction accuracy.

To assess SBM's role in tuning HPC configurations, we analyze the RAJA Performance Suite (RAJAPerf) [6], featuring microbenchmarks for optimizing kernels in the RAJA model. RAJAPerf includes varied kernels representing (i) compute-bound, with high arithmetic intensity, (ii) memory-bound, focusing on data movement, and (iii) hybrid-bound characteristics. We gather detailed performance data from three key kernels on heterogeneous platforms. By leveraging SBM, we show that a limited number of samples can effectively map the performance landscape, lowering prediction costs compared to full sampling.

2 Capturing Diverse Performance Patterns

The main resource constraints in HPC applications fall into three categories: compute-bound, memory-bound, and hybrid-bound workloads. RAJAPerf provides benchmark kernels that demonstrate these constraints in programming models such as OpenMP, CUDA, HIP, and SYCL, and classifies them accordingly [6].

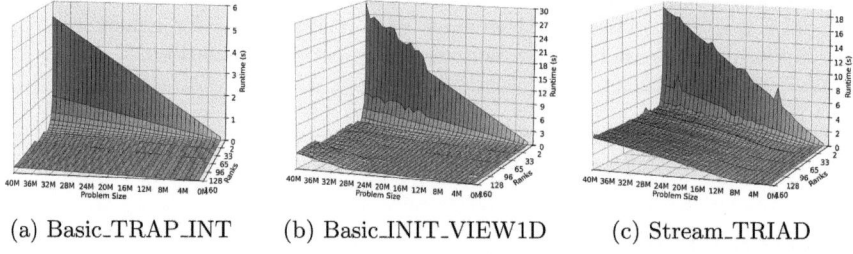

(a) Basic_TRAP_INT   (b) Basic_INIT_VIEW1D   (c) Stream_TRIAD

Fig. 1. Ground truth performance surfaces for compute-bound, memory-bound, and hybrid-bound kernels on CPU across problem sizes and ranks.

Compute-bound kernels, such as Basic_TRAP_INT, are limited by floating-point operations (FLOPs) and benefit significantly from architectures with powerful compute units. Memory-bound kernels such as Stream_TRIAD are constrained by memory bandwidth and show strong performance gains on systems with high bandwidth memory (HBM). Hybrid-bound kernels, such as Basic_INIT_VIEW1D, require a balance of compute power and memory bandwidth, benefiting from both high FLOPs and fast memory. We use RAJAPerf to sample performance across hardware and programming models, generating performance patterns from one kernel per category. We run our tests using Benchpark[7], a reproducible benchmarking repository, with performance metrics collected using Caliper[1] and analyzed using Thicket [2], an open-source Python tool. We evaluated the performance on individual nodes of the Lassen system at LLNL, focusing on CPU-based configurations. Using IBM POWER9 CPUs, we vary the MPI ranks from 2 to 160 and the problem size from 1M to 40M in 1M steps, resulting in 3,200 unique data points. Figure 1 shows performance surfaces for the three selected kernels: Basic_TRAP_INT (compute-bound) maintains flat runtimes due to negligible memory bottlenecks; Stream_TRIAD (memory-bound) shows increasing runtime slopes with problem size; and Basic_INIT_VIEW1D (hybrid-bound) displays mixed behavior where runtime increases at large problem sizes due to memory pressure. Across all kernels, we observe performance cliffs caused by simultaneous multithreading (SMT) imbalance, especially when MPI ranks exceed multiples of the 40 physical cores per node. This effect is most pronounced in Basic_TRAP_INT, indicating greater sensitivity to SMT contention.

3 Generating Performance Approximations with SBM

Building a surrogate model involves approximating the performance landscape using a sampled subset of the parameter space. This data fits a functional model that captures the relationship between parameters and performance. The model is refined iteratively to improve prediction accuracy while balancing computational cost and efficiency.

Selecting the Surrogate Model. When building a surrogate model, choosing the functional form to approximate the performance landscape is critical. In this study, we choose **Polynomial Regression** due to its balance between accuracy and computational efficiency when modeling the performance landscape of RAJAPerf kernels.

Constructing the Polynomial Model. We build our SBM by fitting a **polynomial surface** to sampled data points using **Least Squares Regression**. Consider a scenario with n observed runtimes (data points) corresponding to different parameter settings. These parameters (e.g., number of ranks, number of threads per rank, or problem size) act as independent variables and are represented as vectors $\vec{x}_1, \ldots, \vec{x}_n$. The performance metric, serving as the dependent variable (e.g., application runtime or I/O bandwidth), is denoted as $z_1, \ldots, z_n$. Each data point is modeled by a polynomial function $z(\vec{x})$, expressed as:

$$z(\vec{x}) = \beta_1 + \beta_2 x + \beta_3 y + \beta_4 x^2 + \beta_5 xy + \beta_6 y^2 + \cdots + \beta_D y^d,$$

where β_i are the polynomial coefficients, $\vec{x}$ contains the independent variables (e.g., x, y), and d is the polynomial degree. The total number of monomials is $d_m = \frac{(d+1)(d+2)}{2}$. In matrix form, the system is represented as $Z = X\beta$.

The **Least Squares Problem** is then solved as: $X^\top Z = X^\top X \beta$, where the solution is obtained using the **LU Decomposition** of the matrix $X^\top X$. We use the **MAGMA** library [5,8], optimized for dense matrix computations on heterogeneous architectures. This enables efficient and scalable performance, effectively allowing the surrogate model to handle large datasets and higher polynomial degrees.

Model Validation and Selection. We validate our surrogate models using two key scoring metrics: **Mean Squared Error (MSE)** and **R-squared (R^2)**. These metrics are the foundation for determining the near-optimal polynomial degree and number of sampled points required for the **k-fold cross-validation** procedure and the ensemble agreement approach upon which our assessment is based. They ensure the model maintains high accuracy and remains stable across application parameter settings. We use **k-fold cross-validation** with $k = 10$ to evaluate the performance of the surrogate model for each degree of polynomial d and a number of randomly sampled data points representing the performance measurements of a kernel. These data points are randomly shuffled and divided into k subsets, each subset serving as the validation set to prevent overfitting. For each degree d, given a certain number of sample points, we calculate the **MSE** and R^2 values.

We incorporate **Ensemble Agreement** in our approach to assess the stability of our models. This technique involves training multiple surrogate models with slightly different configurations or data subsets and comparing their predictions. We calculate average **MSE** and R^2 to measure the consistency across different models. Stable metrics (i.e., low standard deviation) are a proxy for high agreement between models and suggest that the surrogate model can generalize well across different computing platforms in HPC environments. We assess prediction stability using k-fold cross-validation and ensemble agreement, ensuring consistent validation and model generalization.

After performing **Ensemble Agreement**, we use the calculated **MSE** and R^2 to determine the near-optimal degree and number of sampled points. We select the degree d_{best} that minimizes the average **MSE** and maximizes the average R^2 as the near-optimal degree for the surrogate model. This process ensures the model generalizes well across different data subsets and avoids overfitting, improving its performance across the application parameter settings (i.e., the problem size and ranks). To determine the optimal number of sample points to train the surrogate model, we use $\mathbf{MSE'(x)}$, which is **the first derivative** of **MSE** with respect to the number of sampled points x. This metric measures the rate of change in **MSE** with respect to the number of points sampled and can find the optimal point where the model accuracy improves without overfitting. This derivative tracks how the **MSE** changes as the number of sampled points increases. We look for the point where the slope of the **MSE** curve transitions from negative to positive. This point represents the **optimal number of sampled points** where adding more data points no longer leads to significant improvements in the model. With the optimal polynomial degree and sample size, the validated model provides a computationally efficient approach to predicting performance metrics, significantly reducing the sampling cost.

4 Performance Predictions

Using the data from Sect. 2 and the methodology in Sect. 3, we predict the performance of the RAJAPerf kernels. We determine the near-optimal polynomial degree, identify the near-optimal number of sampled points required, and compare the performance of SBM with kNN under the same conditions.

Optimal Polynomial Degree Selection. Opting for the right polynomial degree is crucial for accurate SBM performance prediction, aiming to minimize MSE while maximizing R^2. Figure 2 shows the MSE and R^2 for SBM using the Basic_TRAP_INT kernel on a CPU, examining degrees 1 to 13. Increasing the degree reduces MSE, indicating better data fit until it levels off around degree 10. Beyond this, further increases offer little error decrease. Higher degrees also raise R^2 values, indicating better variance explanation. For the kernel, R^2 approaches 1 by degree 10, highlighting improved predictive capacity. The findings underscore that degree 10 offers a balance of complexity and accuracy, while higher degrees risk overfitting and inconsistent prediction.

Similar MSE and R^2 patterns were seen for the Basic_INIT_VIEW1D and Stream_TRIAD kernels. Table 1 presents average metrics for 100 models with degrees ranging from 2 to 12. Bold cells indicate the optimal degree's MSE and R^2 for each kernel. Detailed figures showing the surfaces created by models with these optimal degrees are not shown due to space constraints, but the main observations from these figures are as follows. For Basic_TRAP_INT (CPU) at degree 10, runtime consistency across problem sizes in the predicted surface reflects the kernel's compute-bound nature, with predictions closely matching the actual data. For Basic_INIT_VIEW1D (CPU) at degree 10, the model identifies the hybrid-bound trend but fails to model sharp transitions due to SMT imbalance. For Stream_TRIAD (CPU) at degree 10, as expected, predicted runtime increases with problem size, accurately reflecting its memory-bound nature in the degree 10 SBM model.

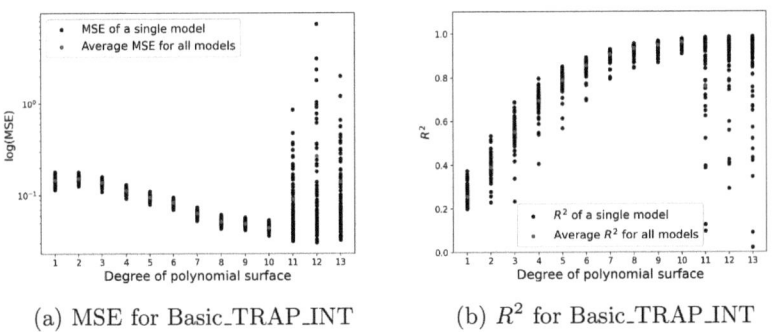

(a) MSE for Basic_TRAP_INT (b) R^2 for Basic_TRAP_INT

Fig. 2. MSE and R^2 for SBM on CPU using full dataset for polynomial degrees from 1 to 13 for Basic_TRAP_INT and Stream_TRIAD.

Table 1. Average MSE and R^2 for SBM degrees ranging from 1 to 13 using all data points for k-fold. Bold cells indicate the MSE and R^2 for the best degree.

Degree	Basic_TRAP_INT CPU		Basic_INIT_VIEW1D CPU		Stream_TRIAD CPU	
	MSE	R^2	MSE	R^2	MSE	R^2
2	0.1521	0.3907	0.7247	0.4616	0.4045	0.7407
4	0.1128	0.6939	0.5873	0.6932	0.3721	0.8363
6	0.0833	0.8599	0.4304	0.8592	0.2723	0.9253
8	0.0520	0.9335	0.2891	0.9316	0.1837	0.9656
10	**0.0443**	**0.9623**	**0.2299**	**0.9613**	**0.1374**	**0.9819**
12	0.2678	−5.1395	1.3101	−4.988	0.8093	−2.1725

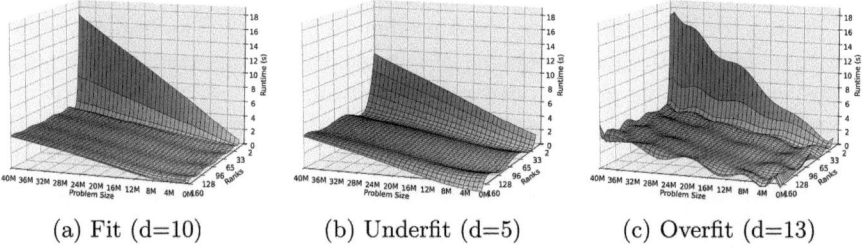

(a) Fit (d=10) (b) Underfit (d=5) (c) Overfit (d=13)

Fig. 3. Overfitting and underfitting on SBM predictions for Stream_TRIAD showing the impact of polynomial degree d on model accuracy.

Figure 3 illustrates the impact of selecting different polynomial degrees than the optimal degree 10 on the accuracy of the SBM for the Stream_TRIAD kernel on CPU. Figure 3 shows the model with the optimal degree of 10, striking the right balance between flexibility and generalization. In Fig. 3, the model is underfitted with a polynomial degree of 5, resulting in an overly simplistic surface that fails to capture the complex memory-bound behavior of Stream_TRIAD. On the other hand, Fig. 3 shows overfitting with a polynomial degree of 13, where the model captures noise and irregularities in the data, leading to unrealistic fluctuations in the predicted surface. Similar trends were observed in the other kernels and are not reported because of space constraints.

Near-Optimal Number of Sampled Points. Accurate performance predictions in HPC applications require efficient sampling strategies to minimize computational costs while maintaining model accuracy. We measure the derivatives of MSE with respect to the number of sampled points for each kernel's optimal SBM degree. We identify the point where the derivative changes sign, indicating the near-optimal number of sampled points beyond which adding more points yields a low accuracy gain while increasing the cost of the model generation. This point represents the number of sampled points where the SBM accuracy plateaus, signifying no significant improvement in model performance with additional data.

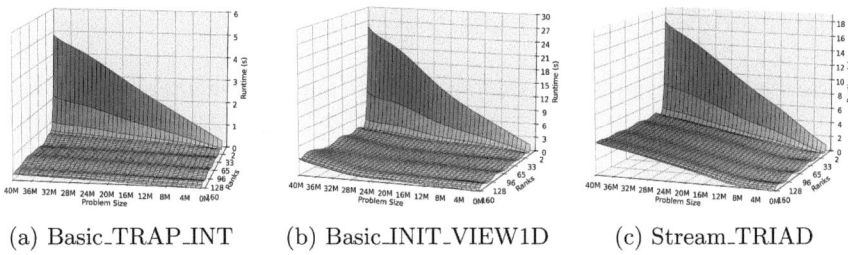

(a) Basic_TRAP_INT   (b) Basic_INIT_VIEW1D   (c) Stream_TRIAD

Fig. 4. Predicted surfaces generated by SBM models using the best degree and the best number of sampled data points for each kernel

Table 2. Performance of the best SBM and kNN models (best value in bold).

Kernel	Model	Degree	k	# Pts	μ(MSE)	σ(MSE)	$\mu(R^2)$	$\sigma(R^2)$
Stream_TRIAD (CPU)	SBM	10	N/A	2200	**0.1453**	**0.0121**	**0.9814**	**0.0046**
	kNN	N/A	3	2700	0.1982	0.1391	0.9415	0.0312
Basic_INIT_VIEW1D (CPU)	SBM	10	N/A	2200	**0.2403**	**0.0210**	**0.9585**	**0.0167**
	kNN	N/A	3	2700	0.5189	0.4148	0.8900	0.0560
Basic_TRAP_INT (CPU)	SBM	10	N/A	2200	0.0465	**0.0037**	**0.9586**	**0.0186**
	kNN	N/A	3	2700	**0.0188**	0.0143	0.8874	0.0547

Across the considered CPU kernels, this optimal number of sampled points is approximately 2200, which is lower than the original 3200 points used. Figure 4 shows the surfaces generated by SBM using the best degree (10) and the best number of data points for each kernel (2200). Using the selected degree and the reduced number of sampled points identified through the MSE derivative analysis, the predicted surfaces in the figure closely match the patterns observed in Fig. 1, effectively capturing performance trends while significantly reducing sampling cost.

Comparison with kNN. To evaluate the effectiveness of SBM, it is crucial to compare it with existing commonly used methods, with k-Nearest Neighbors (kNN) being a popular choice for performance modeling. We compare SBM's capability to capture kernel patterns with kNN under the same circumstances. We use the best degree and number of sampled points identified earlier in the paper for SBM, and, for kNN, we use an optimal k and number of sampled points determined using the validation and selection methodology described in Sect. 3. Table 2 shows each kernel's average and standard deviation of MSE and R^2 for SBM and kNN. These results show that SBM consistently demonstrates better average accuracy and stability (i.e., lower standard deviation) for more complex surfaces, such as those observed with CPU kernels, while requiring fewer sampled points. SBM predictions are up to 54% more accurate and require up to 33% fewer sampled points than kNN.

5 Related Work

Diverse strategies for performance modeling are found in the literature, utilizing methods from surrogate-based models to data-driven insights. Travis et al. [3] optimized MapReduce job settings with polynomial surrogate models. Machine learning frameworks assist in performance modeling. Scikit-learn provides multiple regression models. An observation-based black-box method [9] predicts performance metrics using short partial executions that capitalize on predictable behavior after startup. Our research extends previous work by employing SBM on various platforms like CPUs and GPUs, and on different loop-based computational kernels from the RAJA Performance Suite. Our approach improves scalability over large parameter spaces by integrating polynomial regression with the MAGMA library via SBM [5,8].

6 Conclusions and Future Work

Our SBM approach, which strategically integrates MSE, R^2, k-fold cross-validation, Ensemble Agreement, and $MSE'(x)$, optimizes surrogate models with computational efficiency and high accuracy. It boosts accuracy by up to 54% and reduces sampling needs by 33% compared to traditional methods, significantly cutting sampling costs while ensuring stability and generalization. These advancements enhance the ability to predict performance across various application parameters, establishing SBM as an essential tool for optimizing HPC configurations.

Acknowledgments. This work was supported by the National Science Foundation (NSF) under grant numbers 2331152, 2334945, and 2103845. The work was also performed under the auspices of the U.S. Department of Energy by Lawrence Livermore National Laboratory under Contract DE-AC52-07NA27344 and was supported by the LLNL-LDRD Program under Project No. 24-SI-005 (LLNL-CONF-2003169).

References

1. Boehme, D., et al.: Caliper: performance introspection for HPC software stacks. In: Proceedings of SC16 (2016)
2. Brink, S., et al.: Thicket: seeing the performance experiment forest for the individual run trees. In: Proceedings of HPDC (2023)
3. Johnston, T., Mohammad, A., Cicotti, P., Taufer, M.: Performance tuning of MapReduce jobs using surrogate-based modeling. In: Proceedings of ICCS (2015)
4. Johnston, T., Zanin, C., Taufer, M.: HYPPO: a hybrid, piecewise polynomial modeling technique for non-smooth surfaces. In: Proceedings of SBAC-PAD (2016)
5. Nath, R., Tomov, S., Dongarra, J.: Accelerating GPU kernels for dense linear algebra. In: Proceedings of VECPAR (2010)
6. Pearce, O., et al.: RAJA performance suite: performance portability analysis with caliper and thicket (2024)

7. Pearce, O., et al.: Towards collaborative continuous benchmarking for HPC. In: Proceedings of the SC23 Workshops (2023)
8. Tomov, S., Dongarra, J., Baboulin, M.: Towards dense linear algebra for hybrid GPU accelerated manycore systems. Parallel Comput. **36**(5–6), 232–240 (2010)
9. Yang, L., Ma, X., Mueller, F.: Cross-platform performance prediction of parallel applications using partial execution. In: Proceedings of ACM/IEEE SC (2005)

Cattle Identification Using 2D Mask Retention Network

Niraj Kumar(✉), Sakshi Ranjan, and Sanjay Kumar Singh

Indian Institute of Technology (BHU), Varanasi, Varanasi 221005, India
{nirajkumar.rs.cse22,sakshiranjan.rs.cse21}@itbhu.ac.in,
sks.cse@iitbhu.ac.in

Abstract. Accurate identification of individual cattle is vital for herd management, disease control, and traceability, yet traditional methods like ear tags and RFID are labor-intensive and unreliable for large-scale use. Leveraging advances in computer vision, we propose a novel cattle recognition framework combining Vision Transformers with two-dimensional masked Retention Networks. Evaluated on a self-collected video dataset of 50 cattle, focused on muzzle features, our model efficiently handles high-resolution frames and achieves 91.5% accuracy outperforming state-of-the-art methods. The Retention Network enhances scalability by reducing computational overhead, making the system robust under challenging conditions like occlusions and variable lighting. Our approach provides a practical and high-performing solution for automated cattle identification in precision livestock farming.

Keywords: Cattle identification · Convolutional Neural Networks · Vision Transformers · Retention Networks

1 Introduction

Accurate cattle identification plays a pivotal role in modern livestock management, facilitating health monitoring, resource optimization, and traceability within the food supply chain. Traditional methods, such as ear tags, branding, and RFID chips, remain prevalent but face limitations including labor intensity, risk of loss or damage, and the need for specialized equipment [1,8,12]. These shortcomings have motivated the exploration of biometric-based approaches, which offer a non-invasive and potentially more reliable alternative. Recent studies, including Li *et al.* [6], have proposed multi-modal biometric systems combining muzzle patterns, facial features, and ear tags to improve identification accuracy in diverse farm conditions.

Among biometric features, muzzle pattern recognition has emerged as a particularly robust modality due to the uniqueness and lifelong permanence of the pattern, similar to human fingerprints. As shown in Fig. 1, each cattle's muzzle exhibits a distinct arrangement of beads and ridges that remains unchanged over time, enabling consistent and precise identification. Unlike RFID or visual tags,

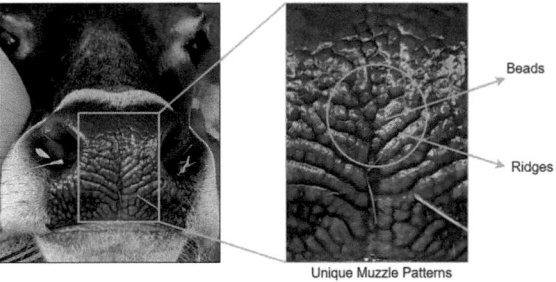

Fig. 1. Muzzle patterns in cattle nose that makes every cattle unique

muzzle-based systems eliminate the risk of physical degradation or detachment, offering a highly dependable solution for automated cattle verification without additional hardware. This makes it well-suited for scalable deployment in open-farm settings.

The integration of deep learning with computer vision has significantly advanced automated cattle recognition. While early approaches based on hand-crafted features and shallow classifiers showed limited robustness, modern models such as Convolutional Neural Networks [10] have improved performance by learning complex visual patterns. However, CNNs predominantly capture local features and often fail to encode global context, a crucial aspect for distinguishing between visually similar individuals [9,11]. To overcome this, we introduce a novel architecture combining Vision Transformers (ViTs) with Retention Networks. ViTs excel at capturing global dependencies through self-attention, but their quadratic complexity hinders real-time use. We address this by introducing a two-dimensional masking strategy to extract both spatial and temporal cues as well as reduced computational overhead from video data.

2 Proposed Approach

2.1 Data Acquisition and Preprocessing

We collected high-definition muzzle-focused cattle videos from Banaras Hindu University farms, covering 50 cattle across diverse conditions. Videos were segmented into frames, resized and normalized to reduce lighting and contrast variation. The resulting dataset was partitioned in an 80:20 ratio, where 80% of the images were used for training and the remaining 20% were reserved for validation.

2.2 Proposed Model

We extend ViTs with Retention Networks to capture spatial and temporal features using a 2D masking approach. Figure 2 shows the full architecture.

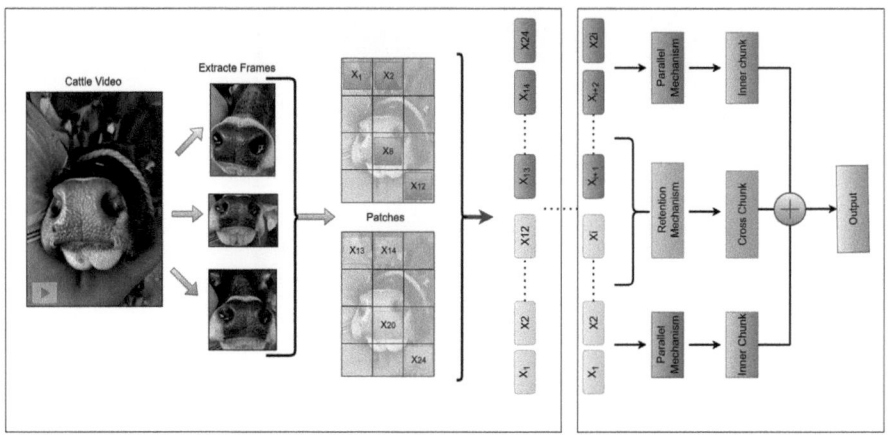

Fig. 2. Architecture of the frame-wise processing and cross-chunk integration in Retention Networks.

Division of Images into Frames. Videos are divided into frames $\{x_1, x_2, \ldots, x_T\}$ enabling frame-wise spatial analysis and efficient temporal modeling. Each frame was divided into non-overlapping patches of size $P \times P$, which were then flattened and linearly projected into D-dimensional embeddings. Positional encoding was added to retain spatial information:

$$x_p = \text{Flatten}(x_p) \cdot W_e \tag{1}$$

where x_p is the p-th patch and W_e is the learnable embedding matrix. To retain spatial positional information, a positional encoding vector e_p was added to each patch embedding, further resulting in a sequence of embedded patches

$$z_p = x_p + e_p \tag{2}$$

$$Z = [z_1, z_2, \ldots, z_N] \tag{3}$$

where N is the total number of patches per frame.

Parallel Mechanism for Frame-Wise Processing. Each frame x_i is processed independently to extract spatial features using ViT self-attention, mathematically, for a given frame x_i, the feature representation is computed as:

$$\text{InnerChunk}(x_i) = Q_i K_i^\dagger V_i, \tag{4}$$

where Q_i, K_i, V_i represent the query, key, and value matrices computed for the frame x_i.

Cross-Chunk Integration for Temporal Dependency. Temporal relationships across frames are captured via a retention mechanism for a chunk of frames $\{x_{i+1}, x_{i+2}, \ldots, x_{2i}\}$, the cross-chunk output is computed as

$$\text{CrossChunk}(x_i) = Q_i R_{i-1} + K_i^\dagger V_i, \tag{5}$$

where R_{i-1} represents the retention state from the previous chunk. This mechanism allows the model to integrate both local (intra-frame) and global (inter-frame) information.

Combining Results. Final sequence representation combines spatial and temporal outputs:

$$\text{Output} = \text{InnerChunk}(x_i) + \text{CrossChunk}(x_i). \tag{6}$$

Algorithm 1 Model Processing Pipeline

1: **Input:** Video $V \to$ Clips $\{C_n\} \to$ Frames $\{f_t\} \in \mathbb{R}^{H \times W \times 3}$
2: **Patching:** Split each f_t into $P = \frac{H}{h} \cdot \frac{W}{w}$ patches $\{x_p^t\}$, then flatten to sequence $\{x_1, \ldots, x_{T \cdot P}\}$
3: **CNN Embedding:**
4: Apply 3×3 conv $\phi_1(x) \to e_p^t \in \mathbb{R}^{d_1}$
5: Normalize: $\hat{e}_p^t = \text{BatchNorm2d}(e_p^t)$, activate: $\text{GELU}(\hat{e}_p^t)$
6: Refine via 1×1 conv $\phi_2(x) \to e_p^t \in \mathbb{R}^{d_2}$
7: **Retention Block (Depth D):**
8: **for** $d = 1$ to D **do**
9:     $z_1 = \text{LayerNorm}(z)$
10:    $z_2 = \text{Dropout}(QK^\top V) + z$ ▷ Self-retention + residual
11:    $z_3 = \text{LayerNorm}(z_2)$
12:    $z = \text{Linear}(\text{GELU}(\text{Linear}(z_3))) + z_2$
13: **end for**
14: **Classification:** $y = \text{Linear}(z) \to \mathbb{R}^C$

2.3 Retention Network with 2D Mask

1D masks fail to capture 2D spatial dependencies. We introduce a 2D decay-based spatial mask to address this. The 2D mask uses a decaying weight α to capture dependencies across spatial neighbors

$$D_{2d}^{nm} = \gamma^{|x_n - x_m| + |y_n - y_m|} \tag{7}$$

For example, the 2D decay matrix for a small patch grid can be represented as:

$$\begin{bmatrix} 1 & \alpha & \alpha & \alpha^2 \\ \alpha & 1 & \alpha^2 & \alpha \\ \alpha & \alpha^2 & 1 & \alpha \\ \alpha^2 & \alpha & \alpha & 1 \end{bmatrix}$$

Incorporating this mask into the attention mechanism, the retention operation is computed as

$$S_n = \gamma S_{n-1} + K^\top V \tag{8}$$
$$X_n = QS_n \tag{9}$$

where γ is the decay factor that determines the contribution of previous retention states. The working mechanism of our method is shown in Algorithm 1.

3 Experimental Results and Discussions

3.1 Experimental Setup

The experiments were performed on a Windows 11 system with an Intel Core i5-8265U CPU (1.60 GHz, 5 cores), 8 GB RAM, and Intel UHD Graphics 620. The models were implemented using Keras 2.11 with TensorFlow 2.11, providing a stable environment for training and evaluation.

3.2 Results and Analysis

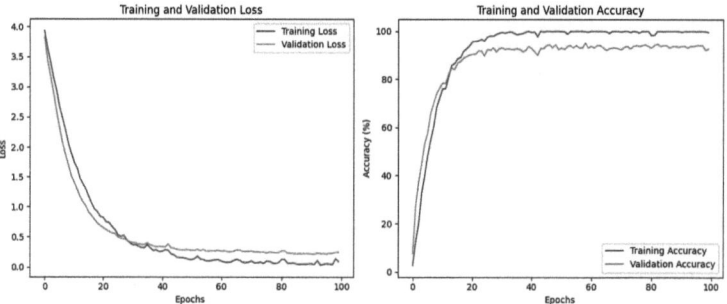

Fig. 3. Training and Validation Performance Curves.

As illustrated in Fig. 3, both training and validation losses steadily decline, and accuracies rise, stabilizing above 90% after 40 epochs. The close alignment of the curves indicates effective learning and strong generalization, with minimal signs of overfitting.

As shown in Table 1, our model achieves 91.5% accuracy, outperforming ViViT, MViT, and TAN. It also leads in Precision (91.0%), Recall (90.8%), and F1 Score (90.9%), owing to its 2D-Mask Retention Network and Transformer-based attention mechanisms that enhance spatiotemporal representation.

Optimal performance is achieved with a learning rate of 0.1 and dropout of 0.2, as shown in Table 2. Lower rates slow convergence, while higher ones reduce generalization. Excessive dropout also harms accuracy, confirming the selected configuration balances learning and regularization effectively.

Table 1. Comparison with State-of-the-Art Models for Video Classification

Model	Acc. (%)	Pre. (%)	Rec. (%)	F1 (%)
TAN (Temporal Attention Networks) [7]	90.5	89.8	89.5	89.6
ViViT (Video Vision Transformer)	91.3	90.5	90.0	90.2
TimeSformer [2]	90.0	89.0	88.5	88.7
MViT (Multiscale Vision Transformers) [4]	91.0	90.2	89.8	90.0
X3D [5]	90.8	90.0	89.5	89.7
I3D (Inflated 3D ConvNet) [3]	89.7	88.9	88.4	88.6
Proposed Model	**91.5**	**91.0**	**90.8**	**90.9**

Table 2. Hyperparameter tuning results for various learning rates(LR) and dropout rates. The table shows the training and validation accuracy(%) for each combination of values.

LR	Drop Rate	Fold-1		Fold-2		Fold-3	
		Train Acc.	Val Acc.	Train Acc.	Val Acc.	Train Acc.	Val Acc.
0.01	0.2	92.3	88.0	92.5	88.3	92.4	88.2
	0.4	91.8	87.5	92.0	87.7	91.9	87.6
	0.5	91.2	86.8	91.4	87.0	91.3	86.9
0.1	0.2	94.1	89.0	94.2	89.2	94.1	89.1
	0.4	93.5	88.4	93.6	88.7	93.5	88.6
	0.5	92.9	87.8	93.0	88.0	92.9	87.9
0.2	0.2	94.5	87.5	94.7	87.7	94.6	87.6
	0.4	94.0	88.0	94.2	88.2	94.1	88.1
	0.5	93.4	88.5	93.6	88.7	93.5	88.6

Table 3. Ablation study on various class subsets and image resolutions, showing Top-1 and Top-5 accuracy(%) using K-Fold cross-validation.

Image Size	Classes	2-Fold		3-Fold		4-Fold	
		Top-1	Top-5	Top-1	Top-5	Top-1	Top-5
64^2	50	85.3	86.7	85.8	87.5	86.1	87.9
	20	86.5	88.2	87.1	88.9	87.3	89.2
	10	88.4	89.7	88.8	90.2	89.0	90.6
128^2	50	87.2	88.5	87.6	89.2	87.9	89.5
	20	88.7	89.9	89.1	90.6	89.4	90.8
	10	90.3	91.1	90.6	91.4	90.8	91.5

From Table 3, higher resolution (128 × 128) images yield superior accuracy across all folds and class subsets. Increasing class count decreases accuracy due to task complexity, but the model maintains stability. More folds improve generalization, reinforcing the trade-off between complexity and performance.

Table 4. Comparison of loss functions used in the model training with Top-1 and Top-5 accuracy.

Loss Function	Top-1 Accuracy (%)	Top-5 Accuracy (%)
Label Smoothing	88.9	90.7
NLLLoss	87.2	89.5
Focal Loss	89.5	91.0
Triplet Loss	89.0	90.5
CrossEntropy	**90.3**	**91.8**

Table 4 shows that Cross Entropy yields the best Top-1 (90.3%) and Top-5 (91.8%) accuracy. While other loss functions (e.g., Focal, Triplet, Label Smoothing) offer specific benefits, none outperform Cross Entropy in balancing precision and robustness for classification.

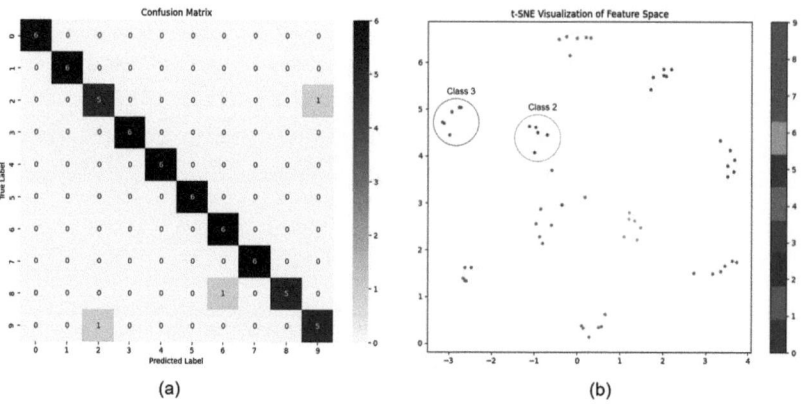

Fig. 4. (a) Confusion Matrix (b) t-SNE graph

Figure 4(a) shows high classification accuracy with minor errors in classes 2 and 8. The t-SNE plot in Fig. 4(b) confirms well-separated feature clusters, validating that the model captures discriminative representations suitable for robust cattle identification.

4 Conclusion and Future Work

This study presents a cattle identification method combining Vision Transformers with a 2D mask-based Retention Network, achieving 91.5% testing accuracy on a self-collected muzzle video dataset of 50 cattle. The 2D mask effectively captures spatial and temporal features, addressing inter-class similarity and intra-class variability. Compared to CNNs, the model improves feature extraction while reducing ViTs' complexity from quadratic to linear, enabling efficient real-time deployment. ViTs' self-attention mechanism enhances robustness under occlusions and lighting variations, making the system suitable for precision livestock management. Despite strong results, limitations include a small, less diverse dataset and training under semi-controlled conditions, which may affect real-world performance. Future work will involve dataset expansion, domain adaptation, unsupervised learning, and multimodal biometric integration. Edge-based deployment will also be explored for real-time on-farm applications.

Acknowledgements. We gratefully acknowledge the support from the PARAM Shivay Facility under the National Supercomputing Mission, Government of India, at IIT (BHU), Varanasi, and the funding from the NASF project 'Artificial Intelligence & IoT-based Smart Vet Ecosystem for Animal Health, Patient Care, and Precision Livestock Farming' (Grant No. NASF/PA-9028/2022-23).

Disclosure of Interests. The authors have no competing interests.

References

1. Bergman, N., Yitzhaky, Y., Halachmi, I.: Biometric identification of dairy cows via real-time facial recognition. Animal **18**, 101079 (2024). https://doi.org/10.1016/j.animal.2024.101079
2. Bertasius, G., Wang, H., Torresani, L.: Is space-time attention all you need for video understanding? In: Proceedings of the International Conference on Machine Learning (2021)
3. Carreira, J., Zisserman, A.: Quo vadis, action recognition? A new model and the kinetics dataset. In: Proceedings of the IEEE Conference on Computer Vision and Pattern Recognition (2017)
4. Fan, H., et al.: Multiscale vision transformers. In: Proceedings of the IEEE/CVF International Conference on Computer Vision (2021)
5. Feichtenhofer, C.: X3D: expanding architectures for efficient video recognition. In: Proceedings of the IEEE/CVF Conference on Computer Vision and Pattern Recognition (2020)
6. Li, G., Sun, J., Guan, M., Sun, S., Shi, G., Zhu, C.: Cattle identification based on multiple feature decision layer fusion. Sci. Rep. **14**(1), 2464 (2024)
7. Lin, S., Zhou, X., Jiang, X., Wang, J.: Temporal attention networks for action recognition. arXiv (2020). https://arxiv.org/abs/2002.12530
8. Lu, Y., Weng, Z., Zheng, Z., Zhang, Y., Gong, C.: Algorithm for cattle identification based on locating key area. Expert Syst. Appl. **228**(C) (2023). https://doi.org/10.1016/j.eswa.2023.120365
9. Srivastava, Y., Murali, V., Dubey, S.R.: PSNet: parametric sigmoid norm based CNN for face recognition, pp. 1–4 (2019).https://doi.org/10.1109/CICT48419.2019.9066169
10. Szegedy, C., et al.: Going deeper with convolutions (2014). https://arxiv.org/abs/1409.4842

11. Li, X., Xiang, Y., Li, S.: Combining convolutional and vision transformer structures for sheep face recognition. Comput. Electron. Agric. **205** (2023)
12. Yang, L., Xu, X., Zhao, J., Song, H.: Fusion of RetinaFace and improved FaceNet for individual cow identification in natural scenes. Inf. Process. Agric. (2023). https://doi.org/10.1016/j.inpa.2023.09.001

Scaling Dynamics of the Electricity Utility Sector:
Assessing the Role of Agglomeration Externalities and Sensitivity to Population Cutoffs in Spatial Dynamics Across European Regions

Vidit Kundu[1(✉)], Debraj Roy[2], Michel Ehrenhard[3], and Karin Pfeffer[4]

[1] Faculty of Geo-Information Science and Earth Observation (ITC), Department of Urban and Regional Planning and Geo-Information Management, University of Twente, Enschede, The Netherlands
v.kundu@utwente.nl

[2] Faculty of Science, Informatics Institute, University of Amsterdam, Amsterdam, The Netherlands

[3] Faculty of Behavioural, Management and Social Sciences (BMS), Hightech Business and Entrepreneurship Department, University of Twente, Enschede, The Netherlands

[4] Faculty of Geo-Information Science and Earth Observation (ITC), Department of Urban and Regional Planning and Geo-Information Management, University of Twente, Enschede, The Netherlands

Abstract. Urban scaling studies have gained popularity in the last two decades, summarising urban attributes' variation with population. Recent research, however, highlights scaling exponents' sensitivity to industry-specific dynamics, population cut-offs, and data distribution. Despite this, few studies systematically examine industry scaling using plant-level data while accounting for sector-specific externalities. This study addresses that gap by analysing longitudinal data on green electricity firms across 968 NUTS (Nomenclature of Territorial Units for Statistics)-3 regions in 14 European countries (1985–2023). We assess how scaling exponents for firm entry and concentration vary across population cutoffs, both with and without controls for agglomeration externalities. Our findings reveal predominantly sublinear scaling, suggesting that population size alone does not drive green energy growth. Concentration consistently scales more strongly than entry, indicating that large cities are more conducive to firm survival than to the creation of new firms. When agglomeration externalities are not controlled for, scaling exponents are systematically underestimated. While variability is observed in regions at population extremes, results remain robust across cutoffs, especially when using inverse thresholds. Comparative analysis with high-tech service and manufacturing sectors confirms sublinear scaling in entry across all sectors, with green electricity showing the lowest exponent, reinforcing its maturity and low innovation intensity. These findings align with the Smart Specialization framework, emphasizing the importance of targeted institutional support, supplier networks, and sector-specific strategies. They also highlight the potential for smaller or lagging regions to take a more active role in the green transition, particularly within cohesion policy efforts.

Keywords: Scaling Dynamics · Energy Sector · Agglomeration Externalities · Electricity Utilities · Population Cut-Offs

1 Introduction

Originally developed to explain biological scaling, the concept of scaling has been increasingly applied in urban studies over the past two decades (Cottineau et al., 2017). Urban scaling theory offers a unified framework to describe how city attributes, such as GDP, wages, and innovation, scale with population, typically in a superlinear manner due to intensified socio-economic interactions (Bettencourt, 2013). However, recent work challenges the focus on aggregate measures, showing that specific industrial concentrations contribute significantly to observed scaling patterns (Sarkar et al., 2020). Despite this, most scaling studies have only indirectly addressed industry emergence and concentration, often using employment or patent data. Some evidence suggests that high-tech and complex economic activities scale superlinearly, while lower-tech sectors like manufacturing and utilities exhibit sublinear scaling (Arcaute et al., 2015; Balland et al., 2020; Cottineau et al., 2017). Traditionally, such questions have fallen within the domain of agglomeration literature, which attributes industry concentration to co-location benefits from similar, related, or diverse firms (Jacobs, 1970; Marshall, 1920). In contrast, the scaling literature treats these externalities as endogenous outcomes of increasing size, typically omitting industry-specific controls and overlooking policy relevance at the sectoral level. Importantly, scaling benefits are not static. Industry life cycles and innovation intensity influence spatial patterns, with young industries concentrating in large cities and mature sectors dispersing over time industries (Frenken et al., 2015; Pumain et al., 2006). Yet, most scaling analyses are cross-sectional and rarely incorporate temporal dynamics or granular plant-level data, which are essential for understanding medium- and low-tech sectors with limited patent activity (De Groot et al., 2016). Moreover, scaling estimates are highly sensitive to population cutoffs and underlying data distribution, raising further concerns when applying them for policy purposes (Cottineau et al., 2017; Leitao et al., 2016). This paper addresses these gaps by focusing on the green energy sector, whose spatial distribution has become a key interest for policymakers aiming to support sustainable transitions and regional job creation. Traditionally seen as a low-tech, regulated industry, the sector has been transformed by liberalisation (Bolton, 2021; Hancher & De Hauteclocque, 2010) and sustainability transitions such as decentralised production and demand-side management (de Gooyert et al., 2016; Tayal, 2016). Innovation has increasingly shifted downstream—from hardware to grid integration—giving the sector a hybrid character: technologically sophisticated but relatively mature in terms of firm entry (Huenteler et al., 2016). Both supply-side externalities and demand-side factors (Bednarz & Broekel, 2020; Geels & Schot, 2007) are critical to its evolution, making it a valuable case for assessing the interaction between population size, industrial dynamics, and agglomeration effects. This paper investigates three core questions: *How do entry and concentration of green energy firms scale with population size over time? How sensitive are these scaling patterns to population cut-offs and industry-specific controls? How does the green energy sector compare to other sectors in terms of scaling behavior, innovation intensity, and maturity?* The analysis covers electricity utilities involved in green energy production, transmission, distribution, and trading across 968 NUTS-3 regions in 14 European countries from 1985 to 2023. For comparison, we utilize one high-tech service sector and three manufacturing sectors with varying knowledge intensities. Our results show that green energy firm entry and concentration generally scale sublinearly, suggesting that population size alone does not drive the sector's growth. Concentration shows higher scaling exponents than entry, indicating that larger regions better support firm survival than foster new ones. Sensitivity to population extremes raises caution regarding the interpretation of edge cases, but results remain robust across a range of population thresholds, particularly when using inverse cutoffs. Moreover, without accounting for agglomeration effects, scaling exponents are systematically underestimated. Compared to benchmark sectors, green energy firms display the lowest scaling exponents for entry, reinforcing the sector's classification as mature and relatively low in innovation intensity. The remainder of the paper is structured as follows: Sect. 2 reviews the sensitivity of scaling estimates. Section 3 outlines data and methods. Section 4 presents the results. Section 5 discusses the implications, followed by conclusions in Sect. 6.

2 Scaling and Its Sensitivity

Scaling laws offer a concise way to describe how system attributes change with size, typically expressed as a power-law:

$$Y = \alpha X^\beta, \tag{1}$$

where Y is the attribute of interest (e.g., GDP), X is population, β is the scaling exponent, and α is a constant. In urban contexts, these laws help summarise how characteristics vary across cities. Depending on the value of β, scaling can be sublinear, linear, or superlinear (Bettencourt, 2013). Scaling laws can be seen as a generalisation of the Cobb-Douglas production function used in economics:

$$Y = \alpha L^\beta K^{1-\beta}, \tag{2}$$

where L and K denote labor and capital and the exponents are assumed to sum to 1, implying constant returns to scale. Unlike Cobb-Douglas, scaling laws relax the assumption of constant returns, replacing labour and capital with population (Lobo et al., 2013; Ribeiro et al., 2019). However, population alone may not fully capture productivity dynamics. Early work (Hyclak, 1986; Moomaw, 1981) showed its influence weakens when accounting for capital and labour inputs. More recent studies use disaggregated data: Sarkar et al. (2020) found superlinear scaling in knowledge-intensive sectors (localisation externalities) and linear returns across broader industry categories (urbanisation externalities) in Australian cities. Cottineau et al. (2017), examining French cities, reported that scaling exponents vary by industry type and population thresholds—high-tech sectors exhibit higher exponents in larger cities, while manufacturing and utilities often scale sublinearly. Crucially, the choice of model affects results (Shalizi, 2011), and high heteroskedasticity and fat-tailed distribution of city sizes can distort results (Leitao et al., 2016).

3 Data and Methods

The analysis includes 962 NUTS-3 regions in 14 European countries: Austria, Belgium, Denmark, Finland, France, Germany, Greece, Italy, Ireland, the Netherlands, Norway, Portugal, Spain, and Sweden, from 1985 to 2023. We use Orbis Bureau van Dijk (BVD) for firm entries and exits, and Eurostat for population, GDP, and employment. We use the North American Industry Classification System (NAICS) code 2211 to query electricity generation, transmission, and distribution utilities. The 6-digit NAICS codes identify firms active in solar (221114), wind (221115), geothermal (221116), and nuclear (221113) electric power generation. For comparison of emergence and concentration dynamics, we also query firms in four other sectors: one high-tech sector, Scientific Research and Development Services (5417), and three manufacturing sectors with varying knowledge complexity: Semiconductor and Other Electronic Component Manufacturing (3344), Plastics Product Manufacturing (3261), and Iron and Steel Mills and Ferroalloy Manufacturing (3311). We assess scaling patterns by estimating the following power-law relationship:

$$\log_{10} y = \log_{10} c + \beta \log_{10} x + \varepsilon \qquad (3)$$

where y is either the number of entries of green energy companies in that region or the total number of active green energy companies in that region, x is the population of a NUTS 3 region, and β is the scaling exponent. To evaluate robustness, we estimate β across a series of population thresholds—both increasing and decreasing in increments of 10,000—by subsetting the data accordingly. This allows us to assess the sensitivity of scaling behavior due to potential distortions caused by extreme values in small or large regions. After calculating scaling exponents without controlling for industry-specific effects, we proceed to do so using a log-additive function similar to the one used by Shalizi (2011):

$$\log_{10} y = \log_{10} c + \beta \log_{10} x + \sum_{j=1}^{n} f_j(x_j) + u_{it}, \qquad (4)$$

where each $f_j(x_j)$ denotes log-linear control terms and u_{it} captures the spatial (country-level) and temporal (annual) fixed effects, as scaling exponents vary over time (Fig. 1). Controls include Marshallian and Jacobean externalities, related variety, GDP per capita, and employment rate—computed following the method of Kundu et al. (2025). Finally, we compare firm entry scaling across the green energy sector and the four benchmark sectors. Due to data limitations (no reliable firm exit information), comparisons are limited to firm entry dynamics and do not include active firm concentrations.

4 Results

4.1 Without Controlling for Industry-Specific Factors

Figure 1a shows that firm concentration exhibits a gradual increase in scaling exponents, rising from ~0.4 (no cutoff) to ~1 at a 1.3 million population cutoff, eventually plateauing around 4.5 million. For firm entry (Fig. 1b), the exponent increases from ~0.2 to ~1 by a 2 million cutoff, with a similar plateau. Notably, concentration rises more steeply than entry, suggesting better survival prospects for firms in larger cities. Using inverse population cutoffs (i.e., excluding larger cities), scaling exponents are more stable. For concentration, the exponent hovers around 0.35, dropping slightly when cities over 1 million are excluded. Entry follows a similar trend, remaining around 0.2 and dipping to 0.1 after the same cutoff.

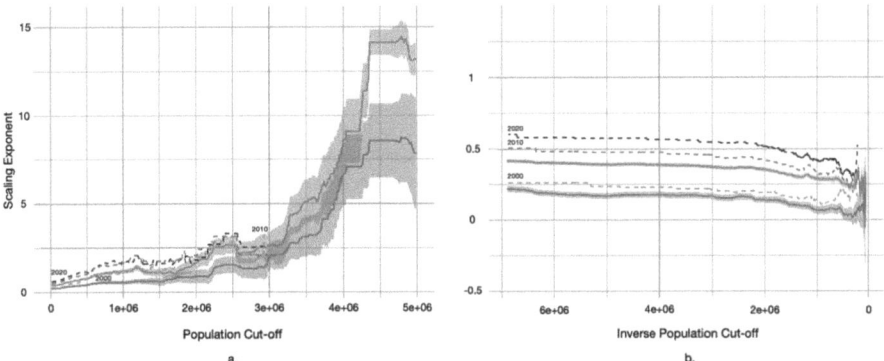

Fig. 1. Varying scaling exponents of firm concentration (orange) and firm entry (green) with a) positively incremental population cut-offs and b) inverse population cut-offs. Grey dashed lines indicate the upward trend in industry concentration scaling exponents over the past three decades. In panel (a), the trend line is truncated at a population threshold of 3 million due to instability in results caused by limited data beyond that point. (Color figure online)

4.2 Controlling for Industry-Specific Factors

After adjusting for agglomeration externalities, GDP per capita, and employment rate, firm concentration (Fig. 2a) starts superlinear (~1.1) and stabilizes near this level, dipping briefly between 2.5–3 million before sharply rising and becoming unstable due to limited data. Entry (Fig. 2b) follows a similar shape but with lower exponents. Inverse cutoff results are again more stable—concentration stabilizes around 0.8, and entry near 0.2—remaining sublinear in all cases.

4.3 Comparison with Other Sectors

Without controls (Fig. 3a), all sectors show similar trends until diverging around a 3.5 million population cutoff. After adjusting for industry factors (Fig. 3b), Scientific R&D consistently shows the highest scaling (0.75), followed by Plastics (0.55), Semiconductors (0.5), and Iron & Steel (0.35). All sectors exhibit a slight drop at higher inverse cutoffs before becoming unstable.

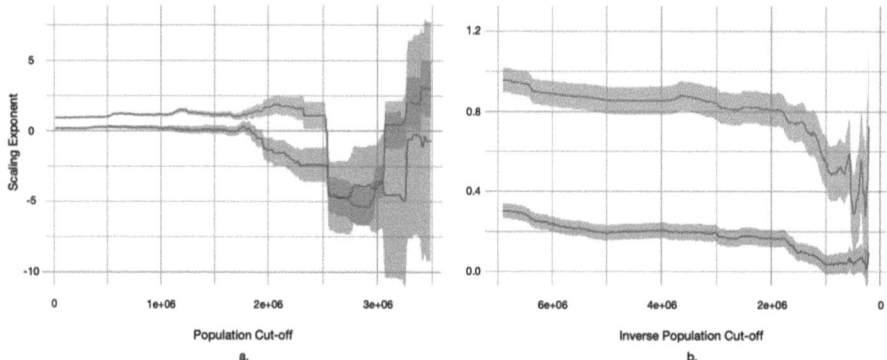

Fig. 2. Varying scaling exponents of firm concentration (orange) and firm entry (green) with a) positively incremental population cut-offs and b) inverse population cut-offs when controlling for agglomeration externalities, GDP per capita, and employment per capita.

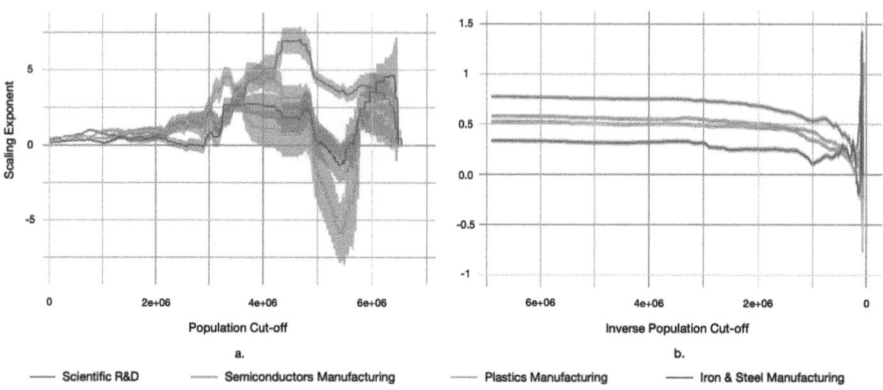

Fig. 3. Varying scaling exponents of firm entry with positively incremental population cut-offs, a) without controls, and b) with inverse population cut-offs when controlling for industry-specific factors for the four different sectors.

5 Discussion

This study examined the scaling of green electricity utilities, with and without industry-specific controls, across a range of population cutoffs. Using plant-level data, we found that both firm entry and concentration generally scale sublinearly, except in highly populous regions where scaling can appear superlinear. Electricity utility concentration consistently exhibits higher scaling exponents than entry, indicating population size supports firm survival more than new firm formation. Our results highlight the sensitivity of scaling estimates to both extremes of the regional size distribution. This pattern in terms of firm entry appears consistent across sectors, raising questions about the validity of scaling results in edge cases. Importantly, results suggest that without agglomeration externalities, size alone does not drive green sector growth, particularly for new entrants. While transition literature often emphasizes demand-side factors for green energy sector niche creation, our findings underscore the continued importance of supply-side dynamics like knowledge spillovers. Additionally, scaling exponents are often underestimated without industry-specific controls. Even with rising exponents over time, low-tech sectors continue to exhibit higher scaling than green electricity, suggesting that recent claims of increased sectoral complexity may be overstated. Despite recent criticisms of sensitivity issues, scaling literature remains a valuable framework for cross-sector comparisons. However, categorizing sectors as sublinear, linear, or superlinear based solely on firm entry may be unrealistic. While edge cases pose a challenge, claims of population cut-off sensitivity appear overstated, as robust results were observed with inverse population cut-offs.

6 Conclusion

This paper explored the scaling behavior of green electricity utilities, contrasting the assumptions of urban scaling and agglomeration literature. The scaling literature, based on a framework of interaction networks, treats externalities, including diversity and specialization, as endogenous outcomes of increasing size. The interest in scaling lies not at the sectoral level but rather in the aggregate behavior of cities. In contrast, agglomeration literature considers agglomeration externalities—such as Marshallian, Jacobean, and related externalities—distinct from size-derived benefits, which are categorized as urbanization externalities. Our findings show that accounting for industry-specific agglomeration externalities yields distinct scaling results, raising questions about the scaling approach. However, we suggest that concerns regarding the sensitivity of scaling exponents may be slightly overstated when excluding edge cases. Green electricity utilities largely scale sublinearly, with lower exponents than most other sectors, supporting their characterization as mature and relatively low in innovation intensity. The study focuses on green electricity utilities using the location of company headquarters rather than subsidiaries, as these are more likely to be knowledge centers driven by supply-side knowledge externalities and are less spatially constrained by energy production demands. These insights are especially relevant for regional transition and cohesion policies. Rather than relying solely on population size or general economic diversification, regions should focus on fostering targeted institutional support, supplier ecosystems, and industry-specific capabilities. This aligns well with the Smart Specialization approach and suggests the potential of smaller or lagging regions playing a more strategic role in the green transition. Future research could incorporate firm-level heterogeneity, such as differences in absorptive capacity and whether a firm is a typical spinoff or a new startup, to understand scaling dynamics better, as agglomeration externalities may affect these firms differently. Finally, due to limited data on employee numbers, we are constrained in assessing consolidation in large regions, which may underrepresent the sector's presence. Future studies could address this gap by combining granular occupation data with plant-level data.

Disclosure of Interests. The authors have no competing interests to declare that are relevant to the content of this article. **Declaration of AI-assisted technologies in the writing process**: During the preparation of this work the authors used ChatGPT large language model in order to enhance the clarity of the text. After using this tool, the authors reviewed and edited the content as needed and take full responsibility for the content of the publication.

References

Arcaute, E., Hatna, E., Ferguson, P., Youn, H., Johansson, A., Batty, M.: Constructing cities, deconstructing scaling laws. J. R. Soc. Interface. **12**(102), 20140745 (2015)

Balland, P.-A., Jara-Figueroa, C., Petralia, S.G., Steijn, M.P.A., Rigby, D.L., Hidalgo, C.A.: Complex economic activities concentrate in large cities. Nat. Hum. Behav. **4**(3), 248–254 (2020)

Bednarz, M., Broekel, T.: Pulled or pushed? The spatial diffusion of wind energy between local demand and supply. Industr. Corp. Change (2020)

Bettencourt, L.M.A.: The origins of scaling in cities. Science **340**(6139), 1438–1441 (2013)

Bolton, R.: Making Energy Markets: The Origins of Electricity Liberalisation in Europe. Springer (2021)

Cottineau, C., Hatna, E., Arcaute, E., Batty, M.: Diverse cities or the systematic paradox of urban scaling laws. Comput. Environ. Urban Syst. **63**, 80–94 (2017)

de Gooyert, V., Rouwette, E., van Kranenburg, H., Freeman, E., van Breen, H.: Sustainability transition dynamics: towards overcoming policy resistance. Technol. Forecast. Soc. Chang. **111**, 135–145 (2016)

De Groot, H.L.F., Poot, J., Smit, M.J.: Which agglomeration externalities matter most and why? J. Econ. Surv. **30**(4), 756–782 (2016)

Frenken, K., Cefis, E., Stam, E.: Industrial dynamics and clusters: a survey. Reg. Stud. **49**(1), 10–27 (2015)

Geels, F.W., Schot, J.: Typology of sociotechnical transition pathways. Res. Policy **36**(3), 399–417 (2007)

Hancher, L., De Hauteclocque, A.: Manufacturing the EU energy markets: the current dynamics of regulatory practice. Compet. Regul. Netw. Industr. **11**(3), 307–334 (2010)

Huenteler, J., Ossenbrink, J., Schmidt, T.S., Hoffmann, V.H.: How a product's design hierarchy shapes the evolution of technological knowledge—evidence from patent-citation networks in wind power. Res. Policy **45**(6), 1195–1217 (2016)

Hyclak, T.: Productivity and city size: some historical evidence. East. Econ. J. **12**(1), 45–51 (1986)

Jacobs, J.: The Economy of Cities. Vintage, New York (1970)

Kundu, V., Roy, D., Ehrenhard, M., Pfeffer, K.: Emergence of electricity utilities: agglomeration externalities and industry dynamics. Zenodo (2025). https://doi.org/10.5281/zenodo.14957718

Leitao, J.C., Miotto, J.M., Gerlach, M., Altmann, E.G.: Is this scaling nonlinear? Roy. Soc. Open Sci. **3**(7), 150649 (2016)

Lobo, J., Bettencourt, L.M.A., Strumsky, D., West, G.B.: Urban scaling and the production function for cities. PLoS ONE **8**(3), e58407–e58407 (2013)

Marshall, A.: Industrial organization, continued. The concentration of specialized industries in particular localities. In: Principles of Economics, pp. 222–231. Springer (1920)

Moomaw, R.L.: Productive efficiency and region. Southern Econ. J. 344–357 (1981)

Pumain, D., Paulus, F., Vacchiani-Marcuzzo, C., Lobo, J.: An evolutionary theory for interpreting urban scaling laws. Cybergeo: Eur. J. Geogr. (2006)

Ribeiro, H.V., Rybski, D., Kropp, J.P.: Effects of changing population or density on urban carbon dioxide emissions. Nat. Commun. **10**(1), 3204 (2019)

Sarkar, S., Arcaute, E., Hatna, E., Alizadeh, T., Searle, G., Batty, M.: Evidence for localization and urbanization economies in urban scaling. Roy. Soc. Open Sci. **7**(3), 191638 (2020)

Shalizi, C.R.: Scaling and hierarchy in urban economies. ArXiv Preprint ArXiv:1102.4101 (2011)

Tayal, D.: Disruptive forces on the electricity industry: a changing landscape for utilities. Electr. J. **29**(7), 13–17 (2016)

Pollution Simulations and in-Field Measurements Performed in March at Longyearbyen, Spitzbergen

Albert Oliver-Serra[2], Leszek Siwik[1], Natalia Leszczyńska[1], Maciej Sikora[1], Tomasz Maciej Ciesielski[3], Eirik Valseth[4,5,6], Jacek Leszczyński[1], Anna Paszyńska[1], and Maciej Paszyński[1(✉)]

[1] AGH University of Krakow, Kraków, Poland
{siwik,maciejsikora,jale,apaszynska,paszynsk}@agh.edu.pl,
s82950@365.sum.edu.pl
[2] The University of Las Palmas de Gran Canaria, Las Palmas de Gran Canaria, Spain
albert.oliver@ulpgc.es
[3] The University Centre in Svalbard, Longyearbyen, Norway
tomaszc@unis.no
[4] Oden Institute for Computational and Engineering, Austin, USA
[5] Simula Research Laboratory, Oslo, Norway
[6] Norwegian University of Life Sciences, Ås, Norway

Abstract. Spitzbergen is Norway's largest island, located in the Svalbard archipelago of the Polar Circle region. The island's population is currently about two and a half thousand, and it works mainly in natural resource extraction, tourism, and scientific research. The subject of our interest is the town of Longyearbyen, the capital of Spitzbergen, with a population of about 1,800. It has an airport and, until recently, the only power plant in Norway that generated electricity by burning coal. The coal has been replaced recently with diesel generators to reduce the pollution output, but the problem is still there. In this paper, we present the hypergraph grammar-based model of mesh generation and finite element pollution propagation simulations at Svalbard at Spitzbergen. We also perform in-field pollution measurements with snowmobiles. The simulations and measurements have been performed in March 2024. We compare our simulation results with in-field measurements.

Keywords: Outdoor air quality at Longyearbyen · Spitzbergen · Finite element simulations · Global Wind Atlas · Hypergraph transformations

1 Introduction

The island of Spitzbergen, located in the Svalbard archipelago, has two permanently inhabited towns, Longyearbyen and Barentsburg. The archipelago residents mostly engage in scientific research, coal mining, and tourism. The capital of Svalbard is the town of Longyearbyen, with a population of about 2,000. It is home to an airport and formerly Norway's only coal-fired power plant, recently replaced by diesel generators. In Longyearbyen, there is also the University Centre in Svalbard (UNIS), the northernmost university in the world, where Arctic Biology, Geology, Geophysics, Technology, and Arctic Safety are the subjects of teaching and research. The Longyearbyen region is also home to the famous seed vault [3]. Longyearbyen is located in the bay of Isfjorden, the second-largest fjord in Svalbard. The town has a location in the valley (see Fig. 1) favorable for a specific local microclimate. There are only two unique seasons on Svalbard. The first is the polar night, which lasts about two and a half months, during which the sun does not rise at all, and night reigns throughout. The polar night begins with a long twilight and ends with a long morning, during which the sun is below the horizon line for days. Between March and October, the sun rises for several hours a day, for about 2 h in March and October, to a maximum of 8 h in August.

Fig. 1. Topography of Longyearbyen at Spitzbergen. Source: https://toposvalbard.npolar.no, permission to use in article granted in terms of service at courtesy of Norwegian Polar Institute. Snowmobiles and the air quality sensor used to take the ($PM_{2.5}$) concentration measurements in Svalbard on 17/03/2024.

Air pollution in Longyearbyen was generated by a coal-fired power plant, recently replaced by diesel generators. During a normal day, due to thermal inversion effects in the morning, the air and pollutants are kept at ground level. When the rising sun subsequently warms the air mass, the layer of fog and pollution is lifted up and dispersed. This air convection process is important for this region's (broadly understood) inhabitants since their exposure to dangerous pollutants is significantly decreased.

In this paper, we perform computer simulations of the propagation of pollution generated by power plants, and we compare them with in-field measurements from snowmobiles, which were performed in March 2024. For the simulations of the pollution propagation from the power plant, we employ a finite element solver with the advection-diffusion model [11] stabilized with the Streamlined-Upwind-Petrov-Galerkin (SUPG) method [2]. The generation and processing of the computational grids for 3D finite element method simulations is a computationally intense task. We employ the hypergraph-grammar model of the 3D longest edge refinement algorithm. The longest-edge refinement algorithm has been initially proposed for 2D meshes by Cecilia Rivara [12,13]. The hypergraph grammar-based mesh refinements for 2D grids have been employed and discussed in [11], and in [7–10] with the hanging nodes version. In this paper, we present a novel hypergraph grammar-based model of mesh transformations expressing the 3D longest-edge refinement algorithm.

2 In-Field Measurements

Snowmobiles are the second most important source of air pollutants (and of ($PM_{2.5}$) in particular) in the Svalbard area. The negative impact of snowmobiles on (local) air quality has been considered by some researchers and environmental agencies, in particular concerning arctic regions [5]. There are 3000 snowmobiles registered at Svalbard between 1974–2024. To understand the impact of these snowmobiles in Svalbard, we need to know the concentration of air pollutants generated by a single snowmobile in an effort to evaluate the impact of all vehicles registered and used in Svalbard on local air quality. We focus on the emissions generating ($PM_{2.5}$), of which a two-stroke snowmobile ranges from 0.5 to 1.0 g per kilometer, whereas four-stroke models lead to 0.1 to 0.2 grams per kilometer see [16]. We performed in-field measurements of ($PM_{2.5}$) generated by snowmobiles used in Svalbard (see Fig. 1). The measurements were taken on 17/03/2024 using a Yamaha Venture Lite 600cc snowmobile and Airly air quality sensor (see Fig. 1) starting at 10:30 AM of Svalbard local time in 10-s intervals. The Airly sensor, used in our experiment, is a multi-pollutant air quality monitoring device that provides real-time measurements of gases (CO, NO_2, O_3 and SO_2) and $PM_{1.0}$, ($PM_{2.5}$) and PM_{10} mass concentrations, and environmental parameters such as pressure, temperature, and relative humidity. These sensors count suspended particles of 0.3, 0.5, 1.0, 2.5, 5.0 and 10 µm [15]. The results of the measurements are presented in Fig. 2. As one may see, there is a significant peak while the snowmobile was speeding up when the concentration reached up to 250 µg/m^3. After that, when the vehicle was driving with constant velocity, the concentration of ($PM_{2.5}$) measured oscillated between 0.9 and around 5 µg/m^3. Finally, the concentration increased up to around 20–30 µg/m^3, which was related to keeping the working vehicle in one place.

3 Hypergraph Grammar-Based Mesh Refinements

In this section, we express the 3D longest-edge refinement algorithm by hypergraph-grammar productions. We consider a 3D mesh with tetrahedral elements. We employ the concept of a hypergraph to model the computational mesh. In the hypergraph, we have the hyperedges that connect multiple vertices. The tetrahedra in the hypergraph are represented by vertices modeled as hypergraph nodes attributed by v. The tetrahedral edges are represented as hyperedges connecting two vertices, and they are attributed by E. The tetrahedral faces are modeled as hyperedges connecting three vertex nodes, and they are attributed by F. The

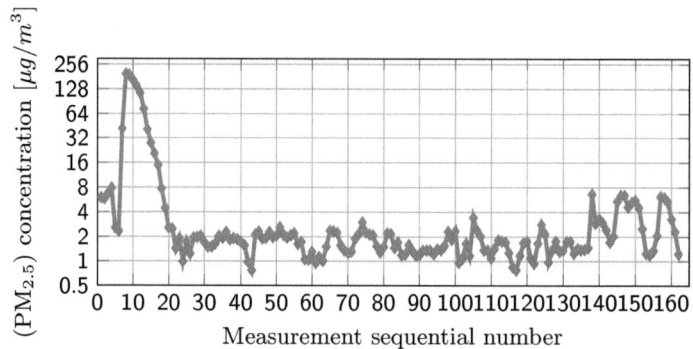

Fig. 2. (PM$_{2.5}$) concentration generated by snowmobile by 160 measurements performed in 10 s intervals in Svalbard on 17/03/2024.

tetrahedral interiors are modeled as hyperedges connecting four vertex nodes, and they are attributed by I. The hyperedges denoting the mesh edges have attributes $L = L_i$ denoting the length of the edge.

The rules for transforming the computational mesh are described by hypergraph transformations, called hypergraph grammar productions. While a large hypergraph represents the entire mesh, we have twelve hypergraph transformations presented in Table 1. For the simplicity of the presentation, we only plot vertices and hyperedges related to mesh edges (we omit the hyperedges related to faces and interiors). We identify the sub-hypergraph from the left-hand side of the transformation, and the right-hand side replaces it. Since the only connections between sub-hypergraphs and the rest of the hypergraph are the vertices, the replacement is straightforward. All our sub-hypergraphs hang on the mesh vertices, denoted by v. Identification and replacement involve locating the affected vertex nodes and replacing the local hypergraphs connecting them. The computationally expensive part is the identification of the left-hand side since many hyperedges are connected to the hypergraph vertices. We have introduced twelve hypergraph grammar productions. The first hypergraph production (**P1**) breaks a tetrahedral element along the longest edge (the edge with attribute $L = L3$ that is the longest edge). It breaks the edge into two new edges, breaks the face into two new faces. It introduces a new internal face, and it breaks an interior into two new interiors. The hypergraph transformations are isomorphic with respect to transformations; that is why we assume we can apply the same hypergraph production for sub-hypergraphs oriented in different ways. All the other hypergraph transformations break tetrahedral elements along the longest edge to remove the hanging edges (to eliminate the situation when a tetrahedral element has a broken face). All the transformations break elements along the longest edge to ensure the proper proportions of the elements. The hypergraph productions (**P2**) and (**P3**) consider the longest edge of a tetrahedral element with one face already broken. The hypergraph productions (**P4**), (**P5**), (**P6**), (**P7**) and (**P8**) consider the longest edge of a tetrahedral with two faces already

Table 1. Hypergraph grammar productions

Production	Hypergraph transformation	Production	Hypergraph transformation
(P1)		(P2)	
(P3)		(P4)	
(P5)		(P6)	
(P7)		(P8)	
(P9)		(P10)	
(P11)		(P12)	

broken. The hypergraph productions (**P9**), (**P10**), (**P11**), and (**P12**) consider the longest edge of a tetrahedral with three faces already broken. There is no case of the tetrahedral with four faces already broken since this case reduces to two tetrahedrons, one with three faces broken or two with two faces broken. Other productions have already expressed these cases.

4 Numerical Simulations

In this section we describe computer simulations aimed at modeling the phenomenon of propagation of pollutants generated by the power plant in March 2024. For the wind direction and intensity, as well as vertical profiles of the temperature, we refer to the High-Resolution Operational Forecasts dataset obtained from the National Science Foundation [6] as well as the Global Wind Atlas [4]. We simulated the pollution propagation using the advection-diffusion model with the source located at the power plant, on the top of the chimney, assuming the average wind direction and velocity as for the winter season. We have generated the computational mesh using a sequence of hypergraph grammar productions

(**P1**) breaking selected finite elements, followed by executions of hypergraph grammar productions (**P2**)–(**P12**) removing the hanging edges. We generate the computational mesh with tetrahedral elements covering the topography of Svalbard from the GMRT data [14]. To model the power plant chimney accurately, it has been manually incorporated into the initial mesh. The criterion of the tetrahedral element refinement implemented by the hypergraph transformation (**P1**) is the presence of the cross-section of the element with the terrain topography or the chimney. The other hypergraph transformations (**P2**)-(**P12**) implement the process of the closure of the mesh, to remove hanging edges. We use the weak form of the advection-diffusion equations with Crank-Nicolson time integration scheme stabilized with the SUPG stabilization method [2]:

$$(u_h^{t+1}, v) - \frac{dt}{2} \left(\beta_x \left(\frac{\partial u_h^{t+1}}{\partial x}, v_h \right) + \beta_y \left(\frac{\partial u_h^{t+1}}{\partial y}, v_h \right) \beta_z \left(\frac{\partial u_h^{t+1}}{\partial z}, v_h \right) \right.$$

$$\left. + K_x \left(\frac{\partial u_h^{t+1}}{\partial x}, \frac{\partial v_h}{\partial x} \right) + K_y \left(\frac{\partial u_h^{t+1}}{\partial y}, \frac{\partial v_h}{\partial y} \right) + K_z \left(\frac{\partial u_h^{t+1}}{\partial z}, \frac{\partial v_h}{\partial z} \right) \right)$$

$$+ (R(u_h^{t+1}), \tau \beta \cdot \nabla v_h) = \frac{dt}{2} \left(\beta_x \left(\frac{\partial u_h^{t}}{\partial x}, v_h \right) + \beta_y \left(\frac{\partial u_h^{t}}{\partial y}, v_h \right) \beta_z \left(\frac{\partial u_h^{t}}{\partial z}, v_h \right) \right.$$

$$\left. + K_x \left(\frac{\partial u_h^{t}}{\partial x}, \frac{\partial v_h}{\partial x} \right) + K_y \left(\frac{\partial u_h^{t}}{\partial y}, \frac{\partial v_h}{\partial y} \right) + K_z \left(\frac{\partial u_h^{t}}{\partial z}, \frac{\partial v_h}{\partial z} \right) \right) + (u_h^t, v) \quad (1)$$

where $(u, v) = \int_\Omega u(x,y,z;t)v(x,y,z;t)dxdydz$ denotes the L^2 scalar product over the computational domain Ω computed for the time moment t, $R(u) = \frac{\partial u}{\partial x} + \frac{\partial u}{\partial y} + \frac{\partial u}{\partial z} + \epsilon \Delta u$, and $\tau^{-1} = \left(\frac{\beta_x}{h_x} + \frac{\beta_y}{h_y} + \frac{\beta_z}{h_z} \right) + 3\epsilon \frac{1}{h_x^2 + h_y^2 + h_z^2}$.

In our simulation, we assume $K_x = K_y = 1.0$ and $K_z = 0.01$. As illustrated in Fig. 3 the pollution propagates into the valley where Longyearbyen is located. Without strong winds and thermal inversion effects, pollutants cannot be dispersed over large areas, contributing to their stagnation in the valley. Four hours after the chimney starts producing the pollution, the whole valley is filled with pollution (see Fig. 3).

Conclusions. Computer simulations discussed in the paper show that the combination of emissions of air pollutants with the specific atmospheric and/or terrain conditions, like the one observed in the Longyearbyen region, may result in a stagnation of the pollutants in the valley of Longyearbyen.

The diesel power plant in Longyearbyen, Svalbard, generates approximately 11 megawatts (MW) of electricity. According to data from the North American Commission for Environmental Cooperation [1], PM2.5 emission rates for oil-fired power plants typically range from 0.05 to 0.10 kg per megawatt-hour (kg/MWh). Assuming high-quality filters, the diesel power plant in Longyearbyen is estimated to emit an average of $4 \times 0.05 = 0.2$ (kg/MWh) kilograms of PM2.5 per four hours under continuous full-load operation. We selected a four hours interval since this is the time when the pollution distributes uniformly over the entire valley. Due to the topography and wind conditions, this PM2.5 spreads in the valley contributing to an average of 200×10^7 grams per m^3 which

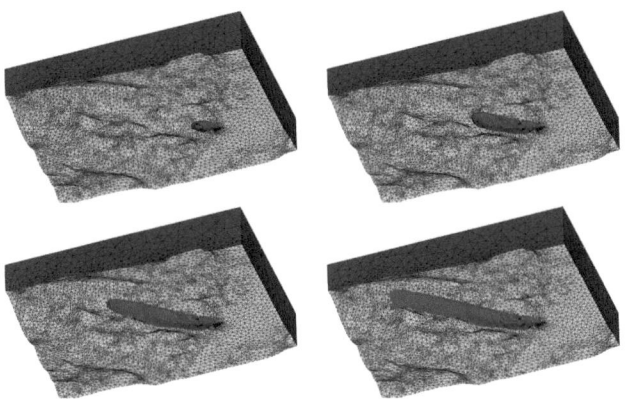

Fig. 3. The front views of the smoke propagated from the chimney into the valley. Half on hour, 90 min, 150 min, and 210 of power plant operation.

results in $200 \times 10^7/10^9 = 2$ ($\mu g/m^3$). This estimate coincides with the measurements from the snowmobiles, oscillating between 0.9 and 5 $\mu g/m^3$. We can also conclude that the pollution generated from snowmobiles is of the same order as the pollution propagating from the new diesel engines power plant.

Acknowledgements. Research project supported by the program "Excellence Initiative - Research University" for the AGH University of Krakow. The research presented in this paper was partially supported by the funds of Polish Ministry of Science and Higher Education assigned to the AGH University of Krakow. The work of Albert Oliver-Serra was supported by "Ayudas para la recualificación del sistema universitario español" grant funded by the ULPGC, the Ministry of Universities by Order UNI/501/2021 of 26 May, and the European Union-Next Generation EU Funds. The Authors gratefully acknowledge the support and assistance of The Polish Polar Station Hornsund for help with data collection.

References

1. North American Power Plant Air Emission. https://www.cec.org/sites/default/napp/en/particulate-matter-emissions.php
2. Brooks, A.N., Hughes, T.J.: Streamline upwind/Petrov-Galerkin formulations for convection dominated flows with particular emphasis on the incompressible Navier-Stokes equations. Comput. Methods Appl. Mech. Eng. **32**(1), 199–259 (1982)
3. Charles, D.: A 'forever' seed bank takes root in the Arctic. Science **312**(5781), 1730–1731 (2006)
4. Davis, N.N., et al.: The global wind atlas: a high-resolution dataset of climatologies and associated web-based application. Bull. Am. Meteorol. Soc. **104**(8), E1507–E1525 (2023)
5. Dekhtyareva, A., et al.: Springtime nitrogen oxides and tropospheric ozone in Svalbard: results from the measurement station network. Atmos. Chem. Phys. **22**(17), 11631–11656 (2022)

6. European Centre for Medium-Range Weather Forecasts: ECMWF IFS High-Resolution Operational Forecasts (2016). https://doi.org/10.5065/D68050ZV, https://rda.ucar.edu/datasets/d113001/. Accessed 25 Nov 2024
7. Paszyńska, A., Paszyński, M., Grabska, E.: Graph transformations for modeling hp-adaptive finite element method with triangular elements. In: Bubak, M., van Albada, G.D., Dongarra, J., Sloot, P. (eds.) ICCS 2008, Part III. LNCS, vol. 5103, pp. 604–613. Springer, Heidelberg (2008). https://doi.org/10.1007/978-3-540-69389-5_68
8. Paszyńska, A., Paszyński, M., Grabska, E.: Graph transformations for modeling hp-adaptive finite element method with mixed triangular and rectangular elements. In: Allen, G., Nabrzyski, J., Seidel, E., van Albada, G.D., Dongarra, J., Sloot, P. (eds.) ICCS 2009. LNCS, vol. 5545, pp. 875–884. Springer, Heidelberg (2009). https://doi.org/10.1007/978-3-642-01973-9_97
9. Paszyński, M.: On the parallelization of self-adaptive hp-finite element methods. Part I. Composite Crogrammable Graph Grammar model. Fundam. Inform. **93**(4), 411–434 (2009)
10. Paszyński, M.: On the parallelization of self-adaptive hp-finite element methods. Part II. Partitioning Communication Agglomeration Mapping (PCAM) analysis. Fundam. Inform. **93**(4), 435–457 (2009)
11. Podsiadło, K., et al.: Parallel graph-grammar-based algorithm for the longest-edge refinement of triangular meshes and the pollution simulations in lesser Poland area. Eng. Comput. **37**, 3857 3880 (2021)
12. Rivara, M.C.: Algorithms for refining triangular grids suitable for adaptive and multigrid techniques. Int. J. Numer. Methods Eng. **20**(4), 745–756 (1984)
13. Rivara, M.C.: Mesh refinement processes based on the generalized bisection of simplices. SIAM J. Numer. Anal. **21**(3), 604–613 (1984)
14. Ryan, W.B.F., et al.: Global multi-resolution topography synthesis. Geochem. Geophys. Geosyst.**10**(3) (2009)
15. South Coast Air Quality Management District: Airly sensor detail (2024). https://www.aqmd.gov/aq-spec/sensordetail/airly. Accessed 25 Nov 2024
16. U.S. Department of the Interior, National Park Service: Scientific assessment of Yellowstone National Park winter use March 2011. https://npshistory.com/publications/yell/winter-use-sci-assessment-2011.pdf

Reversed Model Verification by Inferring Conceptual Models from Simulation Code

Rumyana Neykova[✉] and Derek Groen

Department of Computer Science, Brunel University London, London, UK
rumyana.neykova@brunel.ac.uk

Abstract. Extracting high-level conceptual models from simulation code can benefit model validation and verification, system optimisation, and cross-disciplinary communication. However, conceptual models are often embedded within implementation details, making them difficult to access and interpret. This paper explores the feasibility of using Large Language Models (LLMs) to infer conceptual models from simulation code. We conduct a preliminary investigation on an agent-based simulation (Flee), demonstrating how LLMs can extract key structural, behavioural, and temporal elements. Our results suggest that LLMs can generate meaningful conceptual representations that align with expert-created models, offering potential support for model verification. However, we also identify limitations such as omissions and misinterpretations, highlighting the need for human oversight. While our study is based on a single example, it provides initial insights into the role of LLMs in conceptual model inference and their potential integration into simulation validation workflows.

Keywords: Conceptual Model · LLM · Inference

1 Introduction

Modern simulation systems rely on complex codebases that encapsulate intricate interactions and dynamic behaviors. Despite their importance, the high-level conceptual models underpinning these simulations often remain implicit within implementation details, hindering model verification, optimization efforts, and interdisciplinary collaboration. Traditionally, conceptual models are either manually extracted by experts—a time-consuming and inconsistency-prone process—or neglected entirely, leaving simulations unverified against their conceptual foundations. Recent advancements in Large Language Models (LLMs) offer new opportunities for automating conceptual model inference from simulation code. This paper investigates whether LLMs can extract accurate conceptual models from simulation code, what types of insights these models can capture, and what limitations need to be addressed for effective integration into validation workflows.

To explore these questions, we apply LLM analysis to infer conceptual models from a real-world agent-based simulation (Flee), focusing on temporal structure, entity behavior, and system architecture. We choose Flee because of its wide uptake, and because the authors understand the code deeply, allowing for manual evaluation of results. We evaluate alignment between inferred models and expert-created representations to assess their applicability for model verification. Our findings suggest that LLMs can extract meaningful conceptual elements that support validation, though with limitations requiring human oversight. While based on a single case, this work demonstrates the potential of LLMs as tools for conceptual model inference and outlines directions for integrating these approaches into simulation validation workflows.

2 Related Work

Conceptual Modeling in Simulation. Conceptual modeling constitutes the process of abstracting a model from the real world [12]. Simulations represent implemented conceptual models, typically through computer programs, though exceptions exist [5]. Conceptual models are popular because they are easy to understand and interpret for non-developing researchers [7] and because they are essential in the early stages of simulation development [4].

The various conceptual model diagrams found in literature differ in scope and design, often focusing on specific aspects such as time loop dynamics [3,14], behavioral elements [10,15], or system architecture [6,8], yet these are typically created manually rather than extracted from implementation code. Model verification—comparing the conceptual model with its implementation—commonly relies on targeted tests that ensure specific mechanisms adhere to established rules [13]. For example, one might check whether floating point representations in computer code could lead to unintended instabilities in a fluid dynamics algorithm. However, this approach is limited in scope, as many conceptual models encompass multiple mechanisms, while academic literature frequently presents diagrams representing substantial portions of the overall model (e.g., Figure one in Mehrab et al. [9]).

Within this work we attempt to extract these higher level conceptual models directly from simulation code using a LLM-based approach. To our knowledge, we are the first to do so.

LLMs for Software Analysis. Recent advances in large language models have led to innovative approaches in automated code analysis and conceptual modeling. Ali et al. [1] demonstrated how LLMs support conceptual modeling by creating, updating, and visualizing UML diagrams through natural language interactions. Similarly, Nicola et al. [2] explored how LLMs can assist business process modeling by extracting process models from text and identifying interaction patterns that inform modeling best practices. Nam et al. [11] showed that LLM-based tools can significantly improve code understanding through contextual explanations of unfamiliar code.

While these studies show promising results, they have not been applied to simulation modeling specifically, nor do they focus on autonomous extraction

Time-Loop Model	Entity Behaviour Model	System Architecture Model
Represents the simulation's temporal structure, including: - The main simulation loop and key phases. - The sequence of events and their dependencies.	Describes the decision-making and interactions of entities: - Identifies primary entity types and their roles. - Captures entity states, decision logic, and interactions.	Illustrates the structural organisation of the simulation: - Identifies key system components and their interfaces. - Represents data flow, resource management, and dependencies.

Fig. 1. Types of Conceptual Models Extracted from Simulation Code.

from existing code. Most prior work targets general code documentation or supports human modelers through interactive workflows rather than direct model inference. The unique challenges of simulation code—with its complex temporal structures, entity behaviors, and system interactions—remain unaddressed by current LLM applications.

3 Conceptual Model Extraction: Approach and Case Study

To structure conceptual model inference, we define three conceptual model types that capture distinct aspects of a simulation (Fig. 1). Temporal structure focuses on the time-loop mechanics of the simulation, capturing key phases, event sequences, and dependencies. Understanding the temporal structure is important for validating whether system processes align with expected time-driven behaviours. Behavioural modelling examines entity decision-making and interactions, identifying key agent types, their roles, and the logic governing their state transitions. Behavioural models are particularly relevant for agent-based simulations, where emergent behaviour depends on individual decision rules. System architecture analyses the structural organisation of the simulation, identifying key components, their interfaces, and data flows. Architectural models provide a high-level view of system composition and modular interactions, which is needed for understanding dependencies and debugging. These three dimensions align with established practices in conceptual modelling and provide complementary perspectives on a system's operation. By extracting models across these levels, we assess the extent to which LLMs can infer useful abstractions from raw simulation code.

Case Study: Inferring Conceptual Models from the Flee Simulation.
To evaluate LLM-based conceptual model inference, we conducted experiments on Flee [3], an agent-based simulation framework designed to model forced

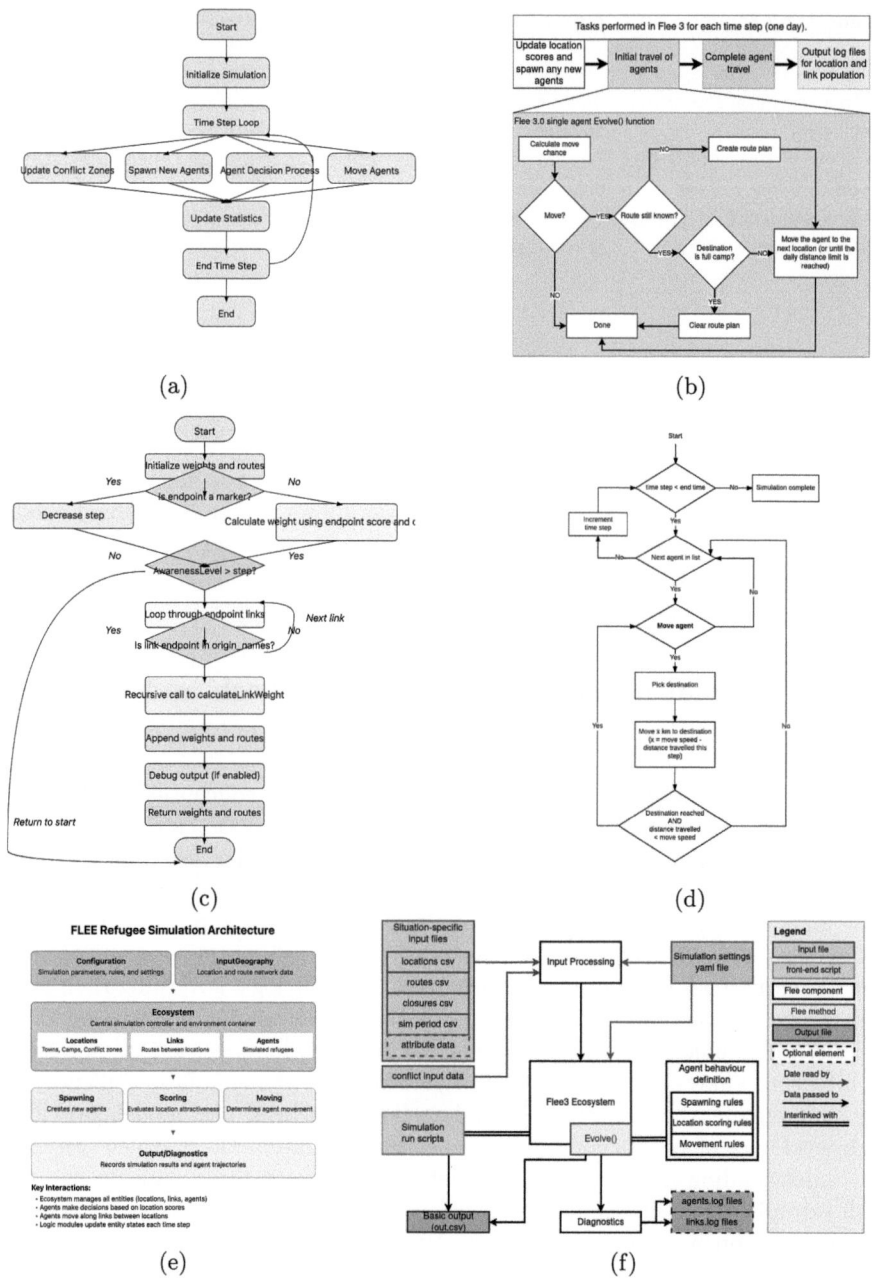

Fig. 2. Comparison of LLM-generated diagrams with literature: (a) LLM-generated time-loop process diagram, (b) time loop process detail from literature [3], (c) LLM-generated agent decision process for calculating link weights, (d) closest corresponding Flee flowchart from literature [15], (e) LLM-generated architectural diagram, and (f) its equivalent in recent literature [3].

migration patterns. The simulation uses discrete time-stepping to track population movements between locations based on conflict dynamics, decision-making heuristics, and resource availability.

Our analysis was conducted on an 18-file, 5,000-line code base, processed using Claude Sonnet 3.5. The entire source code was provided to the LLM as project knowledge, enabling a holistic understanding of the system. The outputs were generated through zero-shot prompting, explicitly requesting simplified high-level overviews. In the case of Fig. 2e, the initial response contained excessive detail, prompting us to refine our request for a more abstract representation. Additionally, we specified visualization in React.js format rather than Mermaid, both being supported by Claude. To extract conceptual models, we employed structured queries to guide the LLM's analysis. For instance, to obtain the system architecture model shown in Fig. 2e, we used the following initial prompt: *Visualise a high-level architectural diagram of the simulation code in your project knowledge. Showing the main components and their interactions.*

The results are summarised in Fig. 2, which presents examples of inferred time-loop, entity behaviour, and system architecture models. Section 4 evaluates these models by comparing them with the original simulation code and conceptual models from the literature.

4 Evaluation

Within this section, we present three LLM generated diagram examples, and provide a brief reflection on their quality. Our evaluation assesses the LLM-generated conceptual models based on accuracy (alignment with code), completeness (coverage of key elements), and interpretability (clarity of presentation). Table 1 summarizes the quantitative assessment of shortcomings across all three model types.

We present two time loop representations from the Flee ABM in Fig. 2a–2b. The LLM-generated diagram effectively captures most conceptual elements with correct terminology, but shows four specific inconsistencies: (i) omitted score updating for non-conflict zones, (ii) mislabeled "update statistics" instead of output writing, (iii) incomplete agent decision process missing the finish travel task, and (iv) incorrect sequencing of conflict zone updates. Table 1 indicates only one missing concept element, suggesting high fidelity. Both LLM and literature diagrams show strong consistency, with the former highlighting the flow from beginning to end (the literature diagram focuses only on the loop iterations), and the latter providing more zoom-in detail on the agent decision-making process.

We present agent decision representations in Fig. 2c–2d. The LLM diagram captures the link weight calculation algorithm comprehensively despite formatting issues, with three specific shortcomings: (i) the loop arrow is inverted, (ii) the "Yes" part in the endpoint link loop seems to be placed one arrow too high, (iii) the arrows around the "is link endpoint in origin_names" are incorrectly arranged. Table 1 shows arrow labeling as the most significant issue, with four incorrect labels. Interestingly, the closest conceptual diagram we could locate in

Table 1. Evaluation overview of LLM-extracted conceptual diagrams

Shortcoming	# of occurrences for each diagram		
	time_loop	agent_decision	architecture
Wrong arrows	1	2	1
Wrong arrow labels	0	4	0
Non-diagram elements	0	0	4
Missing relevant concept element	1	0	0
Overly vague concept element	2	0	1

the literature is not similar at all, as all the logic in the LLM-generated figure is only represented by a single box ("pick destination"). Therefore, despite its shortcomings, the LLM-generated diagram exposes important conceptual elements of the Flee code, particularly because the generation and weighting of possible agent routes is a fundamental aspect of the agent decision-making. In addition, the LLM has successfully picked up the concept and functionality of marker locations (including the need to adjust step numbers), which is something that has not been clearly covered in any of the existing literature.

Figure 2f shows architectural diagrams of Flee. The LLM version depicted in Fig. 2e identifies essential components—inputs (top), ecosystem objects (top middle), dynamical components (middle bottom), and outputs (bottom)—connected via triangular indicators. The literature diagram (Fig. 2f) offers more detailed relations and explicit file type names, though both share clear commonalities in ecosystem and agent behavior modules. Key shortcomings in the LLM diagram include: (i) missing backlinks from behavioral definitions to Ecosystem objects and (ii) vague configuration elements. Table 1 shows this diagram contains fewer issues than the other conceptual models.

Across all three model types, we observe that LLMs excel at identifying main components and general flows, but struggle with precise relationship details (particularly directional relationships) and specific technical terminology. The time loop model exhibits the highest overall accuracy, likely due to its explicit sequential structure in the code, while the agent decision model presents the most challenges in accurately representing complex conditional logic.

5 Discussion and Conclusions

Conceptual model extraction has not been widely explored for model verification because manual extraction is often tedious and sometimes intractable. The ease with which LLMs can generate various levels of conceptual model details from implementations urges revisiting the role these extractions can play in understanding and verifying simulation behavior. What was previously considered impractical due to the effort involved can now potentially become an integral part of simulation development and validation workflows.

Our preliminary investigation demonstrates that LLMs offer potential for conceptual model inference across multiple dimensions of simulation systems. This allows for simulation verification at multiple levels: from time-stepping mechanics to agent decision logic to system composition. LLMs can identify relationships that would otherwise require manual analysis, provide readable abstractions for domain experts without programming knowledge, and surface inconsistencies between implementation and intended behavior. We have demonstrated that it is possible to create high-fidelity conceptual models automatically. The quality of the extracted models in our case study suggests that the potential benefits substantially outweigh the current limitations, warranting further exploration of this approach.

Our study raises important questions about how code commenting style, documentation quality, and naming conventions influence conceptual model extraction. While we haven't explicitly tested this influence, we suspect that well-structured comments and semantically meaningful filenames likely enhance LLM extraction performance significantly. For simulation developers, this suggests an opportunity to strategically instrument code with conceptual markers, standardized documentation patterns, or explicit annotations that could guide more accurate model extraction. The optimal format and extent of such instrumentation remains unexplored, presenting a promising direction for developing standards that maximize verification benefits while minimizing documentation overhead.

In terms of limitations, we have shown that inferred models can exhibit imprecise event sequencing, oversimplified interactions, and miss subtle conceptual elements. Expert validation therefore remains necessary to refine the LLM outputs. In addition, our explorative study examines one simulation code base (Flee), so assessing broader applicability requires further research. In general, reproducibility presents a significant challenge for LLMs. As their outputs can vary between runs and across different model versions, the conceptual models generated may not be consistent.

In recent times, developers increasingly rely on LLMs, instead of programming skills, for application development. Our approach could help these developers to check whether LLM-guided implementations align with their conceptual vision. Furthermore, our findings suggest several integration directions: implementing regular conceptual extraction during development, analyzing models from different implementations of the same system, linking conceptual elements to specific code segments, and creating interfaces for expert review.

Rather than a complete solution, our work serves as a call to action for the simulation community to explore these capabilities further. By investigating across diverse simulation domains and developing specific methodologies for extraction and verification, we can improve the reliability and transparency of simulation-based research.

Acknowledgments. This work has been supported by the SEAVEA ExCAL-IBUR project, which has received funding from EPSRC under grant agreement EP/W00771/1.

Disclosure of Interests. The authors have no competing interests to declare that are relevant to the content of this article.

References

1. Ali, S.J., Reinhartz-Berger, I., Bork, D.: How are LLMs used for conceptual modeling? An exploratory study on interaction behavior and user perception. In: Maass, W., Han, H., Yasar, H., Multari, N. (eds.) Conceptual Modeling, pp. 257–275. Springer (2025). https://doi.org/10.1007/978-3-031-75872-0_14
2. De Nicola, A., Formica, A., Mele, I., Missikoff, M., Taglino, F.: A comparative study of LLMs and NLP approaches for supporting business process analysis. Enterp. Inf. Syst. **18**(10), 2415578 (2024). https://doi.org/10.1080/17517575.2024.2415578
3. Ghorbani, M., et al.: Flee 3: flexible agent-based simulation for forced migration. J. Comput. Sci. **81**, 102371 (2024)
4. Groen, D., Suleimenova, D., Jahani, A., Xue, Y.: Facilitating simulation development for global challenge response and anticipation in a timely way. Journal of Computational Science **72**, 102107 (2023)
5. Holmberg, E.: On the clustering tendencies among the nebulae. II. A study of encounters between laboratory models of stellar systems by a new integration procedure. Astrophys. J. **94**, 385 (1941)
6. Hussan, J.R., Hunter, P.J.: Comfort simulator: a software tool to model thermoregulation and perception of comfort. J. Open Res. Softw. **8**(1), 16 (2020). https://doi.org/10.5334/jors.288
7. Liu, J., Yu, Y., Zhang, L., Nie, C.: An overview of conceptual model for simulation and its validation. Procedia Eng. **24**, 152–158 (2011)
8. Luo, L., et al.: Agent-based human behavior modeling for crowd simulation. Comput. Anim. Virtual Worlds **19**(3–4), 271–281 (2008). https://doi.org/10.1002/cav.238
9. Mehrab, Z., et al.: An agent-based framework to study forced migration: a case study of Ukraine. PNAS Nexus **3**(3), pgae080 (2024)
10. Molina, T., Ortega, J., Mu oz, J.: HELMpy, open source package of power flow solvers. J. Open Res. Softw. **9**(1), 23 (2021). https://doi.org/10.5334/jors.310
11. Nam, D., Macvean, A., Hellendoorn, V., Vasilescu, B., Myers, B.: Using an LLM to help with code understanding. In: Proceedings of the IEEE/ACM 46th International Conference on Software Engineering, pp. 1–13. ACM (2024). https://doi.org/10.1145/3597503.3639187
12. Robinson, S.: Conceptual modeling for simulation. In: 2013 Winter Simulations Conference (WSC), pp. 377–388. IEEE (2013)
13. Sargent, R.G.: Verification and validation of simulation models. In: Proceedings of the 2010 Winter Simulation Conference, pp. 166–183. IEEE (2010)
14. Schmieschek, S., et al.: LB3D: a parallel implementation of the Lattice-Boltzmann method for simulation of interacting amphiphilic fluids. Comput. Phys. Commun. **217**, 149–161 (2017). https://doi.org/10.1016/j.cpc.2017.03.013
15. Suleimenova, D., Bell, D., Groen, D.: A generalized simulation development approach for predicting refugee destinations. Sci. Rep. **7**(1), 13377 (2017)

GPU-Accelerated Out-of-Core HMM Inference with Concurrent CUDA Streams

MohammadReza HoseinyFarahabady[(✉)] and Albert Y. Zomaya

School of Computer Science, Center for Distributed and High Performance Computing, The University of Sydney, Sydney, NSW, Australia
{reza.hoseiny,albert.zomaya}@sydney.edu.au

Abstract. Hidden Markov Models (HMMs), along with their extension, Hidden Semi-Markov Models (HSMMs), are powerful tools for modeling complex systems with multivariate states, but scaling them to ultra-large state spaces presents significant computational and memory challenges. This paper proposes a hybrid CUDA-based implementation to overcome these limitations, enabling efficient processing of HMMs with massive state spaces and extended observation sequences. Key optimizations include log-space computations for numerical stability, memory-efficient data partitioning with sparse matrix representations, and asynchronous data transfers using CUDA streams to overlap host-device communication with GPU kernel execution. Our approach achieves significant performance improvements, demonstrating up to 8.5× speedup over multi-threaded CPU implementations for HMM processing.

Keywords: Complex Systems with Large State Spaces · Hidden Markov Models · GPU Acceleration · Performance Optimization · Memory Coalescing · Asynchronous CUDA Streams

1 Introduction

Hidden Markov Models (HMMs), introduced by Baum and Petrie [1], and their extension, Hidden Semi-Markov Models (HSMMs), are powerful tools for modeling complex systems with multivariate states and hidden structures [5]. Despite their effectiveness, traditional algorithms like Baum-Welch and Viterbi struggle with computational and memory demands in large-scale scenarios. HMMs, defined as doubly stochastic processes with hidden states inferred from observable sequences, are widely applied in fields such as speech recognition [10], bioinformatics [8], and finance [9].

An HMM is characterized by five key components:

- A set of N hidden states $\mathcal{S} = \{S_1, S_2, \ldots, S_N\}$ represents the underlying stochastic process, where each state is not directly observable. The state at time t is denoted by the variable q_t, where $q_t \in \mathcal{S}$.
- A set of M observable symbols $\mathcal{V} = \{v_1, v_2, \ldots, v_M\}$.

- A state transition probability matrix $\boldsymbol{A} \in \mathbb{R}^{N \times N}$ with elements a_{ij}, where $a_{ij} = P(q_{t+1} = S_j | q_t = S_i)$ represents the probability of transitioning from state S_i at time t to state S_j at time $t+1$.
- An emission probability matrix $\boldsymbol{B} \in \mathbb{R}^{N \times M}$ with elements $b_j(m)$, where $b_j(m) = P(O_t = v_m | q_t = S_j)$ represents the probability of observing symbol v_m at time t given that the hidden state is S_j.
- An initial state probability vector $\boldsymbol{\pi} \in \mathbb{R}^N$, where $\pi_i = P(q_1 = S_i)$ represents the probability of starting in state S_i.

These components, collectively denoted as $\lambda = (\boldsymbol{A}, \boldsymbol{B}, \boldsymbol{\pi})$, enable HMMs to address three fundamental problems [4,7]:

- *Evaluation*: To estimate the likelihood of an observed sequence $O = (O_1 \ldots O_T)$ given the model λ, denoted as $P(O|\lambda)$.
- *Decoding*: To determine the most probable sequence of hidden states $Q = (q_1, q_2, \ldots, q_T)$ that generated the observed sequence O, i.e., maximizing $P(Q|O, \lambda)$.
- *Learning*: To estimate the model parameters, λ, to maximize the likelihood of a training set of observation sequences.

This paper introduces a hybrid CPU-GPU out-of-core algorithm designed to efficiently process HMMs with massive state spaces and longer observation sequences. Our approach integrates several optimization techniques that leverages the parallel processing capabilities of modern GPUs while overcoming memory constraints through innovative optimizations. First, we employ log-space computations to mitigate numerical underflow, ensuring stable and accurate results during algorithm execution in the GPU space. Second, we implement a memory-efficient data partitioning strategy that utilizes sparse matrix representations to optimize memory utilization of device. Large matrices are partitioned into blocks, with dimensions dynamically determined by the available GPU memory which enable efficient processing of large-scale HMMs. These blocks are processed sequentially on the GPU to enhance computational throughput. We also leverage asynchronous data transfers via CUDA streams and a triple-buffered pipeline to overlap data transfers between the host and device with kernel execution on GPU to hide the data-transfer latencies. We further optimize memory access patterns through memory coalescing technique for both dense and sparse matrix computations.

The remainder of this paper is organized as follows: Sect. 2 provides an overview of key background concepts, including the fundamentals of Hidden Markov Models, the Baum-Welch algorithm, and the challenges associated with scaling HMMs for large state spaces. Section 3 presents our proposed hybrid CPU-GPU out-of-core algorithm, detailing the optimizations implemented for efficient GPU utilization. Section 4 describes the experimental setup and performance evaluation methodology used to assess the effectiveness of our GPU implementation compared to CPU-based approaches. Finally, Sect. 5 concludes the paper with a discussion of the findings and outlines directions for future research in optimizing HMMs on GPU architectures.

2   Background

Baum-Welch Algorithm is an Expectation-Maximization (EM) technique used for parameter estimation in HMMs. It is an iterative algorithm designed to optimize such parameters when the hidden states are unknown, based only on observed data. The key essential part for this algorithm parameter lies in the *forward-backward procedure* [2].

The Forward-Backward Procedure. A central problem in HMMs is efficiently computing the probability of an observed sequence $O = (O_1, O_2, \ldots, O_T)$ given the model parameters λ. A naive approach of enumerating all possible state sequences of length T has a computational complexity of $\mathcal{O}(TN^T)$, where N is the number of hidden states and T is the number of observations [3,6]. This exponential complexity renders the direct computation intractable even for moderately sized sequences and models. The forward-backward procedure [1] provides a computationally efficient solution by leveraging dynamic programming. The forward pass computes the forward probabilities $\alpha_t(j) = P(O_1, \ldots, O_t, q_t = S_j | \lambda)$, representing the probability of observing the sequence up to time t and being in state S_j at that time. The recursion step calculates $\alpha_t(j)$ by summing over all previous states $S_{i \leq j-1}$, weighted by the transition probabilities A_{ij} and the emission probability of the current observation O_t in state S_j. The algorithm's time complexity is $\mathcal{O}(TN^2)$, significantly more efficient than a direct computation. The backward pass computes the probability of the remaining observation sequence $(O_{t+1} \ldots O_T)$ given state S_j at time t, denoted by $\beta_t(j) = P(O_{t+1}, O_{t+2}, \ldots, O_T | q_t = S_j, \lambda)$. The recursion step calculates $\beta_t(j)$ by summing over all possible next states S_k, weighted by the transition probability A_{jk}, emission probability $B_k(O_{t+1})$, and the subsequent backward probability $\beta_{t+1}(k)$. The algorithm has a time complexity of $\mathcal{O}(TN^2)$.

Parameter Estimation with Baum-Welch. The Baum-Welch algorithm act as an unsupervised learning technique that iteratively trains HMM parameters using the forward-backward procedure to find the parameters of the HMM, *i.e.*, λ, that maximize the probability of a given observation sequence. The algorithm comprises two main steps:

1. **E-Step (Expectation Step):** Compute the expected number of state transitions and emissions using forward and backward probabilities for all time steps t and states S_j. We define the variable $\xi_t(i,j)$ as the probability of being in state S_i at time t and in state S_j at time $t+1$ given the observation sequence and the model, *i.e.*, $P(O | \lambda)$. So, the expected number of transitions from S_i to S_j can be calculated as:

$$\xi_t(i,j) = \frac{\alpha_t(i) A_{ij} B_j(O_{t+1}) \beta_{t+1}(j)}{\sum_{j=1}^{N} \alpha_t(j) \beta_t(j)}. \tag{1}$$

2. **M-Step (Maximization Step):** Re-estimate the model parameters based on the E-step's expected values:

$$A_{ij} = \frac{\sum_{t=1}^{T-1} \xi_t(i,j)}{\sum_{t=1}^{T-1} \gamma_t(i)}, \quad B_j(k) = \frac{\sum_{t=1}^{T} \gamma_t(j) \cdot \mathbb{I}(O_t = k)}{\sum_{t=1}^{T} \gamma_t(j)}, \quad (2)$$

where $\mathbb{I}(O_t = k)$ is an indicator function (1 if $O_t = k$, 0 otherwise), and $\gamma_t(j)$ is the posterior probabilities that estimate the likelihood of being in state S_j at time t given the entire observation sequence and the HMM model. It can be computed using $\gamma_t(j) = \frac{\alpha_t(j)\beta_t(j)}{P(O|\lambda)}$. Summing $\gamma_t(j)$ over t gives the expected number of transitions made from state S_j. The algorithm iterates these steps until convergence (e.g., no significant change in the log-likelihood or a maximum number of iterations is met).

3 Our Approach

We explored the potential of GPU acceleration to enhance the performance of Hidden Markov Models with large state spaces by implementing the Baum-Welch algorithm using the CUDA cuBLAS library. This implementation was specifically designed to leverage the massive parallelism of modern GPUs, enabling highly efficient matrix operations, which are central to HMM computations.

However GPU acceleration is fundamentally limited by the finite capacity of GPU memory. This limitation becomes particularly challenging when processing Hidden Markov Models with ultra-large state spaces or extended observation sequences, where the required data structures, such as the forward and backward probability matrices, exceeds the available GPU memory. To overcome these memory constraints and enable scalable processing of large-scale HMMs on GPUs, we propose a hybrid CUDA implementation incorporating the following key techniques:

- *Log-Space Implementation for Numerical Stability*: Mitigates numerical underflow issues by performing computations in the log domain, ensuring stable results.
- *Memory-Efficient Data Partitioning and Sparse Representations*: Optimizes memory utilization by partitioning large data structures into smaller blocks and leveraging sparse matrix representations and memory coalescing where applicable.
- *Asynchronous Data Transfer for Computation-Communication Overlap*: Maximizes GPU utilization by overlapping data transfers between the host and device memories with kernel execution to minimize the idle time.

These strategies address both the memory capacity limitations of GPUs that enables efficient computation of HMMs with large state spaces and long observation sequences. The following sections detail these techniques.

Dynamic Memory Allocation and Data Partitioning

The algorithm begins by analyzing the HMM parameters: the number of states, transition probabilities, and the length of the observation sequence. Based on these parameters, the memory required for the key data structures, *i.e.*, transition matrix, emission matrix, and forward (α) and backward (β) probability matrices, is dynamically calculated. If the total memory required to store these matrices exceeds the available GPU memory, a data partitioning strategy is employed. The matrices are divided into logical blocks, which are then processed sequentially on the GPU. The dimensions of these blocks are determined based on the available GPU memory and the dimensions of the original matrices. Specifically:

- The transition matrix A is partitioned into blocks of size $b_A \times b_A$.
- The emission matrix B is partitioned into blocks of size $b_B \times M$.
- The forward and backward matrices α and β are each partitioned into blocks of size $N \times b_T$.

The block sizes b_A, b_B, and b_T are chosen to ensure that the combined memory size of active sparse matrix blocks residing in GPU memory remains lower than the available GPU memory capacity. These blocks are also transferred asynchronously between the host memory and the device memory using CUDA streams. The asynchronous transfer mechanism overlaps data transfers with computation to ensure minimizing idle GPU time while maximizing the utilization of GPU core cycles.

Sparse Matrix Representations

In many real-world scenarios, HMMs have sparse transition and emission matrices with numerous near-zero elements. To exploit this sparsity, we use *Compressed Sparse Row (CSR)* or *Compressed Sparse Column (CSC)* formats depending on the computation. These formats minimize data transfer overhead and optimize GPU memory usage by excluding redundant near-zero values, enabling more efficient processing and avoiding unnecessary computations. For a sparse matrix $X \in \mathbb{R}^{N \times M}$, the CSR representation of X is defined by three arrays:

- *Values:* An array containing all non-zero elements of X.
- *Column Indices:* An array storing the column index of each corresponding non-zero value.
- *Row Pointers:* An array of size $N + 1$, where the i-th element specifies the index in the *Values* array where the i-th row begins. The last element indicates the end of the data in the *Values* array.

To enable efficient sparse matrix computations, we utilize the *optimized sparse matrix-vector multiplication (SpMV)* routines from the CUDA cuSPARSE library. These routines are highly optimized for parallelism, ensuring efficient access to the CSR/C data structure, minimizing warp divergence by aligning threads with non-zero elements, and maximizing memory coalescing for accessing the *Values* and *Column Indices* arrays.

Memory coalescing is another GPU optimization technique that maximizes GPU memory bandwidth utilization by ensuring that threads within a warp access contiguous memory locations. This contiguous access pattern minimizes the number of separate memory transactions required to fetch data for the warp. Instead of multiple scattered accesses, coalescing combines them into a single transaction that can leverage the GPU's memory bus width. This reduction in number of transactions directly translates to improved performance for memory-bound applications. In our implementation, we leverage memory coalescing to optimize access patterns for both dense and sparse matrix representations. For dense matrices, we structure the data layout to ensure that threads within a warp access consecutive elements within a row or column, depending on the matrix's layout. For example, if the computation processes rows, threads in a warp access elements as `matrix[row][thread_id + warp_base_index]`, where `warp_base_index` is aligned to a multiple of the warp size. This alignment guarantees that all the data needed for an entire warp is fetched in a single memory transaction to minimize memory access overhead. For sparse matrix representations, we structure the layout of the *Values* and *Indices* arrays so that threads within a warp process consecutive non-zero elements in a row (or column) and their corresponding column (or row) indices in a contiguous manner.

Asynchronous Data Transfers and Kernel Execution
One of the key bottlenecks in GPU-based computation is the data transfer between the host and device memory. To address this issue, we implement asynchronous data transfers using CUDA streams that enable concurrent data movement and kernel execution for increased computational throughput. This technique facilitates computation-communication overlap which effectively reduces GPU idle time. This optimization is particularly important for dense and sparse matrix-vector multiplication operations that form the core of HMM forward, backward, and Baum-Welch training algorithms.

Our asynchronous execution strategy is implemented using a triple-buffered pipelined approach managed by CUDA streams. The pipeline allows us to overlap data transfers with computation to hide the latency of data transfers behind computation. The process is detailed below:

- *Memory Pinning*: Host memory buffers are allocated using `cudaHostAlloc()` with the `cudaHostAllocDefault` flag. This ensures the allocated memory is pinned which allows the GPU to directly access the data via Direct Memory Access during transfers.
- *Stream Creation*: Three distinct CUDA streams are created. These streams operate independently and concurrently which enable the pipelined execution.
- *Triple-Buffered Pipelined Execution*: A three-stage pipeline is implemented using the three streams.
  - *Data Transfer to Device (Stream 0)*: `cudaMemcpyAsync()` is used within stream 0 to transfer a data block from the pinned host memory to the

GPU's global memory. This transfer is non-blocking which allows subsequent operations to begin immediately. The transfer size is determined by the size of the data required for the target HMM computation step.
- *Kernel Execution (Stream 1)*: While the data transfer in stream 0 is in progress, stream 1 can start execution of the computational kernel on a previously transferred data block residing in GPU memory.
- *Result Transfer to Host (Stream 2)*: Concurrently with the kernel execution in stream 1 and the data transfer in stream 0, stream 2 transfers the computed results from GPU global memory back to a separate pinned host memory buffer using asynchronous transfer primitives.

4 Experimental Results

This section evaluates the performance of our GPU-accelerated HMM implementation against a multi-threaded CPU baseline. Hidden state sizes (N) range from 2^{18} to 2^{25} and observation lengths (T) from 2^{10} to 2^{13}. For each (N, T) pair, 10 randomly initialized HMMs were tested with average matrix sparsity of 1%. We used the 99th-percentile execution time to ensure robust comparisons and analyzed speedup and throughput (in TFLOPS) across CPU and GPU implementations. Experiments ran on a workstation system with an Intel Core i7-14700K (28 logical cores), 128 GB DDR5 RAM, and an NVIDIA A10 GPU (24 GB, 600 GiB/s bandwidth). CPU tests used Intel OneAPI with TBB and C BLAS, while GPU tests employed CUDA 12.6.85 and cuBLAS with nvcc optimizations.

Figure 1 shows the p99 execution time (log scale) of the Baum-Welch algorithm (100 iterations, sequence length 1024) as the number of hidden states increases. Due to large matrix sizes, our GPU implementation employed custom memory management techniques described earlier. Single-threaded CPU execution time grew linearly from 334.6 s at $N = 2^{18}$ to 47,350 s at $N = 2^{25}$. The 28-threaded CPU reduced this to 5625.0 s (8.4× speedup). In comparison, our GPU implementation achieved 993.7 s at $N = 2^{25}$, a 5.6× speedup over the multi-threaded CPU, demonstrating both scalability and the effectiveness of GPU acceleration for large-scale HMM inference.

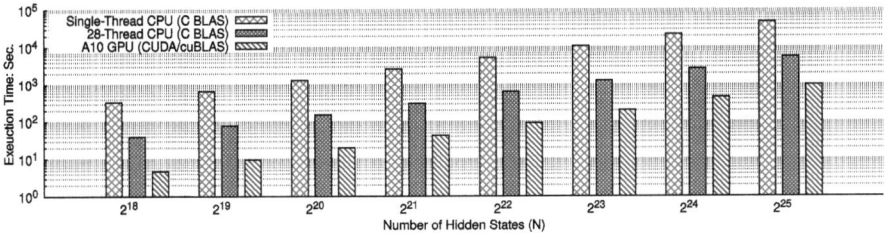

Fig. 1. Execution times (log scale) for Baum-Welch algorithm scaling (100 iterations, 1024 sequence length) with increasing hidden states for single-thread CPU, 28-thread CPU, and our proposed GPU implementations.

5 Conclusion

This work presents a GPU-accelerated Baum-Welch algorithm using CUDA cuBLAS to efficiently scale large Hidden Markov Models. Achieving up to an 8.5× speedup over multi-threaded CPU approaches, our method leverages log-space computation, memory-efficient sparse partitioning, and asynchronous data transfer. Despite notable gains, challenges in host-device transfer and memory access persist. Future directions include adaptive memory management, hardware-aware optimizations, and support for next-gen GPU architectures to enable real-time processing of large-scale HMMs.

Acknowledgements. Professor Albert Y. Zomaya would like to acknowledge the support of the **ONI Grant NI220100111**. Dr. MohammadReza HoseinyFarahabady acknowledges the continued support of The Center for Distributed and High-Performance Computing at The University of Sydney for giving access to advanced high-performance computing and cloud facilities, digital platforms, and necessary tools.

References

1. Baum, L.E., Petrie, T.: Statistical inference for probabilistic functions of finite state Markov chains. Ann. Math. Stat. **37**(6), 1554–1563 (1966). https://doi.org/10.1214/aoms/1177699147
2. Bouguila, N., Fan, W., Amayri, M.: Hidden Markov Models and Applications. Springer (2023). https://doi.org/10.1007/978-3-030-99142-5
3. Dymarski, P.: Hidden Markov Models, Theory and Applications. IntechOpen (2011). https://doi.org/10.5772/1532
4. Elliott, R.J., Aggoun, L., Moore, J.B.: Hidden Markov Models: Estimation and Control. Springer (2008). https://doi.org/10.1007/978-0-387-84854-9
5. Ibe, O.C.: Markov Processes for Stochastic Modeling, 2 edn. Elsevier (2009). https://doi.org/10.1016/B978-0-12-374451-0.X0001-7
6. Ibe, O.C.: Markov Processes for Stochastic Modeling, 2 edn. Elsevier (2013). https://doi.org/10.1016/C2012-0-06106-6
7. Mamon, R.S., Elliott, R.J.: Hidden Markov Models in Finance. Springer (2007). https://doi.org/10.1007/0-387-71163-5
8. Miller, D.R., Leek, T., Schwartz, R.M.: A hidden Markov model information retrieval system. In: Proceedings of the 22nd Annual International ACM SIGIR Conference on Research and Development in Information Retrieval, pp. 214–221 (1999). https://doi.org/10.1145/312624.312680
9. Rabiner, L.R.: A tutorial on hidden Markov models and selected applications in speech recognition. Proc. IEEE **77**(2), 257–286 (1989). https://doi.org/10.1109/5.18626
10. Schuller, B., Rigoll, G., Lang, M.: Hidden Markov model-based speech emotion recognition. In: 2003 IEEE International Conference on Acoustics, Speech, and Signal Processing, vol. 2, p. II-1 (2003). https://doi.org/10.1109/ICME.2003.1220939

Predicting Future Collaborations in a Scientific Community Using Graph Neural Networks

Nachyn Dorzhu[1], Tatiana Sukhomlinova[1], Lijing Luo[2], and Sergey Kovalchuk[1](✉)

[1] ITMO University, Saint Petersburg, Russia
kovalchuk@itmo.ru
[2] Amsterdam, The Netherlands

Abstract. Graph-based machine learning models have gained significant attention in predicting the emergence of new relationships in evolving networks. In this work, we present a study on forecasting scientific collaborations using a Graph Attention Network (GAT) with L2 regularization and dropout. We construct yearly co-authorship graphs based on historical publication data and analyze the evolution of these graphs over time with the International Conference on Computational Science (ICCS) as an example of a living scientific community. Our approach involves training on past yearly graphs to predict the formation of new edges in future graphs. We assess the model's performance by varying the prediction window and evaluating results using link prediction metrics. The proposed method demonstrates the feasibility of utilizing deep learning techniques for predicting future collaborations based on past scientific interactions.

Keywords: graph neural networks · graph attention network · link prediction · complex networks · coauthors graphs · temporal graph

1 Introduction

Scientific collaboration drives innovation, and co-authorship networks reveal how ideas spread [1, 2]. Predicting future researcher collaborations, such as grants and lab space, is key for optimizing resource allocation. Traditional link-prediction methods (e.g., common neighbors, Jaccard similarity) use static features and ignore changes in research interests and trends [3]. Most graph neural network methods also fail to capture temporal dynamics, which is essential for predicting new collaborations [4]. Graph Attention Networks (GATs) overcome this by dynamically weighing evolving neighbor influences [5]. In this study, we build temporal co-authorship graphs from the International Conference on Computational Science (ICCS)[1] publications since 2001 and use GAT-based link prediction with varying forecast windows. Our results show how temporal attention mechanisms enhance prediction accuracy and provide actionable insights for strategic academic planning.

L. Luo—Independent researcher.

[1] https://www.iccs-meeting.org/

2 Modelling Collaboration in Scientific Events

We model collaboration as a link prediction task in a co-authorship network. While focused on ICCS, the approach generalizes to other scientific domains. Here we consider the following definition of emergent scientific collaboration. Two researchers are labeled as target emergent collaboration pair appeared after an event in year Y if they: (1) don't have any publications in common prior to this event in any source (no co-authorship recorded during the years before the event ($y \leq Y$)); (2) come to the event presenting own research (publication in the event proceeding); (3) publish together in any source after the year (co-authorship recorded in $y > Y$).

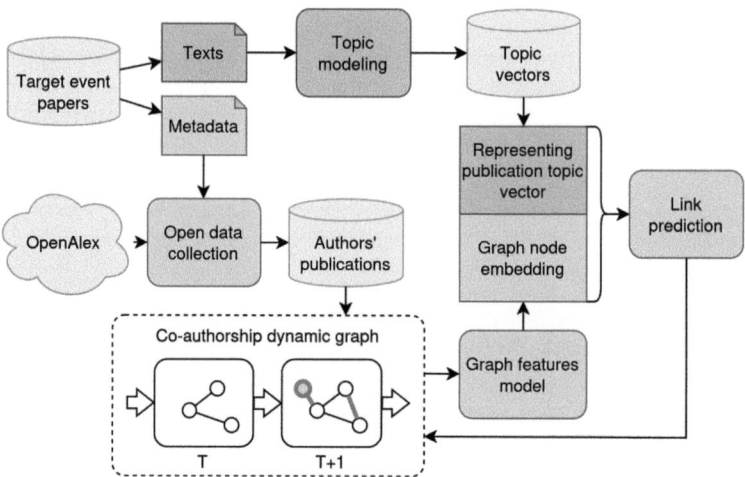

Fig. 1. General pipeline of the collaboration prediction process: (1) data acquisition from ICCS + OpenAlex, (2) yearly co-authorship graph construction, (3) emergent collaboration labeling, (4) GAT-based link prediction, and (5) evaluation.

The definition used in this study may be biased, as it does not account for other forms of collaboration, such as meetings or situations where researchers plan collaborations without publishing together (e.g., sharing affiliations). Additionally, presenting at an event under a single author's name may also obscure collaborations. However, the initial experiment assumes the primary collaboration trends are captured by this definition.

Our approach uses two main data sources (see Fig. 1). First, we utilize the ICCS publication corpus, which is structured and vectorized through topic modeling and manuscript vectorization based on topic relevance [7]. Second, we supplement this with metadata from OpenAlex[2], which provides structured information about authors and their works. We also extend topic analysis using OpenAlex topics. By using these data sources, we reconstruct a yearly co-authorship graph from ICCS history, forming a dynamic graph for link prediction. Both data sources provide features for the GNN-based model. Detailed implementation steps are provided in the following subsections.

---

[2] https://openalex.org/

The construction of co-authorship graphs is a crucial step in modeling scientific collaborations. This process involves extracting data from publication metadata and representing it as a graph, where nodes are authors and edges represent co-authored publications. Authorship data is extracted from the dataset, and co-authorship graphs are constructed yearly using NetworkX,[3] where nodes are authors and edges are shared publications.

The G_{all} graph represents a network of scientific collaborations across all years, where each node corresponds to an author, and edges between nodes denote co-authored publications. Multiple edges between the same nodes reflect repeated collaborations over different years. This graph serves as the foundation for tasks such as link prediction, temporal analysis, and network evolution modeling.

To analyze the evolution of scientific collaborations, yearly co-authorship graphs are constructed, allowing us to study how co-authorship patterns change over time and serving as the basis for link prediction experiments. The starting point for creating temporal graphs is the full co-authorship graph Gall, which includes all collaborations from all years. For each year, from 2023 to 2001, we construct cumulative graphs by excluding all publications after year Y. Each graph G_Y (for Y from 2001 to 2023) represents the cumulative collaboration network at the end of year Y. These graphs allow us to analyze the growth, stability, and evolution of the co-authorship network, including the emergence of new research clusters and future collaboration predictions based on past trends. These temporal graphs serve as the foundation for evaluating link prediction models, assessing how well historical data can forecast new scientific partnerships.

To train a machine learning model for link prediction, we need to create a dataset of positive and negative examples. Positive examples are new edges that appear in the co-authorship network from year Y to Y + 1. For each yearly graph G_Y, we compare it with G_{Y+1} to identify new edges. Each tuple (author1, author2, year) in positive examples represents a new collaboration.

Negative examples represent potential collaborations that didn't happen but could have. We sample pairs of authors who haven't collaborated yet but are in close network proximity. The dataset is balanced to ensure an equal number of positive and negative examples. These datasets will train and evaluate the graph-based neural network model (GAT) for predicting future collaborations.

The GAT model architecture captures structural and relational patterns within the co-authorship network. The input graph $G = (V, E)$ consists of nodes (authors) and edges (co-authorship relationships). The model uses an attention mechanism to assign varying importance to neighboring nodes, enhancing the representation of author interactions.

For link prediction, node embeddings are processed through multiple GAT layers to capture high-order dependencies. The final step involves combining node embeddings for a given author pair (e.g., through concatenation, element-wise multiplication, or absolute difference), followed by a fully connected layer with a sigmoid activation to generate a probability score for a future collaboration.

The training uses a binary cross-entropy loss function, comparing predicted probabilities with actual labels (positive for new collaborations, negative for non-existent links). Dropout and L2 regularization are applied to prevent overfitting and improve

---

[3] https://networkx.org/

generalization. Dropout is applied to both the attention coefficients and the hidden node representations, ensuring the model does not overly depend on specific nodes or features. Through experimentation, we found that a dropout rate of 0.3 and a weight decay of 5e-4 effectively balance overfitting and model capacity, contributing to stable and robust training.

The attention mechanism in GAT enables the model to assign different importance scores to neighboring nodes during message passing. In co-authorship networks, recurring co-authors with shared topics offer a more predictive signal. By learning attention weights, the model amplifies informative connections and suppresses noisy ones, improving link prediction performance.

We use binary cross-entropy loss and evaluate performance with ROC-AUC and F1-score. Figure 2 shows the dynamics of the selected metrics during the model training. All three curves in Fig. 2 show rapid improvement during the first 20 epochs, after which the loss stabilizes and both ROC-AUC and F1-score plateau. This indicates that the model converges early and maintains stable generalization throughout training, with no evidence of overfitting. The model achieves a stable ROC-AUC of ~0.86 and F1-score of ~0.78, indicating a reliable balance between precision and recall. These results suggest that the GAT effectively captures both thematic similarity and structural proximity in co-authorship graphs. The training loss steadily converges without overfitting, validating the regularization strategy. By leveraging these loss and evaluation metrics, we ensure that the model learns meaningful representations and generalizes well to unseen author pairs, effectively predicting future scientific collaborations.

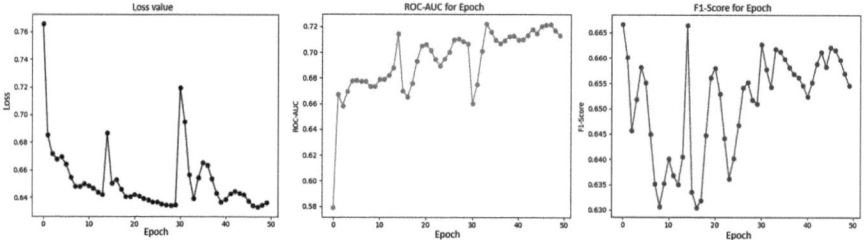

Fig. 2. Loss value (left), ROC-AUC (center), and F1 score (right) during model training.

Abstracts from ICCS proceedings are processed by cleaning, converting to TF-IDF vectors, and reducing them to 128-dimensional semantic embeddings using SVD. Simultaneously, topic identifiers for each paper are retrieved from OpenAlex, and a frequency count of each topic across an author's publications is compiled. The low-dimensional abstract embedding is then combined with this topic-frequency profile to form a rich feature vector for each author.

Co-authorships are represented as a weighted graph, where authors are nodes, and edges reflect the frequency and recency of collaborations. To predict new collaborations, a two-layer Graph Attention Network (GAT) is employed. In the first layer, each author's feature vector is projected into an intermediate representation and combined with those of neighboring authors through parallel attention mechanisms. This enables the model to focus on relevant collaborators by assigning higher weights to similar neighbors. After applying a nonlinear activation function and dropout for regularization, the enriched information flows into a second attention layer, generating a final 32-dimensional embedding for each author.

For link prediction, the embeddings of two authors are fed into a small multi-layer perceptron, which outputs the likelihood of future collaboration. The model is trained on known emerging collaborations versus random author pairs, optimized using binary cross-entropy loss. This end-to-end pipeline—combining TF-IDF reduction, topic counting, graph attention, and a lightweight classifier—provides a clear framework for understanding how shared thematic interests drive scientific collaboration.

3 Prediction of Collaboration in Computational Science Community

We retrieved metadata for all ICCS proceedings via OpenAlex and collected 16178 authors and 321848 papers. We identified 3623 "emerging" collaborations by finding author pairs who first met at ICCS (i.e., had no prior joint publications) and later co-authored any paper. The co-authorship network contains a core of 1,687 authors (65.6%) and multiple smaller clusters. The top 30 contributors show distinct collaboration patterns over time. Figure 3 visualizes this by plotting, for each year, the count of new co-author links each author brought into the network, with larger bubbles indicating more partnerships. Most core contributors join ICCS, spark fresh collaborations, and then sustain a steady rate with gradual tapering in new ties, while a few exhibit intermittent yet significant bursts of activity. Beyond this core lies a multitude of smaller components – 186 isolated author pairs and 109 clusters of 3–16 authors – representing collaborations that emerged largely independently of the main ICCS hub (Fig. 4).

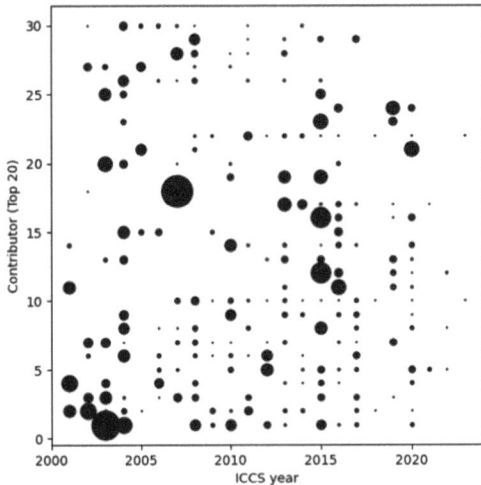

Fig. 3. Top contributors to the collaborating process.

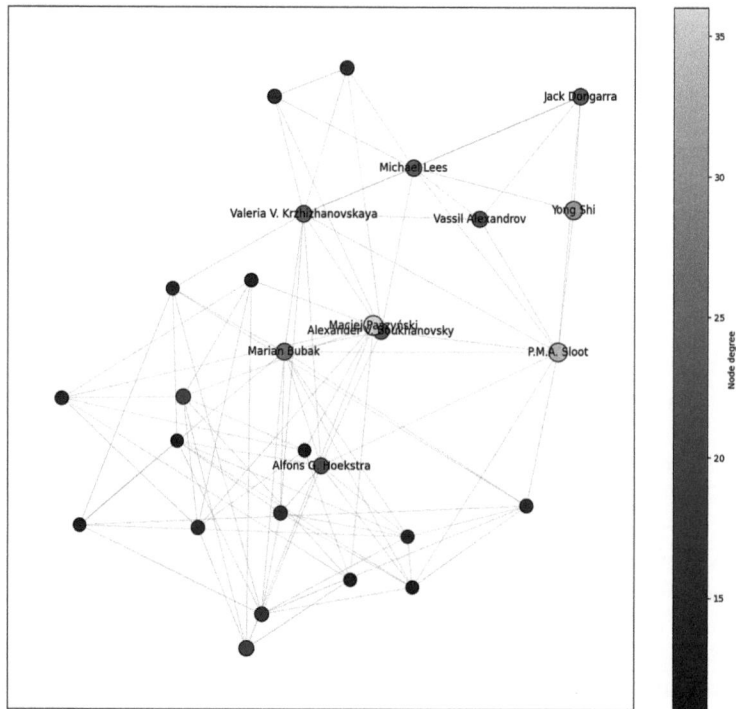

Fig. 4. Co-authorship graph: key authors and significant connections.

Topic similarity correlates with collaboration frequency: dense clusters form around Machine Learning and Data Science, while niche topics form smaller, isolated components.

A key aspect of our study is the effect of varying the time window parameter on predicting future scientific collaborations. The time window defines how much historical data is used when constructing the training graph. We experiment with different values of N, the number of years of past data included, before predicting the next year's collaborations.

By varying N, we can assess how historical data impacts prediction accuracy, such as whether a longer collaboration history improves performance or adds noise. This analysis also explores whether recent collaborations are more predictive than older ones, highlighting temporal dynamics in co-authorship patterns. The optimal time window strikes a balance between using enough historical data and avoiding outdated collaborations (see Fig. 5). A shorter N may capture recent trends more sharply, while a longer N captures long-term relationships at the cost of outdated interactions. We compare the performance of multiple training graphs using different values of N.

Evaluating the predictive performance of a model is crucial for ensuring its reliability in forecasting future collaborations. In this study, we assess the GAT-based model by comparing predicted links with actual collaborations that emerge in subsequent years. A

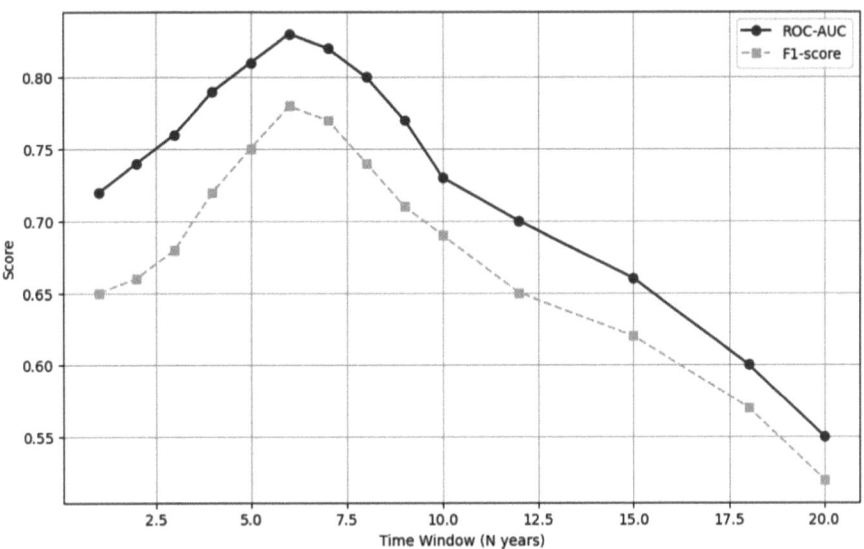

Fig. 5. Model predictive performance depending on the time window.

robust evaluation framework is used to validate the model's ability to identify meaningful connections and distinguish between likely and unlikely co-authorships.

To measure performance, multiple key evaluation metrics are employed. By considering domain-specific variations, the analysis offers insights into the model's strengths and areas for improvement, guiding refinements in model architecture, feature selection, and data preprocessing to enhance predictive accuracy across scientific networks.

The choice of the N-year time window is vital for predicting future collaborations. We conducted experiments with varying N values, from short-term (1–3 years) to long-term (over 10 years) data, and trained models on these windows. Based on AUC results, we identified 5–7 years as optimal for balancing recency and historical depth.

The results show distinct trends depending on the time window. Performance drops for very long histories ($N > 10$), even with controlled regularization. This decline is primarily due to noise from outdated collaborations, not overfitting, as no instability or divergent training loss was observed.

Using a short-term history (1–3 years) captures recent trends but overlooks long-term relationships, yielding higher precision but lower recall. A medium-term history (4–7 years) strikes a balance, incorporating both recent and long-term patterns, and provides the best performance across evaluation metrics. A long-term history (>10 years) includes outdated collaborations, reducing precision as older ties become less relevant.

Figure 5 illustrates the trade-off: short windows focus predictions narrowly, while excessive history dilutes the signal with irrelevant past interactions. An optimal time window of 5–7 years offers the best balance between recency and depth. These findings suggest that recent collaborations play a dominant role in predicting future co-authorship, while very old collaborations contribute less to accuracy.

4 Conclusions

Compared to classic heuristics, the attention mechanism in GAT more effectively captures higher-order patterns in scientific collaboration. Key limitations of the current approach include incomplete publication records and the reliance on a fixed historical window. Our experiments show that a 5–7-year time window provides the best overall performance; however, the optimal span is likely to vary across different fields. Further enhancements could consider integrating author-level features (e.g., citation metrics, transformer-based abstract embeddings), employing adaptive time-window selection, and expanding the dataset by incorporating external sources such as Google Scholar, ORCID, ArXiv, or citation networks. These improvements would help reduce data sparsity and enhance coverage, leading to better prediction accuracy and generalization.

Acknowledgments. The research was supported by the Russian Science Foundation, agreement No. 24-11-00272. https://rscf.ru/project/24-11-00272/.

References

1. Luo, L., et al.: Evolution of Computational Science Community: The Dynamics of Topics Analysis and Authors Collaboration in 24 Years of Iccs and Jocs Publications (2024). https://www.ssrn.com/abstract=5060728, https://doi.org/10.2139/ssrn.5060728
2. Fortunato, S., et al.: Science of science. Science **359**, eaao0185 (2018). https://doi.org/10.1126/science.aao0185
3. Arrar, D., Kamel, N., Lakhfif, A.: A comprehensive survey of link prediction methods. J. Supercomput. **80**, 3902–3942 (2024). https://doi.org/10.1007/s11227-023-05591-8
4. Zare, G., Jafari Navimipour, N., Hosseinzadeh, M., Sahafi, A.: Network link prediction via deep learning method: a comparative analysis with traditional methods. Eng. Sci. Technol. Int. J. **56**, 101782 (2024). https://doi.org/10.1016/j.jestch.2024.101782
5. Gu, W., Gao, F., Lou, X., Zhang, J.: Link Prediction via Graph Attention Network (2019). https://arxiv.org/abs/1910.04807, https://doi.org/10.48550/ARXIV.1910.04807
6. Soundarajan, S., Hopcroft, J.: Using community information to improve the precision of link prediction methods. In: Proceedings of the 21st International Conference on World Wide Web. pp. 607–608. ACM, Lyon France (2012). https://doi.org/10.1145/2187980.2188150
7. Luo, L., et al.: Trends in computational science: natural language processing and network analysis of 23 years of ICCS publications. In: Franco, L., de Mulatier, C., Paszynski, M., Krzhizhanovskaya, V.V., Dongarra, J.J., Sloot, P.M.A. (eds.) Computational Science – ICCS 2024. ICCS 2024. LNCS, vol. 14833, pp. 19–33. Springer, Cham (2024). https://doi.org/10.1007/978-3-031-63751-3_2

Data-Centric Parallel Programming Abstractions for High Performance Computations

Domenico Talia(✉)

University of Calabria, Via P. Bucci 41C, 87036 Rende, Italy
domenico.talia@unical.it

Abstract. This short paper describes the programming paradigm and the main constructs of the DCEx programming model designed for the implementation of data-centric large-scale parallel applications. The DCEx programming paradigm exploits private data structures and limits the amount of shared data among parallel threads in HPC applications. The key idea of DCEx is structuring programs into *data-parallel blocks* mapped on computing elements and managed in parallel by a large number of parallel tasks. Data-parallel blocks are the units of shared- and distributed-memory parallel computations and communications in the memory/storage hierarchy. Tasks execute close to data using near-data synchronization according to the PGAS model. Two use cases implemented using DCEx constructs are also outlined and performance measures on different parallel machine configurations are shown.

Keywords: Parallel programming · data-parallel applications · data-centric computational science · HPDA

1 Introduction

Computational science applications use advanced computing capabilities to model and solve complex scientific problems. To reach this goal, appropriate technologies and tools are needed. In particular, parallel computing systems and scalable data management techniques are vital. Nowadays, data-intensive scientific computing systems are widely used for many computational science applications in several domains. The ever more complex nature of the underlying computing infrastructure necessary to run large-scale use cases asks for data-oriented solutions that simplify the development, deployment, and scalable execution of complex computational tasks. Among these solutions, the scientific workflow model is a leading approach for designing and executing data-intensive applications in high-performance computing infrastructures [1].

When data-intensive applications are targeted, as occurs in high-performance data analysis (HPDA), programming frameworks need to limit task synchronization, reduce communication and remote memory access. Although traditional parallel programming tools and libraries, such as MPI, OpenMP and HPF, are being adapted to manage large datasets, we argue that the best approach is to develop parallel programming paradigms specifically designed according to a data-driven style, especially for supporting for big data analysis and machine learning on high-performance computing (HPC) systems [2]. According this approach, new languages such as X10, Legion, and Chapel, have been defined by exploiting a data-centric parallel programming approach.

This paper introduces the main features and the programming mechanisms of the Data-Centric programming model for Exascale systems (DCEx) [3] designed for the implementation of data-centric parallel applications. DCEx include programming mechanisms to improve the performance of data-intensive computations by reducing accessing, exchanging, and processing of data through the computing nodes of a parallel system. DCEx provides a workflow-based model where tasks are executed closed to input data and computation is distributed where data was generated/stored to limit data transfer overhead.

The DCEx functions are based upon data-aware operations specifically designed for data-intensive applications supporting the scalable use of a massive number of processing elements run in parallel for solving computational science applications. The DCEx model is based on private data structures and associated constructs. The goal is to exploit parallelism starting from data artifacts and limit the amount of shared data among parallel threads.

Instead of starting from parallel operations, we argue that starting from distributing data abstractions specifically defined to be operated in parallel is more appropriate in today data-intensive computations. Therefore, the basic idea of DCEx is structuring programs into *data-parallel blocks* (DPBs) that are the basic units of distributed-memory parallelism, like Resilient Distributed Datasets (RDDs) in Apache Spark, around which computation, communication, and scheduling are accomplished. Computation tasks execute close to data, using near-data synchronization based on the partitioning of data on different processing elements where tasks run in parallel. Using the data-parallel blocks, in DCEx, three main styles of parallelism are exploited: data parallelism, SPMD parallelism, and task parallelism. A prototype API based on the DCEx model has been implemented and some experimental evaluations have been performed.

The remainder of this paper is structured as follows. Section 2 presents the main features of the parallel data model used in DCEx. The parallel data block concept is introduced, and data access and processing operations are illustrated together with the associated types of parallelism. Section 3 briefly illustrate two real use cases developed by means of the programming mechanisms of DCEx, showing performance results. Finally, Sect. 4 concludes the paper.

2 A Data-Centric Parallel Model

Scientific and business applications are becoming more and more data intensive; therefore, data are increasingly playing a centrale role in complex applications. For this reason, the function of data is considered fundamental in the DCEx programming model. In fact, as mentioned in the introduction, the data-centric model used in DCEx is based on the *data-parallel block* (DPB) concept, which defines data structures that are partitioned and distributed on different computing nodes where they are handled in parallel.

Data-parallel blocks are more general data-oriented mechanisms and provide a higher abstraction level with respect to Spark RDDs.

2.1 Data-Parallel Blocks

Data blocks and their message queues are mapped onto tasks to be managed in parallel and are placed in memory/storage units by the DCEx runtime. A DPB is manipulated for managing a composite data element in the main memory of multiple computing nodes. Decomposing a program in terms of block-parallelism, instead of process-parallelism, enables mapping blocks during the program execution among different processors in a parallel computer and execute tasks where data partitions are. This is the main idea that lets us integrate in- and out-of-core programming in the same model. In particular, a DPB **datapb** can be composed of multiple partitions:

```
datapb = [part0][part1][part2]...[partn-1]
```

where each partition is assigned to a given computing node.

Notation **datapb[i]** refers to the i-th partition of DPB **datapb**. However, when a DPB is simply referred by its name (e.g., **datapb**) in a computing node (e.g., the i-th node), it is intended as a reference to the locally available partition (e.g., **datapb[i]**).

A DPB can be created using the **data.get()** operation, which loads into main memory some existing data from secondary storage. This operation is specified as follows:

```
datapb = data.get(source, [format], [part|repl]) at [Cnode|Carea];
```

where the data source, its format, the optional partitioning or replication on a computing node (**Cnode**) or, better, on a computing area (**Carea**) composed of a set of computing nodes, are its parameters.

DCEX also includes the **data.declare()** operation, used to declare a DPB that will be created at run time as a result of a task execution. A DPB can be also written from main memory of processors/cores, where the block has been managed, to secondary storage using the **data.set()** operation, which syntax is as follows:

```
data.set(datapb, dest, [format]);
```

In this case, all partitions are collected in parallel form the processor memories and moved to compose the secondary storage object.

Using these three basic operations, partitions can be mapped on different processing nodes where each task will work in parallel on a given partition. This approach allows computing nodes to manage in parallel the data partitions at each core/node using a set of operations or library APIs that hide the complexity of the underlying actions.

2.2 Parallel Operations

The two main types of computing abstractions defined in DCEx to be coupled with parallel data blocks are:

- *computing nodes* and *computing areas* that specify single processing elements or regions of processors of a parallel machine where to store data and run tasks.
- *tasks* and *task pools* that embody the units of parallelism in the model.

The corresponding DCEx constructs defined to refer to these abstractions are: **Cnode** for a single computing node and **Carea** for a region including a set of computing elements. For instance, a **Cnode** can be declared as follows:

```
nod = Cnode([hardware annotation parameters]);
```

where as an example of **Carea** composed of a two-dimensional array of 30 × 40 computing nodes can be defined as follows:

```
nar = Carea(30,40);
```

where 30 is the number of rows and 40 is the number of columns of the computing area. A single element of this area can be referred as **nar[8][10]** and a sub-area can be also defined like, for example, **na2 = Carea(nar,6,6);** which extracts a 6 × 6 matrix of computing nodes from the **Carea** defined by the variable **nar**.

Concerning tasks management constructs, the programming model allows for expressing parallelism using two concepts: *Task* and *Task pool*. A task can be defined as follows:

```
t=Task(func,func_params) [at Cnode|Carea] [on failure ignore|retry];
```

to execute a function on a given computing node or on a node of a computing area. The **on failure** is an optional directive for specifying an action (for instance, **ignore** or **retry** with it) to be performed in case of task failure. Task pool abstraction is defined to implement SPMD parallelism representing a set of tasks that execute the same function. The basic syntax for declaring a pool of tasks is as follows:

```
tp = Task_Pool([size]);
```

where **tp** identifies the task pool and **size** is an optional parameter specifying the number of tasks in the pool. This statement declares a task pool but does not spawn its execution. Tasks in the pool can be activated explicitly using a **for** loop as in the following example:

```
N = 40;
nodes = Carea(N);
for (i=0; i<N; i++) {
    func_param_1 = x;
    tp[i] = Task(func_name, func_param_1) at nodes[i];
}
```

By exploiting the parallel abstractions of DPBs, tasks, and computing areas, DCEx implements three main types of parallelism that can be combined to develop complex parallel applications: *Task parallelism*, *data parallelism*, and *SPMD parallelism*.

Task parallelism is exploited when different tasks that compose an application run in parallel. It is data driven since data dependencies are used to decide when tasks can be spawn in parallel. As input data of a task are ready its code can be executed. Data parallelism is achieved when the same code is executed in parallel on different data blocks or on partitions of a data block. In SPMD parallelism, differently from data parallelism, tasks cooperate to exchange partial results during execution. In DCEx, these three types of parallelism are combined with the features of the Partitioned Global Address Space (PGAS) model [5] that supports the definition of several execution contexts based on separate address spaces that compose a global address space. For any given task, this allows for the exploitation of memory locality and affinity, providing programmers with a well-define way to distinguish between private and shared memory blocks.

To give a quick example of how locality can be exploited in DCEx using data parallel blocks and computing areas, we can define the computing area `ca`, which includes 8 computing nodes where two data parallel blocks (`df1` and `df2`), storing data coming from two files `f1` and `f2`, can be mapped by splitting them in four partitions each (see Fig. 1).

3 Two Real-World Use Cases

Different real-world applications have been developed by using the DCEx abstractions integrated with the GrPPI language [4], such as urban computing dynamics, parallel neuroimaging, and deep learning for anomaly detection in electric vehicles [6]. The two we discuss here have been developed to analyze diffusion-weighted magnetic resonance imaging data and trajectory discovery from social data analysis.

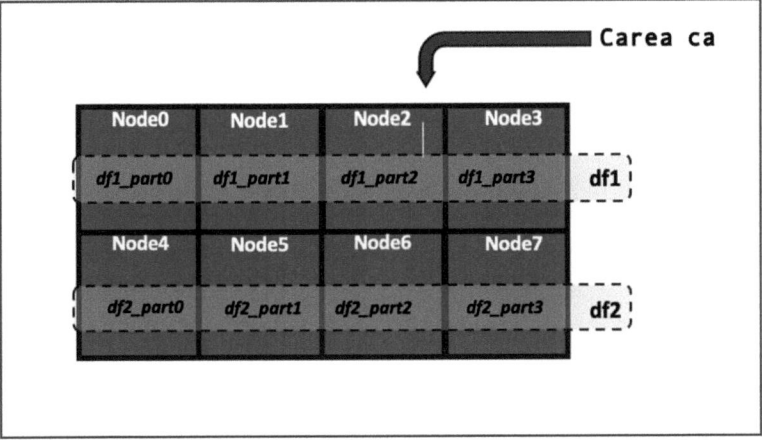

Fig. 1. Two data-parallel blocks storing two files (*f1* and *f2*) partioned on 8 computing nodes.

Diffusion-weighted magnetic resonance imaging (DWI) aims to obtain unique metrics for the study of brain white matter microstructure and structural connectivity. DWI data are four-dimensional images composed from tens to hundreds of three-dimensional brain images. Each image acquisition is composed of the signal of around 1 million volumetric pixels (voxels). This leads to DWI images ranging from hundreds to thousands of MB for each patient. The DWI workflow has been implemented using the DCEx abstractions to orchestrate the execution of different parallel processing steps consisting of Python scripts [4]. The processing steps of the DCEx workflow included processing commands from different state-of-the-art neuroimaging toolboxes.

The trajectory discovery application has been developed for extracting frequent patterns from large volumes of geotagged data gathered from social media. The main steps of the application are as follows: (a) parallel crawling, (b) parallel data filtering, (c) automatic keywords extraction and data grouping, (d) Regions of Interest (RoIs) extraction through a parallel clustering algorithm, and (e) Trajectory mining. This final step is based on a highly parallel versions of the FP-Growth (frequent itemset analysis) and Prefix-Span (sequential pattern mining) algorithms.

3.1 Experimental Evaluations

The experiments carried out to evaluate the DWI application have been performed on an Intel-based cluster with Xeon processors with 128 GB of RAM memory each. Data have been shared using NFS and GlusterFS filesystems with a 10 Gbps network. Performance results are calculated by averaging five consecutive runs and are compared with a baseline implementation based on the use of the Python package Nipype [7], the SLURM cluster and multithreaded execution.

Table 1 summarizes the execution times using four different parallel configurations ranging from 48 to 648 cores. The DCEx results are compared with those obtained with the Nipype/Slurm implementation. The DCEx based solutions are significantly faster than the one based on Nipype under Slurm.

Table 1. DCEx/ Nipype execution time (in minutes) using different parallel machine setting.

Cores	Nypype/NFS	Nypype/GlusterFS	DCEx/NFS	DCEx/GlusterFS
48	74.00	74.12	57.50	54.37
192	18.87	18.56	12.41	11.29
336	12.72	12.53	8.89	7.13
684	8.21	8.09	6.74	5.38

From this performance experiments, we also noted the effect of data locality in DCEx under NFS and GlusterFS. Strong data caching at data nodes benefits the overall execution time of a workflow execution of the same subject. In the best case, data locality results on 25% of improvement using 684 cores.

Concerning the trajectory discovery application, a set of experiments has been run for evaluating turnaround time and speedup using different number of cores and different sizes of datasets to be analyzed. Figure 2 shows the main results.

Figure 2(a) shows the turnaround times (in seconds) of the application for the three datasets we considered (D1, D2, D4), using from 8 to 64 cores. For the smallest dataset

(D1) that contains 1.2 GB of data, the turnaround time decreases from 4 h 10 m using 8 cores to 58 min using 64 cores. For D2 (2.4 GB) the turnaround time decreases from 6 h 21 m to about 1 h 5 m minutes. For D4 (4.8 GB) the turnaround time decreases from 9 h 32 m to 1 h 20 m.

Figure 2(b) reports the speedup obtained by analyzing the different datasets from 8 up to 64 CPU cores. For dataset D1 the speedup is close to linear up to 16 cores, while it slightly decreases to 3.5 and 4.6 with 32 and 64 cores, respectively. For datasets D2 and D4, the application reached a better speedup, which is close to linear. It is worth to note that the speedup increases as the size of the dataset increases. This means that CPU cores are better used when larger datasets are analyzed.

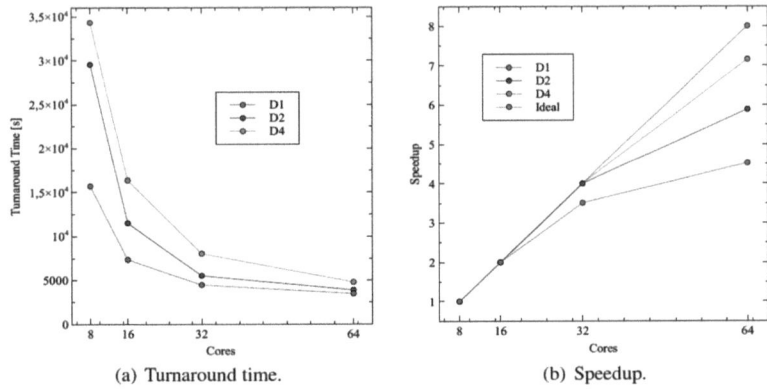

Fig. 2. Turnaround time and speedup of the trajectory discovery application in DCEx.

4 Conclusions

Traditional parallel programming languages are not specifically designed for developing data-intensive applications in science and engineering. With the advent of Big Data, machine learning, and generative artificial intelligence, new programming models, languages, and APIs are needed to combine parallel data abstractions with scalability and performance for extreme data processing [8].

To streamline the development of computational science applications in HPC systems, large-scale data- and task-parallelism techniques have to be developed on top of the data-parallel abstractions divided into many partitions mapped on different computing elements where local tasks process them. This approach allows for processing in parallel the data partitions at each core/node using a set of statements/library calls that hide the complexity of the underlying operations. Since data dependency in this scenario may limit scalability, data-centric abstractions help programmers to avoid or limit it to a local/neighbors scale. Scalability of large data analysis, machine learning and AI applications are closely related to the management of parallelism in the data-driven operations needed in the applications and the limitation of overhead created by data processing mechanisms and techniques.

In this paper we illustrated DCEx, as a data-aware parallel programming paradigm for data intensive computational science applications. The designed DCEx programming model includes data-parallel blocks and data-driven parallel tasks for the implementation of scalable algorithms and applications on top of HPC computers, with a special emphasis on the support of massive data analysis applications.

DCEx provides a workflow-based programming model that enables to set up a data-oriented life cycle management, allowing parallel data locality and data affinity. Moreover, the DCEx implementation offers a runtime system that controls and optimizes the execution of the component-based use-cases and applications. We described here the language features and reported on the experimental evaluation of two significant use cases. Results show that DCEx achieves good performance and scalability.

Machine learning and generative AI are going to play a key role in computational science applications as the analysis of scientific data is integrating traditional simulation approaches. AI and machine learning algorithms are becoming powerful tools to accelerate simulations, extract patterns from data, and enhance scientific discovery by bridging data-driven methods with traditional modeling strategies. To support this new trend, scalable parallel programming models and tools using a data-centric approach for exploiting parallelism in data analysis are vital.

Acknowledgments. This study was partially funded by the EU ASPIDE project (grant number 801091) and the "National Centre for HPC, Big Data and Quantum Computing", (grant number CN00000013- CUP H23C22000360005). Author also acknowledges the contribution of colleagues who worked on the DCEx model in the framework of the ASPIDE project.

Disclosure of Interests. The author has no competing interests to declare that are relevant to the content of this article.

References

1. Ejarque, J., et al.: Enabling dynamic and intelligent workflows for HPC, data analytics, and AI convergence. Futur. Gener. Comput. Syst. **134**, 414–429 (2022)
2. Talia, D., Trunfio, P., Marozzo, F., Belcastro, L., Cantini, R., Orsino, A.: Programming Big Data Applications: Scalable Tools and Frameworks for Your Needs. World Scientific Press, London (2024)
3. Garcia-Blas, J., et al.: Convergence of HPC and Big Data in extreme-scale data analysis through the DCEx programming model. In: IEEE 34th International Symposium on Computer Architecture and High Performance Computing (SBAC-PAD) 2022, IEEE, pp. 130–139, Bordeaux (2022)
4. Garcia-Blas, J., del Rio, D., Garcia, J.D., Carretero, J.: Exploiting stream parallelism of MRI reconstruction using GrPPI over multiple back-ends. In: Workshop on Clusters, Clouds and Grids for Life Sciences, CCGRID-Life 2019, CCGRID 2019, IEEE, Larnaca, Cyprus (2019)
5. Stitt, T.: An introduction to the partitioned global address space programming model. CNX.org (2010)
6. Marchesi, A., et al.: Paradigms and Models at Run-time - Final Report D2.6. ASPIDE Project (2021)

7. Gorgolewski, K., et al.: Nipype: a flexible, lightweight and extensible neuroimaging data processing framework in python. Front. Neuroinform. **5**, 13 (2011)
8. Talia, D.: A view of programming scalable data analysis: from clouds to exascale. J. Cloud Comput. **8**, 4 (2019)

Neural Network for Evaluating the Operational Range of Antennas with Randomly Generated Designs

Bartosz Czaplewski

Department of Teleinformation Networks, Faculty of Electronics, Telecommunications and Informatics, Gdansk University of Technology, Narutowicza 11/12, 80-233 Gdansk, Poland
bartosz.czaplewski@pg.edu.pl

Abstract. This paper introduces a novel machine learning-based methodology to determine the operational range of planar microstrip antennas of randomly generated designs, removing the need for electromagnetic (EM) simulations or expert knowledge. Framed as a multi-label classification task, the proposed approach addresses the inefficiencies of traditional methods, which are prone to high computational cost and engineer's bias. The method quickly identifies promising designs, paving the way for subsequent optimization. This advancement represents a significant step toward automating antenna design processes.

Keywords: Machine Learning · Neural Networks · Planar Microstrip Antennas

1 Introduction

Machine learning (ML) techniques have been widely used in communications, including antenna selection, malicious event detection, and mobility prediction. Applications such as SVM-based speech recognition and context-aware Internet of Things (IoT) show their flexibility. Deep learning has also impacted UAVs [1], THz communication [2], Wi-Fi [3], GPS [4], satellites [5], and IoT [6, 7].

Antennas are essential for efficient signal transmission. Traditional design involves manual, iterative steps—substrate selection, shape definition, and parametric tuning—which are time-consuming, suboptimal, and prone to human bias. Automated, specification-driven methods use optimization algorithms [8–11] to improve designs but face challenges like geometry selection, dimensionality, and computation cost.

Automatically generated antennas follow two main models: compositions of basic shapes (e.g., rectangles or triangles) [12, 13] and coordinate-based representations (e.g., splines or line segments) [14–16]. This work focuses on the latter, which allows flexibility but introduces issues: self-intersecting shapes, initializing geometry, and high dimensionality. Metaheuristics [16, 17] partially address these but remain limited by electromagnetic (EM) simulation costs.

With no exact formulae for antenna topology, initial designs rely on heuristic or literature-based shapes—again introducing bias. Accurate evaluation requires EM solvers, which are costly. Surrogate models [14, 18] help, but dimensionality and bias persist [19]. More on this topic can be found in [20–24].

ML can accelerate designing process by reducing simulation time and predicting antenna behavior. This paper introduces an ML-based framework to evaluate random designs, identify promising geometries for the desired antenna operating range for further optimization, and minimize reliance on EM simulations. It handles designs with 103 parameters, going beyond traditional methods and engineering bias.

The goal is to build an AI-driven system for antenna design and analysis. The system aims to: (1) evaluate random designs without EM simulations, (2) generate designs via generative models, (3) predict antenna responses, and (4) optimize designs via AI. This will enable high-performance, unbiased, fully automated antenna design.

This article addresses the first part of the full system: determining an antenna's operating range from a random design, without EM simulation or expert input. Random generation can yield novel, effective designs. However, most are unsuitable, and filtering them traditionally requires expert knowledge or costly EM simulations. This work proposes a fast ML-based method to estimate the operational range of random designs, enabling quick identification of viable candidates for further optimization.

The contributions are: (1) a methodology for estimating operational range of antennas with random designs via multi-label classification, (2) dataset preparation steps, (3) two neural network architectures—one high-accuracy, one lightweight, (4) training process details to facilitate reproducibility, (5) extensive experimental results.

2 Related Work

In [25, 26], ML was applied to optimize antenna design, assuming an initial design based on expert knowledge, with the operational frequency range predetermined. This paper addresses the earlier stage—automating the search for an initial design without expert knowledge, which can later be optimized using methods like those in [25, 26].

The work in [27, 28] introduced an ML-based antenna synthesis method in three stages: parameter prediction, antenna type classification (e.g., rectangular, horn), and design synthesis. A decision tree classifier was used in [27], while stacking ensemble learning was applied in [28]. The success of neural networks in classification influenced the approach of this paper.

In [29], a survey addressed the regression problem of estimating antennas' frequency responses from designs, using MSE for evaluation. This paper reframes it as a multi-label classification problem, making direct comparison infeasible. In [29], linear regression, support vector regression, polynomial regression, neural networks, and genetic algorithms, were compared demonstrating that neural networks are the most promising approach, which guided the choice of techniques used in this research.

An overview of methods for optimizing antenna designs through regression analysis for various types, including microstrip and patch antennas, was provided in [30]. While direct comparisons with the proposed method are not possible, methods such as support vector machines, Bayesian regularization, and neural networks discussed therein may prove valuable for future research.

Comprehensive reviews of ML methods applied to antenna design and optimization are available in [31, 32], offering broad insights into this field.

3 Dataset

Planar microstrip antennas are analog devices, and their reflection characteristics over a given frequency range are obtained via EM simulations. An antenna is considered suitable for further optimization if its reflection coefficient is below $\theta_R = -3$ dB and its relative bandwidth exceeds $\theta_W = 0.1$ This study introduces a dataset of 106,351 pseudo-random antenna designs with labels indicating whether each sub-band (ten 0.5 GHz sub-bands within 3–8 GHz range) meets both conditions, marking it as suitable for operation and further optimization. Examples are presented in Fig. 1.

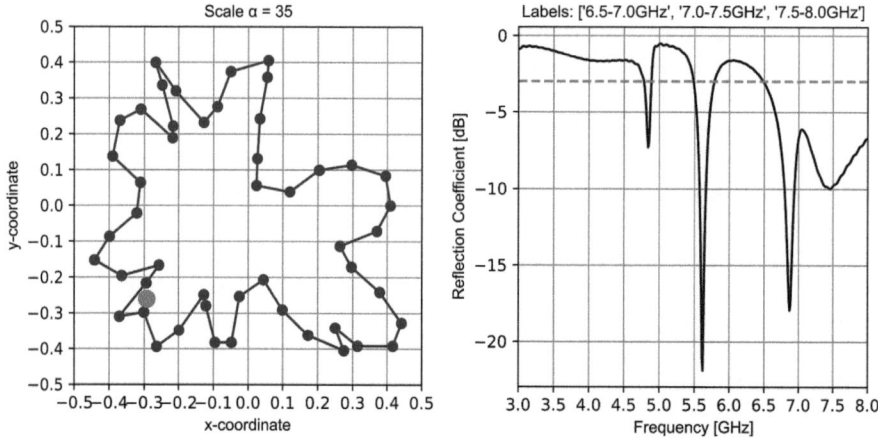

Fig. 1. Example of antenna design (blue – vertices, red – feed point) and antenna response. (Color figure online)

A design is a 103-element vector representing an antenna: $\mathbf{D} = [\alpha, q_x, q_y, \mathbf{v}_x, \mathbf{v}_y]$, where $\alpha \in \langle 24, 36 \rangle$ is an integer scaling factor, q_x, q_y are feed point coordinates, and $\mathbf{v}_x$, $\mathbf{v}_y$ are 50-element vectors of normalized vertex coordinates ($\in <-1, +1>$). Physical dimensions are given by $\alpha \cdot [\mathbf{v}_x, \mathbf{v}_y]$ and $\alpha \cdot [q_x, q_y]$ in mm. Scaling by α shifts resonance frequencies [18, 34]. Designs were generated quasi-randomly [10, 11], ensuring no self-intersections and a valid connection between the feed point and the shape.

A label vector is a 10-element vector indicating the antenna's operational range:
$\mathbf{B} = [b_g \in \{0,1\}]$, $g = 1,\ldots,10$, where $b_g = 1$ indicates the assignment of the g-th label. The set of labels represents ranges from 3.0 GHz to 8.0 GHz, in 0.5 GHz intervals.

To assign labels to the designs, antenna reflection responses $\mathbf{R}$ [dB] were computed for the frequency range $\mathbf{F}$ [GHz] (see Fig. 1) using EM simulations for 20,000 designs $\mathbf{D}$ with $\alpha = 30$. Simulations were performed using CST Microwave Studio [33], an EM solver based on the Finite Integration Technique [9–11]. To avoid additional costly simulations, responses for other α values were estimated using scaling [18, 34]. This can be seen as data augmentation. The computational cost and slight errors in the response amplitude of this interpolation method are negligible. A label vector $\mathbf{B}$ for a given design $\mathbf{D}$ was then assigned using the following algorithm.

1. Compute the set of frequency ranges $\mathbf{Z}$ for which $r_i \leq \theta_R$:

$$\mathbf{Z} = \{z_k = [f_{Lk}, f_{Uk}]\}, \ k = 1, \ldots, m, \tag{1}$$

$$L_k = \min\{i | r_i \leq \theta_R \wedge (i = 1 \vee r_{i-1} > \theta_R)\}, \ i = 1, \ldots, n, \ k = 1, \ldots m, \tag{2}$$

$$U_k = \max\{i | r_i \leq \theta_R \wedge (i = n \vee r_{i+1} > \theta_R)\}, \ i = 1, \ldots, n, \ k = 1, \ldots m, \tag{3}$$

where f_{Lk} and f_{Uk} are the lower and upper bounds for the k-th frequency range.

2. Compute the set of central frequencies $\mathbf{C}$ for each range $[f_{Lk}, f_{Uk}]$ as:

$$\mathbf{C} = \{f_{Ck} = (f_{Lk} + f_{Uk})/2, \forall [f_{Lk}, f_{Uk}] \in \mathbf{Z}\}, k = 1, \ldots m. \tag{4}$$

3. Compute the set of relative widths $\mathbf{W}$ for each frequency range in $\mathbf{Z}$ as:

$$\mathbf{W} = \{w_k = |f_{Uk} - f_{Lk}|/f_{Ck}, \forall [f_{Lk}, f_{Uk}] \in \mathbf{Z}, f_{Ck} \in \mathbf{C}\}, \ k = 1, \ldots m. \tag{5}$$

4. Define the label set $\mathbf{A}$ as frequency ranges:

$$\mathbf{A} = \{a_g = [a_{Lg}, a_{Ug}]\}, g = 1, \ldots, 10 = \{[3.0, 3.5], \ldots, [7.5, 8.0]\} \ [\text{GHz}], \tag{6}$$

where a_g is the g-th label, a_{Lg} and a_{Ug} are the bounds for the g-th label.

5. Compute the set of labels $\mathbf{T} \subseteq \mathbf{A}$ that the design must be tagged as:

$$\mathbf{T} = \{a_g | w_k \geq \theta_W \wedge ((a_{Lg} < f_{Ck} < a_{Ug}) \vee (f_{Lk} < a_{Lg} \wedge a_{Ug} < f_{Uk}))\}, \tag{7}$$

where $\theta_W \in\ <0,1>$ is the minimal acceptable relative width threshold.

6. Compute the binary label vector $\mathbf{B}$ as: $\mathbf{B} = [b_g = \mathbf{1_T}(a_g)], g = 1, \ldots, 10$, where $\mathbf{1_T}(b_g)$ is the indicator function.

An antenna can operate across multiple subranges or have several disjoint ranges; thus, it is a multi-label classification problem, not a multi-class one. Each design in the dataset can be tagged with multiple labels. The dataset is available in [35].

4 Learning Parameters and Model

The proposed method takes a 103-parameter antenna design vector $\mathbf{D}$ as input and produces a 10-parameter label prediction vector $\mathbf{P}$ as output. Before entering the model, the design parameters are normalized to the $<0,1>$ range. The model outputs probability scores $\mathbf{S} = [s_g \in\ <0,1>], g = 1,\ldots,10$, where s_g represents the likelihood that the design corresponds to the g-th label. Label predictions $\mathbf{P}$ are obtained by thresholding $\mathbf{S}$ at θ_P = 0.5: $\mathbf{P} = [p_g = 1 \text{ if } s_g > \theta_P \text{ else } 0], g = 1, \ldots, 10$.

The dataset was split into training set (90%) and test set (10%). The training set was used for 10-fold cross-validation. Metrics such as training loss, training accuracy, validation loss, and validation accuracy were recorded, and average values across all folds were computed. The best model, with the lowest validation loss, was selected and tested on the test set, where overall and label-specific accuracy were calculated.

Early stopping, based on validation loss, was applied to prevent overfitting and reduce training time by halting when no improvement was observed. L2 regularization with weight decay was also used to discourage large weights, promoting generalization and simpler models. A learning rate scheduler dynamically adjusted the learning rate to ensure stable convergence and avoid overshooting.

Initial learning parameters were chosen based on the author's experience, and later fine-tuned using grid search. The parameters in Table 1 yielded the best results. The experiments were run on Python 3.11.0 with PyTorch 2.0.1+ cu118, utilizing Nvidia GeForce GTX 1080 Ti GPU and CUDA driver 12.2.

Table 1. Learning parameters.

Learning parameter	Value	Learning parameter	Value
Batch size	64	LR scheduler	ReduceLROnPlateau
Loss function	BCE	LR scheduler factor	0.1
Solving algorithm	Adam	LR scheduler patience	5
Initial LR	1e−3	Max epochs	1000
Weight decay	1e−7	Early stopping patience	15

Table 2. Structures of the proposed models.

NN#1 (129,327,626 parameters)	NN#2 (29,814,794 parameters)
Fully Connected (FC) layer, 103 neurons	Fully Connected (FC) layer, 103 neurons
FC, 8192 neurons, LeakyReLU, 0.06 dropout	FC, 4096 neurons, LeakyReLU, 0.06 dropout
FC, 8192 neurons, LeakyReLU, 0.06 dropout	FC, 4096 neurons, LeakyReLU, 0.06 dropout
FC, 4096 neurons, LeakyReLU, 0.06 dropout	FC, 2048 neurons, LeakyReLU, 0.06 dropout
FC, 4096 neurons, LeakyReLU, 0.06 dropout	FC, 2048 neurons, LeakyReLU, 0.06 dropout
FC, 2048 neurons, LeakyReLU, 0.06 dropout	FC, 10 neurons, Sigmoid
FC, 1024 neurons, LeakyReLU, 0.06 dropout	
FC, 512 neurons, LeakyReLU, 0.06 dropout	
FC, 10 neurons, Sigmoid	

Due to the small number of input and output parameters and the multi-label classification nature, a classical neural network (NN) model, specifically a multi-layer perceptron, was chosen. Two models were trained: NN#1 aimed for the highest accuracy, while NN#2 sought similar accuracy with fewer learnable parameters. Various NN architectures were tested, differing in layers, neurons, activations, weights initialization, dropout, and more. The best results were obtained with the structures in Table 2.

The model architecture was designed with input and output layers matching the dataset features. The number of hidden layers and neurons balanced model complexity and performance, as more layers and neurons increase capacity but also the risk of overfitting and computational cost. Various activation functions were tested, with LeakyReLU showing the best performance. Weights were initialized using the Kaiming uniform distribution. Dropout layers were added to reduce overfitting by randomly deactivating neurons during training, promoting diverse feature extraction. The output layer used a sigmoid activation for class probability interpretation.

5 Results

Table 3 compares results for models NN#1 and NN#2. Key metrics include label-specific accuracy (per-label performance) and overall accuracy (their average). The table also lists average runtime per sample, covering the entire pipeline: preprocessing, tensor conversion, GPU transfer, classification, and result extraction. For reference, a single EM simulation takes ~ 60–90 s per design.

NN#1 reached 93.55% overall accuracy, slightly outperforming NN#2 (93.31%). Despite this, NN#2 uses 100 million fewer parameters, offering similar accuracy with lower memory usage and computational cost—beneficial for large-scale use. It should be noted that the models assess whether designs are suitable starting points for optimization. Refining their frequency responses remains outside this work's scope.

Table 3. Results of the experiments.

Model	NN#1	NN#2
Avg. Training accuracy	91.76%	87.82%
Avg. Validation accuracy	89.54%	85.36%
Training accuracy (best fold)	**97.33%**	**97.56%**
Validation accuracy (best fold)	**93.36%**	**93.04%**
Test label-specific accuracy	[96.91, 94.77, 94.80, 95.51, 95.90, 93.50, 91.18, 91.39, 91.93, 89.65] %	[96.79, 95.04, 95.20, 95.75, 95.83, 93.45, 90.52, 90.39, 91.00, 89.15] %
Test overall accuracy	**93.55%**	**93.31%**
Avg. Runtime per sample	0.0026 s	0.0012 s

6 Conclusions

This work introduced a method for estimating operational frequency ranges for planar microstrip antennas with randomly generated designs, bypassing EM simulations and expert input. Framing the task as multi-label classification enables fast, low-cost identification of designs suitable for further optimization.

Future work includes testing on diverse datasets, exploring various random design algorithms, and developing a 2D-CNN classifier due to the spatial nature of antenna topologies. Additional plans involve creating a label-to-design generator via an autoencoder with classification support, predicting antenna responses without EM simulations, and building ML-based optimization methods to replace numerical approaches.

Acknowledgments. This work was supported in part by the National Science Centre of Poland (Grant 2021/43/B/ST7/01856).

Disclosure of Interests. The author has no competing interests to declare that are relevant to the content of this article.

References

1. Kouhdaragh, V., Verde, F., Gelli, G., Abouei, J.: On the application of machine learning to the design of UAV-based 5G radio access networks. Electronics **9**(4), 689 (2020)
2. Hung, Y.C., Yang, S.H.: Terahertz deep learning computed tomography. In: 2019 44th International Conference on Infrared, Millimeter, and Terahertz Waves (IRMMW-THz), pp. 1–2. IEEE (2019)
3. Shah, S.A., et al.: Privacy-preserving non-wearable occupancy monitoring system exploiting Wi-Fi imaging for next-generation body centric communication. Micromachines **11**(4), 379 (2020)
4. Orabi, M., Khalife, J., Abdallah, A.A., Kassas, Z.M., Saab, S.S.: A machine learning approach for GPS code phase estimation in multipath environments. In: 2020 IEEE/ION Position, Location and Navigation Symposium (PLANS), pp. 1224–1229. IEEE (2020)
5. Lei, L., Lagunas, E., Yuan, Y., Kibria, M.G., Chatzinotas, S., Ottersten, B.: Deep learning for beam hopping in multibeam satellite systems. In: 2020 IEEE 91st Vehicular Technology Conference (VTC2020-Spring), pp. 1–5. IEEE (2020)
6. Wagih, M., Hilton, G.S., Weddell, A.S., Beeby, S.: Millimeter-wave power transmission for compact and large-area wearable IoT devices based on a higher order mode wearable antenna. IEEE IoT J. **9**(7), 5229–5239 (2022)
7. Gaya, S., Hamza, A., Sokunbi, O., Sheikh, S.I.M., Attia, H.: Electronically switchable frequency and pattern reconfigurable segmented patch antenna for Internet of Vehicles. IEEE IoT J. **11**(10), 17840–17851 (2024)
8. Bod, M., Hassani, H.R., Taheri, M.M.S.: Compact UWB printed slot antenna with extra Bluetooth, GSM, and GPS bands. IEEE Ant. Wirel. Prop. Lett. **11**, 531–534 (2012)
9. Bekasiewicz, A., Kurgan, P., Koziel, S.: Numerically efficient miniaturization-oriented optimization of an ultra-wideband spline-parameterized antenna. IEEE Access **10**, 21608–21618 (2022)
10. Bekasiewicz, A., Dzwonkowski, M., Dhaene, T., Couckuyt, I.: Specification-oriented automatic design of topologically agnostic antenna structure. In: Computational Science – ICCS 2024, LNCS, vol. 14834, pp. 11–18. Springer (2024)
11. Bekasiewicz, A., Askaripour, K.: Performance comparison of automatically generated topologically agnostic patch antennas. In: 2024 25th International Microwave and Radar Conference (MIKON) (2024). https://doi.org/10.23919/MIKON60251.2024.10633953
12. Jacobs, J.P.: Accurate modeling by convolutional neural-network regression of resonant frequencies of dual-band pixelated microstrip antenna. IEEE Ant. Wirel. Prop. Lett. **20**(12), 2417–2421 (2021)

13. Wu, G.-B., Zeng, Y.-S., Chan, K.F., Chen, B.-J., Qu, S.-W., Chan, C.H.: High-gain filtering reflectarray antenna for millimeter-wave applications. IEEE Trans. Ant. Prop. **68**(2), 805–812 (2020)
14. Bekasiewicz, A., Koziel, S., Plotka, P., Zwolski, K.: EM-driven multi-objective optimization of a generic monopole antenna by means of a nested trust-region algorithm. App. Sci. **11**(9), art no. 3958 (2021)
15. Whiting, E.B., Campbell, S.D., Mackertich-Sengerdy, G., Werner, D.H.: Dielectric resonator antenna geometry-dependent performance tradeoffs. IEEE Open J. Ant. Prop. **2** (2021)
16. Lizzi, L., Viani, F., Azaro, R., Massa, A.: Optimization of a spline-shaped UWB antenna by PSO. IEEE Ant. Wirel. Prop. Lett. **6**, 182–185 (2007)
17. Alroughani, H., McNamara, D.A.: The shape synthesis of dielectric resonator antennas. IEEE Trans. Ant. Prop. **68**(8), 5766–5777 (2020)
18. Koziel, S., Bekasiewicz, A.: Multi-objective design of antennas using surrogate models. World Scientific (2016)
19. Koziel, S., Ogurtsov, S.: Multi-objective design of antennas using variable-fidelity simulations and surrogate models. IEEE Trans. Ant. Prop. **61**(12), 5931–5939 (2013)
20. Koziel, S., Bekasiewicz, A.: Fast multi-objective design optimization of microwave and antenna structures using data-driven surrogates and domain segmentation. Eng. Comput. **37**, 735–788 (2019)
21. Koziel, S., Bekasiewicz, A.: Domain segmentation for low-cost surrogate-assisted multi-objective design optimization of antennas. IET Microw. Antennas Propag. **12**(10), 1728–1735 (2018)
22. Koziel, S., Bekasiewicz, A.: Comprehensive comparison of compact UWB antenna performance by means of multiobjective optimization. IEEE Trans. Antennas Propag. **65**(7), 3427–3436 (2017)
23. Koziel, S., Bekasiewicz, A.: Rapid simulation-driven multiobjective design optimization of decomposable compact microwave passives. IEEE Trans. Microw. Theory Tech. **64**(8), 2454–2461 (2016)
24. Koziel, S., Bekasiewicz, A.: Multi-objective optimization of expensive electromagnetic simulation models. Appl. Soft Comput. **47**, 332–342 (2016)
25. Gampala, G., Reddy, C.J.: Fast and intelligent antenna design optimization using machine learning. In: 2020 International Applied Computational Electromagnetics Society Symposium (ACES) (2020). https://doi.org/10.23919/ACES49320.2020.9196193
26. Wu, Q., Chen, W., Yu, C., Wang, H., Hong, W.: Machine-learning-assisted optimization for antenna geometry design. IEEE Trans. Antennas Propag. **72**(3) (2024). https://doi.org/10.1109/TAP.2023.3346493
27. Ramasamy, R., Bennet, M.A.: An efficient antenna parameters estimation using machine learning algorithms. Prog. Electromagn. Res. C **130**, 169–181 (2023). https://doi.org/10.2528/PIERC22121004
28. Shi, D., Lian, C., Cui, K., Chen, Y., Liu, X.: An intelligent antenna synthesis method based on machine learning. IEEE Trans. Antennas Propag. **70**(7) (2022). https://doi.org/10.1109/TAP.2022.3182693
29. ElAbbasi, M.K., Madi, M., Kabalan, K.Y.: Optimization of antenna dimensions for improved frequency response using machine learning: a survey. In: 2024 6th International Conference on Communications, Signal Processing, and their Applications (ICCSPA) (2024)
30. El Misilmani, H.M., Naous, T.: Machine learning in antenna design: an overview on machine learning concept and algorithms. In: 2019 International Conference on High Performance Computing & Simulation (HPCS) (2019)
31. Khan, M.M., Hossain, S., Mozumdar, P., Akter, S., Ashique, R.H.: A review on machine learning and deep learning for various antenna design applications. Heliyon **8**(4), e09317 (2022). https://doi.org/10.1016/j.heliyon.2022.e09317

32. Sarker, N., Podder, P., Mondal, M.R.H., Shafin, S.S., Kamruzzaman, J.: Applications of machine learning and deep learning in antenna design, optimization, and selection: a review. IEEE Access **11** (2023). https://doi.org/10.1109/ACCESS.2023.3317371
33. CST Microwave Studio, Dassault Systems, 10 rue Marcel Dassault, CS 40501, Vélizy-Villacoublay Cedex, France (2015)
34. Koziel, S., Yang, X.S. (Eds.): Computational Optimization, Methods and Algorithms. Studies in Computational Intelligence, Springer-Verlag (2011)
35. Czaplewski, B.: 1Ddesign-to-label-1.8.3 - Antenna Designs with Operating Frequency Range Labels (1.8.3). Gdańsk University of Technology, Faculty of Electronics, Telecommunications and Informatics (2025). https://doi.org/10.5281/zenodo.14982975

Optimizing U-Net Architecture Using Differential Evolution for Brain Tumor Segmentation

Shoffan Saifullah[1,2(✉)] and Rafał Dreżewski[1]

[1] Faculty of Computer Science, AGH University of Krakow, 30-059 Krakow, Poland
`{saifulla,drezew}@agh.edu.pl`
[2] Department of Informatics, Universitas Pembangunan Nasional Veteran Yogyakarta, 55281 Yogyakarta, Indonesia
`shoffans@upnyk.ac.id`

Abstract. Accurate brain tumor segmentation is essential for effective diagnosis and treatment planning. This study proposes DE-UNet, an enhanced U-Net architecture optimized using Differential Evolution (DE) to improve segmentation of multimodal MRI scans. The model was evaluated on two benchmark datasets: Figshare Brain Tumor Segmentation (FBTS) and BraTS 2021 datasets, focusing on whole tumor segmentation across four MRI modalities: FLAIR, T1, T1-CE, and T2. DE-UNet outperformed state-of-the-art methods, achieving Dice Similarity Coefficient (DSC) and Jaccard Index (JI) scores of 0.9160/0.8472 on FBTS and 0.9094/0.8371 on BraTS 2021. DE effectively optimized key hyperparameters—learning rate, dropout, batch size, and filter sizes—enhancing the model generalization across tumor types and imaging conditions. Visual analysis confirmed accurate tumor boundary delineation. These results highlight the potential of DE-UNet as a robust and precise tool for clinical brain tumor segmentation.

Keywords: Brain Tumor Segmentation · Differential Evolution · MRI Modalities · U-Net Optimization · Medical Image Analysis

1 Introduction

Accurate brain tumor segmentation is crucial for diagnosis, treatment planning, and prognosis [15]. MRI is widely used due to its superior soft tissue contrast and non-invasive nature [3], but manual segmentation is time-consuming, subjective, and inconsistent [9]. Tumor heterogeneity further complicates the task, emphasizing the need for reliable automated methods [7].

U-Net and its variants have shown strong performance in medical image segmentation [19,21], but limitations persist in hyperparameter tuning, handling class imbalance, and adapting to multimodal MRI. Recent models like DeepLabV3+ [23] and U-Net extensions [22] improve performance but remain suboptimal for multi-class and multi-modal cases. Metaheuristics such as

PSO [20] and GA [10] aid tuning but often converge prematurely [6], whereas Differential Evolution (DE) offers more robust and adaptive optimization [12,16].

This study introduces DE-UNet, a U-Net architecture optimized via DE to improve segmentation of multimodal MRI brain tumor images. Evaluated on the FBTS and BraTS 2021 datasets, it targets whole tumor segmentation across classes and modalities, achieving superior Dice Similarity Coefficient (DSC) and Jaccard Index (JI) compared to state-of-the-art methods. Section 2 details the proposed framework and datasets, Sect. 3 presents results and comparisons, and Sect. 4 concludes the study.

2 Methods

The proposed DE-UNet was evaluated on two benchmark datasets (Fig. 1): the Figshare Brain Tumor Segmentation (FBTS) [5] and BraTS 2021 [2], selected for their diversity in tumor types and MRI modalities. FBTS includes 3064 slices labeled as Meningioma (708), Glioma (1426), and Pituitary (930), each with expert-annotated binary masks. BraTS 2021 comprises 1251 multimodal slices across T1, T1-CE, T2, and FLAIR, with corresponding whole tumor masks.

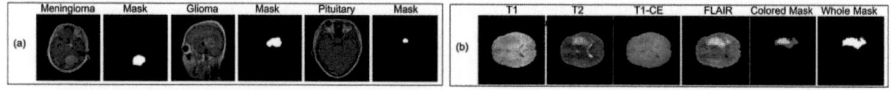

Fig. 1. Sample images from the datasets: (a) FBTS and (b) BraTS 2021.

All images were resized to 256 × 256 pixels using bicubic interpolation [21], and intensities were normalized to [0, 1] to reduce scanner-induced variability. Segmentation masks were binarized to separate tumor from background. This standardized preprocessing enhanced contrast, reduced noise, and ensured consistent inputs for training and evaluation.

DE-UNet integrates a U-Net backbone with Differential Evolution (DE) to optimize four hyperparameters: learning rate, dropout rate, batch size, and number of filters. The architecture follows an encoder-decoder structure with skip connections and bottleneck dropout [20], as shown in Fig. 2.

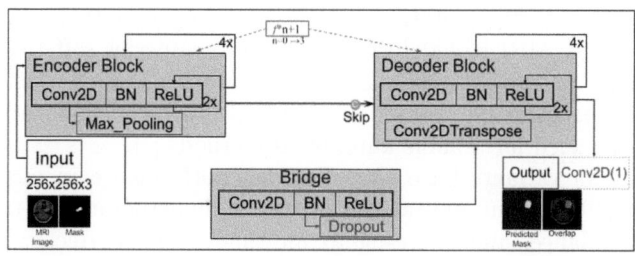

Fig. 2. DE-UNet: Encoder-decoder architecture with skip connections.

Algorithm 1 DE Algorithm for U-Net hyperparameter optimization.

Require: NP, F, CR, G
Ensure: Optimized hyperparameter configuration x^*
1: Initialize a population of NP candidate solutions within predefined bounds.
2: **for** generation $g = 1$ to G **do**
3:     **for** each candidate solution x_i in the population **do**
4:        **Mutation:** Randomly select 3 distinct solutions (x_a, x_b, x_c) from the population.
5:        Generate mutant vector: $v_i = x_a + F \cdot (x_b - x_c)$
6:        **Crossover:** Generate trial vector u_i as:

$$u_{ij} = \begin{cases} v_{ij} & \text{if rand}(0,1) < CR \text{ or } j = j_{\text{rand}} \\ x_{ij} & \text{otherwise} \end{cases}$$

7:        **Selection:** Replace x_i with u_i if $f(u_i) < f(x_i)$
8:     **end for**
9: **end for**
10: Return the best-performing solution x^*

DE was configured with a population size of $NP = 5$, mutation factor $F = 0.8$, crossover probability $CR = 0.9$, and a maximum of $G = 10$ generations. Continuous parameters (learning rate: $[10^{-5}, 10^{-2}]$, dropout: $[0.1, 0.5]$) were mutated within bounds, while discrete parameters (batch size: 8, 16, 32, 64; filters: 16, 32, 64, 128) were rounded post-mutation. The hyperparameter ranges were selected based on prior studies [4,20] and validated empirically. Optimization was performed separately on the FBTS and BraTS 2021 datasets.

The search minimized a composite loss: $\mathcal{L}_{\text{total}} = 1 - \alpha \cdot \text{DSC} - (1 - \alpha) \cdot \text{JI}$, where $\alpha = 0.5$ balances overlap accuracy (Dice Similarity Coefficient, DSC) and boundary agreement (Jaccard Index, JI). This objective encourages robust segmentation across tumor types and modalities. The DE operations—mutation, crossover, and selection—are detailed in Algorithm 1, which iteratively returns the best-performing configuration x^* for final training and evaluation.

DE-UNet was evaluated on the FBTS and BraTS 2021 datasets using a server with 8 NVIDIA A100-SXM4-40GB GPUs. All images and masks were resized to 256 × 256 pixels, converted to three-channel format, and normalized to [0, 1]. An 80/20 training-validation split was applied. During the DE search, each candidate configuration was trained for 10 epochs to evaluate validation performance.

The model was trained using the Adam optimizer and binary cross-entropy loss, with DE optimizing the learning rate, dropout rate, batch size, and filter size. After selecting the best configuration, final training and evaluation were performed on both datasets.

Model performance was assessed using Accuracy, Dice Similarity Coefficient (DSC), and Jaccard Index (JI) [18–20]. Higher DSC and JI indicate stronger overlap between predicted and ground truth masks. Evaluation was performed per tumor class in FBTS (Meningioma, Glioma, Pituitary) and per modality in BraTS 2021 (FLAIR, T1, T1-CE, T2) to assess robustness and generalization.

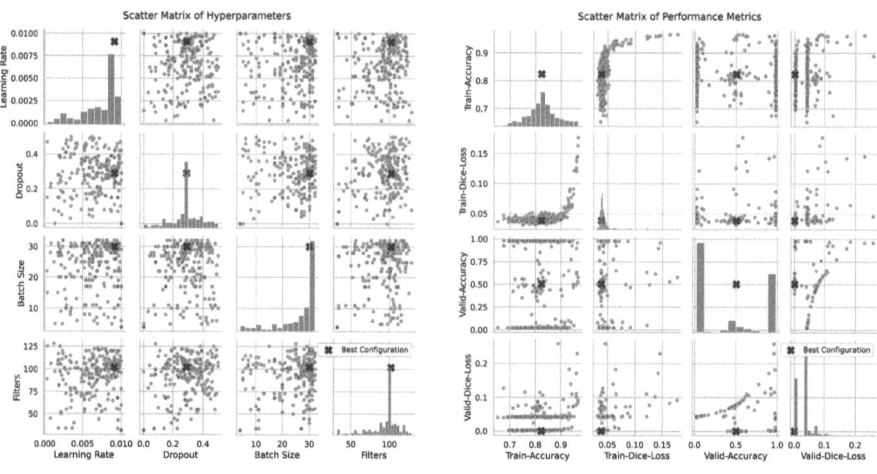

(a) Scatter matrix of hyperparameters.   (b) Scatter matrix of evaluation metrics.

Fig. 3. Effect of DE-tuned hyperparameters on model performance (red: best). (Color figure online)

3 Results and Discussion

Differential Evolution (DE) was applied to optimize four U-Net hyperparameters: learning rate, dropout rate, batch size, and initial filter count. Performance was assessed using training and validation accuracy and Dice-based loss.

Figure 3 visualizes the influence of these hyperparameters on the performance. The hyperparameter matrix (Fig. 3a) shows that learning rate and dropout were the most sensitive, directly impacting the convergence. Batch size and filter count affected the training stability and efficiency. The performance matrix (Fig. 3b) highlights how small hyperparameter shifts can yield notable gains in validation accuracy and Dice scores.

The optimal configuration—learning rate 0.009094, dropout 0.286, batch size 30, and 102 filters—achieved the training accuracy of 0.9302, the validation accuracy of 0.9788, and the validation Dice loss of 0.0024. Among all parameters, the learning rate and the dropout had the strongest impact on generalization, while batch size and filter count mainly influenced the training dynamics.

The DE-optimized U-Net was evaluated on the FBTS and BraTS 2021 datasets to assess segmentation performance across diverse tumor types and MRI modalities. Table 1 presents the results on the FBTS dataset. The model achieved consistently high accuracy, Dice Similarity Coefficient (DSC), and Jaccard Index (JI) across all tumor classes. Meningioma achieved the best DSC (0.9348), while Glioma—despite its irregular morphology—maintained strong performance. Pituitary tumors exhibited stable, near-perfect accuracy and overlap.

Table 1. Performance metrics on the FBTS dataset.

Tumor Type	Training				Validation			
	Accuracy	Loss	DSC	JI	Accuracy	Loss	DSC	JI
Meningioma	0.9983	0.0042	0.9286	0.8677	0.9984	0.0038	0.9348	0.8784
Glioma	0.9971	0.0070	0.9023	0.8231	0.9968	0.0080	0.8943	0.8103
Pituitary	0.9991	0.0022	0.9183	0.8509	0.9991	0.0021	0.9200	0.8539

Table 2. Performance on BraTS 2021 before and after extended training.

Modality	Training				Validation			
	Accuracy (pp)	Loss (pp)	DSC (pp)	JI (pp)	Accuracy (pp)	Loss (pp)	DSC (pp)	JI (pp)
FLAIR	0.9956 (+0.18)	0.0110 (−0.47)	0.8941 (+4.31)	0.8103 (+7.21)	0.9961 (+0.15)	0.0095 (−0.37)	0.9068 (+3.31)	0.8304 (+5.69)
T1	0.9935 (+0.32)	0.0157 (−0.79)	0.8464 (+7.58)	0.7353 (+12.12)	0.9930 (+0.39)	0.0168 (−0.96)	0.8327 (+9.07)	0.7154 (+14.3)
T1-CE	0.9950 (+0.34)	0.0119 (−0.82)	0.8823 (+8.05)	0.7900 (+13.86)	0.9940 (+0.43)	0.0147 (−1.06)	0.8576 (+10.37)	0.7524 (+17.34)
T2	0.9946 (+0.28)	0.0135 (−0.72)	0.8707 (+6.65)	0.7714 (+11.1)	0.9942 (+0.34)	0.0147 (−0.89)	0.8602 (+7.97)	0.7550 (+13.23)

Table 2 summarizes the performance on BraTS 2021 before and after extended training. Improvements are shown as percentage point (pp) gains. T1 and T2 exhibited the largest relative gains, particularly in DSC and JI. T1-CE achieved the highest overall segmentation accuracy (DSC: 0.9613, JI: 0.9258), highlighting the benefit of contrast-enhanced imaging. These results validate DE-UNet's effectiveness across MRI modalities and tumor structures. These findings are consistent with prior studies showing that modality contrast significantly influences segmentation quality [20].

The DE-optimized U-Net was evaluated on the FBTS and BraTS 2021 test sets using Accuracy, Loss, Dice Similarity Coefficient (DSC), and Jaccard Index (JI). Table 3 summarizes the model's final test performance. On the FBTS dataset, Meningioma achieved the highest DSC (0.9410) and JI (0.8895), while Pituitary showed the best accuracy (0.9991). Glioma remained more challenging due to its irregular morphology, resulting in the lower DSC (0.8922).

Table 3. Test performance of the DE-optimized U-Net model.

Class	Accuracy	Loss	DSC	JI	Modality	Accuracy	Loss	DSC	JI
Meningioma	0.9985	0.0035	0.9410	0.8895	FLAIR	0.9976	0.0058	0.9447	0.8956
Glioma	0.9966	0.0083	0.8922	0.8069	T1	0.9969	0.0073	0.9277	0.8657
Pituitary	0.9991	0.0020	0.9148	0.8451	T1-CE	0.9983	0.0041	0.9634	0.9297
					T2	0.9976	0.0058	0.9399	0.8873

For BraTS 2021, the T1-CE modality achieved the highest segmentation performance (DSC: 0.9634, JI: 0.9297), benefiting from enhanced tumor contrast. FLAIR and T2 also performed well in delineating edema regions. T1 showed comparatively lower overlap metrics but remained effective in less complex cases.

Figure 4 presents qualitative segmentation results, comparing ground truth (red contours) with predicted masks (green contours). The model produced high-quality segmentations across all tumor types and modalities, with T1-CE and FLAIR showing the closest alignment. Despite Glioma complexity, prediction masks demonstrated strong overlap with expert annotations. These results confirm the model's robustness in handling modality-specific and tumor-type variations. DE-UNet adapts well to anatomical complexity, achieving reliable segmentations suitable for clinical diagnostic support.

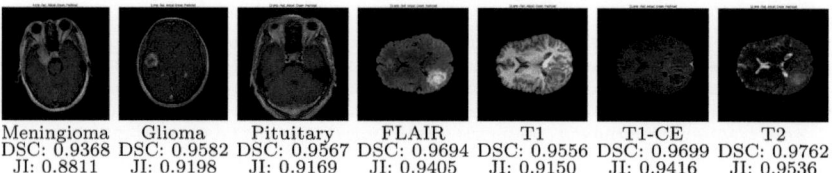

Fig. 4. Qualitative results of DE-UNet (red: ground truth, green: prediction). (Color figure online)

Table 4. Comparison of DE-UNet with state-of-the-art models.

Method	FBTS Dataset		Method	BraTS 2021	
	DSC	JI		DSC	JI
Proposed DE-UNet	**0.9160**	**0.8472**	**Proposed DE-UNet**	**0.9094**	**0.8371**
DeepLabV3+Xception [23]	0.8115	0.8018	UNet [8]	0.8600	0.7807
KFCM-CNN [14]	0.8884	0.8204	U-Net base [25]	0.9080	–
U-Net based [1]	0.8900	0.8100	SPPNet-2 [25]	0.9040	–
MST-based [13]	0.8469	0.7443	UNCE-NODE [17]	0.8949	–
U-Net with ResNet [11]	0.9011	–	nnU-Net [24]	0.8900	–

Table 4 compares the proposed DE-UNet against state-of-the-art (SOTA) methods on the FBTS and BraTS 2021 datasets. DE-UNet outperformed all baselines across Dice Similarity Coefficient (DSC) and Jaccard Index (JI), demonstrating superior segmentation accuracy.

On the FBTS dataset, DE-UNet achieved the DSC of 0.9160 and the JI of 0.8472, outperforming the previous best model, U-Net with ResNet [11] (DSC: 0.9011). For BraTS 2021, DE-UNet scored 0.9094 in DSC and 0.8371 in JI, exceeding prior models such as UNet [8] and modular approaches such as UNCE-NODE [17]. These improvements reflect the impact of differential evolution (DE) in dynamically tuning learning rate, dropout, batch size, and filter count, unlike traditional static or manually configured models.

DE-UNet's modality-agnostic architecture and deeper encoder-decoder design contribute to its robust performance across MRI modalities (FLAIR,

T1, T1-CE, T2). The model captures fine tumor boundaries without requiring modality-specific preprocessing or architectural modifications. These strengths enhance generalization to diverse tumor characteristics and support clinical applicability, positioning DE-UNet as a competitive and practical solution for brain tumor segmentation.

4 Conclusions

This study proposed DE-UNet, a U-Net architecture enhanced with Differential Evolution (DE) for optimized brain tumor segmentation. By automatically tuning key hyperparameters—learning rate, dropout, batch size, and filters—DE-UNet achieved superior performance on the FBTS and BraTS 2021 datasets, with DSC of 0.9160 and 0.9094, respectively. The model demonstrated strong generalization across MRI modalities (FLAIR, T1, T1-CE, T2) and tumor types, outperforming several state-of-the-art methods. Visual results confirmed accurate tumor boundary alignment with expert annotations. These findings highlight the effectiveness of DE-driven optimization in medical image segmentation. Future work will explore hybrid metaheuristics and validation on larger, multimodal datasets to enhance clinical applicability.

Acknowledgement. Research funding was provided by AGH University of Krakow (Program "Excellence initiative – research university"), ACK Cyfronet AGH (Grant no. PLG/2024/017503), and Polish Ministry of Science and Higher Education funds assigned to AGH University of Krakow.

References

1. Akter, A., et al.: Robust clinical applicable CNN and U-Net based algorithm for MRI classification and segmentation for brain tumor. Expert Syst. Appl. **238**, 122347 (2024). https://doi.org/10.1016/j.eswa.2023.122347
2. Baid, U., et al.: RSNA-ASNR-MICCAI-BraTS-2021 Dataset (2023). https://doi.org/10.7937/jc8x-9874
3. Batool, A., Byun, Y.C.: Brain tumor detection with integrating traditional and computational intelligence approaches across diverse imaging modalities – challenges and future directions. Comput. Biol. Med. 108412 (2024). https://doi.org/10.1016/j.compbiomed.2024.108412
4. Biswas, S., et al.: Improving differential evolution through bayesian hyperparameter optimization. In: 2021 IEEE Congress on Evolutionary Computation (CEC), pp. 832–840. IEEE (2021). https://doi.org/10.1109/CEC45853.2021.9504792
5. Cheng, J.: brain tumor dataset. Figshare (2017). https://doi.org/10.6084/m9.figshare.1512427.v5
6. Gad, A.G.: Particle swarm optimization algorithm and its applications: a systematic review. Arch. Comput. Methods Eng. **29**(5), 2531–2561 (2022). https://doi.org/10.1007/s11831-021-09694-4
7. Ghadimi, D.J., et al.: Deep learning-based techniques in glioma brain tumor segmentation using multi-parametric MRI: a review on clinical applications and future outlooks. J. Magn. Reson. Imaging (2024). https://doi.org/10.1002/jmri.29543

8. Hernandez-Gutierrez, F.D., et al.: Brain tumor segmentation from optimal MRI slices using a lightweight U-Net. Technologies **12**(10), 183 (2024). https://doi.org/10.3390/technologies12100183
9. Kaifi, R.: A review of recent advances in brain tumor diagnosis based on AI-based classification. Diagnostics **13**(18), 3007 (2023). https://doi.org/10.3390/diagnostics13183007
10. Khouy, M., Jabrane, Y., Ameur, M., Hajjam El Hassani, A.: Medical image segmentation using automatic optimized U-Net architecture based on genetic algorithm. J. Personal. Med. **13**(9), 1298 (2023). https://doi.org/10.3390/jpm13091298
11. Kumar Sahoo, A., Parida, P., Muralibabu, K., Dash, S.: Efficient simultaneous segmentation and classification of brain tumors from MRI scans using deep learning. Biocybern. Biomed. Eng. **43**(3), 616–633 (2023). https://doi.org/10.1016/j.bbe.2023.08.003
12. Kuş, Z., et al.: Differential evolution-based neural architecture search for brain vessel segmentation. Eng. Sci. Technol. Int. J. **46**, 101502 (2023). https://doi.org/10.1016/j.jestch.2023.101502
13. Mayala, S., et al.: Brain tumor segmentation based on minimum spanning tree. Front. Signal Process. **2** (2022). https://doi.org/10.3389/frsip.2022.816186
14. Rao, S.K.V., Lingappa, B.: Image analysis for MRI based brain tumour detection using hybrid segmentation and deep learning classification technique. Int. J. Intell. Eng. Syst. **12**(5), 53–62 (2019). https://doi.org/10.22266/ijies2019.1031.06
15. Rasool, N., Bhat, J.I.: A critical review on segmentation of glioma brain tumor and prediction of overall survival. Arch. Comput. Methods Eng. (2024). https://doi.org/10.1007/s11831-024-10188-2
16. Ren, L., et al.: Multi-level thresholding segmentation for pathological images: optimal performance design of a new modified differential evolution. Comput. Biol. Med. **148**, 105910 (2022). https://doi.org/10.1016/j.compbiomed.2022.105910
17. Sadique, M.S., et al.: Brain tumor segmentation using neural ordinary differential equations with UNet-Context encoding network. In: Bakas, S., et al. (eds.) Brainlesion: Glioma, Multiple Sclerosis, Stroke and Traumatic Brain Injuries. BrainLes 2022, pp. 205–215. LNCS, Springer, Cham (2023). https://doi.org/10.1007/978-3-031-33842-7_18
18. Saifullah, S., Dreżewski, R.: Brain tumor segmentation using ensemble CNN-transfer learning models: deepLabV3plus and ResNet50 approach. In: Franco, L., et al. (eds.) Computational Science – ICCS 2024, LNCS, vol. 14835, pp. 340–354. Springer, Cham (2024). https://doi.org/10.1007/978-3-031-63772-8_30
19. Saifullah, S., Dreżewski, R.: Improved brain tumor segmentation using modified U-Net based on Particle Swarm Optimization image enhancement. In: Proceedings of the Genetic and Evolutionary Computation Conference Companion, pp. 611–614. GECCO '24 Companion, Association for Computing Machinery, New York, NY, USA (2024). https://doi.org/10.1145/3638530.3654339
20. Saifullah, S., et al.: Automatic brain tumor segmentation using convolutional neural networks: u-net framework with PSO-tuned hyperparameters. In: Affenzeller, M., et al. (eds.) Parallel Problem Solving from Nature – PPSN XVIII, LNCS, vol. 15150, pp. 333–351. Springer, Cham (2024). https://doi.org/10.1007/978-3-031-70071-2_21
21. Saifullah, S., et al.: Modified U-Net with attention gate for enhanced automated brain tumor segmentation. Neural Comput. Appl. (2024). https://doi.org/10.1007/s00521-024-10919-3

22. Saifullah, S., et al.: Optimizing brain tumor segmentation through CNN U-Net with CLAHE-HE image enhancement. In: Proceedings of the 2023 1st International Conference on Advanced Informatics and Intelligent Information Systems (ICAI3S 2023), pp. 90–101 (2024). https://doi.org/10.2991/978-94-6463-366-5_9
23. Saifullah, S., et al.: Advanced brain tumor segmentation using DeepLabV3Plus with Xception encoder on a multi-class MR image dataset. Multimed. Tools Appl. (2025). https://doi.org/10.1007/s11042-025-20702-8
24. Sørensen, P.J., et al.: Repurposing the public BraTS dataset for postoperative brain tumour treatment response monitoring. Tomography **10**(9), 1397–1410 (2024). https://doi.org/10.3390/tomography10090105
25. Vijay, S., et al.: MRI brain tumor segmentation using residual spatial pyramid pooling-powered 3D U-Net. Front. Public Health **11** (2023). https://doi.org/10.3389/fpubh.2023.1091850

Simulation Modeling of Clinical Decision Making for Personalized Policy Identification

Ashish T. S. Ireddy(✉) and Sergey V. Kovalchuk

ITMO University, Saint Petersburg, Russia
{ireddy,kovalchuk}@itmo.ru

Abstract. With Human – Artificial Intelligence (AI) collaboration booming in all fields, the pace of task-based cooperation is ever-expanding. Yet, in most applications, AI induction is sidelined to test beds and is perceived skeptically as a competitor rather than a collaborator. The healthcare domain is one field where AI support is viewed as theoretical and far from practical. While most focus is directed towards developing and training AI models, the human expert and their interactions with the AI model are often overlooked. We present an experiment that incorporates, the personalization of human experts into the AI's model training, aiming to improve collaboration and overall outcome. Using a simulation-based approach, we optimize the AI learning policy of a domain expert's behaviour when evaluating decision support data of a patient's risk of acquiring type 2 diabetes mellitus (T2DM). With Linear and Maximum Entropy inverse reinforcement learning (IRL) algorithms, we analyze various learning strategies by including context, rewards and sampling rates to show personalized expert characteristics with optimal policies and effective reward functions respectively. Our results provide insights into experts' personalized evaluation policy and the AI model's learning behaviour in various environmental scenarios and further the implicit difference in the evaluation of domain experts.

Keywords: Inverse Reinforcement Learning · Decision Support Systems · Behaviour Optimization · Policy Simulation · Diabetes Mellitus

1 Introduction

Artificial intelligence - the world of recent times partially revolves around it. From minor hints to complex solutions, the use of AI assistance in the general domain workflow is ever-expanding. The healthcare domain is one of many fields that has seen a rise of clinical decision support systems (CDSS) incorporated into a quasi-practical state where recommendations provide valuable insight to the decision maker therefore aiding in achieving a better outcome. While most decision support systems are based on offline training using historical observations, few approaches account for the remaining aspects of decision support, i.e. interaction with experts, environment variables, information context, perceptional states and the human expert [7]. Context is one aspect that is overlooked during training, the inclusion of which can improve the outcome with perceptional states [4]. However, in instances where we have only an expert's demonstration of performing a task, IRL can be an effective approach to recover the behaviour and implicit decision-making policy of the human user [1]. Together with the inclusion of context, background and perceptional states, we can improve recommendations from CDSS to be more aligned with the domain expert's solution and scenario. Further, the type of training for the AI model (i.e. offline or online) has been given little attention when working with expert demonstrations. Our work aims to simulate the AI model's learning curve and investigate the difference in policy behaviour of doctors from various specializations via IRL in a clinical decision-making scenario. We present a simulation-based approach to model the domain expert's (medical professional) personalized policy when evaluating recommendation data of a prediction model, to assess the patient's risk of acquiring type 2 diabetes mellitus (T2DM). Medical professional's evaluation is acquired through subjective metrics (understandability, agreement and usability). We perform personalization, using Maximum entropy (MaxEnt) and Linear Inverse Reinforcement Learning (IRL) to extract the underlying reward functions and the real optimal policies. We simulate the personalization of policies on three groups of data, comprising individual doctors, by specialization and a global dataset, to show the sensitivity of respective reward functions via an entropy measure. Using context, strategy of evaluation and information levels, we scrutinize our observations to reveal behavioural patterns of medical professionals on real-world data. Our results provide a collective insight into identifying personalized policies based on expert behaviour. Further, the paper is structured as follows: Sect. 2 introduces the methodology of modelling CDSS for personalized policies. Section 3 shows the interpretation of policies from IRL, Sect. 4 investigates results after simulation. Section 5 is the conclusion.

2 Modeling Personalization in Clinical Decision Making

This section introduces our approach to simulating and modelling expert personalization using IRL algorithms. We define notations that will be used throughout the paper. A set of n (finite) **expert trajectories** $E_T = \{\tau_1; \tau_2; ....\tau_n\}$ constitute to a combination of **states** $S = \{s_1, s_2, ...s_n\}$ and **actions** $A = \{a_1, a_2, ...a_n\}$ that an agent can take in E_T where $T_{PA}(.)$ is the state **transition probabilities** of moving to state s' from s upon taking action a (i.e. $T(s, a, s')$. A **discount factor** $\gamma \in [0, 1)$ dictates the weightage for long-term-short-term reward strategy. $L1 \in [0, 1)$ is the **regularization factor**. π is the **policy** function that defines the action to be taken in each state i.e. $(\pi : S \rightarrow A)$, π^* is the **optimal policy** that defines the optimal actions to take in each state such that the generated reward is maximum. $\tau = \{(s_0, a_1, s_1); (s_1, a_2, s_2); ....(s_{n-1}, a_n, s_n)\}$ is a **trajectory** describing one complete iteration of the agent in the MDP. $R(s_n, a_n) \in R_f$:

is the **reward** received for reaching state s_n by taking action a_n where R_f is the collective reward function for all policies π in trajectories in E_T. To identify personalization, we have used a combination of linear and maximum entropy algorithms to generate individual reward functions, optimal policies and analyze the impact across doctors and specializations extending our prior experiment [3]. Our linear IRL [6] is based on the approach of using RL inside IRL to iterate across all policies and identify the maximal reward using the assumed optimal policy π^* over trajectories E_T. This procedure provides us with a complete overview of all possible rewards in the state space. The maximum entropy IRL is implemented as in [8] i.e. maximizing the reward function relative to their weights $\theta^* = \underset{\theta}{argmax} \sum_{E_T} log\, P(E_T|\theta, E_T)$. Further, using the maximum entropy algorithm we recover the real optimal policy π^* therefore describing true expert behaviour. Figure 1 showcases our approach. We use the setup of the IRL algorithm by [2] and customized Python scripts to perform simulations.

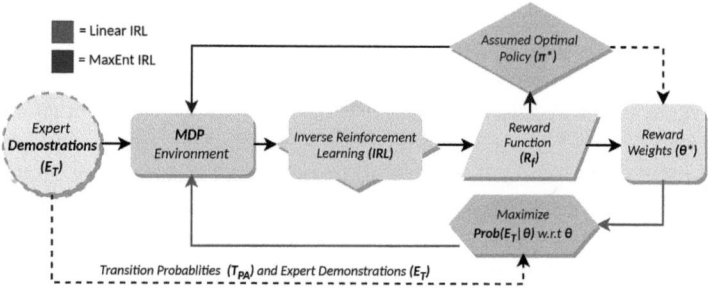

Fig. 1. Linear and MaxEnt IRL in our experiment. An MDP is setup using E_T. A tuple of trajectories is fed to IRL, generating reward function R_F. Linear IRL iterates through all policies π, while MaxEnt maximizes reward weights θ^* relative to T_{PA} and E_T

CDSS Data: The dataset used in our experiment is from an experimental survey [5] conducted at a medical research center where the authors analyze the effect of having decision makers (doctors) supported by information from a prediction model, a FINDRISK measure and case-explanation to assess the perceptional state through subjective metrics of patients suffering from T2DM. Physicians were provided with the patients' basic information (age, BMI etc.), one of three prediction information and asked to assess the data via three subjective perception measures on a Likert scale from 1 (strongly disagree) to 5 (strongly agree). The measure being *Understandability* denoting interpretation, *Agreement* of model's prediction and *Usability* of prediction data in diagnosis. A total of 541 cases of patient assessment data were found to be usable for our experiment. We include internal context (as doctors' experience, specialization), external context (as patient risk) to generate inferences.

MDP Setup and Policy Iteration: In [3], we used CDSS data to model the internal perceptional state of the medical experts using IRL. We extend the

MDP of $S:5$ **states** $= \{End,\ Understandability,\ Agreement,\ Usability,\ Completion\}$, $A:2$ **actions** $= \{Continue,\ Terminate\}$ with T_P: Transition probabilities of moving from state s to s' extracted from E_T using $T_P(s,a,s') = \frac{\#\ of\ times\ (s\ \to s')\ occurs\ in\ T_E}{Total\ \#\ of\ occurences\ in\ T_E}$. Where the agent is initialized at understandability state and takes actions based on the scored evaluation from the physician. A deterministic function compares the doctor's assessment to a metric threshold M_T (i.e. reflecting lenient, strict evaluation) and is used to decide the action of the agent. We thresholds of $M_T = [2, 3, 4]$ relating to intensity of assessment.

3 Analysis of Reward Functions and Maximum Entropy Policies

To derive doctor's personalized evaluation policy, we first assume that all doctors have universal behaviour and run IRL on all 541 cases of CDSS data (i.e. global demonstrations). We select a metric threshold $M_T = 2$ to ensure the inclusion of a broad range of evaluations and create a subset of CDSS data trajectories. We feed the linear IRL a Linear IRL algorithm with a tuple $(S, A, T_P, [\pi], R_{max}, \gamma, L1)$, where $[\pi]$ is a set of all policies extracted from the trajectories, $L1$ and γ were set to $(0.9, 0.9)$ aimed at long term rewards. We obtain the reward function as shown in Table 1 representing reward/penalties for reaching each state respective to the policy, here the assumed optimal policy π^* is $[0,1,1,1,0]$ as per reward-penalty distribution. Here we encounter optimal ambiguity where the same rewards are obtained for multiple policies. Therefore, we introduce entropy to reflect the collective behaviour of individual policies. It is defined as the probability of reward-penalty received during the resolution of trajectories $H(R_f) = \sum_{R(s_n,a_n) \in R_f} P(R(s_n,a_n))\ log(P(R(s_n,a_n)))$. Next, we use maximum entropy IRL to recover the real optimal policy using the same trajectory subsets.

Table 1. The reward function of global demonstrations generated over policies π and with reward terms and respective entropy $H(R_F)$. For all other π the reward and entropy were 0

Policy	End	Understand	Agree	Use	Complete	Entropy
[0,0,1,1,0]	0.0	0	0	0	10.00	0.217
[0,1,1,1,0]	0.0	0	0	10.00	10.00	0.293
[0,1,1,0,0]	0.0	0	0	10.00	0	0.218
[0,1,0,1,0]	0.0	0	0	0	10.00	0.218

Assessment by Specializations: On running linear IRL and MaxEnt for the specialization subset of CDSS data, we identify collective reward functions and optimal policies shown in Table 2. We analyse the variance of policy-reward space and compare it with MaxEnt policies. To ensure an unbiased analysis, we selected specializations where more than 3 individual doctors' evaluations were available i.e. endocrinologists, cardiologists and general medicine [E, C, GM] while sparsely involving other doctors since they had less than 2 samples or their behaviour was erroneous. Table 3 shows the reward function for the three specializations [E, C, GM] at $M_T = [2, 3]$. On analysing the MaxEnt policies and respective R_F, we observe the π^* for [E, C, GM] doctors to be relatively the same across M_T with the entropy of endocrinologists being consistent compared to all specializations. The average score of evaluation metrics for [E, C, GM] were (4.30, 3.72, 3.75), (4.9, 4.2, 4.7) and (4.27, 4.30, 4.0) respectively. This shows the volatility in the optimal policy followed and given the high metric scores, the IRL assumes that the amount of information does not impact the evaluation thus moderately saturating the reward function. The infectionsists and behaviour follow closely with endocrinologists. Gynaecologists have a peculiar deviation of policy at $M_T = 2$ due to a higher number of evaluations at low M_T. For neurologists and rheumatologists, the R_F remained the same throughout all M_T since they had only trajectory. Overall, at a higher M_T, the optimal policy for all specializations follows similar behaviour, and at MT = 2,3, the policy reveals individual assessment tendencies (i.e. strict/lenient) which is reflected in the average evaluation scores. At $M_T = 4$ the behaviour converges as in Table 1 with rewards awarded for reaching usability states across specializations.

Table 2. The maximum entropy optimal policies π^* for all doctors per specialization representing their personalized decision policy at various levels of assessment M_T

Specialization	No of Docs	MT = 2	MT = 3	MT = 4
Endocrinologist	5	[0, 0, 0, 0, 0]	[0, 0, 0, 1, 0]	[0, 0, 1, 1, 0]
Cardiologist	3	[0, 0, 0, 0, 0]		[0, 0, 1, 1, 0]
General Medicine	4	[0, 0, 0, 0, 0]	[0, 0, 0, 1, 0]	[0, 0, 1, 1, 0]
Gynaecologist	2	[0, 1, 1, 0, 0]	[0, 0, 0, 1, 0]	[0, 0, 1, 1, 0]
Ophthalmologist	1	[0, 0, 0, 0, 0]	[0, 0, 0, 1, 0]	[0, 1, 1, 1, 0]
Infection Specialist	2	[0, 1, 0, 0, 0]	[0, 1, 1, 0, 0]	[0, 1, 1, 1, 0]
Rheumatologist	1		[0, 1, 1, 1, 0]	
Neurologist	1		[0, 1, 1, 1, 0]	

4 Simulating Personalized Learning via Policy Iteration

The simulation setup for personalized learning behaviour uses the same MDP and IRL as in Sect. 3. To simulate behaviour, we extend the IRL cycle to randomly resample the trajectories from the subsets of specialization and individual doctors to train the model over multiple steps. To investigate optimal policies we select subsets of three specializations as per CDSS data, i.e. 5 endocrinologists, 4 general medicine specialists and 3 cardiologists since their sample trajectories are large enough for simulating individual and groupwise behaviour. For uniformity in experiments, we selected our IRL parameters to be focused on having a long-term reward strategy of $(L1 : 0.9, \gamma : 0.9)$ and $M_T = 2$ to cover all possible scenarios. Figure 2 gives an overview of our simulation in three cases.

Case 1: Simulation using Individual Demonstrations. The linear IRL algorithm is initialized with a doctor's full trajectory set and trained. On concluding one iteration of training, we randomly select a single sample from the original set of doctors' trajectories and append it to the training set. The process is continued for N Sample additions and the entropy scores for all sample addition steps up to are averaged over M cycles. To make the computing process more efficient, we selected two combinations of sample addition and randomization (M, N) i.e. (25, 600) and (100, 100) since at smaller intervals, the behaviour was incomplete and larger intervals tended to have constant variance while also taking immense computing power and time. We evaluate the training by assessing the reward entropy $H(R_f)$ across N and M randomization cycles.

Table 3. The reward functions and entropy of Endocrinologists (E), Cardiologists (C) and General medicine therapists (GM) at $M_T = [2, 3]$ when running IRL on subsets of specialization. The rewards are constant for states of **End**, **Understandability** (*Und*) and **Agreement** (*Ag*) while varying for **Usability** (*Use*) and **Complete**

Policy	End	Und	Agree	Use		Complete		Entropy	
Specializations	[E, C, GM]			[E]	[C, GM]	[E]	[C, GM]	[E]	[C, GM]
[0,0,0,0,0]	0	0	-10	0	-10	0	-10	0.217	0.292
[0,0,0,1,0]	0	0	-10	0	-10	0	10	0.217	0.458
[0,0,1,1,0]	0	0	-10	0	10	10	10	0.412	0.458
[0,1,1,1,0]	0	0	10	10	10	10	10	0.292	0.292
[0,1,1,0,0]	0	0	10	10	10	0	0	0.292	0.292
[0,1,0,0,0]	0	0	10	0	0	0	0	0.217	0.217
[0,0,1,0,0]	0	0	-10	0	10	0	0	0.217	0.412
[0,1,0,1,0]	0	0	10	0	0	10	10	0.292	0.218

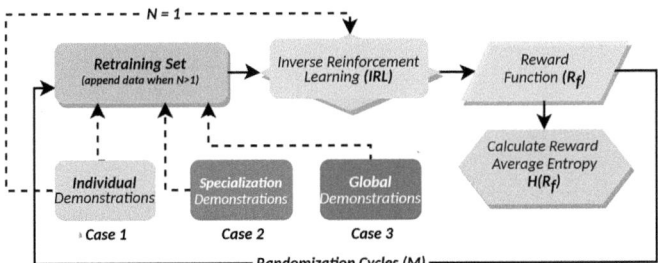

Fig. 2. Our simulation setup to learn personalized policies of doctors in various training scenarios. Given a single doctors demonstrations at $N = 1$ we run IRL on the current set of trajectories to obtain R_F and its entropy. The initial set of trajectories is appended with one sample from; Case 1: doctor undergoing personalization; Case 2: doctor's specialization; Case 3: global demonstrations; This cycle is continued for M steps.

Across all specializations, we observe spikes in the entropy with the addition of new trajectories reflecting new unlearned behaviour. After cycles of random addition, the entropy gradually saturates. Figure 3A shows the learning behaviour of cardiologists (A, B, C) trained with 600 iterations and 25 randomization cycles. On comparing with respective MaxEnt policies and R_F, we observe that the physicians have a strict policy of evaluation at lower M_T with penalties. Whereas for general medicine therapists trained with 100 iterations of sampling and 100 randomization cycles, (Fig. 3B), the reward entropy does not stabilize for specialists (A, C) until 100 samples. We attribute this behaviour to higher metric scores. Specialist (D) has a constant entropy due to erroneous evaluation scores. We observe that an increase in sample additions (N) elongates the learning behaviour with sharp deviations in R_F, while an increase in randomization (M) results in smooth learning behaviour with fewer steps to adopt a policy.

Case 2: Resampling using Specialization Trajectories: Here, we initialize the MDP and IRL algorithm as in the previous case, however, our re-sampling data is sourced from the data of specializations (i.e. grouped data of doctors from the same specialization except the doctor being assessed). We consider the case of endocrinologists, where we feed specialization data to all doctors individually. Figure 4A shows the results of the endocrinologists. Unlike in case 1, the reward entropy takes comparatively longer to converge and stabilize while there are sections of constant trajectory in between. This behaviour is the impact of deviating policies given the breadth of added samples (i.e. trajectories of four endocrinologists were added during training) therefore requiring more cycles of randomization to stabilize.

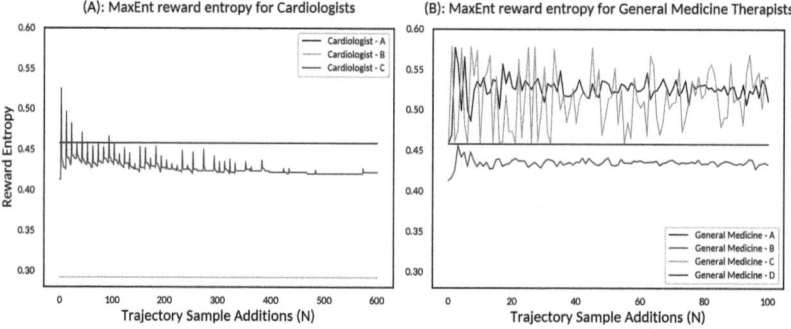

Fig. 3. The personalized learning curve of reward entropy at Maximum Entropy policy [0, 0, 1, 1, 0] for; (A): cardiologists after training with (25,600) randomized -sampling iterations; (B): Personalized learning curve for general medicine specialists after training with (100, 100) randomized-sampling iterations

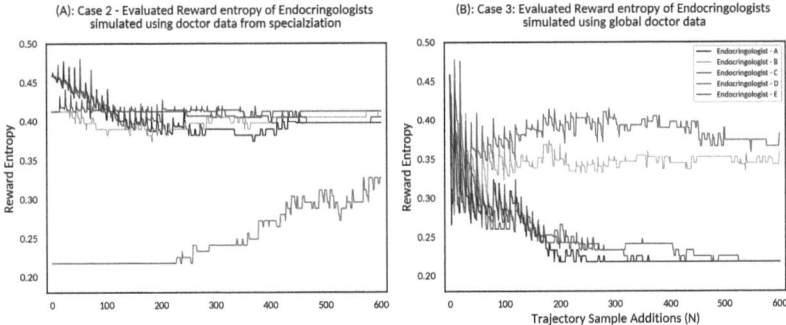

Fig. 4. Learning curve of endocrinologist's reward entropy of MaxEnt policy [0,0,1,1,0] when simulated at (25,600) cycles of randomization; (A): Sampling data from endocrinologists specialization; (B): Sampling data from global doctors

Case 3: Resampling using Global Trajectories: The final case is aimed at providing a broad perspective of the learning behaviour as the trajectories are added from global demonstrations. The setup of MDP and IRL follows case 1. We experiment with endocrinologists again as they provide a wider base for analysis. At 25 randomization and 600 iterations, we observe the behaviour as in Fig. 4B. Compared to case 1 and case 2, we notice the magnitude of reward entropy is much lower than the latter, when analyzed, the reward functions are observed to vary frequently with changes in penalty and reward terms across all policies. The variance in behaviour does not subside despite 600 additions and 25 randomization cycles, we believe this is attributed to the nature of the dataset since it consists of 541 samples consisting of doctor trajectories of all specializations, the ideal number of iterations required to stabilize the reward function for collective policies should be much larger. Hence, the number of trajectories required to reach a constant policy is much larger.

5 Conclusion and Future Work

Overall, the results of our experiment provide insights into modelling and simulating clinical decision-making data using inverse reinforcement learning to identify personalized policies of doctors when evaluating prediction data of patients' risk of type 2 diabetes mellitus. We perform simulations to evaluate the personalized learning of doctors' individual policies in three cases of trajectory sampling. From our investigation, we observed an increase in evaluated entropy when trained by experts of different specializations whereas when trained using one's own data, the identification of policy is faster (i.e. less than 600 steps of sampling). We see these results as a crucial step towards the development of personalization in DSSs, the inclusion of which can improve the AI model's alignment with domain experts enabling a more efficient collaboration. In future works, we plan to extend AI model simulation with online and offline training constructed around theory of mind and feedback learning to improve human-AI collaboration in universal domains.

Acknowledgments. The research was supported by The Russian Science Foundation, agreement No. 24-11-00272, https://rscf.ru/project/24-11-00272/.

References

1. Abbeel, P., Ng, A.Y.: Apprenticeship learning via inverse reinforcement learning. In: Proceedings of the 21st International Conference on Machine Learning, p. 1 (2004)
2. Alger, M.: Inverse reinforcement learning (2017)
3. Ireddy, A.T., Kovalchuk, S.V.: Modelling information perceiving within clinical decision support using inverse reinforcement learning. In: International Conference on Computational Science, pp. 210–223. Springer (2024)
4. Kovalchuk, S., Ireddy, A.T.S.: Prediction of users perceptional state for human-centric decision support systems in complex domains through implicit cognitive state modeling. In: Proceedings of the Annual Meeting of the Cognitive Science Society, vol. 46 (2024)
5. Kovalchuk, S.V., Kopanitsa, G.D., Derevitskii, I.V., Matveev, G.A., Savitskaya, D.A.: Three-stage intelligent support of clinical decision making for higher trust, validity, and explainability. J. Biomed. Inform. **127**, 104013 (2022)
6. Ng, A.Y., Russell, S., et al.: Algorithms for inverse reinforcement learning. In: ICML, vol. 1, p. 2 (2000)
7. Schmidt, P., Biessmann, F.: Calibrating human-ai collaboration: impact of risk, ambiguity and transparency on algorithmic bias. In: International Cross-Domain Conference for Machine Learning and Knowledge Extraction, pp. 431–449. Springer (2020)
8. Ziebart, B.D., Maas, A.L., Bagnell, J.A., Dey, A.K., et al.: Maximum entropy inverse reinforcement learning. In: AAAI, vol. 8, pp. 1433–1438. Chicago, IL, USA (2008)

Verified Eigenvalue Calculation for the Laplace Operator

Jijing Zhao and Shuyu Sun$^{(\boxtimes)}$

School of Mathematical Sciences, Tongji University, Shanghai, China
suns@tongji.edu.cn

Abstract. This paper proposes a method for estimating eigenvalues and eigenvectors based on a computable residual. We provide guaranteed upper and lower bounds for eigenvalues and establish first-order and second-order error estimates. Additionally, error bounds for eigenvectors are provided, ensuring precise estimates for both eigenvalues and eigenvectors. Numerical experiments have validated the results. Furthermore, the interval algorithm is used for the calculation of residual function, and the resulting eigenvalue bounds are expected to be mathematically accurate.

Keywords: Error bounds · eigenvalue problem · self-adjoint operators

1 Introduction

The solution of eigenvalue problems for operators arises in numerous applications in numerical analysis and scientific computing. These problems are central to diverse fields such as quantum mechanics, elasticity theory, and stability analysis. Given a linear self-adjoint operator A, its eigenvalues and eigenvectors are often of primary interest.

A standard approach for computing eigenvalues and eigenvectors of large matrices is the Rayleigh-Ritz method, which approximates the eigenvalue problem within a subspace. For the Laplace operator, we typically employ finite element discretization. Following this discretization, iterative methods—such as the power method, inverse power method, and Lanczos method—are commonly applied. The numerical solution process introduces several types of errors, including discretization errors and floating-point arithmetic errors. When using the conforming finite element method, the computed eigenvalue is usually larger than the exact eigenvalue due to the min-max principle. Consequently, determining a precise interval for the analytical eigenvalue, or its rigorous upper and lower bounds, remains a significant challenge.

In recent decades, guaranteed upper and lower bounds for eigenvalue problems have been a research focus. Methods for estimating these bounds generally fall into two categories: a priori estimates and a posteriori estimates [1]. Based on prior error estimates, Liu [5] provided a lower-order estimate for the Laplace eigenvalue problem by accurately estimating the constants in the projection operator. The theory in [2,4] applies to arbitrarily coarse meshes and provides convincing numerical results in various test cases. Hu et al. [3], Luo et al. [6] and Yang et al. [8] also derived (guaranteed) eigenvalue estimates for the nonconforming finite element method.

In this paper, we propose a method for estimating eigenvalues and eigenvectors of linear operators. The approach is based on a computable residual function that quantifies the discrepancy between approximate and exact eigenvalues. Using this residual, we derive explicit upper and lower bounds for the eigenvalues, which are independent of unknown constants. The residual function is fully computable, and we further establish first- and second-order error estimates for the eigenvalues. In addition to the approximation error arising from the finite-dimensional discretization of the original infinite-dimensional problem, and the iterative solution error arising from numerical computation, we also account for rounding errors in floating-point arithmetic. The IEEE Standard enables the estimation of rounding errors through the use of interval arithmetic [7].

The structure of this paper is as follows: In the Sect. 2 we introduce the eigenvalue problem. The Sect. 3 provides the upper and lower bound estimates, while the Sect. 4 presents numerical validation of the theoretical results.

2 Eigenvalue Problem

Consider the eigenvalue problem for the Laplace operator Δ on a domain $\Omega \subset \mathbb{R}^n$, defined as:
$$-\Delta u = \lambda u \quad \text{in} \quad \Omega,$$
where $\Delta = \nabla^2$ is the Laplace operator, λ is the eigenvalue, and u is the corresponding eigenvector (eigenfunction). We also have boundary conditions $u|_{\partial\Omega} = 0$ (Dirichlet boundary conditions), though other boundary conditions can be chosen depending on the specific problem.

Multiply both sides by a test function $v \in H_0^1(\Omega)$ and integrate, obtaining the weak form:
$$\text{Find} \quad (\lambda, u) \quad \text{such that} \quad (\nabla u, \nabla v) = \lambda(u, v) \quad \forall v \in H_0^1(\Omega),$$
where $(\cdot, \cdot)$ represents the standard inner product:
$$(u, v) = \int_\Omega uv\, dx, \quad (\nabla u, \nabla v) = \int_\Omega \nabla u \cdot \nabla v\, dx.$$

The key idea in finite element methods is to approximate the solution by choosing a suitable finite-dimensional space of test and trial functions. Let $V_h \subset H_0^1(\Omega)$ be a finite-dimensional subspace. We can obtain a discrete system of linear equations:
$$S u_h = \lambda_h M u_h,$$
where:
$$S_{ij} = (\nabla \varphi_i, \nabla \varphi_j), \quad M_{ij} = (\varphi_i, \varphi_j),$$
and $u_h = [u_1^h, u_2^h, \ldots, u_N^h]^T$ is the vector of coefficients in the finite element basis.

In this paper, we employ a global spectral method to solve the problem. Compared with traditional finite element methods, spectral methods offer superior accuracy and computational efficiency through the use of globally orthogonal basis functions for solution representation.

After discretizing the problem using global basis functions, we obtain a matrix eigenvalue problem that could conventionally be solved using methods like the power method or inverse power method. However, in this work, we employ a gradient flow approach instead. This method offers unconditional energy stability, making it particularly advantageous for our purposes.

$$\frac{\tilde{u}_h^{n+1} - u_h^n}{\Delta t} = \Delta \tilde{u}_h^{n+1},$$
$$u_h^{n+1} = \tilde{u}_h^{n+1}/(\tilde{u}_h^{n+1}, \tilde{u}_h^{n+1}).$$

One key advantage of using the gradient flow method is that it is well-suited for solving eigenvalue problems in the context of non-linear and large-scale systems, where other methods might fail to converge or require extensive computational resources.

3 Guaranteed Bounds

In this section, we provide upper and lower bound estimates for the eigenvalues. Define $\{(\lambda_i, u_i), i \in \mathbb{N}\}$ are the exact eigen-pairs of the operator $A = -\Delta$, where $\lambda_1 \leq \lambda_2 \leq \lambda_2 \leq \cdots \lambda_{k-1} \leq \lambda_k \leq \cdots$. We assume the numerical scheme provides an approximate u_h intended to approximate the k-th exact eigenvector u_k. The corresponding eigenvalue λ_k is then computed using the following equation. We note that even the numerical scheme gives us also λ_k, we do not use it, and we update λ_k by

$$\lambda_h = \frac{(Au_h, u_h)}{(u_h, u_h)}.$$

And we define the residual vector as

$$r_h = \frac{Au_h - \lambda_h u_h}{\|u_h\|_{L^2(\Omega)}}.$$

First, we present a guaranteed numerical bound for eigenvalues known as D. Weinstein's bound in the literature.

Lemma 1. *We assume among all the exact eigenvalues of A, the eigenvalue λ_k is the one closest to λ_h, that is $|\lambda_k - \lambda_h| \leq |\lambda_i - \lambda_h|, \forall i \in \mathbb{N}$. Then we have*

$$\lambda_h - \delta \leq \lambda_k \leq \lambda_h + \delta,$$

where $\delta = \|r_h\|_{L^2(\Omega)}$.

Proof. We first expand u_h in terms of $\{u_i, i \in \mathbb{N}\}$,

$$u_h = \sum_{i=1}^{\infty} \alpha_i u_i, \quad \alpha_i = (u_h, u_i),$$

We then note that

$$r_h = Au_h - \lambda_h u_h = \sum_{i=1}^{\infty} \alpha_i(Au_i - \lambda_h u_i) = \sum_{i=1}^{\infty} \alpha_i(\lambda_i - \lambda_h)u_i.$$

We can obtain,

$$\|r_h\|_{L^2(\Omega)}^2 = \sum_{i=1}^{\infty} \alpha_i^2(\lambda_i - \lambda_h)^2 \geq \sum_{i=1}^{\infty} \alpha_i^2(\lambda_k - \lambda_h)^2 = (\lambda_i - \lambda_h)^2.$$

□

Remark 1. In our calculation, the first step is to calculate the first eigenpair (λ_1, u_1) using the regular floating numbers, the second step is to obtain a guaranteed interval estimate for $\delta = \|r_h\|_{L^2(\Omega)}$ and then for λ_k by using the Interval Arithmetic (IA) technique: $\delta \in [\delta_m, \delta_M]$ and $\lambda_k \in [\lambda_{k,m}, \lambda_{k,M}]$, where $[\delta_m, \delta_M] = \text{IA}(\|r_h\|_2)$, $\lambda_{k,m} = \text{IA}_m(\lambda_h - \delta_M)$, and $\lambda_{k,m} = \text{IA}_M(\lambda_h + \delta_M)$.

The first step (numerical computation) introduces multiple potential error sources: discretization error from approximating the infinite-dimensional space with a finite-dimensional subspace; domain truncation error, particularly significant for quantum systems like the hydrogen atom requiring unbounded-domain approximation; iterative convergence error inherent in numerical approximation schemes; and floating-point round-off error due to finite-precision arithmetic limitations. The second step yields a guaranteed interval estimate for the eigenvalue λ_k. This overestimation arises from two primary sources: (i) the inherent non-sharpness of our *a posteriori* error estimator (as seen in the bounding estimate $\lambda_h - \delta \leq \lambda_k \leq \lambda_h + \delta$), and (ii) the application of interval arithmetic during the post-processing step. Both factors contribute conservatively to the final error bounds.

Theorem 1. *We assume among all the exact eigenvalues of A, the eigenvalue λ_1 is the one closest to λ_h; that is, $|\lambda_1 - \lambda_h| \leq |\lambda_i - \lambda_h|, \forall i \in \mathbb{N}$. Let $\lambda_{2,L}$ is a lower approximate of λ_2. Then we have*

$$\sin \alpha(u_h, u_1) \leq \frac{\delta}{\lambda_{2,L} - \lambda_h}, \|u_h - u_1\|_{L^2(\Omega)} \leq \left(\frac{\delta}{\lambda_{2,L} - \lambda_h}\right)\sqrt{1 + \left(\frac{\delta}{\lambda_{2,L} - \lambda_h}\right)^2},$$

where $\delta := \|r_h\|_{L^2(\Omega)}$.

Proof. We expand u_h in terms of $\{u_i, i \in \mathbb{N}\}$, $u_h = \sum_{i=1}^{\infty} \alpha_i u_i, \alpha_i := (u_h, u_i)$, we have

$$\sin^2 \alpha(u_h, u_1) = 1 - \cos^2 \alpha(u_h, u_1) = 1 - \alpha_1^2$$

As
$$r_h = Au_h - \lambda_h u_h = \sum_{i=0}^{\infty} \alpha_i(Au_i - \lambda_h u_i) = \sum_{i=0}^{\infty} \alpha_i(\lambda_i - \lambda_h)u_i,$$
we have
$$\|r_h\|_{L^2(\Omega)}^2 = \sum_{i=1}^{\infty} \alpha_i^2(\lambda_i - \lambda_h)^2 \geq \sum_{i=2}^{\infty} \alpha_i^2(\lambda_2 - \lambda_h)^2 = (1 - \alpha_1^2)(\lambda_2 - \lambda_h)^2.$$

For $(1 - \alpha_1)^2 \leq (1 - \alpha_1^2)^2 = \sin^4 \alpha(u_h, u_1)$, we obtain,
$$\|u_h - u_1\|_2^2 = (1-\alpha_1)^2 + \sum_{i \geq 2} \alpha_i^2 = (1-\alpha_1)^2 + 1 - \alpha_1^2 \leq \sin^4 \alpha(u_h, u_1) + \sin^2 \alpha(u_h, u_1),$$
implying
$$\|u_h - u_1\|_{L^2(\Omega)} \leq \sin \alpha(u_h, u_1)\sqrt{1 + \sin^2 \alpha(u_h, u_1)}$$
$$\leq \left(\frac{\delta}{\lambda_{2,L} - \lambda_h}\right)\sqrt{1 + \left(\frac{\delta}{\lambda_{2,L} - \lambda_h}\right)^2}.$$
□

Next, we give a second-order result.

Theorem 2. *We assume among all the exact eigenvalues of A, the eigenvalue λ_1 is the one closest to λ_h; that is, $|\lambda_1 - \lambda_h| \leq |\lambda_i - \lambda_h|, \forall i \in \mathbb{N}$. We assume $\lambda_h = \frac{(Au_h, u_h)}{(u_h, u_h)}$ precisely (without any errors including round-off errors). Then we have*
$$\lambda_h - \frac{\delta^2}{(1 - \sin^2 \alpha(u_h, u_1))(\lambda_2 - \lambda_1)} \leq \lambda_1 \leq \lambda_h,$$
where $\delta := \|r_h\|_{L^2(\Omega)}$. As a result,
$$\lambda_h - \frac{\delta^2}{\lambda_2 - \lambda_1} \leq \lambda_1 \leq \lambda_h.$$

Proof. We expand u_h in terms of $\{u_i, i \in \mathbb{N}\}$, $u_h = \sum_{i=1}^{\infty} \alpha_i u_i, \alpha_i = (u_h, u_i)$. First $|\lambda_1 - \lambda_h| \leq |\lambda_i - \lambda_h|, \forall i \in \mathbb{N}$ implies $\lambda_h \leq \lambda_i, \forall i \geq 2$. We then note that
$$r_h = Au_h - \lambda_h u_h = \sum_{i=1}^{\infty} \alpha_i(Au_i - \lambda_h u_i) = \sum_{i=1}^{\infty} \alpha_i(\lambda_i - \lambda_h)u_i.$$

Due to the definition of $\lambda_h = \frac{(Au_h, u_h)}{(u_h, u_h)}$, we see that r_h is orthogonal to u_h. Thus,
$$0 = \left(\sum_{i=1}^{\infty} \alpha_i(\lambda_i - \lambda_h)u_i, \sum_{i=1}^{\infty} \alpha_i u_i\right) = \alpha_1^2(\lambda_1 - \lambda_h) + \sum_{i=2}^{\infty} \alpha_i^2(\lambda_i - \lambda_h),$$

which implies $\lambda_h \geq \lambda_1$ and

$$\alpha_1^2(\lambda_h - \lambda_1) = \sum_{i=2}^{\infty} \alpha_i^2(\lambda_i - \lambda_h) \geq 0.$$

On the other hand,

$$\begin{aligned}\|r_h\|_{L^2(\Omega)}^2 &= \sum_{i=1}^{\infty} \alpha_i^2(\lambda_i - \lambda_h)^2 \geq \alpha_1^2(\lambda_h - \lambda_1)^2 + \sum_{i=2}^{\infty} \alpha_i^2(\lambda_i - \lambda_h)(\lambda_2 - \lambda_h) \\ &= \alpha_1^2(\lambda_h - \lambda_1)^2 + \alpha_1^2(\lambda_h - \lambda_1)(\lambda_2 - \lambda_h) \\ &= \alpha_1^2(\lambda_h - \lambda_1)(\lambda_2 - \lambda_1).\end{aligned} \quad (1)$$

Noting that
$$\sin^2 \alpha(u_h, u_1) = 1 - \cos^2 \alpha(u_h, u_1) = 1 - \alpha_1^2,$$
we have the desired result. □

We assume among all the exact eigenvalues of A, the eigenvalue λ_1 is the one closest to λ_h; that is $|\lambda_1 - \lambda_h| \leq |\lambda_i - \lambda_h|, \forall i \in \mathbb{N}$. Given u_h to approximate u_1, we obtain $\lambda_h = \frac{(Au_h, u_h)}{(u_h, u_h)}$ using Interval Arithmetic to get $\mathrm{IA}(\lambda_h) = [\lambda_{h,L}, \lambda_{h,U}]$. Let $\mathrm{IA}(\|r_h\|_{L^2(\Omega)}) = [\delta_L, \delta_U]$. Then we have

$$\mathrm{IA}_L\left(\lambda_{h,L} - \frac{\delta_L^2}{\lambda_{2,L} - \lambda_{1,U}}\right) \leq \lambda_1 \leq \lambda_{h,U},$$

where $\lambda_{2,L}$ is a guaranteed lower bound for λ_2, and $\lambda_{1,U}$ is a guaranteed upper bound for λ_1. Both $\lambda_{2,L}$ and $\lambda_{1,U}$ can be obtained from Lemma 1.

Remark 2. When we get $(\lambda_h, u_h) \approx (\lambda_1, u_1)$ eigen-pair, u_h is not exactly u_1 due to round-off error (and other numerical errors like iteration error, truncation error etc.). When employing the gradient flow method to compute the approximate eigenfunction u_h, the resulting solution predominantly captures the first eigenvector u_1 rather than $u_i, i \geq 2$. By subtracting the projection of u_h onto the first eigenmode u_1 from u_h itself, we obtain the residual component orthogonal to u_1: $u_{h,\neq 1} := u_h - (u_h, u_1)u_1$. Intuitively, we see $u_{h,\neq 1}$ contains mainly u_2 rather than $u_i, i \geq 3$. Similarly, because the residual vector $r_h = Au_h - \lambda_h u_h$ is orthogonal to u_h, which is close to u_1, it is also expected that r_h mainly contains u_2 rather than $u_i, i \neq 2$. Consequently, we can use r_h to get $(\lambda_{2,h}, u_{2,h})$, an approximation of (λ_2, u_2), by $u_{2,h} = \frac{r_h}{\|r_h\|_{L^2(\Omega)}}, \lambda_{2,h} = \frac{(Au_{2,h}, u_{2,h})}{(u_{2,h}, u_{2,h})}$. Then we use Lemma 1 to get a guaranteed interval for λ_2, then get $\lambda_{2,L}$.

4 Numerical Examples

In this section, we validate the results using two examples with rectangular domains. We consider the spectral method for rectangular domains. The basis functions are Legendre polynomials, which are orthogonal polynomials defined globally.

$$L_0(x) = 1, L_1(x) = x, L_2(x) = \frac{3}{2}x^2 - \frac{1}{2}, \dots$$

$$(n+1)L_{n+1}(x) = (2n+1)xL_n(x) - nL_{n-1}(x).$$

The k-th basis function and the j-th basis function are orthogonal on the interval $[-1, 1]$,

$$\int_{-1}^{1} L_k(x)L_j(x)dx = \frac{1}{k+\frac{1}{2}}\delta_{kj}.$$

We will use this basis function to solve the Laplace eigenvalue problem on two different rectangular domains $[0, \pi]^2$ and $[0, 1]^2$. The smallest exact eigenvalue for the $[0, \pi]^2$ domain is 2, and the smallest exact eigenvalue for the $[0, 1]^2$ domain is $2\pi^2$. We present the upper and lower bounds of the smallest eigenvalue, as shown in Tables 1 and 2.

Table 1. Results of $[0, 1]^2$.

Poly Degree	λ_h	$|\lambda - \lambda_h|$	Upper Bound	Lower Bound	Upper - Lower
4	2.02642367284675	2.64e-02	2.66724	1.38561	1.28e-00
6	2.00002942777066	2.94e-05	2.04342	1.95664	8.68e-02
8	2.00000000686899	6.87e-09	2.00107	1.99893	2.14e-03
9	2.00000000000054	5.40e-13	2.000013417	1.999986583	2.68e-05
11	2.00000000000001	1.02e-13	2.000000138	1.999999862	2.76e-06
12	1.99999999999992	7.99e-14	2.000000138	1.999999862	2.71e-06
13	1.99999999999998	2.00e-14	2.000000091	1.999999909	1.82e-07

Table 2. Results of $[0, \pi]^2$.

Poly Degree	λ_h	$|\lambda - \lambda_h|$	Upper Bound	Lower Bound	Upper - Lower
4	20	2.61e-01	26.3246	13.6754	1.26e01
6	19.7394992426336	2.90e-04	20.1678	19.3112	8.56e-01
8	19.7392088699728	6.78e-08	19.74975953	19.72865821	2.11e-02
10	19.7392088021836	4.90e-12	19.73934122	19.73907639	2.65e-04
11	19.7392088021788	9.90e-14	19.73921007	19.73920753	2.54E-06
12	19.7392088021789	1.99e-14	19.73920952	19.73920808	1.44E-06
13	19.7392088021791	3.40e-14	19.73920960	19.73920800	1.40E-06

From our numerical experiments, we can see that the residual can effectively provide upper and lower bounds for the true eigenvalues using interval arithmetic. Specifically, interval arithmetic maintains error bounds at computational step, avoiding the numerical uncertainties that may arise in traditional floating-point calculations, thus providing stronger mathematical guarantees for the computation of eigenvalues.

By accurately computing the residuals, we are able not only to obtain approximate eigenvalues but also to provide their exact upper and lower bounds, offering an effective guarantee for the reliability of the eigenvalues.

5 Conclusions

This paper presents an approach for estimating eigenvalues using a computable residual function. The method provides guaranteed upper and lower bounds for eigenvalues, along with first-order and second-order error estimates that enhance the precision of eigenvalue and eigenvector approximations.

Acknowledgments. This study was funded by National Key Research and Development Project of China (No. 2023YFA1011701). The work of J. Zhao was supported by the China Postdoctoral Science Foundation under Grant Number GZB20240552 and 2024M762396.

Disclosure of Interests. The authors have no competing interests to declare that are relevant to the content of this article.

References

1. Cancès, E., Dusson, G., Maday, Y., Stamm, B., Vohralík, M.: Guaranteed and robust a posteriori bounds for laplace eigenvalues and eigenvectors: conforming approximations. SIAM J. Numer. Anal. **55**(5), 2228–2254 (2017)
2. Carstensen, C., Gedicke, J.: Guaranteed lower bounds for eigenvalues. Math. Comp. **83**(289), 2605–2629 (2014). published electronically: April 25, 2014
3. Hu, J., Huang, Y., Lin, Q.: Lower bounds for eigenvalues of elliptic operators: by nonconforming finite element methods. J. Sci. Comput. **61**(2), 196–221 (2014)
4. Liu, X.: A framework of verified eigenvalue bounds for self-adjoint differential operators. Appl. Math. Comput. **267**, 341–355 (2015). the Fourth European Seminar on Computing (ESCO 2014)
5. Liu, X., Oishi, S.: Verified eigenvalue evaluation for the Laplacian over polygonal domains of arbitrary shape. SIAM J. Numer. Anal. **51**(3), 1634–1654 (2013)
6. Luo, F., Lin, Q., Xie, H.: Computing the lower and upper bounds of laplace eigenvalue problem: by combining conforming and nonconforming finite element methods. Sci China Math **55**(5), 1069–1082 (2012)
7. Moore, R., Kearfott, R., Cloud, M.: Introduction to Interval Analysis. SIAM, Philadelphia (2009)
8. Yang, Y., Han, J., Bi, H., et al.: The lower/upper bound property of the crouzeix–raviart element eigenvalues on adaptive meshes. J. Sci. Comput. **62**(2), 284–299 (2015)

A Hybrid Q-Learning Automata Routing Protocol for Wireless Sensor Networks

Jakub Gąsior[✉]

Department of Mathematics and Natural Sciences, Cardinal Stefan Wyszyński University, Warsaw, Poland
j.gasior@uksw.edu.pl

Abstract. This paper proposes a novel hybrid approach to routing in Wireless Sensor Networks that combines Q-learning and Learning Automata models. It is designed to optimize the routing process by leveraging the strengths of both techniques: Q-learning's ability to adapt to dynamic network conditions and Learning Automata's fast adaptation and convergence in stable scenarios. Preliminary analysis indicates feasibility of the proposed approach, showing that it can improve the network lifetime and packet delivery ratio when compared with similar routing protocols.

Keywords: Wireless sensor networks · Routing · Learning automata · Q-learning

1 Introduction

Wireless Sensor Network (WSN) is a distributed network comprising small, battery-powered devices called sensors, capable of sensing and collecting data from their surrounding environment. The sensors are typically low-power and have limited computing capabilities, making energy efficiency a crucial aspect of their design. These sensors communicate with one another wirelessly using radio frequency waves and collaborate to perform specific tasks, such as monitoring environmental parameters like temperature, humidity, or air quality. Data is then collected and sent for further processing via a specialized sink node.

Our work will focus on efficient routing methods, allowing for the optimization of the route taken by data from the source to the sink. To that end, we present a hybrid routing model where Q-learning and Learning Automata (LA) are integrated to leverage their respective strengths while minimizing their weaknesses. This combination allows the algorithm to adapt dynamically to the varying conditions of WSNs, ensuring better performance in terms of latency, packet delivery, and routing stability.

The rest of this paper is organized as follows. Section 2 presents the works related to reinforcement learning-based routing algorithms in WSNs. We introduce the theoretical background of the problem in Sect. 3. Section 4 describes our proposed hybrid Q-LA routing protocol. We present the findings of our experiments in Sect. 5. The last section concludes the paper with future research directions.

2 Related Work

There has been a growing interest in developing reinforcement learning and automata models to tackle the challenges of energy efficiency in WSNs. For example, Manju and Kumar [7] introduced a scheduling algorithm utilizing learning automata to address the target coverage problem. This approach allows sensor nodes to select their operational state autonomously. To validate the efficacy of their scheduling method, comprehensive simulations were conducted, comparing its performance against existing algorithms.

In another study, Lin et al. [5] presented a novel on-demand coverage-based self-deployment algorithm tailored for significant data perception in mobile sensing networks. The authors employed the cellular automata model to accommodate the characteristics of mobile sensing nodes and spatial-temporal node evolution. Subsequently, leveraging learning automata theory and historical node movement data, they proposed a new mobile cellular learning automata model to intelligently and adaptively determine optimal movement directions with minimal energy consumption.

Gudla and Kuda [3] utilized a LA-based model as a routing mechanism for enhanced energy efficiency and reliable data delivery. The approach aims to calculate the selection probability of the next node in a routing path based on various factors such as node score, link quality, and previous selection probability. Furthermore, they proposed an energy-efficient and reliable routing mechanism by combining learning automata with the A-star search algorithm.

Another contribution by Upreti et al. [12] introduced a scheduling technique named Pursuit-LA. Each sensor node in the network was equipped with an LA agent to autonomously determine its operational state to achieve comprehensive target coverage at minimal energy cost. Lastly, Qarehkhani et al. [10] proposed a continuous learning automata-based approach for optimizing sensor angles in Distributed Sensor Networks (DSNs). The method involved continuously adapting sensing angles using LA models. Comparative analysis against a conventional automata-based approach demonstrated the efficacy of the proposed algorithm.

Q-learning, a model-free reinforcement learning algorithm, was also the subject of recent studies in efficient data routing in WSN. Maivizhi and Yogesh [6] employed it to design a routing algorithm for in-network aggregation (RINA) to build a routing tree based on minimal information such as residual energy, the distance between nodes, and link strength.

Gao et al. [1] employed a Q-learning-based routing optimization algorithm for underwater wireless sensor networks. The authors proposed two reward functions based on the average residual energy of the network, transmission delay, and link success rate to balance transmission quality and lifetime better. A similar solution was proposed by Nandyala et al. [8]. The authors employed the QTAR protocol to determine the next-forwarder candidates along the routing path and adopt Q-learning to aid in the optimal global decision-making of next-hop candidates. It showed a lower energy consumption, shorter latency, and longer network lifetime than other state-of-the-art solutions.

Finally, Jain et al. [4] assessed Q-Learning-based routing based on energy depletion rate, node duration, and packet delivery ratio. The authors proved that reinforcement learning schemes can outperform traditional algorithms such as LEACH and K-means by adopting better energy utilization, reduced node mortality, and higher network throughput.

3 Theoretical Background

We consider a WSN comprising N sensors $S = \{s_1, s_2, ..., s_N\}$ randomly deployed over a two-dimensional rectangular area of $x \times y \, [m^2]$. The area contains M targets $T = \{t_1, t_2, ..., t_M\}$ (also called Points of Interest (POI)) that are uniformly distributed with a step of g. All sensors are assumed to have the same sensing range R_s^i and battery capacity b_i. A Boolean disk represents the coverage model of a sensor node [13] and assumes omnidirectional sensing with no random variations.

We use a bipartite graph $G = (V, E)$ to model the Target Coverage Problem (TCP), with $V = S \cup T$, where S represents a set of sensor nodes, T a set of targets and E the set of edges as follows: $\{s,t\} \in E$ if and only if the sensor node s_i detects the target t_j. We define the degree $d(t_j)$ of the target t as the number of sensor nodes that detect the target t_j.

Further, we employ the first-order radio model for the sensors and assume the energy spent for transmitting and receiving a data packet is constant. In addition to this, the energy expenditure is proportional to the distance between two nodes [6].

3.1 Q-Learning

Q-learning is a model-free reinforcement learning algorithm used for sequential decision-making, especially when the environment is uncertain or dynamic. It learns optimal policies over time by adjusting its action-value function based on rewards for selecting specific actions (in our case, selecting next-hop nodes). The core components of the Q-learning algorithm can be defined as follows [8]:

1. Define the State Space ($S = \langle E, L, C, D \rangle$) by assigning the variables that affect routing in the WSN, i.e.:
   - Energy level (E): remaining energy of the node,
   - Link quality (L): measured by the Packet Delivery Ratio (PDR),
   - Congestion level (C): the number of packets currently in the node's buffer or the average delay,
   - Distance to destination (D): distance to the sink or destination node.
2. Define the Action Space (A) corresponding to the set of potential decisions a node can make, i.e., the next-hop nodes;
3. Initialize the Q-values for each state-action pair to an arbitrary value (e.g., zero);
4. Routing Decision (Action Selection) are selected based on the ϵ-greedy policy:

- Exploration: with probability ϵ, the node randomly selects an action from the set of possible actions,
- Exploitation: with probability $1 - \epsilon$, the node selects the action with the highest Q-value (best-known action based on previous experiences).

5. Once the action is chosen (i.e., the next-hop node is selected), the node forwards the packet and observes the reward based on the outcome of the transmission:
   - A reward is given if the packet is successfully transmitted to the next hop,
   - A penalty is assigned if the transmission fails (due to poor link quality, congestion, or node energy depletion).
6. After observing the outcome, the Q-value for the state-action pair is updated using the Bellman equation [1]:

$$Q^{new}(s_t, a_t) = Q(s_t, a_t) + \alpha[R_t + \gamma \times max_{a'}Q(s_{t+1}, a') - Q(s_t, a_t)], \quad (1)$$

where:
- α is the learning rate;
- γ is the discount factor;
- R_t is the reward observed after performing action a_t;
- $max_{a'}Q(s_{t+1}, a')$ is the maximum Q-value of the next state s_{t+1}, which represents the best possible future reward.

This update process gradually refines the Q-values, leading to better routing decisions.

This solution provides several advantages, such as adaptability to dynamic network conditions (e.g., node mobility, varying link quality, energy constraints) by continually learning the optimal routing paths. Due to its decentralized nature, each node can independently learn from its environment without requiring global knowledge, making it scalable and suitable for larger networks [1,4,11].

Some challenges need to be acknowledged. One of the main problems is balancing exploration and exploitation (controlled by the parameter ϵ). Too much exploration can lead to inefficient routing, while too much exploitation can cause the network to get stuck in suboptimal paths. Additionally, Q-learning requires maintaining and updating a Q-value table, which can be computationally expensive, especially in large-scale networks. Thus, converging to the optimal policy may take a long time, especially in highly dynamic environments.

3.2 Learning Automata

A learning automaton is a self-operating mechanism that responds to a sequence of instructions in a certain way to achieve a particular goal. The automaton either responds to a predetermined set of rules or adapts to the environmental dynamics in which it operates [9]. We define the environment influencing the activities of the automaton as a triple $E = \langle A, C, B \rangle$, where:

- $A = \alpha_1, \alpha_2, \ldots, \alpha_r$ is the set of actions;
- $B = \beta_1, \beta_2, \ldots, \beta_m$ is the output set of the environment. When $m = 2$, $\beta = 0$ corresponds a reward and $\beta = 1$ represents a penalty;
- $C = c_1, c_2, \ldots, c_r$ is a set of punishment or penalty probabilities, where $c_i \in C$ corresponds to an input activity α_i.

The learning process involving the LA and a random environment is presented in Fig. 1. Whenever an automaton generates an action α_t, the environment sends a response β_t either penalizing or rewarding the automaton with a specific probability c_i.

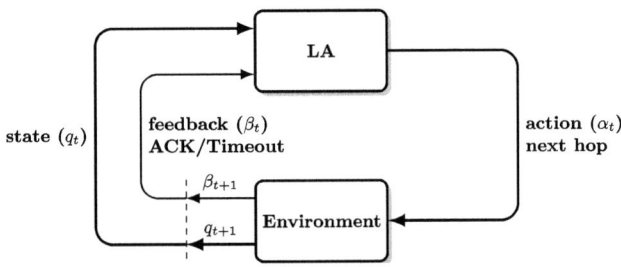

Fig. 1. A feedback loop of learning automata.

Generally, LA can be categorized as a fixed structure LA or a variable structure LA. This paper considers variable structure LA, where the action probability vector is not fixed, and the action probabilities are updated after each iteration. Thus, through interactions with the environment, LAs may adjust their action-selection probabilities by a positive reinforcement (i.e., Reward, Eq. 2):

$$p_i(t+1) = p_i(t) + a(1 - p_i(t)) \quad j = i$$
$$p_j(t+1) = (1-a)p_j(t) \quad \forall j, j \neq i \quad (2)$$

or a negative reinforcement (i.e., Penalty, Eq. 3):

$$p_i(t+1) = (1-b)p_i(t) \quad j = i$$
$$p_j(t+1) = \frac{b}{r-1} + (1-b)p_j(t) \quad \forall j, j \neq i \quad (3)$$

Values $p_i(t)$ and $p_j(t)$ are the probabilities of actions α_i and α_j at time t, r is the number of actions, while a and b are the reward and the penalty parameters, respectively. We employ a learning algorithm called *Linear Reward-Penalty* (L_{R-P}) with $a = b$ in our work [9].

4 A Hybrid Q-LA Routing Protocol

In order to mitigate the inherent challenges present within the Q-Learning algorithm, we propose the hybrid approach, which dynamically combines Q-Learning and Learning Automata to make routing decisions based on network conditions such as energy levels, link quality, and congestion. The idea is to use Q-Learning for adaptive, long-term decision-making (in high-congestion situations) and Learning Automata for short-term adjustments when the network is stable.

The hybrid model leverages Q-Learning when the network is highly dynamic or when energy depletion is significant (for better adaptation). On the other hand, Learning Automata is favored when the network is stable, as it allows quicker convergence with fewer computational overheads [2].

At each decision-making step, nodes switch between Q-Learning and Learning Automata based on network conditions. If the node runs low energy, it will use Q-learning to optimize routing decisions and conserve power. Similarly, if the queue length is high (overloaded network), Q-learning can help find alternative, less congested routes. On the other hand, LA is more efficient when the network does not experience frequent link failures or congestion. It adjusts probabilities of choosing the next hop based on local conditions without needing extensive state exploration.

Every sensor node s_i maintains a Q-table where each entry represents each neighbor's expected reward (path quality). The algorithm then adjusts the Q-values to improve the routing decisions over time. The action with the highest Q-value will be selected (i.e., next-hop with the highest potential for success) with probability $1 - \epsilon$. Additionally, nodes maintain a probability distribution when selecting each neighboring node (Eqs. 2 and 3). These probabilities are dynamically updated based on reward signals from the environment, where successful packet delivery increases the probability of selecting a particular neighbor, and packet loss or failed delivery decreases the probability.

When conditions are stable, the node uses Learning Automata for quicker decision-making and convergence. When unstable conditions occur (e.g., low energy or congestion), it switches to Q-learning for optimal path discovery.

5 Experimental Study

In this section, we aim to evaluate the effectiveness of the proposed hybrid Q-LA routing protocol through multiple computer simulations. To accomplish this, we will employ a custom WSN simulator written in Matlab. We use a fixed network, where sensor nodes are randomly positioned within a 1000: m × 1000: m area alongside a static deployment of $T = 400$ targets. The sensing range of sensors was set at a value of $R_s^i = 175$. The number of nodes will vary in the range $S = \{100, 150, 200, 250, 300\}$ sensors.

We will compare its performance with multiple routing protocols, including LEACH (Low Energy Adaptive Clustering Hierarchy), AODV (Ad hoc On-demand Distance Vector), and RPL (Routing Protocol for Low Power and Lossy Networks). The performance of the basic Q-Learning routing scheme (without LA improvements) will serve as a benchmark against the proposed solution. We will be evaluating the routing efficiency based on performance metrics listed below:

- Packet Delivery Ratio (PDR): the ratio of packets delivered successfully to the destination node (sink) over the total packets sent;
- Latency: the average time a packet takes from the source node to the destination node (sink).
- Energy Consumption: the total energy consumed by the network, considering both the transmission and reception of packets;
- Network Lifetime: the duration for which the network remains operational before the first node runs out of energy.

The experiment results are presented in Table 1. There were averaged over 30 runs to ensure robustness. Through this evaluation, we seek insights into the algorithms performance across various network conditions.

Table 1. Averaged results of comparative performance metrics for tested routing protocols.

Protocol	PDR [%]	Latency [ms]	Energy [%]	Lifetime [t]
Q-Learning	86.23	74.51	**21.64**	63
Hybrid Q-LA	**89.42**	**65.59**	23.66	**68**
AODV	79.23	**75.34**	**24.75**	59
RPL	**83.58**	88.71	29.46	**61**
LEACH	75.43	91.47	32.77	54

As stated before, the hybrid model dynamically selects the most appropriate routing technique based on the network's current state. Q-Learning provides long-term adaptability to changing network conditions (i.e., failures, congestion), while Learning Automata ensures faster, local adjustments during stable conditions.

By dynamically switching between these two approaches, the algorithm provides a higher delivery success rate, lower latency, and longer network lifetime at the cost of a slight increase in overall energy consumption. Regardless of the variant, the reinforcement learning-based protocols offer better overall efficiency than the standard routing solutions.

6 Conclusion

This paper presents a novel Q-Learning and Learning Automata (Q-LA) routing protocol for Wireless Sensor Networks. Our early research findings demonstrate that this hybrid Q-LA approach provides a robust and adaptive solution to wireless sensor networks' dynamic and unpredictable nature. Learning optimal routing policies based on the local conditions of each node improves packet delivery, latency, and network lifetime. Though there are challenges related to convergence and computational overhead, proper parameter tuning could significantly enhance the performance of routing protocols in WSNs. Our future work will include further testing in real-world WSNs, especially in interference-prone environments, by introducing link failures and node mobility.

References

1. Gao, J., Wang, J., Jianlei, G., Shi, W.: Q-learning-based routing optimization algorithm for underwater sensor networks. IEEE Internet Things J. **11**(22), 36350–36357 (2024). https://doi.org/10.1109/JIOT.2024.3398797
2. Gąsior, J.: Learning automata strategies for prolonging lifetime of wireless sensor networks. In: Mathieu, P., De la Prieta,F. (eds.), Advances in Practical Applications of Agents, Multi-Agent Systems, and Digital Twins: The PAAMS Collection, pp. 109–120. Springer Nature Switzerland, 2025. ISBN 978-3-031-70415-4
3. Gudla, S., Kuda, N.R.: Learning automata based energy efficient and reliable data delivery routing mechanism in wireless sensor networks. J. King Saud Univ. Comput. Inf. Sci. **34** (8, Part B), 5759–5765 (2022). ISSN 1319-1578. https://doi.org/10.1016/j.jksuci.2021.04.006, https://www.sciencedirect.com/science/article/pii/S1319157821000926
4. Jain, A., Jain, S., Mathur, G.: Optimizing wireless sensor network routing with q-learning: enhancing energy efficiency and network longevity. Eng. Res. Express **6**, 11 (2024). https://doi.org/10.1088/2631-8695/ad9138
5. Lin, Y., Wang, X., Hao, F., Wang, L., Zhang, L., Zhao, R.: An on-demand coverage based self-deployment algorithm for big data perception in mobile sensing networks. Futur. Gener. Comput. Syst. **82**, 220–234 (2018). ISSN 0167-739X. https://doi.org/10.1016/j.future.2018.01.007. https://www.sciencedirect.com/science/article/pii/S0167739X17313262
6. Maivizhi, R., Yogesh, P.: Q-learning based routing for in-network aggregation in wireless sensor networks. Wirel. Netw. **27**(3), 2231–2250 (2021). ISSN 1022-0038. https://doi.org/10.1007/s11276-021-02564-8
7. Manju, S., Kumar, B.: Target coverage heuristic based on learning automata in wireless sensor networks. IET Wirel. Sens. Syst. **8**(3), 109–115 (2018). https://doi.org/10.1049/iet-wss.2017.0090. https://ietresearch.onlinelibrary.wiley.com/doi/abs/10.1049/iet-wss.2017.0090
8. Nandyala, C.S., Kim, H.W., Cho, H.S.: Qtar: a q-learning-based topology-aware routing protocol for underwater wireless sensor networks. Comput. Netw. **222**,109562 (2023). ISSN 1389-1286. https://doi.org/10.1016/j.comnet.2023.109562. https://www.sciencedirect.com/science/article/pii/S1389128623000075
9. Narendra, K.S., Thathachar, M.A.: Learning Automata: An Introduction. Prentice-Hall Inc, USA, 1989. ISBN 0134855582
10. Qarehkhani, A., Golsorkhtabaramiri, M., Mohamadi, H., Yadollahzadeh-Tabari, M.: Solving the target coverage problem in multilevel wireless networks capable of adjusting the sensing angle using continuous learning automata. IET Commun. **16**(2), 151–163 (2022). https://doi.org/10.1049/cmu2.12323. https://ietresearch.onlinelibrary.wiley.com/doi/abs/10.1049/cmu2.12323
11. Sharma, V.K., Shukla, S.S.P., Singh, V.: A tailored q- learning for routing in wireless sensor networks. In: 2012 2nd IEEE International Conference on Parallel, Distributed and Grid Computing, pp. 663–668, 2012. https://doi.org/10.1109/PDGC.2012.6449899

12. Upreti, R., Rauniyar, A., Kunwar, J., Haugerud, H., Engelstad, P., Yazidi, A.: Adaptive pursuit learning for energy-efficient target coverage in wireless sensor networks. Concurr. Comput. Pract. Exp. **34**(7) (2022). https://doi.org/10.1002/cpe.5975. https://onlinelibrary.wiley.com/doi/abs/10.1002/cpe.5975
13. Wang, B.: Coverage problems in sensor networks: a survey. ACM Comput. Surv. **43**(4) (2011). ISSN 0360–0300. https://doi.org/10.1145/1978802.1978811

Improving Project-Level Code Generation Using Combined Relevant Context

Dmitriy Fedrushkov[(✉)], Denis Tereshchenko, Sergey Kovalchuk, and Artem Aliev

Chebyshev Research Center, Saint Petersburg, Russia
fedrushkov.dmitriy1@huawei.com

Abstract. Within this study, we propose and evaluate an approach to structure and improve context provided in RAG-based solutions for code generation. The approach is based on combination of semantically relevant API and code selection and filtering for better context representation in following LLM prompt. The experimental evaluation performed with CodeGen-350M-mono and several popular benchmarks such as RepoCoder, CoderEval, CoIR show good overall performance (even in comparison to bigger LLMs). Also, the experimental evaluation shows improvement with narrower and more focused context representation (project-scope API instead of popular public API).

Keywords: Artificial Intelligence · Natural Language Processing · Code Generation · Retrieval Augmented Generation

1 Introduction

Retrieval Augmented Generation (RAG) has pushed the boundaries of Large Language Models applications in versatile domains, including software development [5]. Involving the relevant information from external sources to generative models might strongly enhance the outputs in such tasks in software engineering as code generation, code search, question answering and many others. It is also applicable to internal sources and could make language models be useful when working with documents that must not be disclosure [7]. Despite the mentioned advantages, RAG has its own limitations. Inconsistent use of context to be provided to generative AI could lead to reduction of performance which is not resolved with larger size of information provided [2]. This emphasizes the necessity of accurate use of information retrieval techniques when working with LLMs. We believe that small but properly structured pieces of the most relevant information might be more efficient than providing the whole information from relevant documents. Thus, our work is focused on development and investigation of the various approaches of improving RAG prompts by relevant context and their combinations in software development (e.g., code generation task).

2 Context Structuring and Optimization

The main idea of the study is to investigate possible performance improvement in RAG-based solution with smaller yet more structured and more relevant information given as a context in LLM query. We consider text-to-code problem of code generation, i.e. given a natural language query we want LLM to generate a source code in the target language resolving the stated problem. Our approach is based on combination of two main sources of information for query extension: relevant API search and code search. The extracted API and code can be included into the prompt of an LLM (see Fig. 1).

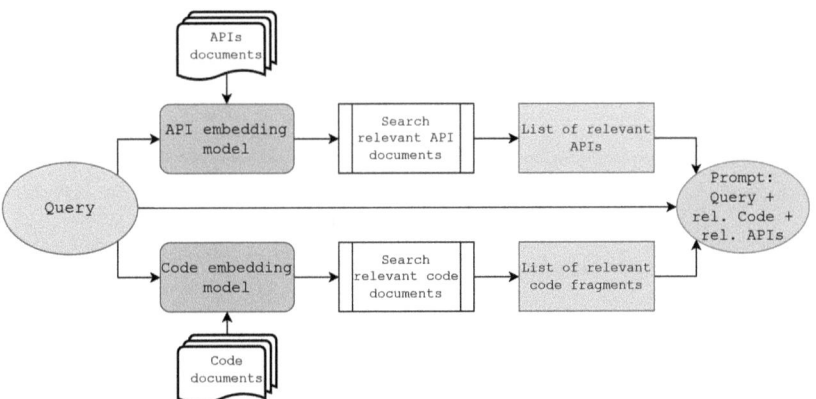

Fig. 1. Prompt creation pipeline

Having the relevant information extracted by retrievers, we've selected the prompt template for code generation with the following structure. Here, RETRIEVED APIS and RELEVANT CODE FRAGMENTS FROM REPO stays for relevant API and pieces of code extracted by search algorithms respectively. FUNCTION SIGNATURE defines target function signature. DOCSTRING includes natural language query for code generation task.

```
# Here are some relevant code fragments from other files of the repo:
# ---------------------------------------------------
%RELEVANT CODE FRAGMENTS FROM REPO%
# ---------------------------------------------------
# Using apis: %RETRIEVED APIS% continue the function:
def %FUNCTION SIGNATURE%:
    %DOCSTRING%
```

Bidirectional Decoder for Code Retrieval. Recent works [1] demonstrated that replacing unidirectional attention masks in decoder-only language models (LLMs) with bidirectional variants enables competitive performance in retrieval tasks. Building upon this, we propose a fine-tuning framework for code retrieval that integrates masked next-token prediction and hard negative contrastive learning, achieving state-of-the-art results on Python code-search benchmarks.

Our approach adapts pre-trained decoder models (e.g., CodeGen 350M-mono) for retrieval tasks through the following key steps:

1. **Structural Tokenization:** Code tokens are augmented with explicit indentation markers to capture syntactic hierarchies.
2. **Bidirectional Training:** Masked next-token prediction with bidirectional attention enables full-sequence context utilization.
3. **Hard Negative Mining:** During training, code snippets are encoded into vectors using the current model checkpoint. The top-k nearest neighbors to each query, excluding the ground truth, are selected as hard negatives for contrastive learning.

Within our experimental study, we fine-tune CodeGen 350M-mono using an EOS token pooling strategy on the AdvTest and CodeSearchNet (CSN) datasets, among others. The training procedure is based on a combination of following methods: Masked Language Modeling (MLM) (30% of tokens are masked, and the model reconstructs them using bidirectional attention); Hard Negative Contrastive Learning (a cross-entropy loss is applied to similarity scores between the query and positive/negative code pairs).

Relevant API Retrieval. We used a machine learning search algorithm to search for semantically relevant APIs. Receiving natural language query as an input, the search system returns the list of relevant APIs, which could be applicable to solve the programming question. While conventional code search methods return the list of code slices, which can help to tackle the task, our approach suggests the functions' qualified names from standard, publuc and local libraries. For instance, this apparatus returns System.out.println java method when receiving "How to print something" question instead of searching for whole code slice. Our experimental setup involves the following data processing: rather than manually collecting documentation of popular APIs, like Pandas, Torch, etc. [9], we downloaded 26728 open-source Python repositories through GitHub API and parsed them to extract the information about public API calls in functions and form a private API usage database [7]. To avoid the insufficient uses of APIs in real code examples, we collected only the repositories with more than 100 stars. Then we used the input queries (function docstring) to match them with the API calls and develop our ML approach to search semantically relevant APIs.

API Database. We used the test part of previously collected public Python repositories as a database to apply searching for the public APIs that would help in developing process according to user necessity. Instructing model to use these APIs seems to improve the quality of generation. To extend the API database, we used an information about project imports and locally installed packages to provide the context about public and/or local APIs that was already used in project. It allowed us to make the context be focused on project scope; this might involve private context to development process and appears to enhance the generated code security owing to use trusted libraries.

Benchmark. We've used two popular repository-level benchmarks for evaluation of our approach, namely RepoCoder [8] and CoderEval [6]. For most of the experiments, we used the Python part of CoderEval benchmark to assess the impact of various contexts in the valid code generation. CoderEval implemented multiple levels of project scopes, which allows us to assess the usefulness of different suggestions from multiple angles. This benchmark also provided the Oracle context that could be used to complete the code, making us able to compare the effectiveness of our retrieved information with the "golden" context. We also used the repositories provided by CoderEval to retrieve the relevant piece of code, which were added to prompts while generating the solutions.

Code Generation Model. We applied CodeGen-350M-mono [4] model to complete the code generation tasks. Since this model was used in original CoderEval approach, we compared the possibilities of each useful context to improve the efficiency of code generation. The model was run in sampling mode with 256 max new tokens and other parameters equal to default values.

3 Results

Code Retrieval. For preliminary evaluation of code retrieval with our training procedure we used CoIR benchmark [3]. For CodeGen 350M-mono, we obtained NDCG@10 metric 0.475 which shows relatively good performance outperforming such models as UniXCoder (0.373), BGE-M3 (0.393), OpenAI-Ada-002 (0.456). The better performance in the benchmark was recorded only by significantly bigger models such as E5-Mistral and Voyage-Code-002.

Next, we evaluated our model on the RepoCoder and CoderEval benchmark for code retrieval and completion tasks for Python code. We were focusing on enhancing bidirectional CodeGen performance but also have fine-tuned UnixCoder-base on AdvTest and CodeSearchNet Python dataset. Exact Match (EM), Edit Similarity (ES), and Edit Tree Distance (ED) were used as quality metrics, with results averaged across repositories (Table 1). The results show that the choice of encoder-decoder architecture significantly impacts performance. For the best results are achieved when the bidirectional CodeGen 350M-mono is used. This configuration outperforms all other combinations, achieving an EM score of 0.321 for API completion and 0.423 for line completion (RepoCoder tasks). The Edit Tree Distance scores for these tasks are 0.690 and 0.769, respectively. In contrast, standalone decoders (without a separate encoder) perform poorly, underscoring the importance of context-aware encoding for code generation tasks.

For CoderEval benchmark we have also tested several decoders together with bidirectional Codegen 350M as well as function completion without RAG (Table 2). The bidirectional CodeGen demonstrated promising results on several benchmarks and demonstrated significant improvement in passing tests for CodeEval function generation benchmark.

The results from CoderEval reveal that bidirectional CodeGen 350M-mono achieves the highest pass@1 score of 0.250, outperforming even the larger 6B model, which achieves a pass@1 score of 0.171. This demonstrates that smaller models with RAG can outperform larger models.

Table 1. Code generation evaluation with RepoCoder

Encoder	Decoder	Compl. type	EM	ES	ED
–	CodeGen 350M-mono	API	0.204	0.496	0.610
UnixCoder-base	CodeGen 350M-mono	API	0.233	0.482	0.600
UnixCoder (fine-tuned)	CodeGen 350M-mono	API	0.310	0.585	0.679
CodeGen 350M-mono	CodeGen 350M-mono	API	**0.321**	**0.597**	**0.690**
–	CodeGen 350M-mono	Line	0.263	0.520	0.690
UnixCoder-base	CodeGen 350M-mono	Line	0.340	0.540	0.712
UnixCoder (fine-tuned)	CodeGen 350M-mono	Line	0.398	0.625	0.762
CodeGen 350M-mono	CodeGen 350M-mono	Line	**0.423**	**0.638**	**0.769**

Table 2. Code generation evaluation with CoderEval

Encoder	Decoder	EM	ES	ED	pass@1
–	CodeGen 350M-mono	0.200	0.512	0.574	12.3%
UnixCoder-base	CodeGen 350M-mono	0.200	0.534	0.59	19.3%
UnixCoder-(fine-tuned)	CodeGen 350M-mono	0.215	0.559	0.618	21.0%
CodeGen 350M-mono	**CodeGen 350M-mono**	**0.222**	**0.547**	**0.606**	**24.8%**
–	CodeGen 6B-mono	0.210	0.532	0.592	17.1%

RAG Context Results. Next, we've evaluated the influence of context given with RAG onto performance of code generation with CoderEval benchmark using pass@1 metric. Table 3 represents the evaluation results with the use of different prompting strategies. Having versatile prompts and scopes of project, we assessed the influence of each relevant information.

Table 3. Code generation results with different contexts (pass@1)

Scope	Query	Oracle	Pub API	Project API	Project Code	Project API+Code	ChatGPT 3.5
plib runnable	4.76%	19.05%	19.05%	19.05%	28.57%	**38.10%**	21.43%
project runnable	3.91%	8.70%	8.70%	8.70%	4.35%	**34.78%**	9.57%
class runnable	5.82%	10.91%	16.36%	20.00%	**39.62%**	26.42%	8.73%
self contained	22.57%	42.86%	42.86%	28.57%	40.00%	40.00%	**52.29%**
file runnable	7.79%	17.65%	19.12%	19.12%	14.71%	19.12%	**21.03%**
slib runnable	12.50%	35.71%	25.00%	**39.29%**	17.86%	28.57%	15.36%
overall	9.48%	21.30%	21.74%	22.17%	24.78%	**28.26%**	21.04%

Providing relevant context that would not overwhelm the model resulted in undeniable upgrades of project-level code generation. We observed significant improvement of small model contribution that overcame ChatGPT generation results. As expected, different prompting strategies influenced various aspects of project-level code completion. Interestingly, combined context improved project runnable by approximately 4 times in comparison with Oracle context, while use of only one of context suggestions did not improve this at all. All these might signalize that the quality of context in project scope is more essential than the quantity of model parameters.

Code Generation Example. We've inspected several randomly chosen tests in the CoderEval experimental evaluation and see that our RAG approach provides reasonable improvement in the code generation process. For example, for the following task (CoderEval 62ece4982e6aefcf4aabbd65) we have the following natural language query:

```
Given an argparse.ArgumentParser instance, return its argument flags in a
    space-separated string.
Args: script: argparse.ArgumentParser instance
Returns: argument flags in a space-separated string
```

Ground-truth solution given by the benchmark was as follows:

```
def parser_flags(parser):
    """
    Given an argparse.ArgumentParser instance, return its argument flags in a
        space-separated string.
    Args:
    script: argparse.ArgumentParser instance

    Returns:
    argument flags in a space-separated string
    """
    return ' '.join(option for action in parser._actions for option in action
        .option_strings)
```

The model without RAG extension gives the following solution:

```
def parser_flags(parser):
    """
    Given an argparse.ArgumentParser instance, return its argument flags in a
        space-separated string.
    Args:
    script: argparse.ArgumentParser instance

    Returns:
    argument flags in a space-separated string
    """
    flags = ""
    for arg in parser.arg_strings:
        flags += " " + arg
    return flags
```

Our approach constructed following RAG-based prompt for code generation:

```
# Here are some relevant code fragments from other files of the repo:
# ----------------------------------------------------
# the below code fragment can be found in:
# borgmatic-commands-completion-parser_flags5.py
# borgmatic-commands-completion-parser_flags0.py
# ----------------------------------------------------
# import argparse
#
# def parser_flags(parser: argparse.ArgumentParser) -> str:
#     return ' '.join(parser.prefix_chars + flag for action in parser.
    _actions for flag in action.option_strings)
#
#
# def test_parser_flags():
# ----------------------------------------------------
# the below code fragment can be found in:
# borgmatic-commands-completion-parser_flags1.py
# ----------------------------------------------------
# import argparse
#
# def parser_flags(parser: argparse.ArgumentParser) -> str:
#     return ' '.join(parser.prefix_chars + option_strings for action in
    parser._actions for option_strings in action.option_strings)
#
#
# def test_parser_flags():
# ----------------------------------------------------
# Using apis: ['borgmatic-commands-completion-parser_flags1.parser_flags', '
    borgmatic-commands-completion-parser_flags2.parser_flags', 'borgmatic-
    commands-completion-parser_flags3.parser_flags', 'borgmatic-commands-
    completion-parser_flags4.parser_flags', 'getattr'] continue the function:
def parser_flags(parser):
    """
    Given an argparse.ArgumentParser instance, return its argument flags in a
        space-separated string.
    Args:
    script: argparse.ArgumentParser instance

    Returns:
    argument flags in a space-separated string
    """
```

Finally, the following code was obtained with much better matching to the original (ground-truth) solution:

```
def parser_flags(parser):
    """
    Given an argparse.ArgumentParser instance, return its argument flags in a
        space-separated string.
    Args:
    script: argparse.ArgumentParser instance

    Returns:
    argument flags in a space-separated string
    """
    return ''.join(flag for action in parser._actions for flag in action.
        option_strings)
```

4 Conclusion

Within the presented study we propose an approach for structuring RAG prompt in code generation with relevant code samples and API retrieved with semantic search in public and project-related scope. Having CodeGen 350M-mono as the main experimental model, our experimental evaluation shows that the proposed approach enables achieving high performance with relatively smaller model. Moreover, the approach gives insight that proper (project-related) selection of RAG context show better performance comparing to direct usage of all public information (e.g., with comparing of public API vs. project-related API). We see this approach as promising starting point for further development of RAG-based code generation and other software development tasks which can give more consistent, comprehensive, and aligned application of RAG solutions. The promising research directions we consider as further step in this approach development are semantic and knowledge-graph involvement into RAG implementation, application of more advanced metrics in code generation evaluation, extending the study to larger number of modern benchmarks, etc. Future work would also include the comparison with other RAG approaches to see the effectiveness of versatile retrieving methods and contexts for generative models that are used when solving software engineering tasks.

References

1. BehnamGhader, P., Adlakha, V., Mosbach, M., Bahdanau, D., Chapados, N., Reddy, S.: Llm2vec: large language models are secretly powerful text encoders (2024). https://arxiv.org/abs/2404.05961
2. Leng, Q., Portes, J., Havens, S., Zaharia, M., Carbin, M.: Long context rag performance of large language models (2024). https://arxiv.org/abs/2411.03538
3. Li, X., et al.: Coir: a comprehensive benchmark for code information retrieval models (2024). https://arxiv.org/abs/2407.02883
4. Nijkamp, E., et al.: Codegen: an open large language model for code with multi-turn program synthesis (2023). https://arxiv.org/abs/2203.13474
5. Yang, Z., et al.: An empirical study of retrieval-augmented code generation: challenges and opportunities. ACM Trans. Softw. Eng. Methodol. (2025). https://doi.org/10.1145/3717061
6. Yu, H., et al.: Codereval: a benchmark of pragmatic code generation with generative pre-trained models. In: Proceedings of the IEEE/ACM 46th International Conference on Software Engineering. ICSE '24, Association for Computing Machinery, New York, NY, USA (2024). https://doi.org/10.1145/3597503.3623316
7. Zan, D., et al.: Private-library-oriented code generation with large language models (2023). https://arxiv.org/abs/2307.15370
8. Zhang, F., et al.: Repocoder: repository-level code completion through iterative retrieval and generation (2023). https://arxiv.org/abs/2303.12570
9. Zhang, K., Zhang, H., Li, G., Li, J., Li, Z., Jin, Z.: Toolcoder: teach code generation models to use api search tools (2023). https://arxiv.org/abs/2305.04032

Author Index

A
Adàlia, Ramon 19
Ahn, Kwangwon 114
Akram, Juniad 225
Albert, Miquel 216
Aliev, Artem 438
Anaissi, Ali 207, 225
Asao, Shinichi 27
Askaripour, Khadijeh 121

B
Balis, Bartosz 97
Bekasiewicz, Adrian 121
Bhowmick, Sanjukta 36
Blachnik, Marcin 89
Bogale, Befikir 327
Borkowski, Piotr 3
Braytee, Ali 207, 225
Brink, Stephanie 327
Bruballa, Eva 11
Bungartz, Hans-Joachim 292

C
Carrillo, Carlos 63
Chang, Victor 283
Chaturvedi, Kunal 207
Chen, Kehao 207
Ciesielski, Tomasz Maciej 353
Cortés, Ana 63
Czachórski, Tadeusz 146
Czaplewski, Bartosz 394
Czarnul, Paweł 130
Czekalski, Piotr 146
Czerski, Dariusz 3, 318

D
Dąbrowski, Daniel 89
Dam, Tien Minh 300
De Giusti, Armando 180
De Lucia, Marco 233
Diakun, Oksana 130

Dinh, Duong Trung 300
Dong, Xueyan 81
Dorzhu, Nachyn 377
Dreżewski, Rafał 403
Dzwonkowski, Mariusz 106

E
Ehrenhard, Michel 345
Epelde, Francisco 11, 46

F
Fedrushkov, Dmitriy 438
Feist, Alexander 275

G
Galeano, Ramona 11
Gąsior, Jakub 429
González, Irene 63
Gratl, Fabio Alexander 292
Groen, Derek 361
Grzesiak-Kopeć, Katarzyna 266
Guo, Yipei 198
Gupta, Amarnath 189

H
He, Chunrong 81
HoseinyFarahabady, MohammadReza 369

I
Ikeda, Takahiro 27
Irany, Fariba Afrin 36
Ireddy, Ashish T. S. 412

J
Jankowski, Norbert 138
Janz, Arkadiusz 154
Jeong, Minhyuk 114
Ji, Yuhao 225
Jung, Leehyun 114

K

Kaczmarek, Sylwester 163
Kazienko, Przemysław 54
Kica, Piotr 257
Kłopotek, Mieczysław A. 3
Kobayashi, Yusei 27
Kobusińska, Anna 283
Kölle, Michael 275
Kordos, Mirosław 89
Kovalchuk, Sergey V. 412
Kovalchuk, Sergey 377, 438
Koziel, Slawomir 121, 171
Kuaban, Godlove Suila 146
Kumar, Niraj 336
Kundu, Turja 36
Kundu, Vidit 345

L

Lang, Zheng 81
Lee, Saiho 225
Lelek, Tomasz 97
Leon, Betzabeth 216
Lerchner, Manuel 292
Leszczyńska, Natalia 353
Leszczyński, Jacek 353
Li, Meisi 207
Li, Shengqi 189
Lichołai, Sabina 257
Ling, Feng 198
Linnhoff-Popien, Claudia 275
Long, Jiang 309
Lübke, Max 233
Lumsden, Ian 327
Luo, Lijing 377
Luque, Emilio 11, 46, 216

M

Malawski, Maciej 257
Margalef, Tomàs 19
Markiewicz, Maciej 54
Martyniak, Remigiusz 106
Mazurek, Szymon 97
Mesas, Francisco 46
Mishra, Manish Kumar 292
Młynarczuk, Magdalena 163
Moska, Julia 154
Motyka, Dawid 154
Mukherjee, Animesh 36

N

Naiouf, Marcelo 180
Newcome, Samuel James 292
Neykova, Rumyana 361
Nguyen, Huynh Hoai 198
Nguyen, Linh Thuy Thi 300
Nguyen, Tuan Anh 300
Nikitin, Nikolay O. 241
Nycz, Monika 146
Nycz, Tomasz 146

O

Ogrodniczuk, Maciej 318
Oliver-Serra, Albert 353
Orzechowski, Michał 257

P

Paszyńska, Anna 353
Paszyński, Maciej 353
Pazar, Abdulkadir 292
Pearce, Olga 327
Pechko, Anastasiya 266
Petri, Stefan 233
Pfeffer, Karin 345
Pietrenko-Dabrowska, Anna 171
Potemkin, Vadim A. 241
Pousa, Adrián 180
Prasad, Mukesh 207

R

Radliński, Łukasz 54
Ranjan, Sakshi 336
Revin, Ilia 241
Rexachs, Dolores 11, 46, 216
Roy, Debraj 345

S

Sac, Maciej 163
Saifullah, Shoffan 403
Sánchez, Paula 63
Sanjuan, Gemma 19
Sanz, Victoria 180
Sarakar, Soumya 36
Sato, Hiroto 27
Schnor, Bettina 233
Seweryn, Karolina 154
Shimokawabe, Takashi 249
Sikora, Maciej 353
Singh, Sanjay Kumar 336

Siwik, Leszek 353
Skarupski, Mateusz 138
Ślusarczyk, Grażyna 266
Song, Jinpei 81
Song, Yena 114
Starosta, Bartłomiej 3
Stein, Jonas 275
Strug, Barbara 266
Sukhomlinova, Tatiana 377
Sukkari, Dalal 327
Sun, Shuyu 421
Suppi, Remo 11, 63
Szczęsny, Aleksander 54
Szerszeń, Krzysztof 72

T
Tabaczyński, Damian 283
Taboada, Manel 46
Takeuchi, Seiichi 27
Takii, Ayato 27
Talia, Domenico 385
Tang, Jia 81
Taufer, Michela 327
Tereshchenko, Denis 438
Tomaszewska, Aleksandra 318
Truong, Long Viet 300

V
Valseth, Eirik 353

W
Walkowiak, Paweł 154
Wang, Liye 207
Wang, Yuji 81
Wielgosz, Maciej 97
Wierzchoń, Sławomir T. 3
Wölckert, Sebastian 275
Wong, Alvaro 11, 46, 216
Wu, Yanjie 225
Wu, Zhe 81

X
Xie, Jie 81

Y
Yamakawa, Masashi 27
Yokelson, Dewi 327
Yu, Guiqing 81
Yuan, Ziheng 249

Z
Zamora, Ismael 19
Zhang, Zhuobin 81
Zhao, Jijing 421
Zhao, Taoxu 207
Zhou, Xucheng 207
Zieniuk, Eugeniusz 72
Zomaya, Albert Y. 369
Żuk, Bartosz 154, 318